The Hippocampus

Volume 1: Structure and Development

THE HIPPOCAMPUS

Volume 1: Structure and Development
Volume 2: Neurophysiology and Behavior

The Hippocampus

Volume 1: Structure and Development

Edited by

Robert L. Isaacson
Department of Psychology
University of Florida

and

Karl H. Pribram
Department of Psychology
Stanford University

PLENUM PRESS · NEW YORK AND LONDON

Library of Congress Cataloging in Publication Data

Main entry under title:

The Hippocampus.

   Includes bibliographies and index.
   CONTENTS: v. 1. Structure and development.—v. 2. Neurophysiology and be-
havior.
   1. Hippocampus (Brain) I. Isaacson, Robert Lee, 1928-   II. Pribram, Karl
H., 1919-   [DNLM: 1. Hippocampus. WL300H667]
QP381.H53   612'.825   75-28121
ISBN 978-1-4684-2978-7   ISBN 978-1-4684-2976-3 (eBook)
DOI 10.1007/978-1-4684-2976-3

© 1975 Plenum Press, New York
Softcover reprint of the hardcover 1st edition 1975
A Division of Plenum Publishing Corporation
227 West 17th Street, New York, N.Y. 10011

United Kingdom edition published by Plenum Press, London
A Division of Plenum Publishing Company, Ltd.
Davis House (4th Floor), 8 Scrubs Lane, Harlesden, London, NW10 6SE, England

Contributors
to this volume

JOSEPH ALTMAN, Laboratory of Developmental Neurobiology, Department of Biological Sciences, Purdue University, West Lafayette, Indiana

PER ANDERSEN, Institute of Neurophysiology, University of Oslo, Oslo, Norway

JAY B. ANGEVINE, JR., Department of Anatomy, College of Medicine, University of Arizona, Tucson, Arizona

SHIRLEY BAYER, Laboratory of Developmental Neurobiology, Department of Biological Sciences, Purdue University, West Lafayette, Indiana

BÉLA BOHUS, Rudolf Magnus Institute for Pharmacology, University of Utrecht, Utrecht, The Netherlands

R. B. CHRONISTER, Department of Neurobiology, University of South Alabama, Mobile, Alabama

CARL W. COTMAN, Department of Psychobiology, University of California, Irvine, California

JOHN L. GERLACH, Rockefeller University, New York, N.Y.

GARTH HINES, Department of Psychology, University of Arkansas, Little Rock, Arkansas

MICHAEL J. KUHAR, Departments of Pharmacology and Experimental Therapeutics, and Psychiatry and the Behavioral Sciences, Johns Hopkins University School of Medicine, Baltimore, Maryland

GARY LYNCH, Department of Psychobiology, University of California, Irvine, California

PAUL D. MacLEAN, Laboratory of Brain Evolution and Behavior, National Institute of Mental Health, Bethesda, Maryland

BRUCE S. McEWEN, Rockefeller University, New York, N.Y.

BRENDA K. McGOWAN-SASS, Department of Physiology-Anatomy, University of California, Berkeley, California

DAVID J. MICCO, Rockefeller University, New York, N.Y.

ROBERT Y. MOORE, Department of Neurosciences, University of California, San Diego, La Jolla, California

SHINSHU NAKAJIMA, Dalhousie University, Halifax, Nova Scotia, Canada

ERVIN WILLIAM POWELL, Department of Anatomy, University of Arkansas School of Medicine, Little Rock, Arkansas

DONALD W. STRAUGHAN, Wellcome Professor and Chairman, Department of Pharmacology, The School of Pharmacy, London, England

PAOLA S. TIMIRAS, Department of Physiology-Anatomy, University of California, Berkeley, California

CAROL VAN HARTESVELDT, University of Florida, Gainesville, Florida

L. E. WHITE, JR., Division of Neuroscience, University of South Alabama, Mobile, Alabama

Preface

These books are the result of a conviction held by the editors, authors, and publisher that the time is appropriate for assembling in one place information about functions of the hippocampus derived from many varied lines of research. Because of the explosion of research into the anatomy, physiology, chemistry, and behavioral aspects of the hippocampus, some means of synthesis of the results from these lines of research was called for. We first thought of a conference. In fact, officials in the National Institute of Mental Health suggested we organize such a conference on the hippocampus, but after a few tentative steps in this direction, interest at the federal level waned, probably due to the decreases in federal support for research in the basic health sciences so keenly felt in recent years. However, the editors also had come to the view that conferences are mainly valuable to the participants. The broad range of students (of all ages) of brain-behavior relations do not profit from conference proceedings unless the proceedings are subsequently published. Furthermore, conferences dealing with the functional character of organ systems approached from many points of view are most successful *after* participants have become acquainted with each other's work. Therefore, we believe that a book is the best format for disseminating information, and that its publication can be the stimulus for many future conferences.

As editors we would like to thank those authors whose contributions appear in these books for their efforts and for sharing our belief in the timeliness of such a publication. We also would like to thank Mr. Seymour Weingarten and Plenum Publishing Corporation for their help and encouragement. They share our hopes and beliefs.

Robert L. Isaacson

Karl H. Pribram

Contents

I. Organization

1. Fiberarchitecture of the Hippocampal Formation: Anatomy, Projections, and Structural Significance

R. B. CHRONISTER AND L. E. WHITE, JR.

2. Septohippocampal Interface

ERVIN WILLIAM POWELL AND GARTH HINES

3. Development of the Hippocampal Region

JAY B. ANGEVINE, JR.

4. Postnatal Development of the Hippocampal Dentate Gyrus Under Normal and Experimental Conditions

JOSEPH ALTMAN AND SHIRLEY BAYER

II. Neurochemistry and Endocrinology

9. Neurotransmitters and the Hippocampus

DONALD W. STRAUGHAN

10. Cholinergic Neurons: Septal–Hippocampal Relationships

MICHAEL J. KUHAR

11. Putative Glucocorticoid Receptors in Hippocampus and Other Regions of the Rat Brain

BRUCE S. McEWEN, JOHN L. GERLACH, AND DAVID J. MICCO

12. The Hippocampus and the Pituitary–Adrenal System Hormones

BÉLA BOHUS

Contents of Volume 2

III.   Electrical Activity

Introduction

ROBERT L. ISAACSON AND KARL H. PRIBRAM

When the editors of these volumes began their research, the hippocampal formation presented a prime example of an enigma wrapped within layers of puzzles. As recently as 1960, Green in the *Handbook of Physiology* could state that "ablation of the hippocampus without serious damage to large areas of the brain has not yet proved possible" and that "up to this time no truly satisfactory theory concerning the role of the hippocampus has been advanced." Still this large structure of the brain was a delight to observe through the microscope. Its simplicity of organization, when viewed in cross-section, was unique and reminded one of a jelly roll whose layering invited electrophysiological analysis; its length and connections through the impressive fornix bundle (one of the largest tracts in the brain) attested to its importance. For some years the cellular architecture of the hippocampus had been worked out thoroughly (e.g., Ramón y Cajal, 1968) and some of its interrelations with adjacent cortical areas were already known. The fornix had been established as the major output channel, although the relationship between the hippocampus and the septum did not become clear until the early 1950s (for review, see Pribram and Kruger, 1954).

Even something of the electrophysiology of the hippocampus was already established. Jung and Kornmüller (1938) had observed a θ rhythm discharge of 4–7 Hz in the rabbit hippocampus. Green and Arduini (1954) and Green and Adey (1956) and also Grastyán (1959) were to follow up this observation in the early 1950s in their classic studies which showed that this rhythm could be elicited by a variety of stimuli; these studies also showed that it was difficult to obtain such rhythms in the cat (see Volume 2 of the present publication) and nearly impossible to find them in the monkey. Yet the significance of the θ rhythm remained difficult to fathom.

Seminal to the prodigious production reported in these volumes were two publications which appeared during the latter 1930s. One was the anatomical report by Papez (1937), who suggested that the hippocampal circuit might serve as the substrate of emotion. Two common misconceptions about Papez's contribution

1

should be noted here, however. First, the view that the entire limbic forebrain is concerned with emotion is sometimes attributed to Papez. However, Papez restricted himself to the hippocampal circuit as it was then known. Second, it has been assumed that Papez held to a James–Lange, visceral feedback theory of emotion. Papez made it clear that he subscribed to an *attitude* theory of emotion, and, in fact, contributed a chapter to Nina Bull's book *The Attitude Theory of Emotion* (1951).

The other major influence on postwar productivity was the publication of the experiments of Klüver and Bucy (1939). Klüver, interested in the hallucinogenic effects of mescaline, had wondered about the role of this remarkable histological structure, the hippocampus, in the production of hallucinations. He asked Bucy, a young neurosurgeon, to remove the structure in order to see whether the hallucinogenic effect would persist in the absence of the structure. Bucy said he could not remove the hippocampus selectively but that he could take out the entire temporal lobe. The resulting dramatic effect on behavior came as a surprise and was interpreted to be a confirmation of Papez's proposal. The effects consisted of hyperorality, "hypermetamorphosis," hypersexuality, and psychic blindness, those characteristics which constitute the now-famous Klüver–Bucy syndrome. Scientific fame is capricious. Brown and Schäfer (1888) had described these consequences of temporal lobectomy fully in the late 1890s and, indeed, Schäfer's textbook of 1900 already carried a comprehensive account of the "Klüver–Bucy" syndrome.

As Bucy's first neurosurgical resident (see Bucy and Pribram, 1943), one of the editors of this book, Karl Pribram, was steeped in these initial attempts to solve the riddle of the hippocampus. He first reproduced the Klüver–Bucy syndrome by making temporal lobectomies to prove to Karl Lashley that the effects on behavior were not due to invasion of the hypothalamus (no histological verification of the lesions accompanied Klüver and Bucy's initial reports). He then went on to map by electrical (Lennox *et al.*, 1950) and chemical (Pribram *et al.*, 1950; Pribram and MacLean, 1953; MacLean and Pribram, 1953) stimulation the responses evoked (neuronography) in the hippocampal circuit. It soon became evident that the cortex on the orbital surface of the frontal lobe, the anterior insula, and the temporal pole, centering on the amygdalar complex, had intimate interconnections with the hippocampal circuit (Fulton *et al.*, 1949; Kaada *et al.*, 1949) and that, except for the psychic blindness, the entire Klüver–Bucy syndrome could be reproduced by resecting these areas which are closely associated with the amygdala (Pribram and Bagshaw, 1953). Psychic blindness, i.e., visual agnosia, was produced when the temporal isocortex (especially that of the inferior temporal convolution) was removed (Blum *et al.*, 1950; Mishkin and Pribram, 1954).

As it became clear that most of the symptoms making up the Klüver-Bucy syndrome were due to intrusions of the amygdala circuit, the functions of the hippocampal circuit remained enigmatic as ever. As noted by Green (1960), few investigators had found any behavioral effect from bilateral hippocampal destruction. In addition, Kaada *et al.* (1949) and Kaada (1951) had shown that electrical stimulation of the hippocampus was without discernible effect on autonomic activities. This paucity of information about the behavioral contributions of the hippocampus was comparable to a similar lack of knowledge about the electrophysiological and neurochemical aspects of hippocampal activity. What was best known 20 years ago was the anatomy of the structure based on the pioneering work of Ramón y Cajal (e.g., 1968) and Lorente de Nó (1934). But even here, as the chapters in Volume 1 by Chronister and White, Angevine, and Powell and Hines will indicate, considerable advances have been made.

Consistent evidence about the behavioral contributions of the hippocampal circuit had to await the use of quantitative techniques for behavioral testing with lesioned animals and the reports which followed the inadvertant production of an astounding and bizarre memory defect when the medial portions of the temporal lobe were resected bilaterally (Penfield and Milner, 1958; Scoville and Milner, 1957). Quantitative information began to pour in during the 1950s (Pribram et al., 1952; Pribram and Fulton, 1954; Pribram, 1954; Pribram and Weiskrantz, 1957; MacLean et al., 1955–1956).

One of the editors of this book, Robert Isaacson, began his research into hippocampal function by combining quantitative behavioral studies with measurements of electrical rhythms of the hippocampus (Isaacson, 1958). In these studies, a leg-lift avoidance task was used with dogs fitted with chronically implanted electrodes in the hippocampus and surrounding areas. Isaacson noted a burst of fast electrical potentials occurring just after the conditioned stimulus and before the conditioned response. These electrical changes were apparent even in well-trained dogs given a narcotic or a major tranquilizer. Under these conditions, the spontaneous electrical rhythms became less desynchronized and the animals sometimes made mistakes. However, a period of fast electrical activity always preceded each conditioned response. Working with a Romanian scholar, Dr. Cornell Guirgea, who was visiting the laboratories of R. W. Doty, Isaacson also studied the learning of the leg-lift avoidance response in dogs with bilateral surgical destruction of the hippocampus. On the basis of the reports from the patients described by Milner in association with Scoville and Penfield, it was anticipated that the dogs would have severe difficulties with learning and retention. It was a considerable surprise, therefore, for them to find that the lesioned dogs learned and retained the response better than intact animals. It was this observation that led to the more elaborate study of two-way active avoidance learning in rats with bilateral hippocampal damage (Isaacson et al., 1961). Once again, acquisition was accomplished faster by the animals with hippocampal damage than by control animals. It is interesting, therefore, to compare these more recent reports with earlier ones that attempted to elucidate hippocampal functions. For example, Schäfer, in Volume II of his 1900 *Textbook of Physiology,* describes the results of his own and others' experiments as follows:

> Ferrier was led by the results of his experiments upon the hippocampal region of the hemisphere to the conclusion that tactile sensibility is in every case impaired or abolished, in proportion to the destruction of the hippocampal and inferior temporal region. It was, however, found by Horsley and myself that extensive lesions might be made in the hippocampus major without producing hemianaesthesia, although in other cases, when the lesion was more extensive and involved the greater part of the hippocampal gyrus, and a large part of the adjacent under surface of the temporal lobe, hemianaesthesia was very marked. The same result was obtained on destruction of the gyrus fornicatus, so that we were led to the conclusion that these convolutions, which form the greater part of the limbic lobe (especially the gyrus fornicatus),

were the seat of perception of cutaneous sensations, and especially of tactile sensitility. Our experiments upon the gyrus fornicatus were confirmed by Munk, who was, however, of opinion that the anaesthesia is the effect of unavoidable injury to the neighbouring Rolandic (motor) region of the cortex, in which alone he localises tactile perceptions. Neither did we ourselves, however, nor Ferrier, ever obtain anaesthesia on directly injuring the [sensory] motor cortex solely. Nevertheless, it must be borne in mind that in endeavouring to excise such deeply lying portions of the cortex as the gyrus fornicatus and hippocampus gyrus, it is necessary to raise up or draw aside the whole mass of the brain, and it is possible that the hemianaesthesia we obtained in the experiments upon the hippocampus and gyrus fornicatus were due not to the actual injury to those parts, but to a general disturbance in the functions of the whole hemisphere (perhaps of the optic thalamus), which the manipulation of the brain might produce. (Schäfer, pp. 766–767)

Schäfer also presents the possibility that the anterior portions of the hippocampal formation are concerned with olfaction, but decided the evidence was inconclusive:

In Ferrier's earlier experiments by this method, which were admittedly not very exact, he found, with destruction of the lower temporal regions on both sides, indications of impairment or abolition of the senses of smell and taste. In monkeys operated upon by myself, in conjunction with Sanger Brown, we were unable to detect such impairment, even when the whole of both temporal lobes was removed. We did not, it is true, remove the hippocampus major as well, and in some animals small pieces of the hippocampal convolution were left. Nevertheless, in spite of the very extensive lesion in this region, the animals unquestionably gave very distinct indications of still possessing both smell and taste. Ferrier has since repeated the operation upon another monkey, and obtained evidence of the loss of both senses for a period of about two months, after which time both senses appeared gradually to undergo recovery. There is, therefore, not complete localization of those senses to these parts of the brain. Munk has recorded the case of a dog, rendered blind by destruction of the occipital cortex, which seemed also to have lost the sense of smell, and in which it was found postmortem that the whole of the hippocampal gyrus upon each side was converted into a thin-walled cyst. Luciani got no evident deficiency of smell after extirpation of the temporal lobe in dogs, but found olfactory disorders to follow lesions of the gyrus hippocampi and of the hippocampus major combined. (Schäfer, p. 766)

Thus matters stood 50 years, except for the lone attempts by Allen (1940, 1941) to relate hippocampal function to conditioning. Allen found, despite rather gross destruction of the hippocampus and surrounding tissue, that it was possible for dogs to obtain conditioned responses to odors and, indeed, these conditioned responses were often established more readily than in intact dogs. Allen also noted that the animals often had difficulty in withholding responses to inappropriate stimuli and thus was the first to describe behavioral difficulties that seem to reflect an impairment of inhibitory control.

Several decades have passed since many of these early studies were undertaken and during this period new techniques and methods have been applied to the study of the hippocampus. These include studies of the activities of single cells as recorded from within and without the cell membrane, assays of presumed neurotransmitters including their regional distribution and functional systems, improved procedures for the analysis of endocrine activities, and improved methods for the analysis of behavior. With the information derived from these improved techniques, the reader can now gauge for himself or herself how well the earlier formulations have stood the tests of time.

References

ALLEN, W. F. Effect of ablating the frontal lobes, hippocampi, and occipito-parieto-temporal (excepting piriform areas) lobes on positive and negative olfactory conditioned reflexes. *American Journal of Physiology,* 1940, **128,** 754–771.

ALLEN, W. F. Effect of ablating the piriform-amygdaloid areas and hippocampi on positive and negative olfactory conditioned reflexes and on conditioned olfactory differentiation. *American Journal of Physiology,* 1941, **132,** 81–92.

BLUM, J. S., CHOW, K. L., AND PRIBRAM, K. H. A behavioral analysis of the organization of the parieto-temporo-preoccipital cortex. *Journal of Comparative Neurology,* 1950, **93,** 53.

BROWN, S., AND SCHÄFER, E. A. An investigation into the functions of the occipital and temporal lobes of the monkey's brain. *Philosophic Transactions of the Royal Society of London (B),* 1888, **179** (London: Harrison and Sons), 303–327.

BUCY, P. C., AND PRIBRAM, K. H. Localized sweating as part of a localized convulsive seizure. *Archives of Neurology and Psychiatry,* 1943, **50,** 456–461.

BULL, N. *The attitude theory of emotion.* New York: Nervous and Mental Disease Monograph No. 81, 1951.

FULTON, J. F., PRIBRAM, K. H., STEVENSON, A. F., AND WALL, P. D. Interrelations between orbital gyrus insula, temporal tip and anterior cingulate. *Transactions of the American Neurological Association,* 1949, 175.

GRASTYÁN, E. The hippocampus and higher nervous activity. In Mary A. B. Brazier (Ed.), *Central nervous system and behavior.* New York: Josiah Macy, Jr., Foundation, 1959.

GREEN, J. D. The hippocampus. In *Handbook of physiology. Vol. II: Neurophysiology.* Williams and Wilkins, 1960, pp. 1373–1389.

GREEN, J. D., AND ADEY, W. R. Electrophysiological studies of hippocampal connections and excitability. *Electrencephalography and Clinical Neurophysiology,* 1958, **8,** 245–262.

GREEN, J. D., AND ARDUINI, A. A. Hippocampal electrical activity in arousal. *Journal of Neurophysiology,* 1954, **17,** 533–577.

ISAACSON, R. L. An electrographic study of the dog during avoidance learning. *American Psychologist,* 1958, **13,** 375.

ISAACSON, R. L., DOUGLAS, R. J., AND MOORE, R. Y. The effects of radical hippocampal ablation on acquisition of avoidance responses. *Journal of Comparative and Physiological Psychology,* 1961, **54,** 625–628.

JUNG, R., AND KORNMÜLLER, A. E. Eine Methodik der Abkitung lokalisierter Potentialschwankugen aus subcorticalen Hirngebieten. *Archiv für Psychiatrie und Nervenkrankheiten,* 1938, **109,** 1–30.

KAADA, B. R. Somato-motor, autonomomic and electrocorticographic response to electrical stimulation of rhinencephalic and other structures in primates, cat and dog. *Acta Physiologica Scandinavica Supplement,* 1951, **23,** 83.

KAADA, B. R., PRIBRAM, K. H., AND EPSTEIN, J. A. Respiratory and vascular responses in monkeys from temporal pole, insula, orbital surface and cingulate gyrus. *Journal of Neurophysiology,* 1949, **12,** 347.

KLÜVER, H., AND BUCY, P. C. Preliminary analysis of functions of the temporal lobes in monkeys. *Archives of Neurology and Psychiatry,* 1939, **42,** 979–1000.

LENNOX, M. A., DUNSMORE, R. H., EPSTEIN, J. A., AND PRIBRAM, K. H. Electrocorticographic effects of stimulation of posterior orbital, temporal and cingulate areas of *Macaca mulatta. Journal of Neurophysiology,* 1950, **13,** 383.

LORENTE DE NÓ, R. Studies on the structure of the cerebral cortex. II. Continuation of the study of the ammonic system. *Journal für Psychologie und Neurologie,* 1934, **46,** 113.

MACLEAN, P. D., AND PRIBRAM, K. H. A neuronographic analysis of the medial and basal cerebral cortex. I. Cat. *Journal of Neurophysiology,* 1953, **16,** 312–323.

MACLEAN, P. D., FLANIGAN, S., FLYNN, J. P., STEVENS, J. R., AND KIM, C. Hippocampal function: Tentative correlations of conditioning, EEG drug and radioautographic studies. *Yale Journal of Biology and Medicine,* 1955/1956, **28,** 380–395.

MISHKIN, M., AND PRIBRAM, K. H. Visual discrimination performance following partial ablations of the temporal lobe. I. Ventral vs. lateral. *Journal of Comparative and Physiological Psychology*, 1954, **47**, 14–20.

PAPEZ, J. W. A proposed mechanism of emotion. *A.M.A. Archives of Neurology and Psychiatry*, 1937, **38**, 725.

PENFIELD, W., AND MILNER, B. Memory deficit produced by bilateral lesions in the hippocampal zone. *A.M.A. Archives of Neurology and Psychiatry*, 1958, **79**, 475–497.

PRIBRAM, K. H. Concerning three rhinencephalic systems. In: The rhinencephalon (a symposium). *Electroencephalography and Clinical Neurophysiology*, 1954, **6**, 708–808.

PRIBRAM, K. H., AND BAGSHAW, M. H. Further analysis of the temporal lobe syndrome utilizing frontotemporal ablations in monkeys. *Journal of Comparative Neurology*, 1953, **99**, 347–375.

PRIBRAM, K. H., AND FULTON, J. F. An experimental critique of the effects of anterior cingulate ablations in monkeys. *Brain*, 1954, **77**, 34–44.

PRIBRAM, K. H., AND KRUGER, L. Functions of the "olfactory brain." *Annals of the New York Academy of Sciences*, 1954, **58**, 109–138.

PRIBRAM, K. H., AND MACLEAN, P. D. A neuronographic analysis of the medial and basal cerebral cortex. II. Monkey. *Journal of Neurophysiology*, 1953, **16**, 324.

PRIBRAM, K. H., AND WEISKRANTZ, L. A comparison of the effects of medial and lateral cerebral resections on conditioned avoidance behavior of monkeys. *Journal of Comparative and Physiological Psychology*, 1957, **50**, 74–80.

PRIBRAM, K. H., LENNOX, M. A., AND DUNSMORE, L. H. Some connections of the orbito-fronto-temporal, limbic and hippocampal areas of *Macaca mulatta*. *Journal of Neurophysiology*, 1950, **13**, 127.

PRIBRAM, K. H., MISHKIN, M., ROSVOLD, H. E., AND KAPLAN, S. J. Effects on delayed response performance of lesions of dorsolateral and ventromedial frontal cortex of baboons. *Journal of Comparative and Physiological Psychology*, 1952, **45**, 565.

RAMÓN Y CAJAL, S. *The structure of Ammon's horn*. Springfield, Ill.: Charles C Thomas Publishers, 1968.

SCHÄFER, E. A. *Textbook of physiology*. Edinburgh: Young J. Pentland, 1898, 1900.

SCOVILLE, W. B., AND MILNER, B. Loss of recent memory after bilateral hippocampal lesions. *Journal of Neurology, Neurosurgery and Psychiatry*, 1957, **20**, 11–21.

I
Organization

1

Fiberarchitecture of the Hippocampal Formation: Anatomy, Projections, and Structural Significance

R. B. CHRONISTER AND L. E. WHITE, JR.

1. Introduction

Despite advances in the neurosciences, many specialized terms are being used in different ways: first, to delimit anatomical structure and, second, to conceptualize function. In many instances, the two usages are incompatible. Examples are found where the term "reticular formation" is used as though it meant a morphological entity, while what is really being referred to is a physiological area. The same statement can be directed toward another functional conceptualization which has become increasingly popular, the "limbic system," which refers to a functional system but derives its name from the border (i.e., "limbus") structure of the cerebrum around the foramen of Monro. Indeed, it can be said that terms such as "reticular formation" and "limbic system" are becoming meaningless, especially with regard to morphology. Evaluation of papers concerning these topics often leaves one in a state of confusion as to the subject matter being covered. This type of confusion has prompted the suggestion to eliminate the concept of "limbic" in its entirety (Brodal, 1969). The orientation of this chapter will be *strictly morphological* in order not to confuse structure with function.

R. B. CHRONISTER AND L. E. WHITE, JR. • Department of Neurobiology and Division of Neuroscience, University of South Alabama, Mobile, Alabama.

Problems of conceptual nomenclature have been previously noted (Broca, 1878a) and remain obvious in the name of the relatively circumscribed area of the medial temporal lobe, the hippocampus. The difficulties in the name of this cortical region are partly descriptive—i.e., different nomenclature (Lewis, 1923; Tilney, 1938)—and partly a result of differences of opinion concerning the nature of the structure so named (Hill, 1894; Crosby et al., 1962). It is the purpose of this chapter to examine the structure and fiber connections within the hippocampal formation (hippocampus and contiguous structures) and their interrelationships with the remaining limbic area of the cerebral cortex.

2. Limbic Lobe

Meynert (1872) in describing the cerebral hemispheres wrote:

> The cerebral hemispheres originate in the form of two laterally situated, lenticular, and hollow processes, which are budded off from the anterior cerebral vesicle. The entire superficies develops into cortical substance. The external surface of the lens is convex, like a shield; the internal surface, turned towards the constricted base or peduncle is annular, and the lumen of the ring forms the aperture of communication between the first or anterior cerebral vesicle and the vesicle of the hemisphere [Meynert is here referring to the foramen of Monro]. The perfora-tion of the vesicles by the trabecular (callosal) fibers divides off a portion of the upper periphery of the ring of the median surface as the septum pellucidum. Besides this, the ring of the internal surface of the hemispheres is divisible into an anterior smaller and a posterior larger segment (or, as they may be termed, two semi-circles). The "posterior" semi-circle forms the "gyrus for-nicatus" which curves around the corpus callosum; the "anterior," presenting an angle opening posteriorly, constitutes the "olfactory lobe." The apex of the angle dilates to form the bulbus olfactorius; the internal and at the same time upper limb runs as the "internal" olfactory convo-lution into the frontal extremity of the gyrus fornicatus; the external and at the same time lower limb of the angle, on the other hand, runs, as the external convolution of the olfactory lobe, into the temporal extremity of the gyrus fornicatus (the hooked convolution, gyrus uncinatus). The temporal portion of the gyrus fornicatus, the uncinate convolution or subiculum cornu Ammonis, with the cornu Ammonis, is . . . a portion of the hemispheres, within the lumen of which the cortex terminates by a free border. (pp. 378–393)

In this description, Meynert notes that the cerebrum is comprised of a ring whose center is the foramen of Monro. (He divided this ring into an inner portion and an outer portion. This inner ring is comprised of the cingulate gyrus, the hip-pocampal formation, and the prepyriform cortex.) Furthermore, Meynert stated that it is in this inner ring that the cortex terminates. Broca (1878b) reexamined the area (i.e., the inner ring) visualized descriptively by Meynert in man and stated that it remains "absolutely constant across the whole series of mammals" (p. 404). He named the area "la grande lobe limbique" or limbic lobe. The limbic lobe has been described in great detail by numerous investigators and simplified in model form by one of us (White, 1965a). Figure 1 shows the model of the classical view of the limbic lobe as seen from a medial position. In a two-dimensional rendering, symmetry is achieved by distorting the convexity of the medial hemisphere and removing the com-missures (which Meynert, 1885, also stipulated as essential to understanding the cerebrum). In this model, the limbic lobe is therefore structurally depicted as that

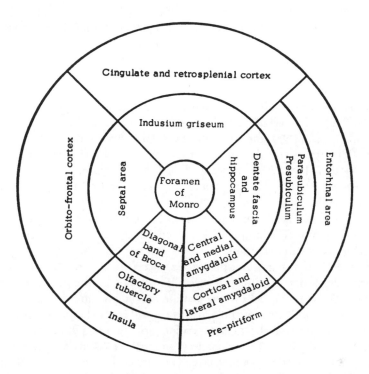

Fig. 1. The classic view of the limbic lobe. The limbic lobe is composed of circumferential rings of cortex centered on the foramen of Monro. From White (1965a) with permission of Academic Press.

cerebral structure which surrounds the foramen of Monro and forms the limbus or border of the cerebral hemispheres. This morphological concept has historical, ontogenetic, and phylogenetic support (White, 1965a).

It should be stated that "limbic lobe" as defined here is not to be considered synonymous with "rhinencephalon." The latter term has historically confused concept with structure and now commonly refers to more than the primary olfactory areas originally described by Sir Richard Owen (1868) in the first anatomical use of the term. "Rhinencephalon" (functionally referring to the sense of smell) should be structurally utilized in a very narrow and restricted manner; therefore, it is defined as the region immediately posterior to the olfactory bulb (olfactory tract, tubercle prepyriform area), as Sir Richard suggested.

3. Anatomy of the Hippocampal Formation

Cortex is, in general, divided into two basic types depending on ontogenetic development. These types are allocortex and isocortex. During development, allocortex fails to cleave completely from the mantle layer (called cortex incompletus by Filimonoff, 1947), while isocortex completely separates from the mantle layer (cortex completus of Filimonoff). The hippocampal formation is comprised of two types of cortex: (1) allocortex (allocortex primitivus or allocortex simplex) and (2) cortex immediate to the allocortex—periallocortex (from the Greek prefix "peri" meaning "around").

The allocortex consists of the dentate fascia (fascia dentata or dentate gyrus), the hippocampus, and nearly all of the subiculum (the allocortex proper terminates at "7" in Figs. 2–6). The periallocortex consists of the presubiculum (area 27), the area retrosplenialis e (area 29e), the parasubiculum (area 49), and the entorhinal region (area 28). The region of the subiculum lying adjacent to the presubiculum is probably also periallocortex (Lorente de Nó, 1934). Because of the complexities of the input systems, it is necessary to examine in detail the fine anatomy of the various regions listed above.

The allocortical dentate fascia (see Figs. 2–6, "DF") and hippocampus are relatively simple in structure. The primary cell of the dentate region is granular while that of the hippocampus is pyramidal. (Indeed, it appears that, through growth distortion and specific cell migration, the normal structure of cortex has been shifted so that the primary cortical granule cell layers lie perched on the primary cortical pyramidal cells.) The granular cells of the dentate region exist in a single layer with the apical dendrites oriented toward the pial surface within the hippocampal fissure. Because of the curvature of the dentate region around the pyramidal layer of the hippocampus (see Figs. 2–6), the dentate granular layer is divided into a suprapyramidal layer and an infrapyramidal layer (i.e., in horizontal sections of the rodent brain the suprapyramidal layer is adjacent to the subiculum and initial regions of the hippocampus while the infrapyramidal region is intraventricular—in Fig. 2, "DF" is pointing to the infrapyramidal dentate fascia).

Next to the granular cells are several layers (Ramón y Cajal, 1893; Lorente de Nó, 1934) of polymorphic cells. The initial layer of polymorphic cells is within the concavity of the granule layer. These cells form the region known as the hilus of the dentate fascia (see Fig. 2, "12"). Great discrepancies can be found in the literature concerning the boundaries of the dentate region and the hippocampus within the hilus of the dentate fascia. Indeed, Blackstad (1956) has suggested that the region bounded by the two layers of the dentate fascia should be included, along with the dentate, into a region termed the area dentata (all the region from Fig. 4, "11" to "hf"). While there is great merit to this attempt at simplicity, some important features of the organization of the dentate tend to be minimized. The polymorphic cells (the limitant and polymorph layers of Ramón y Cajal) are part of the dentate region, as are the adjacent and the initial part of the modified pyramidal cells within the hilus proper. This parcellation was suggested by Lorente de Nó (1934), but he apparently did not consider it important at that time. There are several types of cells found in this "sub"-granular layer of the dentate that, at least from a morphological point of view, appear to be very important to the normal integrity of the dentate fascia. (This polymorphic layer is a continuation of the stratum oriens of the hippocampus and has cell types similar to those of the stratum oriens.) Perhaps the most important cell is the basket cell, which appears to contact numerous granular cells through a supragranular axon plexus (Ramón y Cajal, 1893). The other cell type of

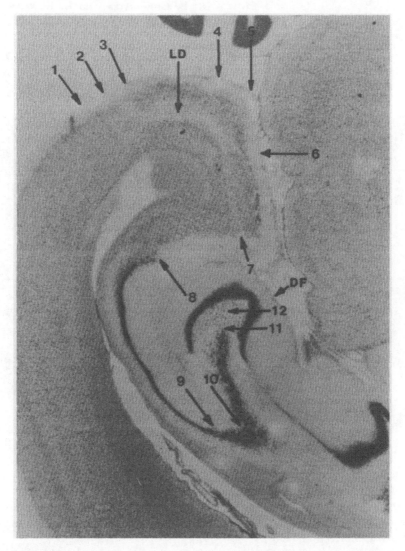

FIG. 2. A cytoarchitectural horizontal section through the rodent hippocampal formations demarcating the various areas and subareas of the hippocampal formation at this dorsal level. For an explanation of the figure numbers, see the list of abbreviations and the text. Cresyl violet stain. ×18.

Abbreviations: LD, Lamina dissecans; Df, dentate fascia; hf, hippocampal fissure. 1, Border between area 35 and isocortex; 2, separation of 35 into its two components; 3, border between the perirhinal area (35) and the entorhinal area (28) dorsally and entorhinal and prepyriform cortex ventrally; 3a, border between lateral entorhinal region (28b) and the intermediate entorhinal area; 3b, border between medial entorhinal region (28a) and the intermediate entorhinal area; 4, region where the layers of the entorhinal area lose discreteness, border between entorhinal area (28) and parasubiculum (49) according to Lorente de Nó (1933); 5, border between the entorhinal area (28) and parasubiculum (49) according to Blackstad (1956); 6, border between parasubiculum and presubiculum (27) except in Fig. 3, which shows area retrosplenialis e ("6" to "6a"); 7, border between presubiculum and subiculum (except transition zone "6b" to "7"); 8, border between subiculum and hippocampus; 9, border between CA1 and CA2; 10, border between CA2 and CA3; 11, border between CA3 and CA4; 12, hilus of the dentate fascia.

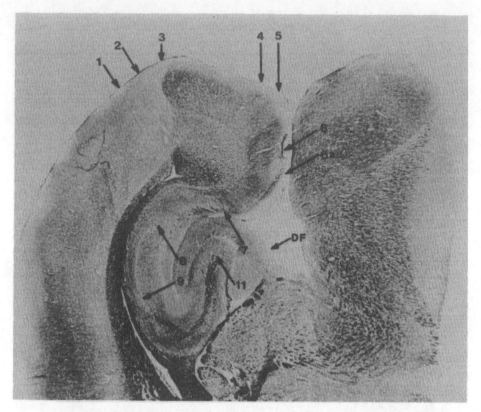

Fig. 3. A fiberarchitectural horizontal section through the rodent hippocampal formation at a level ventral to that in Fig. 2. Note the presence of area retrosplenialis e (between "6" and "6a"), which is not noticeable in Fig. 2. Original Nauta. See abbreviations (Fig. 1) and text. ×18.

possible importance is cells that have axons coursing through the polymorphic layer (i.e., cells with horizontal axons).

For the purposes of this chapter, the fascia dentata will be considered to have three layers: (1) a molecular layer which, with special stains, can be further sub-divided (e.g., Doinikow, 1908), (2) a granular layer, and (3) a layer of polymorph cells which contains basket cells and some modified pyramidal cells. The significance of these three layers will be evident when the fiber connections are reviewed.

The hilus of the dentate fascia contains a considerable number of pyramidal cells comprising area CA4 of Lorente de Nó (1934). Very little is known about the significance of this region, but it is extremely variable in appearance across species (Geneser-Jensen, 1972). Whether this region belongs to the dentate fascia or the hippocampus is unclear at the present time, although it forms an obvious structural transition into hippocampus. Pyramidal cells are seen to stream out of CA4 and form the hippocampus proper. With the utilization of the reduced silver impregnations, it is easy to observe a dense fiberplexus above this pyramidal cell layer. This dense

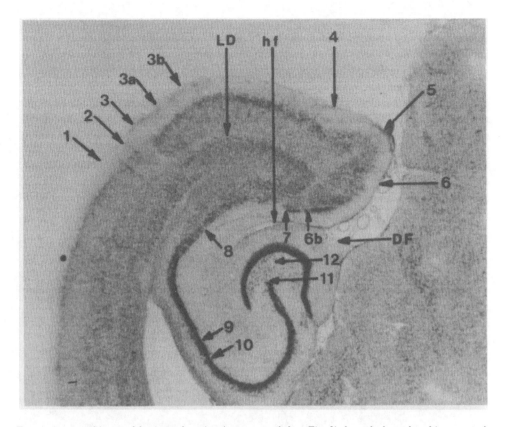

Fig. 4. A cytoarchitectural horizontal section (more ventral than Fig. 3) through the rodent hippocampal formation. Note the clearly discernible lamina dissecans (LD) and its extension to the surface at "3." See abbreviations (Fig. 1) and text. Cresyl violet stain. X21.

plexus of fibers (constituting the stratum radiatum) demarcates region CA3 (Fig. 2, "10" to "11," and Fig. 3, approximately from "9" to "11"). At the dentate end, the plexus terminates abruptly within the two blades of the fascia dentata. The cessation at this point corresponds to the border between CA3 and CA4 (Fig. 3, "11"). The CA3 pyramidal cell has its apical dendrites oriented into this dense plexus. Also characteristic of field CA3 are axons penetrating through the pyramidal cell layer into the stratum radiatum. These axons presumably are collaterals from the giant pyramidal cells of CA3 either terminating within the cell layer or going on to form the Schaffer collaterals.

Near the distal end of the dense plexus (the transition zone away from the dentate), the large pyramidal cells of CA3 are replaced gradually with smaller pyramidal cells. At the same time, the fibers penetrating the pyramidal cell layer disappear. This region of smaller pyramidal cells is CA1. Between CA1 and CA3 (Fig. 2, "9" to "10") is a small zone of transition referred to by Lorente de Nó (1934) as CA2. Blackstad (1956), for other purposes and based on observations with the original Nauta reduced silver method (Nauta, 1950) for fiber impregnation, argued that CA2 was difficult to delineate, thus questioning its existence. However, Nissl stains clearly show its presence (see Figs. 2 and 4), and silver impregnation indicates that the terminating interpyramidal axons may possibly be densest in CA2.

Field CA1 or the region of small pyramidal cells extends from CA2 as one layer of pyramidal cells. The cells appear to begin to lose their clustering; as a result, the region between the pyramidal cells and the overlying alveus (stratum oriens) gradually diminishes and becomes a cell layer (Fig. 2, "8"). This is the border between CA1 and the subiculum (referred to as prosubiculum by Lorente de Nó). CA1 is the region of the hippocampus considered by neuropathologists as being extremely susceptible to anoxia and referred to as Sommer's sector, since Sommer (1880) pointed out that the region is a primary focus for hippocampal damage occurring with epilepsy.

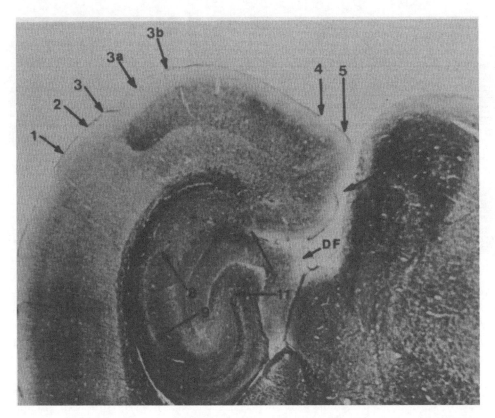

FIG. 5. A fiberarchitectural horizontal section at the approximate level of Fig. 4. Note the dense fiber plexus occupying the external cell layers of the entorhinal region. See abbreviations (Fig. 1) and text. Original Nauta. ×21.

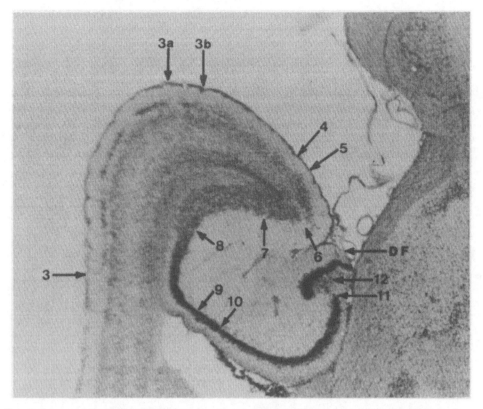

Fig. 6. A cytoarchitectural horizontal section through the ventral hippocampal formation of the rodent. This section is below the rhinal fissure, and, by convention, no perirhinal areas exist. For this reason, there is no "1" or "2." Cresyl violet stain. ×21.

Like the dentate region, the hippocampus has several other cell types that are extremely important to its structural anatomy. Paramount among these cells are the basket cells. The basket cells, like their counterparts elsewhere, tend to synapse with numerous pyramidal cells and are considered to be inhibitory in nature (Andersen *et al.*, 1964). These cells are in the pyramidal layer itself and also deep to it in the stratum oriens. In the stratum oriens are also located cells with horizontally running axons. Thus it can be seen that the organization of the dentate fascia and hippocampus is basically the same. The major difference is the cell type present. Another difference is found in the elaboration of the suprapyramidal cell layer. Instead of a single molecular layer, the suprapyramidal layer is composed of the stratum moleculare, the stratum lacunosum, and the stratum radiatum. These layers have significance in the fiber relations of the hippocampus.

Next to the field CA1 of the hippocampus is the subiculum. This starts out with a simple structure near CA1 (Fig. 2, "8") but becomes progressively more complex as the periallocortex is approached (Fig. 2, "7"). Classically, the subiculum has been regarded as a simple transition cortex, but this is probably an understatement since the subiculum appears to possess rather unique morphological characteristics (Braak, 1972) and functional characteristics (Chronister *et al.*, 1974). Although it has been subdivided into numerous fields (e.g., by Lorente de Nó, 1934), its structure is not especially conducive to discrete parcellation. At its junction with periallocortex, a part of the subiculum extends into the periallocortical presubiculum (Fig. 4, between "7" and "6b"). This junction is rectilinear dorsally (Fig. 2, "7") but becomes trianglelike ventrally (Figs. 4 and 6, "7"). This area can be referred to as either subiculum or presubiculum (Lorente de Nó, 1934, regarded it as subiculum field A). This overlap appears to be the result of an extension to the surface of the lamina dissecans (Figs. 2 and 4, "LD") of the periallocortex.

The allocortical hippocampal formation is comprised of the dentate fascia, the hippocampus, and the subiculum. The former area is primarily granular (afferent) while the latter two areas are primarily pyramidal (efferent). Indeed, all of the hippocampus and subiculum send very heavy fiber projections to the fornix system.

The structure of the periallocortical hippocampal formation is much more complex than that of the allocortical portion. The first and most obvious difference is the presence of the cell-poor layer—the lamina dissecans (Figs. 2 and 4, "LD"). This lamina is the distinctive feature of the periallocortex. Nearest the allocortex, the first periallocortical structure is the presubiculum (or Brodmann's area 27) (Fig. 2 and Figs. 4, 5, and 6, between "6" and "7"). The superficial cell layer (lamina principalis externa) is characterized by the presence of granular-appearing cells. These cells exhibit the appearance of clustering at certain dorsoventral levels and are usually quite distinctive in appearance. At the lateral border of the presubiculum and at relatively dorsal levels, there exists an area, area retrosplenialis e or 29e. Area 29e is a small triangular-shaped region sandwiched between the presubiculum and the parasubiculum. It appears rather indistinct in Nissl stains, but is shown to be a fiber-poor region with reduced silver impregnation techniques (Fig. 3, "6" to "6a"). Whether area 29e has any structural significance is unknown, but it has been described in detail by some (Vaz Ferreira, 1951; Blackstad, 1956; Smith and White, 1964).

Lateral to area 29e dorsally and the presubiculum ventrally lies the parasubiculum. This border is only relatively discrete in the adult brain but extremely sharp in the neonatal brain. The parasubiculum, a newcomer to the hippocampal formation (Rose, 1927; Smith and White, 1964), is comprised of pyramidal cells including a layer of superficial cells. Furthermore, there is a dorsal–ventral variability in the fiber density within the parasubiculum. Blackstad (1956) and White (1959) distinguished a ventral fiber-poor region (49a) and a dorsal fiber-rich region (49b). The significance of this structural heterogeneity is not known. It has been suggested that the parasubiculum arises in conjunction with the elaboration of the orbitofrontal convexity of the frontal lobe (White, 1959). Because of the large concentration of pyramidal cells in this area, the parasubiculum may be a major projection region for the periallocortex—although direct evidence is lacking.

Laterally, the parasubiculum merges with the entorhinal area. The border between the two regions is unclear and arbitrary. According to Lorente de Nó (1933, 1934), the diffuse nature of the layers of the parasubiculum is replaced by the discrete layers of the entorhinal region (which seems to occur at "4" in Figs. 2 and 4). A slightly different border is obtained following silver impregnation (Fig. 3, "4"). This transition area (between "4" and "5" in Figs. 2–6) between the parasubiculum (area 49) and the entorhinal area (area 28) is included in area 49 by Lorente de Nó (1933, 1934) but in area 28 by Blackstad (1956). (Very close examination of the text and figures in the two papers of Lorente de Nó reveals ambiguity concerning where this author placed the area 49/28 border. Most contemporary authors place the border at "5" in figs. 2–6.) Whether or not this transition region should be considered a separate entity (such as CA2 in the hippocampus) is an acceptable structural conjecture.

The entorhinal area (area 28) has been described by numerous authors. It has a dense fiber plexus of such a nature that the plexus is unique in the nervous system (Ramón y Cajal, 1901). Great disagreement has existed in subdividing area 28. M. Rose (1927) showed it as varying between species by as many as 16 subfields. Lorente de Nó (1934), on the other hand, demonstrated four fields in the mouse and five in the monkey (his Figs. 2 and 33). Most recent authors follow the nomenclature of Brodmann (1909) and M. Rose (1912) and divide area 28 into area 28a or pars medialis and an area 28b or pars lateralis (e.g., Krieg, 1946; Vaz Ferreira, 1951; Blackstad, 1956; White, 1959). These parcellations are usually made in a plane perpendicular to the long axis of the hippocampus. Thus, for example, in the rodent the normal manner of demarcating area 28 is in horizontal section, although this may be an oversimplification (Rose, 1927).

The medial region of area 28 adjacent to area 49 is extremely fiber rich and forms the characteristic dense fiber plexus noted by Ramón y Cajal (Fig. 3, "3" to "5"). This region (28a) of the entorhinal cortex is especially pronounced in more dorsal levels of the hippocampal formation. Area 28a has relatively large cells in layer II. Laterally, the cells in layer II get smaller and tend to cluster in islands (prominent at ventral levels of the hippocampus), and the dense fiber plexus is absent. This region is the lateral entorhinal area or area 28b (Fig. 6, "3" to "3d"). Between the two regions is a small transition zone described by Blackstad (1956) (Figs. 4–6, "3a" to "3b"). Golgi preparations discussed by Lorente de Nó (1934) lead to further structural differences, but continue to support a general medial (area 28a) and lateral (area 28b) subdivision of the entorhinal area. At the lateral end of the entorhinal region, the lamina dissecans extends to the surface and terminates. Dorsally, this termination occurs at the rhinal fissure and forms a region referred to by Brodmann as area 35 or perirhinal area. Ventrally, the border of area 28b is not precise and appears to merge rather indistinctly with the periamygdaloid area (Rose, 1926) and prepyriform cortex (Fig. 6, "3"). In this ventral portion of area 28, the lamina dissecans loses its striking appearance.

Area 35 is identified in the depths of the rhinal sulcus and is a special variety of cortex termed perisemicortex by Filimonoff (1947) and proisocortex by Sanides

(1970, 1972). At dorsal levels of the entorhinal region, it is difficult to distinguish precisely the area 28/35 border (Fig. 2, "2" and "3"). There is an abrupt cessation of the lamina principalis externa, but the lamina principalis interna continues for a short distance. Embryonic preparations (see Fig. 7) show that the cortex in the depth of the rhinal sulcus as well as on both sides of the rhinal sulcus is distinct from the cortex in area 28. The precise parcellation of this junction, as seen in Fig. 7, is complex. These complexities were noted many years ago by Lorente de Nó (1934), who stated, "I certainly do not exaggerate in stating, that to ascertain connections and other characteristic traits of the subfields of the area entorhinalis will demand many years' work" (p. 159).

Although considerable change is noted in the evolutionary elaboration of the lateral periallocortical hippocampal formation, in its medial to lateral extent it has basically the same laminar components. The cell layers, although showing morphological modifications, are basically continuations of one another. This point was recognized and emphasized by Lorente de Nó (1934). Therefore, it seems reasonable to conclude that the hippocampal formation is comprised of two discrete units, here termed the medial allocortical and lateral periallocortical components. (With the exception of the septohippocampal fiber system, the fiber systems also are organized into two discrete units.) In review, Table 1 shows the subdivisions of the hippocampal formation into its various subareas.

4.   Hippocampal Formation: External Afferent Supply

Because of the position of the hippocampal formation (i.e., configuration of the formation within and around the lateral ventricle), several sources of external input are possible. The pathways must follow one of three routes: (1) travel toward the lateral border of the hippocampal formation either superficially or in the angular bundle and gain access to the entorhinal region, (2) course supracallosally (i.e., cingulum and stria Lancisii) to enter the hippocampal formation through extremely dorsal regions, or (3) enter the hippocampal formation through the fornix–fimbrial system. Each of these fiber systems will be examined separately.

4.1.   Lateral Input System (Through Entorhinal Area)

Observations of general cortical differentiations suggest that cortical fibers should enter the entorhinal area laterally, especially anterolaterally. (This was suggested as early as 1883 by Flechsig.) Consequently, the entorhinal region should receive fibers from the olfactory cortex. The meticulous research of Ramón y Cajal (1893, 1901, 1911) did not reveal the existence of these fibers. Nevertheless, the idea that the entorhinal region received direct input from the olfactory bulbs seemed logical and became accepted dogma. Numerous investigators (e.g., Hilpert, 1928) described fibers emanating from the amygdala and distributed to the hippocampal gyrus or entorhinal area. Attempts to examine these connections with experimental methods (primarily the Marchi technique) were unsuccessful (Allen, 1948). With the advent of the reduced silver impregnations, different results were obtained.

White (1965b) demonstrated that olfactory bulb efferents passed superficial in the amygdaloid regions (i.e., in the 1a fiber systems) to terminate in what he termed the ventral portion of the lateral entorhinal cortex. This pattern of termination was noticed by Scalia (1966) and by Heimer (1968) but denied by Powell *et al.* (1965). Price and Powell (1971) and Price (1973) have shown rather conclusively that olfactory fibers cross the periamygdaloid region and end in the transition region between the periamygdaloid region and entorhinal region (Price and Powell, 1971). This termination area does seem to be related to the entorhinal region (White, 1965b; Hjorth-Simonsen, 1972; Price, 1973). Kerr and Dennis (1972) found electrophysiologically that the distribution of olfactory fibers might be to the entirety of the ventrolateral entorhinal region. Clearly, olfactory evoked responses have access to the hippocampal formation.

Several investigators have examined the prepyriform lobe (based on strict

TABLE 1

Areas of the Hippocampal Formation

Doinikow (1908)	Brodmann (1909)	Rose (1926)	Lorente de N6 (1934)	Blackstad (1956)
Dentate fascia		Dentate fascia	Dentate fascia	Dentate fascia
End blade		h3, h4, h5	CA3C, CA4	Hilus of the dentate fascia
Transition blade		h2	CA2, CA32, CA3b	Regio inferior (probably including CA1c of Lorente de N6)
Lower blade		h1	CA1	Regio superior
			Prosubiculum a, b, c, subiculum c	Subiculum
	Area 27		Subiculum a and presubiculum	Presubiculum
	Area 29e			Area retrosplenialis e
	Area 49		Parasubiculum (including transition zone)	Parasubiculum
	Area 28a		Entorhinal fields B, C	Medial entorhinal (including transition zone)
	Area 28b		Entorhinal field A (including Lorente de N6's area 35)	Lateral entorhinal
	Area 35		Unlabeled	Perirhinal area

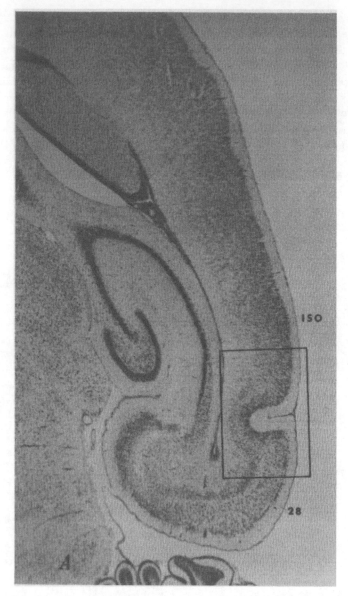

Fig. 7. The structure of the cortex of the rhinal fissure. *A*: A horizontal section through the posterior region of an 80 mm C-R length pouch opossum. Note the similarity to the corresponding area in the rodent (Fig. 4). *B*: Higher power of the area outlined in *A*. Note cell proliferation occurring on both sides of the rhinal fissure (under "ISO" and extending through "D" to "A" in the entorhinal area or area 28). *C*: Similar section to *B* in the adult opossum. The same cytoarchitectural boundaries are demarcated in both *B* and *C*. Note the obscuring of the lamina dissecans of the periallocortical entorhinal region between "A" and "B." The regions between "B" and "D" correspond to area 35 of the proisocortex. The region between "A" and "B" is the initial segment of the entorhinal region. Abbreviations: ISO, progressive elaboration into isocortex; 28, entorhinal area; rf, rhinal fissure. Cresyl violet stain. A—×20, B—×70, C—×80.

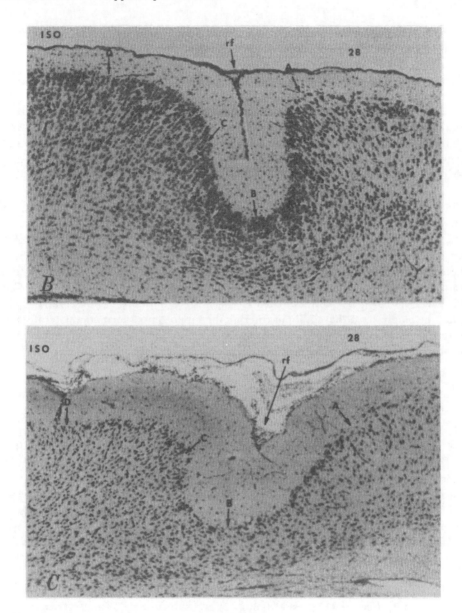

FIG. 7. (*Continued*)

nomenclature, the term "piriform lobe" is synonymous with "entorhinal area") to ascertain whether any portions of it send fibers into the hippocampal formation. Gloor (1955) showed evidence for multisynaptic connections between the amygdaloid regions and the entorhinal region. Cragg (1961) further described two pathways, one superficial and one deep, between the prepyriform lobe and entorhinal areas. These two fiber systems were verified by Powell *et al.* (1965) and also by Valverde (1965). All of these fiber systems probably correspond to the uncal fibers discussed by Flechsig (1883) many years earlier. It would appear, therefore, that both primary olfactory and secondary olfactory fibers enter the hippocampal formation, but only through the lateral entorhinal region.

The lateral entorhinal region appears to be important also in the influx of fibers from other cortical regions. Van Hoesen *et al.* (1972) reported that fibers from the orbitofrontal and temporal cortices projected heavily upon the lateral entorhinal region (28b) and the medial entorhinal region (28a). In addition, they noted terminal degeneration within the banks of the rhinal sulcus. This type of degeneration had been noted previously (White, 1965*a*; Jones and Powell, 1970; Price and Powell, 1971) and appears to be distributed to area 35 and the initial segment of area 28 (subfield Aa of Lorente de Nó, 1934). Terminal degeneration in the rhinal sulcus area results from general cortical lesions, with frontal lobe fibers terminating more dorsally in area 35 of the rodent than do temporal–occipital fibers (White, unpublished observations). The course of these fibers, especially those from the frontal region, is via the uncinate fasciculus in primates (Van Hoesen *et al.*, 1972) and via a homologous path in rodents (Chronister, unpublished observations). It should be emphasized that all input entering the hippocampal formation laterally appears to terminate in the periallocortical entorhinal cortex and the adjacent transitional cortex, area 35. No degeneration gains access to the hippocampus directly.

4.2. Supracallosal Input

The classic literature places great emphasis on medially oriented fiber tracts coursing through the arch of the hemisphere to terminate in the hippocampal formation. An example of this type of tract is the fasciculus marginalis described by Elliot Smith (1898). In callosal animals, this type of fiber tract, which may be efferent with regard to the hippocampus (Simpson, 1952), is related to the induseum griseum, which is extremely rudimentary in many mammals (Humphrey, 1937).

The other medial fiber tract, the cingulum bundle, has strong terminations in the hippocampal formation, as defined by Ramón y Cajal (1911), Adey and Meyer (1952), White (1959), and Domesick (1969), among others. These investigations are consistent in the description of the degeneration patterns. The terminal degeneration is restricted exclusively to the periallocortex with no apparent exceptions. All periallocortical structures receive an input varying only in extent; ranked in terms of heaviest input, they are (1) presubiculum, (2) parasubiculum, (3) area retrosplenialis e, and (4) entorhinal. Like the laterally oriented fibers, the supracallosal system exhibits considerable specificity of termination and does not enter the allocortical hippocampal formation.

The cingulum bundle can be divided into various bundles based on lateromedial and dorsoventral considerations. The lateral aspects of the bundle (White's subdivisions a, b, and c) terminate in the presubiculum. Bundle "d" distributes to the parasubiculum and bundle "e" terminates in the entorhinal area. Although controversy exists as to the origin of these fibers, these areas have been suggested to be the source of the fibers: (1) medial hemisphere, (2) anterior and intralaminar thalamus groups, and (3) the extreme anteromedial portion of the frontal lobe. The last area appears to project solely into the parasubiculum, suggesting that the parasubiculum arises in conjunction with increased development of the frontal lobe. Comparative observations (including embryology) support the recent development of the parasubiculum. For a more complete description of this system, see White (1959).

4.3. Fornix–Fimbrial Fibers

It was accepted for many years that the fornix was afferent to the hippocampus. By this route, olfactory information was reported to reach the hippocampus (e.g., the olfactory radiations of Zuckerkandl, 1888). Ramón y Cajal (1911) emphasized that the fornix was strictly composed of hippocampal efferent fibers. However, Crosby (1917) noted fornix afferents (with respect to the hippocampus) in the alligator, and this was confirmed later in mammals (Gerebtzoff, 1939). Since that time, fornix afferents have been detected and described by many researchers (Ban and Zyo, 1962; Votaw and Lauer, 1963; Raisman et al., 1965; DeVito and White, 1966; Siegel and Tassoni, 1971; Hjorth-Simonsen, 1973; Mellgren and Srebro, 1973). These fibers play a great role in normal hippocampal functioning, particularly in slow-wave electrical activity (Green and Arduini, 1954).

The septohippocampal fibers travel in both the dorsal fornix and the body of the fornix to enter the alveus. Termination appears densest in the hilus of the dentate fascia and the region of CA2/CA3. These endings are primarily in the stratum oriens and stratum radiatum. A very diffuse and scant projection enters the molecular layer of the dentate fascia, presumably with the perforant path (Mellgren and Srebro, 1973). Field CA1 does not seem to receive as many fornix afferents as do CA2 and CA3, but electrophysiological evidence indicates that these fibers are very important (Grantyn et al., 1971). The fornix afferents continue beyond CA1 to the subiculum, presubiculum, parasubiculum, and entorhinal region, terminating in the area 28/35 border. Both the medial and the lateral septal regions send efferents to the hippocampal formation, but are distributed on different dorsal–ventral levels (Siegel and Tassoni, 1971b). Axons from the medial septal area, according to Siegel and Tassoni (1971b), enter and terminate in the dorsal hippocampus, while axons from the lateral septal area project into the ventral hippocampus. Thus it is apparent that the septal area has efferent projections to the totality of the hippocampal formation—both allocortical and periallocortical.

5. Hippocampal Formation: Commissural Connections

The commissural connections of the hippocampal formation have not been mapped as systematically as the ipsilateral systems. Blackstad (1956) has published the most definitive study on this subject and the reader is referred to this report for more detail.

The majority of crossed hippocampal fibers appear to course through the ventral commissure, with only a few traveling through the dorsal commissure. All of the fields of the hippocampal formation have been demonstrated to have commissural connections but not all are of homotypic origin (Gottlieb and Cowan, 1973). In this latter regard, the commissural connection of field CA1 appears to emanate from field CA3 and distributes in the same manner as the Schaffer collaterals (Gottlieb and Cowan, 1973). Similarly, the crossed dentate fibers have been shown to have origin in CA3 of the opposite side. If these fibers are of heterotypic origins as suggested above, they are, in reality, crossed association fibers and not commissural fibers. Such connections have been postulated also by electrophysiological evidence (Andersen *et al.*, 1961).

As mentioned earlier, some of the fibers that cross do so in the dorsal psalterium. Paramount among these are crossed fibers related to the subiculum (Blackstad, 1956; Raisman *et al.*, 1965). According to Sperti *et al.* (1970), there is also a crossed influence exerted on the hippocampus via relays in the dorsal entorhinal region. Although a great deal is known about the mode of termination of the crossed fibers and their "incredible" specificity (Gottlieb and Cowan, 1972), the structural significance of these crossed fiber systems is unclear. For example, it is known that the electrical activity expressed by one hippocampus is independent of the activity of the other hippocampus (Green and Arduini, 1954; Chronister, Zornetzer *et al.*, 1974). That this is the case is even more surprising in view of crossed septohippocampal fibers (Mellgren and Srebro, 1973). Clearly, the commissural system is in need of further definitive investigation.

6.   *Hippocampal Formation: Internal Afferents*

Numerous investigators have shown the entorhinal region to be the major source of afferent input into the hippocampus and dentate fascia (e.g., Ramón y Cajal, 1901; Zuckerkandl, 1900; Lorente de Nó, 1933, 1934; Blackstad, 1958). These fiber systems are so intense that Ramón y Cajal (1901) remarked that the key to the understanding of the hippocampus was in the understanding of the entorhinal region.

Lorente de Nó (1934) stated that the lateral entorhinal region gives rise to the perforant bundle while the medial entorhinal region gives rise to the alvear path. The former tract is distributed to the dentate fascia and hippocampus, while the alvear path terminates in the subiculum and initial segment of the hippocampus. Subsequent degeneration studies have not confirmed these normal observations. Rather, Hjorth-Simonsen (1973) and Hjorth-Simonsen and Jeune (1972) demonstrated that the lateral entorhinal region is the source of fibers terminating close to the hippocampal surface near the hippocampal fissure (i.e., outer portion of the molecular layer) while the medial entorhinal region gives rise to fibers that terminate in the deep half of the stratum moleculare-lacunosum of the hippocampus and the middle portion of the molecular layer of the dentate fascia. These entorhinal afferents to the hippocampus have also been noted by Van Hoesen *et al.* (1972), and Chronister (unpublished observations). For example, Fig. 8 shows the pattern of degeneration following a medial entorhinal lesion (including the presubiculum and parasubiculum). Note the dense band of terminal degeneration in the middle zone of the molecular layer of the dentate fascia ("F"). Heavy degeneration in the hippocampus begins near the terminal third of CA1 ("A") and changes its form abruptly at the CA1/CA2 border ("B"). At this juncture, the degeneration becomes denser (approximately "B" to "C"), gradually diminishing to a very delicate degeneration in CA3. Near the end of CA3 (CA3c), the degeneration again is denser ("D" to "E"). The significance of this latter change in density is probably physically related to the tight orientation of the apical dendrites of CA3c. Conversely, the striking lack of degeneration in the stratum radiatum, stratum oriens, and alveus indicates the importance of the perforant path to a limited area of the hippocampus and dentate fascia.

Lesions restricted to the region of the rhinal fissure in what is referred to as area 35, but also including the initial segment of area 28 called subfield Aa of Lorente de Nó (1934) (i.e., in the depth of the rhinal fissure itself), produces afferent degeneration within the lateral entorhinal area and diffusely within the molecular layer of the dentate fascia (Chronister and White, 1972). Further, Hjorth-Simonsen (1973) demonstrated in the rodent that the perforant bundle courses through the depth of area 35 before entering the hippocampus and dentate fascia. Whether this is the source of the dentate fascia degeneration is unknown; however, superficial lesions of area 35 in primates produces degeneration exclusively in the entorhinal area (Van Hoesen and Pandya, 1973). Either there is a difference in the organization of the afferents between rodents and primates, or area 35 is not equivalent in the two species. Nevertheless, the structural significance of area 35 as a prominent source of ipsilateral afferents to the hippocampal formation seems established.

In all of these studies, no degeneration within the alvear bundle was demonstrated. Clearly, lesions of the entorhinal area do not produce degeneration in this tract, which is in contrast with the general observations of Allen (1948) and Raisman et al. (1965). Similarly, it does not appear that lesions of the presubiculum and parasubiculum produce significant degeneration in or near the alveus; however, lesions of the subiculum result in heavy degeneration throughout the alvear bundle with perhaps a fine diffuse degeneration in the initial segment of CA1 which cannot be distinguished from commissural afferents (Chronister and Zornetzer, 1973). Therefore, the contribution of the parasubiculum, presubiculum, and subiculum to the afferent supply of the allocortical hippocampal formation (hippocampus and dentate fascia) is not established.

Several other interesting intrinsic pathways have been described within the hippocampal formation. It was suggested many years ago (Azoulay, 1894; von Koelliker, 1896; Ramón y Cajal, 1911; Lorente de Nó, 1934) that fibers traveled from the hippocampus to the ipsilateral entorhinal area. These fibers were denied by Russell (1961), Nauta (1966), and Raisman et al. (1965). However, Adey et al. (1956, 1957) and Votaw (1959, 1960) claimed their existence. Utilizing a special preparation, Hjorth-Simonsen (1971) described CA3–entorhinal area fibers. However, possible contamination of the study by "sprouting" cannot be ruled out, although it seems likely that hippocampus to entorhinal area fibers do exist.

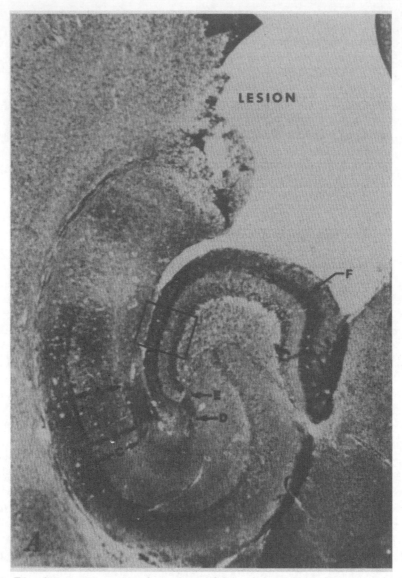

Fig. 8A: Fiber degeneration resulting from a lesion of the medial entorhinal region, parasubiculum, and presubiculum (with moderate subiculum involvement). Note the dense degenerating band ("F") near the middle of the dentate fascia molecular layer. ×35.

It is known that the dentate granule cells give rise to axons that terminate in the hippocampus. These mossy fibers have been mapped rather carefully by Lorente de Nó (1934) and Blackstad et al. (1970). The granule cells of both the infra- and suprapyramidal aspects of the dentate fascia give off mossy fibers. The infrapyramidal fibers course a short distance and terminate close to the dentate fascia mark-

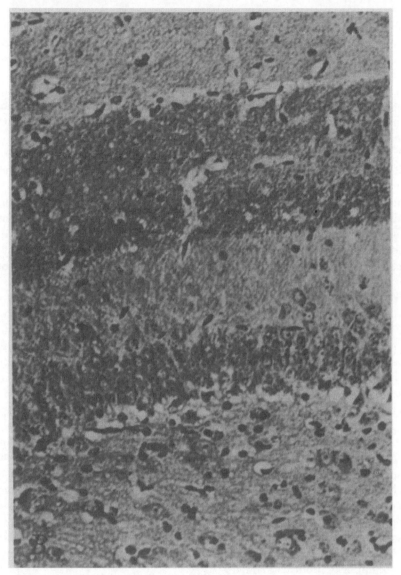

FIG. 8B: High power of region outlined in *A*. The hippocampal fissure is at the top of the figure. Note the dense degeneration band corresponding in "F" in *A*. Fink–Heimer I stain, modified. ×275.

ing the CA3c border of Lorente de Nó. The remaining mossy fiber bundle continues through all of CA3 and ends in CA2 (Blackstad *et al.*, 1970; Lorente de Nó, 1934, p. 144).

Hippocampal association fibers have been described by Schaffer (1892) and Ramón y Cajal (1893). Both of these authors described collaterals of CA3 pyramidal cells (according to Lorente de Nó, 1934, both CA3 and CA4) perforating the pyramidal layer and distributing to the stratum lacunosum of CA1 and (according to Lorente de Nó, 1934) of CA2. Furthermore, Lorente de Nó (1934) found another type of fiber with its primary source in CA2 and coursing with the long axis of the hippocampus. These fibers travel in the stratum radiatum as the longitudinal association path or axial association path. Hjorth-Simonsen (1973) was unable to find any intrahippocampal fibers in stratum lacunosum-moleculare (contrary to Raisman *et al.*, 1965) and considered the two bundles (i.e., Schaffer collaterals and axial bundle) to be synonymous, with only their axes being different.

Hjorth-Simonsen (1973) described fibers from CA1 distributed to the junction between subiculum and presubiculum. These fibers have been noted by others (Chronister and Zornetzer, 1973) and seem to have a component from the subiculum (Chronister and Zornetzer, 1973). It appears that there is a cascading of output from the allocortical dentate fascia to CA3 to CA1 to subiculum and/or presubiculum. This cascading input was described in part by Andersen *et al.* (1966) as a trisynaptic innervation from entorhinal cortex to CA1. Nafstad (1967) found few terminal boutons in CA1 following entorhinal damage. Although his observations were restricted to a small portion of CA1, they support the circuitous interrelations within the hippocampal formation. On the other hand, Segal (1972) showed that blockade of part of trisynaptic pathway does not eliminate CA1 responses to entorhinal stimulation. This demonstrates that the input to CA1 is not well understood, but it seems to be dependent on septal input (Grantyn *et al.* 1971) and perhaps receives afferents from the prepiriform regions (Hjorth-Simonsen, 1972) (see also Fig. 8 and its description in the text).

Little else is known about intrinsic intrahippocampal formation fiber systems. An association bundle within the entorhinal region has been described (Lorente de Nó, 1934), but this plexus has not been examined with degeneration techniques. There are fibers from CA3, CA4, and the polymorph cells within the hilus that are distributed to the initial segment (commissural zone) of the dentate fascia (Ramón y Cajal, 1893; Zimmer, 1971). The structural relationships among the parasubiculum, area retrosplenialis e, presubiculum, and subiculum in the internal organization of the hippocampal formation are unclear. It can be stated that these well-demarcated structures are much more important to normal hippocampal structure than previously thought, at least in relationship to the normal ongoing electrical activity of the hippocampus (Chronister *et al.*, 1974).

7. Hippocampal Efferents

Anatomists have long been intrigued by the efferent distribution of hippocampal fibers. As mentioned earlier, all of the hippocampus and subiculum contribute fibers to the fornix. Fornix fibers have been analyzed by numerous researchers from von Gudden (1881) or earlier to recent investigators. The classic descriptions are by Nauta (1956), Guillery (1956), and Raisman *et al.* (1966). The last authors maintain that there exists subfield specificity with regards to the hippocampal efferents. In other words, the individual fields of the hippocampus demarcated by Lorente de Nó (1934) have individual projection systems and terminations. For example, Raisman *et al.* (1966) stated that the subiculum (their prosubiculum) sent fibers, in the midline, to the regions of the hypothalamic–preoptic system that control pituitary function. Discrete lesions (300–400 μm in diameter) of the subiculum do not confirm the origin of this corticohypothalamic tract (Chronister and Zornetzer, 1973). Instead, subicular fibers were found to terminate in the lateral part of the medial septal nucleus and in nucleus accumbens. Siegel and Tassoni (1971a) claim that there is no subfield contribution to the fornix system. The contribution to the fornix is based on a dorsal–ventral distinction in the hippocampus. Dorsal hippocampus projects to the medial septal regions, while the ventral hippocampus projects to lateral septal regions. The total pattern of the hippocampal efferent distribution to the septal area, etc., is well defined (e.g., Nauta, 1956; Guillery, 1956) and deserves no further comment.

Adey *et al.* (1956) and Koikegami (1964) found that fibers exiting from the entorhinal region and traveling in the angular bundle join the stria medullares thalami to be distributed to the habenular region. Both Price and Powell (1971) and Hjorth-Simonsen (1972) described fibers from the posterior pyriform lobe to the anterior continuation of the hippocampus, which presumably is intimately related to the hypothalamus and septal area. The anterior hippocampus has very heavy fiber contributions to the tuberculum olfactorium and the medial forebrain bundle (Chronister, unpublished observations). Van Hoesen and Pandya (1973) were unable to find area 28 efferents traveling any farther than area 35. Although information from the hippocampus does reach area 28, the much-discussed exit from area 28 (e.g., Votaw, 1959, 1960) does not have strong anatomical verification. By far, the major number of hippocampal formation efferents exit via the fimbrial–fornix system despite strong behavioral arguments to the contrary (Douglas, 1967).

8. Conclusion

In the original postulation of a model in the limbic region of the cerebrum (White, 1965a), great emphasis was placed on the manner in which the allocortical and periallocortical areas were interrelated. Little attention was given to the way in which the limbic lobe proper was related to the remaining parts of the telencephalon. In addition, structural interrelations were not discussed in specific terms. These structural topics are extremely important to our understanding of temporal lobe function. It seems clear that isocortical fibers gain direct access to the limbic lobe through two major routes. The first of these routes is the cingulum fasciculus (Ramón y Cajal, 1911; Adey and Meyer, 1952; White, 1959). This bundle, as discussed earlier, terminates exclusively in the periallocortical hippocampal formation. The origin of the bundle, as also noted earlier, has been disputed in recent years. Classically, the cingulum has been believed to be comprised of both short, arcuate types of fibers and long fibers (see White, 1959, for review). They were assumed to arise from the frontal lobe, anterior thalamus, and cingulate gyrus. Electrophysiological evidence (White *et al.*, 1960) and differential cingulum lesions (White, 1959) demonstrate that frontal fibers, thalamic fibers, and cingulate fibers join the cingulum bundle. These origins are supported by the findings of Adey and Meyer (1952) concerning the frontal lobe and Nauta and Whitlock (1954) with thalamic fibers. The cingulate cortex contribution has been disputed recently (Domesick, 1969), although Ramón y Cajal (1911) argued strongly for its existence. Domesick (1973) claims that the majority of the degeneration in the cingulum can be caused by anterior thalamic nuclei lesions. The reasons for these apparent discrepancies are not clear. Despite the controversy, the cingulum fasciculus is a major extrinsic source of input to the hippocampal formation.

The general mode of access of isocortex information to the limbic lobe is via laterally oriented fibers that terminate either in area 35 or area 28. Little is known about the structural significance of this input system, but it represents, as pointed out by Van Hoesen *et al.* (1972), a mode of access in the isocortical sensory association systems to the hippocampal formation. This isocortical input is reinforced by lateral geniculate body fibers that terminate in a specialized section of the posterior hippocampal gyrus (MacLean and Creswell, 1970). Although distinct from areas 28 and 35, this region of the hippocampal gyrus is a proisocortex area (Vitzthum and Sanides, 1966). Thus proisocortex is important to the transfer of input to the limbic lobe.

The entorhinal region has direct access to the hippocampus proper and the dentate fascia. In primates, isocortical access to the hippocampal formation is probably isocortex to area 28 to allocortical hippocampal formation. In rodents, on the other hand, cortical input appears restricted to area 35, and perhaps the initial segment of area 28. The perirhinal region (35) has a heavy projection to area 28b, and the area between area 35 and area 28 has direct fibers to the dentate fascia. These latter fibers probably represent the system in the perirhinal region that Feldberg and Lotti (1970) found activated the hippocampus in much the same way as did excitation of the entorhinal area. These findings emphasize the significance of this lateral input system to both isocortex and allocortex.

Because of these recent findings, the model (White, 1965a) should be extended to include another associational ring which represents a transition from periallocortex to isocortex (Fig. 9). This border is ill-defined (Yakovlev *et al.*, 1966) but has been examined systematically by Sanides (1970, 1972). In well-documented studies, Sanides (1970, 1972) proposed the concept of proisocortex to refer to the interface between the allocortical type of structure and isocortex. Proisocortex, according to Sanides, is closer allied to isocortex than to periallocortex. This appears to be true in view of its input system and its ontogenetic development (see Fig. 7). Clearly, any cortical input to area 28 must pass at least through area 35. From the model, similar input to the other areas of the limbic lobe can be postulated.

Although difficult to specify, a proisocortex ring does enclose the entirety of the limbic lobe. Several other regions, in addition to area 35, that are included in proisocortex are a substantial area of the insula (which apparently projects to the temporal lobe and cingulate area, Astruc, 1972) and a small zone around the cingulate gyrus.

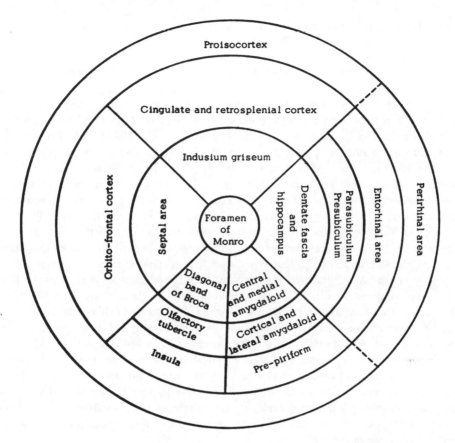

FIG. 9. A proposed revision of FIG. 1 showing the transition ring of proisocortex separating the limbic lobe from the isocortex. This proisocortex is very important to the influx of isocortical information to the limbic lobe.

Again, any isocortical input to the limbic lobe has to gain access to it by crossing the proisocortex. This zone, as Sanides has stressed, is probably structurally significant for the interactions of iso- and allocortex.

The internal organization of the periallocortex and the allocortex has also been ignored. For many years, it has been accepted that the characteristic hippocampal θ rhythm was exclusively septal dependent, despite some evidence to the contrary (Carreras *et al.*, 1955). We recently determined that the classically defined transitional area, the subiculum hippocampi, is crucial to the normal existence of θ rhythm (Chronister *et al.*, 1974). This finding again suggests that greater importance should be assigned to regions of transition. Perhaps the subiculum should be termed a "properiallocortex." Furthermore, the importance of the cascade sequence to the morphology of the cerebrum as suggested by Andersen *et al.* (1966) in the hippocampus–dentate fascia relations cannot be ignored. The need to concentrate on sequential neuron-to-neuron relationships must be examined since the sequences contain meaningful morphological information. (Indeed, this notion of sequence analysis has formed the analysis of the limbic lobe by Pribram and Kruger, 1954.)

Several other interesting aspects of the hippocampus should be emphasized. The concept, voiced originally by Lorente de Nó (1934), that the hippocampus is comprised of a series of interrelated structures arranged along an axis perpendicular to the alveus has been substantiated. Indeed, these chips or "lamellae" are organized very precisely within the hippocampus (Andersen et al., 1971) and also the hippocampus–subiculum complex (Chronister et al., 1974). The second observation is that the hippocampus appears to have an efferent system based not on subfield contribution but rather on its dorsoventral axis. This is certainly true of hippocampal–septal fibers (in both directions). The observations concerning nonseptal fibers are only preliminary (Chronister and Zornetzer, 1973), but the subiculum does not appear to project to the medial hypothalamic areas as suggested earlier by Raisman et al. (1966). The significance of the dorsoventral organization of hippocampal efferents and septohippocampal afferents is unclear at the present time.

It is becoming increasingly evident that the relatively simple regions of the forebrain are complex, especially in their relationships to the surrounding forebrain. That this is true is characterized by the isocortex–proisocortex–allocortex interactions and the septal area–allocortex–subiculum interrelationships. Future research will have to contend with the manner in which areas of the forebrain talk to each other. At the present time, it seems reasonable to postulate (on morphological grounds) the existence of a series of progressive cascades, beginning in the isocortex and terminating in the allocortex. This concept is very similar to that proposed earlier as the result of physiological neuronography studies (Pribram and MacLean, 1953). Additional studies must be done to clarify the significance of these observations to the integrated relations of the forebrain.

9. Addendum

Additional observations on the hippocampal formation have clarified some of the issues in this chapter. The subiculum–CA1 complex appears to undergo considerable elaboration with phylogenetic advance (Chronister and White, 1974). The subiculum also seems to be the source of postcommissural fornix fibers to the mammillary bodies. The presubiculum–parasubiculum area is the apparent source of fibers to the anterior thalamus (Chronister and White, 1975).

Rapid Golgi observations on the septal area (Chronister, DeFrance, Srebro, and White, in preparation) reveal that the midline septal area has bilateral dendritic arborizations. These neurons respond to bilateral medial forebrain bundle stimulation but not to stimulation to the fornix–fimbria. In addition, the septal area has a very complex morphology that is poorly understood. The lateral septal area has a completely different appearance in Golgi preparations than does the medial septal–diagonal band area. Neurons in the lateral septal region have spiny dendrites, whereas neurons in the medial septal region are nearly spine free but have very large and extensive dendritic arborizations. These observations suggest a possible functional dichotomy in the septal area including the relationship of the septal area to the hippocampus.

ACKNOWLEDGMENTS

The authors wish to thank B. Kimball and S. Walker (University of South Alabama) and G. Smith (University of Florida) for histological assistance, W. Lamm and D. Lee (University of South Alabama) for assistance in photography, K. Farnell (University of South Alabama) for technical assistance, and L. Wilks and L. Hollinger for secretarial assistance. Special thanks are expressed to Academic Press for permission to reproduce Fig. 1. This work was supported by NIH Grant No. NS 10809.

10. References

ADEY, W. R., AND MEYER, M. An experimental study of hippocampal afferent pathways from prefrontal and cingulate areas in the monkey. *Journal of Anatomy*, 1952, **86,** 58–74.

ADEY, W. R., MERRILLEES, N. C. R., AND SUNDERLAND, S. The entorhinal area: Behavioral, evoked potential and histological studies of its interrelationships with brain-stem regions. *Brain*, 1956, **79,** 414–439.

ADEY, W. R., SUNDERLAND, S., AND DUNLOP, C. W. The entorhinal area: Electrophysiological studies on its interrelationships with rhinencephalic structures and the brainstem. *Electroencephalography and Clinical Neurophysiology*, 1957, **9,** 309–324.

ALLEN, W. F. Fibre degeneration in the Ammon's horn resulting from extirpation of the pyriform and other cortical areas and from transection of the horn at various levels. *Journal of Comparative Neurology*, 1948, **88,** 425–438.

ANDERSEN, P., BRULAND, H., AND KAADA, B. R. Activation of the dentate area by septal stimulation. *Acta Physiologica Scandanavica*, 1961, **51,** 17–28.

ANDERSEN, P., ECCLES, J. C., AND LØYNING, Y. Location of postsynaptic inhibitory synapses on hippocampal pyramids. *Journal of Neurophysiology*, 1964, **27,** 592–607.

ANDERSEN, P., HOLMQVIST, B., AND VOORHOEVE, P. E. Excitatory synapses on hippocampal apical dendrites activated by entorhinal stimulation. *Acta Physiologica Scandinavica*, 1966, **66,** 461–472.

ANDERSEN, P., BLISS, T. V. P., AND SKREDE, K. K. Lamellar organization of hippocampal excitatory pathways. *Experimental Brain Research*, 1971, **13,** 222–238.

ASTRUC, J. Corticofugal fiber degeneration following lesions of the insular cortex in *Macaca mulatta*. Paper delivered before Society for Neuroscience, Houston, Texas, 1972.

AZOULAY, L. Structure de la corne D'Ammon chez l'enfant. *Comptes Rendus des Séances Societé de Biologie (Paris)*, 1894, **1,** 212–214.

BAN, T., AND ZYO, K. Experimental studies on the fiber connections of the rhinencephalon. 1. Albino rat. *Medical Journal of Osaka University*, 1962, **12,** 385–424.

BLACKSTAD, T. W. Commissural connections of the hippocampal region in the rat, with special reference to their mode of termination. *Journal of Comparative Neurology*, 1956, **105,** 417–537.

BLACKSTAD, T. W. On the termination of some afferents to the hippocampus and fascia dentata: An experimental study in the rat. *Acta Anatomica*, 1958, **35,** 202–214.

BLACKSTAD, T. W. BRINK, K., HEM, J., AND JEUNE, B. Distribution of hippocampal mossy fibers in the rat: An experimental study with silver impregnation methods. *Journal of Comparative Neurology*, 1970, **138,** 433–450.

BRAAK, H. Zur Pigmentarchitektonik der Grosshirnrinde des Menschen. II. Subiculum. *Zeitschrift fur Zellforschung und Mikroskopische Anatomie,* 1972, **131,** 235–254.

BROCA, P. Nomenclature cérébrale dénomination des divisions et subdivisions des hémisphéres et des anfractuosités de leur surface. *Revue D'Anthropologie,* 1878a, **1,** 193–236.

BROCA, P. Anatomie comparée des circonvolutions, cérébrales: Le grande lobe limbique et la scissure limbique dans la série des mammiferes. *Revue d'Anthropologie,* 1878b, 1(3s), 385–498.

BRODAL, A. *Neurological anatomy in relation to clinical medicine,* New York: Oxford University Press, 1969.

BRODMANN, K. *Vergleichende Lokalisationslehre der Grosshirnirnde in ihren Prinzipien dargestellt auf Grund des Zellenbaues.* Leipzig: Barth, 1909.

CARRERAS, M., MACCHI, G., ANGELERI, F., AND URBANI, M. Sull attivita elettrica della formazione ammonica: Effetti determinati dall'ablazione della corticcia entorinale. *Bollettino Della Societa Itoliana Di Biologica Sperimental,* 1955, **31,** 182–184.

CHRONISTER, R. B., AND WHITE, L. E., JR. Ipsilateral afferents to the dentate fascia of rodents. *Anatomical Record,* 1972, **172,** 289–290.

CHRONISTER, R. B., AND WHITE, L. E., JR. Paper delivered at the Richard A. Lende Memorial Symposium. Albany, N.Y., 1974.

CHRONISTER, R. B., AND WHITE, L. E., JR. In *International septal symposium,* Detroit: Wayne State Press, 1975, in press.

CHRONISTER, R. B., AND ZORNETZER, S. F. The fiber connections of allocortex: The subiculum hippocampi. *Anatomical Record,* 1973, **175,** 292.

CHRONISTER, R. B., ZORNETZER, S. F., BERNSTEIN, J. J., AND WHITE, L. E., JR. Hippocampal theta rhythm; intra-hippocampal formation contributions. *Brain Research,* 1974, **65,** 13–28.

CRAGG, B. G. Olfactory and other afferent connections of the hippocampus in the rabbit, rat, and cat. *Experimental Neurology,* 1961, **3,** 588–600.

CROSBY, E. C. The forebrain of *Alligator mississippiensis. Journal of Comparative Neurology,* 1917, **27,** 325–402.

CROSBY, E. C., HUMPHREY, T., AND LAUER, E. W. *Correlative anatomy of the nervous system,* New York: Macmillan, 1962.

DEVITO, J. L., AND WHITE, L. E., JR. Projections from the fornix to the hippocampal formation in the squirrel monkey. *Journal of Comparative Neurology.* 1966, **127,** 389–398.

DOINIKOW, B. Beitrag zur vergleichenden Histologie des Ammonshorns, *Journal für Psychologie und Neurologie,* 1908, **13,** 166–202.

DOMESICK, V. B. Projections from the cingulate cortex in the rat. *Brain Research,* 1969, **12,** 296–320.

DOMESICK, V. B. Thalamic projections in the cingulum bundle to the parahippocampal cortex of the rat. *Anatomical Record,* 1973, **175,** 308.

DOUGLAS, R. J. The hippocampus and behavior. *Psychological Bulletin,* 1967, **67,** 416–442.

ELLIOT SMITH, G. The relation of the fornix to the margin of the cerebral cortex. *Journal of Anatomy and Physiology,* 1898, **32,** 25–58.

FELDBERG, W., AND LOTTI, V. J. Direct and indirect activation of the hippocampus by tubocurarine. *Journal of Physiology,* 1970, **210,** 697–716.

FILIMONOFF, I. N. A rational subdivision of the cerebral cortex. *Archives of Neurology and Psychiatry,* 1947, **58,** 296–311.

FLECHSIG, P. *Plan des menschlichen Gehirns.* Leipzig: von Veit, 1883.

GENESER-JENSEN, F. A. Distribution of acetyl cholinesterase in the hippocampal region of the guinea pig. III. The dentate area. *Zeitschrift fur Zellforschung und mikroskopische Anatomie,* 1972, **131,** 481–495.

GEREBTZOFF, M. A. Sur quelques voies d'association de l'ecore cerebrale. (Recherches anatomo-experimentales). *Journal Belge de Neurologie et Psychiatrie,* 1939, **39,** 205–221.

GLOOR, P. Electrophysiological studies on the connections of the amygdaloid nucleus in the cat. Part 1. The neuronal organization of the amygdaloid projection system. *Electroencephalography and Clinical Neurophysiology,* 1955, **7,** 223–242.

GOTTLIEB, D. I., AND COWAN, W. M. Evidence for a temporal factor in the occupation of available synaptic sites during the development of the dentate gyrus. *Brain Research,* 1972, **41,** 452–456.

GOTTLIEB, D. I., AND COWAN, W. M. Autoradiographic studies of the commissural and ipsilateral association connections of the hippocampus and dentate gyrus of the rat. I. The commissural connections. *Journal of Comparative Neurology*, 1973, **149**, 393–422.

GRANTYN, A., GRANTYN, R., AND HANG, T. L. Hippokampole Einzelzellantworten auf mesenzephale Reizungen nach Septumläsion. *Acta Biologica et Medica Germanica*, 1971, **26**, 985–996.

GREEN, J. D., AND ARDUINI, A. Hippocampal electrical activity in arousal. *Journal of Neurophysiology*, 1954, **17**, 533–557.

GUILLERY, R. W. Degeneration in the post-commissural fornix and the mammillary peduncle of the rat. *Journal of Anatomy*, 1956, **90**, 350–371.

HEIMER, L. Synaptic distribution of centripetal and centrifugal nerve fibers in the olfactory system of the rat: An experimental anatomical study. *Journal of Anatomy*, 1968, **103**, 413–432.

HILL, A. The hippocampus. *Philosophical Transactions of the Royal Society of London (Series B)*, 1894, **184**, 389–429.

HILPERT, P. Der Mandelkern des Menschen. 1. Cytoarchitektonik und Faserverbindungen. *Journal für Psychiatrie und Neurologie*, 1928, **36**, 44–74.

HJORTH-SIMONSEN, A. Hippocampal efferents to the ipsilateral entorhinal area: An experimental study in the rat. *Journal of Comparative Neurology*, 1971, **142**, 417–437.

HJORTH-SIMONSEN, A. Projection of the lateral part of the entorhinal area to the hippocampus and fascia dentata. *Journal of Comparative Neurology*, 1972, **146**, 219–232.

HJORTH-SIMONSEN, A. Some intrinsic connections of the hippocampus in the rat: An experimental analysis. *Journal of Comparative Neurology*, 1973, **147**, 145–162.

HJORTH-SIMONSEN, A., AND JEUNE, B. Origin and termination of the hippocampal perforant path in the rat studied by silver impregnation. *Journal of Comparative Neurology*, 1972, **144**, 215–232.

HUMPHREY, T. The hippocampal vestiges in relation with the dorsal commissural systems in mammals. *University of Michigan Medical Bulletin*, 1937, **3**, 34–35.

JONES, E. G., AND POWELL, T. P. S. An anatomical study of converging sensory pathways within the cerebral cortex of the monkey. *Brain*, 1970, **93**, 793–820.

KERR, D. I. B., AND DENNIS, B. J. Collateral projection of the lateral olfactory tract to entorhinal cortical areas in the cat. *Brain Research*, 1972, **36**, 399–403.

KOIKEGAMI, H. Amygdala and other related limbic structures; experimental studies on the anatomy and function. II. Functional experiments. *Acta Medica et Biologica*, 1964, **12**, 73–266.

KRIEG, W. J. S. Connections of the cerebral cortex. *Journal of Comparative Neurology*, 1946, **84**, 221–225.

LEWIS, F. T. The significance of the term hippocampus. *Journal of Comparative Neurology*, 1923, **35**, 213–230.

LORENTE DE NÓ, R. Studies on the structure of the cerebral cortex. I. The area entorhinalis. *Journal für Psychologie und Neurologie*, 1933, **45**, 381–438.

LORENTE DE NÓ R. Studies on the structure of the cerebral cortex. II. Continuation of the study of the ammonic system. *Journal für Psychologie und Neurologie*, 1934, **46**, 113–177.

MACLEAN, P. D., AND CRESWELL, G. Anatomical connections of visual system with limbic cortex of monkey. *Journal of Comparative Neurology*, 1970, **138**, 265–278.

MELLGREN, S. I., AND SREBRO, B. Changes in acetylcholinesterase and distribution of degenerating fibres in the hippocampal region after septal lesions in the rat. *Brain Research*, 1973, **52**, 19–36.

MEYNERT, T. The brain of mammals. In S. Stricker (Ed.), *Manual of human and comparative histology*. Vol. 2 (trans. by H. Power). London: The New Sydenham Society, 1872, pp. 367–537.

MEYNERT, T. *Psychiatry. A clinical treatise on diseases of the forebrain based upon a study of its structure, functions, and nutrition*. New York: G. P. Putnam's Sons, 1885.

NAFSTAD, P. H. J. An electron microscope study on the termination of the perforant path fibers in the hippocampus and fascia denta. *Zeitschrift für Zellforschung und mikroskipische Anatomie*, 1967, **76**, 532–542.

NAUTA, W. J. H. Ueber die sogenannte terminale Degeneration im Zentralnervensystem und ihre Darstellung durch Silberimprägnation. *Schweizer Archiv für Psychiatrie*, 1950, **66**, 353–376.

NAUTA, W. J. H. An experimental study of the fornix in the rat. *Journal of Comparative Neurology*, 1956, **104**, 247–272.

Nauta, W. J. H. Some brain structures and functions related to memory. In F. O. Schmitt and T. Melnechick (Eds.), *Neurosciences research symposium summaries.* Cambridge, Mass.: M.I.T. Press, 1966, pp. 73–103.

Nauta, W. J. H., and Whitlock, D. G. An anatomical analysis of the non-specific thalamic projection system. In J. F. Delafresnaye (Ed.), *Brain mechanisms and consciousness.* Springfield, Ill.: Charles C. Thomas, 1954, pp. 81–116.

Owen, R. *Anatomy of vertebrates.* 3 vols. London: Longmans and Green, 1868.

Powell, T. P. S., Cowan, W. M., and Raisman, G. The central olfactory connections. *Journal of Anatomy,* 1965, **99,** 791–813.

Pribram, K. H., and Kruger, L. Functions of the "olfactory brain." *Annals of the New York Academy of Sciences,* 1954, **58,** 109–138.

Pribram, K. H., and MacLean, P. D. Neuronographic analysis of medial and basal cerebral cortex. *J. Neurophysiol.,* 1953, **16,** 324–340.

Price, J. L. An autoradiographic study of complementary laminar patterns of termination of afferent fibers to the olfactory cortex. *Journal of Comparative Neurology,* 1973, **150,** 87–108.

Price, J. L., and Powell, T. P. S. Certain observations of the olfactory pathway. *Journal of Anatomy,* 1971, **110,** 105–126.

Raisman, G., Cowan, W. M., and Powell, T. P. S. The extrinsic afferent, commissural and association fibers of the hippocampus, *Brain,* 1965, **88,** 963–996.

Raisman, G., Cowan, W. M., and Powell, T. P. S. An experimental analysis of the efferent projection of the hippocampus. *Brain,* 1966, **89,** 83–108.

Ramón y Cajal, S. Estructura del asta de Ammon. *Anales de la Sociedad Espanola de Historia Natural.* 1893, **22,** 53–114. Translated as *The structure of ammon's horn.* Springfield, Ill.: Charles C. Thomas, 1968.

Ramón y Cajal, S. Estudios sobre la corteza cerebral humana. IV. Estructura de la corteza cerebral ofactiva del hombre y mamiferos. *Trabajos del Laboratorio de Investigaciones Biologicas de la Universidad Madrid,* 1901, **1,** 1–140. Translated as *Studies on the cerebral cortex.* Chicago: New Book, 1955.

Ramón y Cajal, S. *Histologie de Systeme Nerveux de l'Homme et des Vertebres.* Vol. 2. Paris: Maloine, 1911.

Rose, M. Histologische Lokalisation der Grosshirnrinde bei kleinen Säugeticien (Rodentia, Insectivora, Chiroptera). *Journal für Psychologie und Neurologie,* 1912, **19,** 391–479.

Rose, M. Ueber das histogenetische Prinzip der Einteilung der Grosshirnrinde. *Journal für Psychologie und Neurologie,* 1926, **32,** 97–160.

Rose, M. Die sog. Riechrinde beim Menschen und beim Affen, II. "Teil des Allocortex bei Tier und Mensch." *Journal für Psychologie und Neurologie,* 1927, **34,** 261–401.

Russell, G. V. Interrelationships within the limbic and centrencephalic systems. In D. E. Sheer (Ed.), *Electrical stimulation of the brain.* Austin: University of Texas Press, 1961, pp. 167–181.

Sanides, F. Functional architecture of motor and sensory cortices in primates in the light of a new concept of neocortex evolution. In C. R. Noback and W. Montagne (Eds.), *The primate brain.* New York: Appleton-Century-Crofts, 1970, pp. 137–208.

Sanides, F. Representation in the cerebral cortex and its areal lamination patterns. *Structure and function of nervous tissue,* 1972, **5,** 329–453.

Scalia, F. Some olfactory pathways in the rabbit brain. *Journal of Comparative Neurology,* 1966, **126,** 285–310.

Schaffer, K. Beitrag zum Histologie der Ammonshornformation. *Archiv für Mikroskopische Anatomie,* 1892, **39,** 611–632.

Segal, M. Hippocampal unit responses to perforant path stimulation. *Experimental Neurology,* 1972, **35,** 541–546.

Siegel, A., and Tassoni, J. P. Differential efferent projections from the ventral and dorsal hippocampus of the cat. *Brain, Behavior, and Evolution,* 1971a, **4,** 185–200.

Siegel, A., and Tassoni, J. P. Differential efferent projections of the lateral and medial septal nuclei to the hippocampus in the cat. *Brain, Behavior, and Evolution,* 1971b, **4,** 201–219.

Simpson, E. A. The efferent fibers of the hippocampus in the monkey. *Journal of Neurology, Neurosurgery, and Psychiatry,* 1952, **15,** 79–92.

Smith, R. W., and White, L. E., Jr. The fiber architectonics of the cat hippocampal formation. *Journal of Comparative Neurology,* 1964, **123,** 11–28.

Sommer, W. Erkrankung des Ammonshorns als aetiologisches Moment der Epilepsie. *Archiv für Psychiatrie und Nervenkrankheiten,* 1880, **10,** 631–675.

Sperti, L., Gessi, T., Volta, F., and Sanseverino, E. R. Synaptic organization of commissural projections of the hippocampal region in the guinea pig. II. Dorsal psalterium: pre-hippocampal and intrahippocampal relays. *Archivio di Scienze Biologiche,* 1970, **54,** 183–210.

Tilney, F. The hippocampus and its relations to the corpus callosum. *Bulletin of the Neurological Institute of New York,* 1938, **7,** 1–77.

Valverde, F. *Studies on the piriform lobe.* Cambridge, Mass.: Harvard University Press, 1965.

Van Hoesen, G. W., and Pandya, D. N. Afferent and efferent connections of the perirhinal cortex (area 35) in the rhesus monkey. *Anatomical Record,* 1973, **175,** 460.

Van Hoesen, G. W., Pandya, D. N., and Butters, N. Cortical afferents to the entorhinal cortex of the rhesus monkey. *Science,* 1972, **175,** 1471–1473.

Vaz Ferreira, A. The cortical areas of the albino rat studied by silver impregnation. *Journal of Comparative Neurology,* 1951, **95,** 177–243.

Vitzthum, H., and Sanides, F. Entwicklungsprinzipen der menschlichen Sehrinde. In R. Hassler and H. Stephan (Eds.), *Evolution of the forebrain: Phylogenesis and ontogenesis of the forebrain.* Stuttgart: Georg Thieme, 1966, pp. 435–442.

von Gudden, B. Beitrag zur Kenntniss des Corpus mammillare und des sogenannten Schenkel der fornix. *Archiv für Psychiatrie und Nervenkrankheiten,* 1881, **11,** 428–452.

von Koelliker, A. *Handbuch der Gewelbelehre des Menschen.* Vol. 2: *Nervensystem des Menschen und der Tiere.* Leipzig: Wilhelm Engelman, 1896.

Votaw, C. L. Certain functional and anatomical relations of the cornu Ammonis of the macaque monkey. 1. Functional relations. *Journal of Comparative Neurology,* 1959, **112,** 353–382.

Votaw, C. L. Certain functional and anatomical relations of the cornu Ammonis of the macaque monkey. 1. Anatomical relations. *Journal of Comparative Neurology,* 1960, **114,** 283–293.

Votaw, C. L., and Lauer, E. W. An afferent hippocampal fiber system in the fornix of the monkey. *Journal of Comparative Neurology,* 1963, **121,** 195–206.

White, L. E., Jr. Ipsilateral afferents to the hippocampal formation in the albino rat. 1. Cingulum projections. *Journal of Comparative Neurology,* 1959, **113,** 1–41.

White, L. E., Jr. A morphologic concept of the limbic lobe. *International Revue of Neurobiology,* 1965a, **8,** 1–34.

White, L. E., Jr. Olfactory bulb projections of the rat. *Anatomical Record,* 1965, **152b,** 465–480.

White, L. E., Jr., Nelson, W. M., and Foltz, E. L. Cingulum fasciculus study by evoked potentials. *Experimental Neurology,* 1960, **2,** 406–421.

Yakovlev, P. I., Locke, S., and Angevine, J. B. The limbus of the cerebral hemisphere, limbic nuclei of the thalamus, and the cingulum bundle. In D. P. Purpura and M. D. Yahr (Eds.), *The thalamus.* New York: Columbia University Press, 1966, pp. 77–97.

Zimmer, J. Ipsilateral afferents to the commissural zone of the fascia dentata, demonstrated in decommissurated rats by silver impregnation. *Journal of Comparative Neurology,* 1971, **142,** 393–416.

Zuckerkandl, E. Das Riechbundel des Ammonshornes. *Anatomischer Anzeiger,* 1888, **3,** 425–434.

Zuckerkandl, E. Beiträge zur Anatomie des Riechcentrums. *Sitzenberichte der Kaiserlichen Akademie der Wissenschafler im Wien, der Mathematisch-naturwissenschaftlichen Klasse,* 1900, **109,** 459–500.

2

Septohippocampal Interface

Ervin William Powell and Garth Hines

1. Introduction

The hippocampal formation is usually understood to include the hippocampal gyrus (or at least the parahippocampal portion of that gyrus), the hippocampus proper (cornu ammonis), and the fascia dentata (Fig. 1). The hippocampal rudiment, located in the rostral septal area and ventral to the genu of the corpus callosum, was suggested by Hines (1922) as the anlage of the fascia dentata. Furthermore, its cells are cytoarchitectonically similar to the granular cells of the fascia dentata. However, an adequate comparative cell study of the various dense granule cell areas of the brain (e.g., the fascia dentata, olfactory tubercle, hippocampal rudiment, and pyriform cortex) has not been performed, so whether these areas are rudimentary portions of the hippocampus or relatively undifferentiated cell clusters is not known. In primitive animals (Hoffman, 1967), the hippocampus forms a medial archicortex along the major length of the brain, forming a pre- and postcommissural hippocampus. In higher animals, the hippocampus gradually comes to lie entirely in the temporal lobe, as it does in the squirrel monkey. The dorsal hippocampus is prominent in the rat, with very little ventral hippocampus, while the cat and dog have a nearly equal representation of both the dorsal and ventral portions.

The hippocampus forms a semicircle around the thalamus. That part which extends over the thalamus is referred to by many authors as the dorsal hippocampus; that part which extends into the temporal horn of the lateral ventricle is regarded as the ventral hippocampus. These two portions may be further subdivided into anterior

Ervin William Powell • Department of Anatomy, University of Arkansas School of Medicine, Little Rock, Arkansas. Garth Hines • Department of Psychology, University of Arkansas at Little Rock, Little Rock, Arkansas. This research was supported by NSF Grant GB 32170.

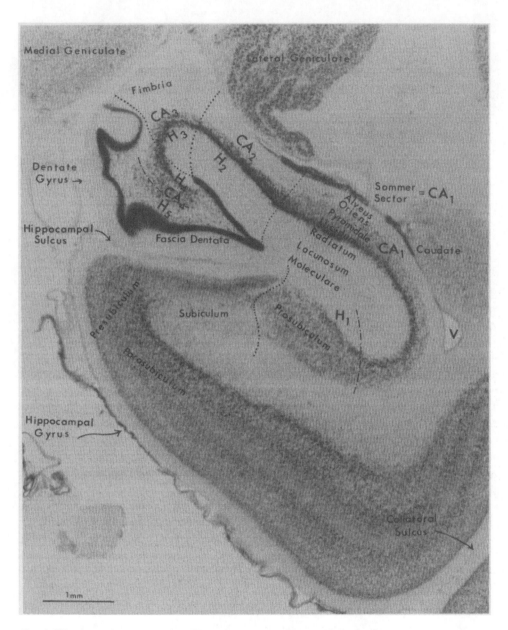

Fig. 1. Hippocampal area in monkey. This coronal section corresponds to the stereotaxic coordinate of A3.5 (Emmers and Akert, 1963). Cresyl violet stain.

and posterior parts: i.e., anterior ventral hippocampus, posterior ventral hippocampus, posterior dorsal hippocampus, and anterior dorsal hippocampus. Elul (1964) has made this type of regional differentiation for the cat. In most primates, the hippocampus is located entirely in the temporal horn of the lateral ventricle, hence making the term "dorsal hippocampus" inappropriate; the terms "anterior hippocampus" and "posterior hippocampus" as used by Simpson (1952) are more useful.

Cytoarchitectonic layers and fields have been delineated for the hippocampus (Ramón y Cajal, 1911, translated by Kraft, 1968; Rose, 1926; Lorente de Nó, 1934). The alveus layer is next to the eipthelium of the lateral ventricle. The subsequent layers are shown in Fig. 1 as oriens, pyramidale, lucidum, radiatum, lacunosum, and moleculare, respectively. The areas designated as CA1, etc., are according to the classification of Lorente de Nó (1934), and the areal markings H_1, etc., are based on the work of Rose (1926, 1927). H_1 corresponds to the subiculum and CA1, H_2 to CA2 and CA3, with H_4 and H_5 included in CA4.

Cytoarchitecturally, the hippocampus merges with the septum rostrally, especially in the region of the hippocampal rudiment (Young, 1936). Working with amphibians and reptiles, Hines (1929) and Hoffman (1967) defined the septum as the medial telencephalic wall ventral to the hippocampal formation and rostral to the anterior commissure. This placed the nucleus accumbens intermediate between the septum and the caudate nucleus. A compreshensive account of the septal area has been made by Young (1936), which is also contained in a practical stereotaxic atlas for the rat (König and Klippel, 1963). The basic nuclear areas are shown in Fig. 2A, and consist principally of the medial and lateral septal nuclei and nucleus of the diagonal band (the last nucleus having an infracommissural location). The septofimbrial and triangular nuclei are considered components of the caudal septum. Raisman (1966) appeared to include the bed nucleus of the stria terminalis as part of the septum. Crosby et al. (1967) included as part of the septal area the bed nucleus of the anterior commissure, the bed nucleus of the stria terminalis, and the nucleus accumbens. Further, they appear to use the terms "paraolfactory region" and "septal area" synonymously. Powell (1963) limited the septum to that area rostral and anterior to the anterior commissure and within the near margin of the corpus callosum—a region that Nauta (1956) refers to as the "supracommissural septum." Finally, considerable cytoarchitectural detail has been worked out for the septal area by Andy and Stephan for a variety of animals including the cat (1964a), monkey (1964b), and man (1968). These authors have classified septal nuclei into four groups—dorsal, ventral, medial, and caudal—which in turn have been divided into subnuclei, as shown in Figs. 2B and 2C for the cat and monkey.

2. Connections Between the Septum and the Hippocampus

2.1. Hippocamposeptal

While attempts to delineate fiber paths from the hippocampus to general septal regions have varied greatly in terms of procedures (see Appendix), species investi-

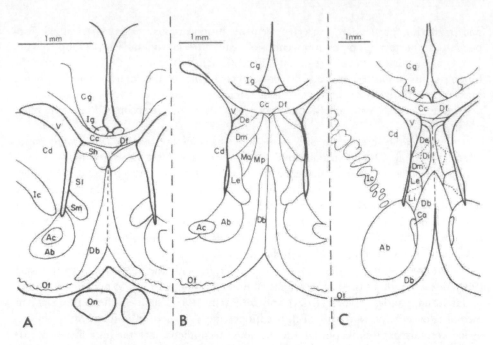

FIG. 2. Midseptal area representative of septal nuclear divisions. A: Rat. B: Cat. C: Monkey. Ab,
Accumbens nucleus (n); Ac, anterior commissure; Ca, island of Calleja; Cc, corpus callosum; Cd, caudate
n; Db, diagonal band n; De, dorsal external septal n; Df, dorsal fornix; Di, dorsal internal septal n; Dm,
dorsal intermediate septal n; Ic, internal capsule; Ig, indusium griseum; Le, lateral external septal n; Li,
lateral internal septal n; Ma, medial anterior septal n; Mp, medial posterior septal n; Sh,
septohippocampal n; Sl, lateral septal n; Sm, medial septal n; On, optic nerve; Ot, olfactory tubercle; V,
ventricle.

gated, and specifics of connections observed, hippocampal projections have been
clearly established to the medial and lateral septal nuclei (Young, 1936; Nauta,
1956, 1958; Johnson, 1957; Cragg, 1961; Ban and Zyo, 1962; Knook, 1966), the
nucleus accumbens (Carman *et al.*, 1963; Knook, 1966), and the nucleus of the
diagonal band (Craigie, 1925; Daitz and Powell, 1954; Knook, 1966). Ban and Zyo
(1962) have further reported connections with the bed nucleus of the stria terminalis,
although Knook (1966) has claimed that these were fibers of passage, terminating in
the anterior thalamic nuclei.

There would appear to be a large degree of topographic specificity in these projection systems, although the nature of the relationships may vary from species to
species. Simpson (1952), using the Adey and Meyer (1952) modification of the Glees
stain, has described the topography of hippocamposeptal projections in the monkey.
In general, lesion of the anterior hippocampus was found to result in degeneration in
the ventral part of the lateral septal nuclei and the medial nucleus accumbens. Lesion
of the central portion of the hippocampus resulted in almost no septal degeneration,
with the small amount seen limited to the ventral part of the lateral septal nuclei. Lesion of the posterior hippocampal region was followed by degeneration in the dorsal

part of the lateral septal nuclei, with some fibers passing horizontally to the diagonal band.

Raisman *et al.* (1966), using Nauta–Gygax stains to trace degeneration in the rat, reported still further topographic specificity. They found fiber pathways through the fimbria (which they indicated receives axons from the posterior CA1 region, from CA3 and CA4, and possibly from CA2) passing through the precommissural fornix to terminate in the septofimbrial nucleus, lateral septal nucleus, medial septal nucleus (except for the dorsomedial quadrant or that portion of the medial region immediately adjacent to the midline), and the nucleus accumbens. Fiber pathways to the septum passing through the dorsal fornix (which receives its axons from the anterior portion of CA1) were found to pass from the precommissural fornix through the septofimbrial nucleus, with rostral fibers terminating in the middorsoventral segment of the medial septal nucleus close to the midline. Some few of these fibers were found leaving the medial septal region to course caudally in the diagonal band. The specificity of these relationships was further delineated by these authors through the placement of lesions in the specific hippocampal regions. Lesion of the anterior portion of CA1 resulted in degeneration in the middorsoventral segment of the medial septal nucleus and the diagonal band, with some evidence of termination in the nucleus of the diagonal band. Lesion of the posterior portion of CA1 resulted in degeneration in the medioventral quadrant of the lateral septal nucleus. Lesion of CA3 and CA4 resulted in terminal degeneration in the dorsolateral quadrant of the lateral septal nucleus, the nucleus accumbens, the nucleus of the diagonal band, and the medial forebrain bundle. Finally, the authors suggest that CA2 has no septal projections, contributing solely to the postcommissural pathway.

Siegel and Tassoni (1971a), using Fink–Heimer and Nauta–Gygax stains as well as electron microscopic examination to study projection pathways in the cat, reported a topography based on a dorsal–ventral hippocampus distinction, rather than on that of hippocampal cytoarchitectonic divisions. Unilateral lesions of the anterior portions of the ventral hippocampus were found to result in fiber path degeneration through the fimbria to the lateral margin of the subcallosal fornix on both sides. Upon reaching the ipsilateral septum, degenerating fibers were found to pass via the precommissural fornix to terminate in the lateral septal nucleus and the nucleus accumbens. There was also a distinct bundle of fibers which joined the diagonal band to terminate in the nucleus of the diagonal band, and to pass on to the olfactory tubercle rostrally and to the ventrolateral preoptic area immediately above the supraoptic nucleus. Very little difference was seen in the projections (which pass through the ventral psalterium) to the contralateral septal region, although lateral septal degeneration was limited to the caudal end of the ventrolateral septum. One cat received a lesion at the posterior crus of the ventral hippocampus. The projections observed were basically the same as those described above, although septal terminations were found only in the midlateral septum.

Lesion of the dorsal hippocampus was found by these authors to result in fiber degeneration in the medial half of the subcallosal fornix. These fibers were traced directly to the septum, where they descended ventromedially in the precommissural fornix to terminate in the medial septum, with a bundle continuing to the nucleus of the diagonal band. No degeneration was found in the lateral septal region or the olfactory tubercle. In opposition to the findings of Raisman *et al.* (1966), these authors found no degeneration in the dorsal fornix following lesion of any portion of the hippocampus.

2.2. Septohippocampal

Again, although procedures and species of investigation have varied greatly, projections from the lateral and medial septal nuclei (Daitz and Powell, 1954; Ban and Zyo, 1962; Votaw and Lauer, 1963; Morin, 1950; Knook, 1966; Powell, 1963) and the nucleus of the diagonal band (Craigie, 1925; Daitz and Powell, 1954) to the hippocampus have been clearly established.

Knook (1966), using Nauta–Gygax stains with rats, described the pathway as originating or gathering in the medial septal nucleus, from which the majority of fibers pass via the fornix body to the hippocampus. A few fibers were reported reaching the hippocampus by the dorsal fornix. He cautioned, however, that these fibers may all originate in the nucleus of the diagonal band, with no fibers reaching the hippocampus from the medial septal nucleus.

Powell (1963), investigating rat septal efferents using Nauta–Gygax staining of fiber degeneration, concluded that septal fibers from midline septal regions enter the body of the fornix, reaching the hippocampus over the alveus. Lesion of the dorsal septum resulted in a greater degeneration density in the dorsal region of the fornix body, while posterior midline regions projected with greater density at the lateral tip of the fornix body. Posterior septal lesions caused large amounts of degeneration in both the body and column of the fornix. These fibers were observed to terminate generally, throughout the hippocampus.

Raisman (1966), using Nauta–Gygax to investigate septal connections in the rat, found that the caudal aspect of the medial septal nucleus and the diagonal band send fibers through the septofimbrial nucleus to lie first in the lateral and then the ventral portion of the ventral hippocampal commissure. From there, they were seen to enter the ventral portion of the fimbria, sending terminals to regions CA2, CA3, CA4, and the fascia dentata. No degeneration was found in CA1, nor did lesion of the lateral septal nucleus result in hippocampal degeneration, although lateral septal lesions did result in degeneration in the nucleus accumbens, rostral olfactory tubercle, and medial forebrain bundle. These results stand in contradiction of the findings reported earlier by Powell (1963), who found that lesion of the dorsal midline region of the septum (anterior and posterior parts) resulted in degeneration in the medial portion of CA1, as well as in CA2, CA3, and CA4. Further, lesions of the lateral septal area were reported to produce degeneration in the fascia dentata. This latter difference between the findings of Powell and those of Raisman may be due either to the more anterior placement of the Powell lesions or to the involvement of fibers of the medial nuclei in the Powell lesion. The latter would not seem to be the case, however, since medial lesions were not seen to involve the fascia dentata.

Siegel and Tassoni (1971*b*), using Fink–Heimer stains to investigate septal efferents in the cat, disagree with Raisman as to the topographical details of septohippocampal projections. Lateral septal lesions were found to result in degeneration through the fornix body to the fimbria, and then directly to all portions of the ventral hippocampus. Hippocampal degeneration was made up of both terminals and preterminals, and was especially marked in the rostral levels of the ventral hippocampus, decreasing in density caudally. There was no regional preference for terminal degeneration within either the cornu ammonis or the fascia dentata. Nor did these lesions result in degeneration in the dorsal hippocampus. Fibers from the medial septal nucleus were found to course through the medial aspect of the fornix body and through the dorsal fornix to terminate in the dorsal hippocampus. Again, both terminals and preterminals were found, with no regional preference seen within the cornu ammonis or the fascia dentata. These results have been more recently replicated by Siegel and Edinger (1973).

That the topographical differences found between the work of Siegel and Tassoni (1971b) and of Raisman (1966) are the results of procedural differences rather than of species differences is suggested by the work of Powell (1963, above) as well as by the work of Anderson et al. (1961), using evoked potential tracing in the rabbit. These authors found that stimulation of the dorsal septal region caused bilateral surface-positive potential changes in CA1, with the primary stimulus area lying in the magnocellular portion of the medial septal nucleus. The septal efferents providing this stimulation were found to be situated between the most medial and lateral parts of the fornix fiber system, then traveling via the alveus to the dorsal CA1 region. Fibers terminating in the lateral CA1 area probably course in the fimbria before merging in the alveus. Stimulation of the posterior medial septal region resulted in potential changes in a wide area of the dorsal hippocampus—principally surface-positive potentials, with fewer, smaller surface-negative potentials observed in CA3. The authors concluded that these probably represented mono- or disynaptic connections, and that some fibers may have been contributed by the nucleus septohippocampalis.

2.3. Indirect Connections

Further septal–hippocampal connections may exist indirectly through the indusium griseum and through the pyriform and subicular regions. Simpson (1952) reported terminal degeneration in the indusium following lesion of the medial fornix column, the central hippocampus, and the posterior hippocampus; while Craigie (1925) had earlier reported fibers transversing corpus callosum to join the dorsal fornix from the indusium and the region of the cingulum. However, Knook (1966) was unable to verify Craigie's findings. In further support of hippocampoindusium pathways, Ban and Zyo (1962) have reported fibers in the fornix coursing anteriorly into the indusium, while Gastaut and Lammers (1961) have even suggested that the indusium griseum may be little more than an extension of the fascia dentata of the hippocampus. Domesick (1969) has reported fibers from the indusium penetrating the corpus callosum to terminate in the septum. A final point of possible indusium–septum interaction is the hippocampal rudiment of the septum, which is continuous with the indusium rostrally.

Septohabenular connections have been firmly established, coursing from the nucleus triangularis and the caudal portions of the medial and lateral septal nuclei via the septohabenular component of the stria medullaris to terminate in the medial habenular nucleus (Gurdjian, 1924/1925; Craigie, 1925; Young, 1936; Nauta, 1958; Cragg, 1961). Furthermore, Knook (1966) observed fibers from the rostral parts of the lateral and medial septal nuclei terminating in the lateral habenular nuclei, while Powell (1968) reported that rostral parts of the septum projected to the lateral habenular nucleus, and posterior parts to the medial nucleus. This latter connection is reciprocal, with the medial habenular nucleus sending efferents to the posterior medial septal region (Akagi and Powell, 1968). Since the habenular nuclei do have connections with the pyriform region (Fox, 1949; Nauta and Valenstein, 1958; Nauta, 1961; Powell *et al.*, 1965), which has been found to interconnect with the hippocampus via the entorhinal cortex and subiculum (Cragg, 1961; Powell *et al.*, 1965), this pathway would appear to be another possible point of septal–hippocampal interaction.

Kemper *et al.* (1972) have indicated that the lateral septal nucleus of the monkey projects to both the anterior and posterior regions of the cingulate gyrus; and Powell (1963) has found fibers passing from the rostral midline septum via the longitudinal stria to the cingulate of the rat. In the cat, Powell (1966) observed that the rostral cingulate and the gyrus rectus received projection from the septum via the dorsal fornix. Finally, Powell *et al.* (1974) observed terminal degeneration in the anterior part of the medial septal nucleus and the nucleus of the diagonal band following lesion of the cingulate region comparable to Brodman's area 23. These fibers passed via the cingulum to the corpus callosum toward the midline, where they joined midline fibers of the septum. While there is no evidence of direct connections between the hippocampus and the cingulate gyrus—Domesick (1969) and Raisman *et al.* (1965) failed to observe hippocampal degeneration following cingulate lesions—there are cingulosubiculum–entorhinal fiber connections (Ramón y Cajal, 1911; Gardner and Fox, 1948; Adey, 1951; Adey and Meyer, 1952), again allowing hippocampal influence.

Finally, Siegel and Tassoni (1971*a,b*) report that the ventral hippocampus projects to subicular and pyriform cortex via the alveus, while septal projections (from both lateral and medial nuclei) travel through the dorsal hippocampus to the cingulum, terminating in the pyriform and entorhinal area, thus allowing an even more direct avenue of interaction at this cortical level than either of the two discussed above.

3. Major Outputs of the Septum and the Hippocampus

Thus far, the picture of septohippocampal interconnection would seem to involve two distinct, although interacting, systems. The first is that involving dorsal hippocampal–medial septal regions; and the second involves the ventral hippocampus and the lateral septal complex, including the nucleus accumbens. Of all the septal nuclei, only the nucleus of the diagonal band receives fibers from both hippocampal divisions (Nauta, 1956; Valenstein and Nauta, 1959; Ban and Zyo, 1962; Knook, 1966; Siegel and Tassoni, 1971a), as well as having efferent connections to the hippocampus generally (Craigie, 1925; Daitz and Powell, 1954; Knook, 1966). The question thus becomes whether this separation is maintained in terms of their other projection systems, as has been suggested by Elul (1964).

3.1. Mammillary Projections

Fiber projections from the hippocampus to the mammillary bodies (Guillery, 1955, 1956; Craigie, 1925; Tsai, 1925; Gurdjian, 1927; Loo, 1931; Allen, 1944; Powell *et al.*, 1957; Nauta, 1958; Valenstein and Nauta, 1959; Cragg, 1961; Ban and Zyo, 1962; Szentagothai *et al.*, 1962), the preoptic region (Rioch, 1931; Young, 1936; Nauta, 1956, 1958; Valenstein and Nauta, 1959; Ban and Zyo, 1962), and the lateral hypothalamic region (Nauta, 1958; Valenstein and Nauta, 1959) have been well established.

Raisman *et al.* (1966) reported fibers from the fimbria projecting via the postcommissural fornix and the medial corticohypothalamic tract to the lateral part of the medial mammillary nucleus and the lateral mammillary nucleus. The fibers of the dorsal fornix were found to project via the postcommissural fornix to "fill" the lateral mammillary nucleus and the medial and lateral parts of the medial mammillary region.

Siegel and Tassoni (1971*a*) found ventral hippocampal lesions to result in terminal degeneration (via the diagonal band) in the ventrolateral preoptic area immediately above the supraoptic nucleus. Lesions of the dorsal hippocampus were not found to result in degeneration in preoptic regions. On the other hand, degeneration was found throughout the ventrolateral hypothalamus, lateral mammillary nucleus, and lateral part of the medial mammillary nucleus following lesion of either the dorsal or the ventral hippocampus. Some topographical distinction was seen in the lateral hypothalamic projections, in that ventral lesions resulted in degeneration of greater magnitude in the anterior portion of the lateral hypothalamus, while dorsal lesions caused greater amounts of degeneration in the posterior portion. The projection of all portions of the hippocampus to the mammillary region supports the earlier findings of Simpson (1952) that lesions of the anterior, central, and posterior hippocampus all resulted in degeneration in the mammillary region (both lateral and medial) of the monkey.

Septal projections to the preoptic region (Zyo *et al.*, 1963; Valverde, 1963), lateral hypothalamus (Gurdjian, 1927; Nauta, 1956, 1958; Szentagothai *et al.*, 1962; Raisman, 1966), and mammillary body (Knook, 1966) have also been well established. Powell (1963) has reported that lesion of the medial septal area of the rat resulted in degeneration via the medial forebrain bundle with terminals distributed to nuclei between the olfactory tubercle and mammillary bodies, with some fibers seen entering the mammillotegmental tract. Lesion of the posterior midline septal region resulted in relatively fewer degeneration granules in the medial forebrain bundle and mammillotegmental tract. The greatest degree of degeneration along these paths was seen following lesion of the ventral septum (including a portion of the nucleus accumbens). This lesion resulted in heavy degeneration in the medial forebrain bundle and mammillotegmental tract, with terminals in the supramammillary and ventral tegmental nuclei. Simmons and Powell (1972) have further reported on septomammillary projections in the squirrel monkey, using Nauta-Gygax stain. They concluded that there are at least three sorts of septomammillary projection systems. The main one occurs through the ipsilateral column of the fornix and is distributed to the ipsilateral medial mammillary nucleus with some termination in the contralateral medial mammillary nucleus. Another crosses the midline in the septal region, and perhaps at more caudal levels, to descend in the contralateral column of the fornix to the medial mammillary nucleus. Third, some of the projections enter the ipsilateral medial forebrain bundle and terminate bilaterally in the medial mammillary nucleus. Finally, Siegel and Tassoni (1971b) observed that projections of either the medial or the lateral septal nuclei entered Zuckerkandl's bundle, traveling to the lateral preoptic region and the lateral hypothalamus.

While these results suggest that the more caudal projections (preoptic, mammillary, and hypothalamic) of the septum and hippocampus are not contributed to differentially along a dorsal, ventral hippocampus or a lateral, medial septum division, it is worth noting that Elul (1964), investigating afterdischarge spread in the cat following stimulation of either the dorsal or the ventral hippocampus, indicated that a notable amount of afterdischarge spread was detected in the medial mammillary nucleus and lateral hypothalamus only following ventral hippocampal stimulation, with little or no spread to these structures observed following stimulation of the dorsal hippocampus. Further complicating the picture is the finding of Sparks and Powell (1966) that the application of a conditioning pulse to either the septum or the hippocampus had but little effect on the response elicited by mammillary body stimulation, while preceding either hippocampal or septal stimulation with stimulation of the mammillary body resulted in the attenuation of all components of the septal or hippocampal response wave.

3.2. Thalamic Projections

Projections of both the septum and the hippocampus to thalamic nuclei have been variously described (Nauta, 1956; Guillery, 1959; Valenstein and Nauta, 1959; Cragg, 1961; Trembly and Sutin, 1961; Zyo et al., 1963). Powell (1963) observed two pathways from the rostral midline septal area to thalamic nuclei in the rat. Following lesion of this area, the medial portion of the stria medullaris showed consistent degeneration and contributed terminals to the anteroventral, anteromedial, anterodorsal, and reticular nuclei. The stria medullaris then contributed posteriorly to the habenula, with no degeneration seen beyond these nuclei. Degenerating axons were also seen leaving the fornix column, distributing terminals to the zona incerta, anteroventral, and anteromedial nuclei. Lesion of the posterior midline septal area resulted in a similar distribution of degeneration, with a relative increase in the number of terminals seen in the anteroventral, reticular, and habenular nuclei. There was also an increase in the number of degenerating fibers seen in the stria medullaris. The dorsolateral septum was found to contribute, almost solely via the stria medullaris, to the anteromedial, anteroventral, reticular, reuniens, and suprachiasmatic nuclei (the last perhaps via medial fasciculi). Finally, deep lesions of the ventral septal area (including some destruction of the nucleus accumbens) resulted in some terminal degeneration in the nucleus medialis dorsalis of the thalamus, traveling in large part via the medial forebrain bundle, with a few contributions from axons coursing in the medial portion of the stria terminalis.

In the cat (Powell, 1966), lesion in the nucleus lateralis interna of the septum, with damage to the nucleus accumbens as well, has been found to result in degeneration via the stria medullaris terminating in the ventral portion of the anteromedial nucleus of the thalamus, the nucleus medialis dorsalis, and the lateral habenular nucleus. Lesion of the medialis anterior nucleus of the septum and ipsilateral diagonal band resulted in fornix column degeneration with terminals appearing in the nuclei centralis and lateralis, and the anteroventral and anterodorsal nuclei. Fewer anterodorsal and anteroventral terminations were seen when the diagonal band was spared in this lesion, although degeneration was reported via the stria medullaris and fornix column. Terminals were also observed in the medialis dorsalis nucleus of the thalamus. The greatest number of anteroventral nucleus terminals were seen following lesion of the medialis posterior nucleus of the septum and the diagonal band, with terminal degeneration also noted in the anteromedial nucleus and the posterior part of the nucleus centromedianum. In the monkey (Powell, 1973, using Fink–Heimer stains), septal efferents were found to join the fornix in the posterior part of the septum and course to the thalamic nuclei via the internal medullary lamina and stria medullaris. Thalamic degeneration in both the anteroventral and anteromedial nuclei was observed to be greater in density than that observed in the mammillary body of the same animal (Fig. 3). Degeneration in the lateral dorsal nucleus was as great as that seen in the mammillary body.

Finally, Sparks and Powell (1966) found an attenuation of the long-latency responses recorded from the anteroventral nucleus following dorsal hippocampal stimulation when sequential stimuli were delivered to the septum and hippocampus. Similar anteroventral response attenuation was obtained when septal stimulation was preceded by stimulation of the hippocampus. That this attenuation was not a simple failure of the anteroventral nucleus to follow sequential stimulation was indicated by the failure of successive septal–septal stimulation to produce any change in the anteroventral response. However, hippocampal–hippocampal sequences did result in response attenuation. The authors speculated that perhaps impulses originating in the septum or other brain areas and reaching the thalamus via the hippocampus would not be transmitted during this depressed period. Similar refractory periods have been observed following stimulation of the guinea pig hippocampus (Liberson, 1962). That some efferent feedback reaches the septum from the thalamus is suggested by the finding by Powell et al. (1968) of fiber connections from the anteroventral nucleus with the nucleus septalis fimbrialis and the nucleus septalis pars anterior coursing via the fornix and the diagonal band.

Nauta (1956, 1958), Guillery (1956), Powell et al. (1957), Valenstein and Nauta (1959), Cragg (1961), and Knook (1966) have all described projections of the hippocampus to such various thalamic nuclei as the anteroventral, anteromedial,

Se 30   Fx 45   Cg 50 ——

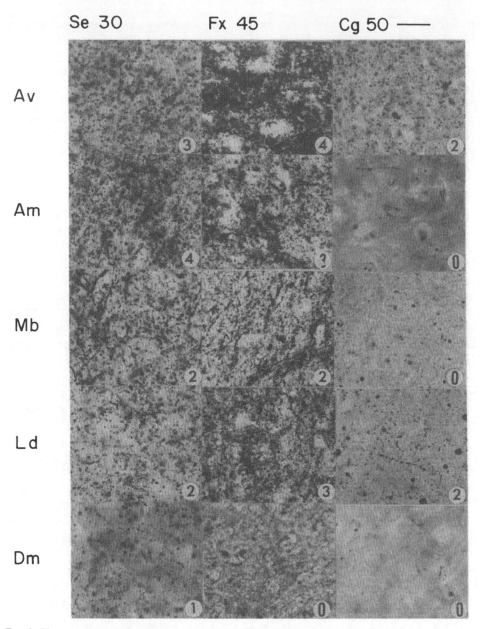

FIG. 3. Photographs of degeneration in thalamus and hypothalamus. The letters on the left indicate the nucleus in which the degeneration was studied. Av, Anteroventral nucleus (n); Am, anteromedial n; Mb, medial mammillary n; Ld, lateral dorsal n; Dm, dorsal medial n. The number in the lower right corner of each frame is an expression of the degeneration observed in the nucleus represented (interreliability ratings equal 0.60–0.84, Simmons and Powell, 1972). The letters at the top indicate the structures damaged by the lesion. Se, Septum; Fx, fornix body; Cg, cingulate gyrus. The numbers refer to specific cases used. The calibration line to the upper right equals 25 μm. Fink–Heimer I stain. Reproduced from Powell (1973), Fig. 2, with the permission of the publisher.

parataenialis, reuniens, and centralis medius. Raisman *et al.* (1966) report that CA1 fibers project via the postcommissural fornix to the anteromedial and anteroventral nuclei, with no thalamic projection from area CA2, CA3, or CA4. Siegel and Tassoni (1971*a*) report that lesion generally of either the dorsal or ventral hippocampus resulted in terminal degeneration in only one thalamic nucleus—the anteroventral. Sparks and Powell (1966) and Powell (1973) indicated hippocampal projection to the anteroventral and lateral dorsal nuclei. Elul (1964) observed spread of hippocampal afterdischarge following stimulation of the dorsal hippocampus only in the dorsal thalamus.

4.   Summary and Conclusions

1.   The primary direct pathways from hippocampus to septum run from the dorsal hippocampus via the medial half of the fornix body to the precommissural fornix, terminating in the medial septum and nucleus of the diagonal band, and from the ventral hippocampus via the fimbria, lateral margin of the fornix body, and precommissural fornix to the lateral septal region, nucleus accumbens, and nucleus of the diagonal band.

2.   The hippocampus also projects via the postcommissural fornix and the diagonal band to the preoptic area, ventrolateral hypothalamus, and mammillary bodies.

3.   Hippocampothalamic fibers course via the fornix column and stria medullaris, primarily to anterior thalamic nuclei.

4.   The septum is seen to project via the dorsal fornix to the hippocampus. While the topographical nature of these projections is still somewhat of an open question, the results generally suggest that the medial septal region projects via the medial aspect of the fornix body and the dorsal fornix to the dorsal hippocampus, while the lateral septal area and the nucleus accumbens project through the lateral portion of the fornix column to the fimbria and the ventral hippocampus.

5.   The septum and hippocampus may interact indirectly via the indusium griseum, and the pyriform and subicular regions.

6.   The septum projects via the medial forebrain bundle and mammilotegmental tract to the preoptic region, lateral hypothalamus, and mammillary nuclei.

7.   The septum also projects via the diagonal band, the stria medullaris, the fornix column, and the internal medullary lamina to the anterior thalamic nuclei.

8.   These pathways are summarized in Fig. 4.

9.   In the light of these pathway interactions, it appears that the major thrust of septal and hippocampal outputs is not, as is traditionally considered, to downstream structures such as the hypothalamus and mammillary bodies, but rather to the anterior thalamic nuclear complex, where it serves to interact with inputs entering via the mammillothalamic tract from the mammillary bodies. Limbic projections downstream to the mammillary body (which may be quantitatively fewer than those to the anterior thalamic nuclei) may simply represent feedback loops to that structure carrying information as to higher limbic states of activity. Since the cingu-

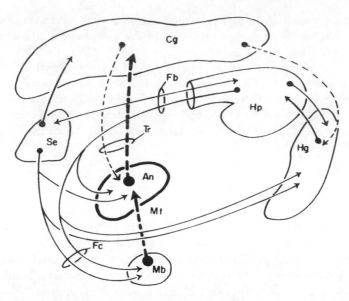

Fig. 4. Principal septal and hippocampal connections are shown by solid lines. Broken lines represent some main secondary connections. An, Anterior nuclei (n); Cg, cingulate gyrus; Fb, fornix body; Fc, fornix column; Hg, hippocampal gyrus; Hp, hippocampus; Mb, mammillary body; Mt, mammillothalamic tract; Se, septum; Tr, thalamic radiations.

late gyrus also has entensive connections with the anterior thalamic complex (Domesick, 1969, 1970, 1972), it becomes feasible to further speculate that the entire thrust of the limbic system in general runs via the thalamus and cingulate gyrus to higher cortical areas. Limbic projections downstream might thus be conceived of as serving primarily to modulate the types and amounts of stimulus information entering the system, much as the septotectal efferents (Powell and Hoelle, 1967) apparently act in the modulation of medial geniculate body activity (Powell *et al.*, 1970).

5. Appendix: Comments on Stains

5.1. Cell Stains

Such stains as cresyl violet (Clark and Clark, 1971) are used to indicate the location, size, and density of cells and their arrangement, especially in the central nervous system. More specifically, these stains are used for localizing lesions, electrode tips, and electrode tracks. They are also used in cytoarchitectonic studies to classify cells and nuclei. They are further used in retrograde studies of anatomical connections as related to their capacity to reveal neuron chromatolysis and gliosis.

5.2. Fiber Stains

a. Weigert, Weil, and Luxol fast blue (Clark and Clark, 1971) are myelin stains which do not indicate the direction of fiber projection in bundles. They may be used to demonstrate, by the absence of myelin, and therefore the absence of staining, large fiber tract degeneration subsequent to injury or lesion.

   b.   Marchi stains (Clark and Clark, 1971) are relatively sensitive in the detection of degenerated myelinated fibers. In skilled hands, single fibers will stain intense black on a light background. This method labels only myelinated degenerating fibers and does not differentiate between fibers of origin and fibers of passage.

   c.   Nauta–Gygax (1954) silver impregnation methods for degenerating axons stain fibers whether they are myelinated or not. There is, however, some question as to whether or not the method labels *all* degenerating axons subsequent to lesion. Terminals are poorly impregnated or do not retain the silver deposited. This method does not distinguish between fibers of origin and fibers of passage.

   d.   Fink–Heimer (1967) procedures are designed to selectively impregnate degenerating axons. In the experience of the senior author, the Fink–Heimer procedure I is the best method presently available for the demonstration of pathway termination in nuclei and cellular areas of the brain. The weakness of this method, as with all of the previous, is that it does not distinguish fibers of origin from fibers of passage. There is also some question as to whether negative data are valid for non-projection.

5.3.   Autoradiography

   In the autoradiographic procedure, tritiated leucine is used to focally mark cells so that their axons, and especially their terminals, are made visible for light and electron microscopic examination. The nuclei and cellular aggregates which have accumulated the radioactive leucine are considered valid target sites of the structure originally injected. This method avoids electrode damage problems as well as those related to fibers of passage since only cell bodies have been reported to absorb the leucine. A major problem with this method is the spread of the labeled amino acid from the prescribed injection site.

5.4.   Histochemical Neuroanatomy

   Histochemical determination of neuroanatomy has been made possible by the histofluorescence methods of Falck *et al.* (1962). These methods have been applied to projection areas of the septum and hippocampus (Powell *et al.*, 1972; Morgane and Stern, 1972; Moore *et al.*, 1971).

ACKNOWLEDGMENTS

   The authors wish to thank Drs. Ed Lucas, Ture Schoultz, and Robert Skinner, Department of Anatomy, University of Arkansas School of Medicine, for discussing and helping with the manuscript, and Geraldine Brown, Jean Galatzan, and Robert Leman for their technical assistance.

6. References

ADEY, W. R. An experimental study of the hippocampal connexions of the cingulate cortex in the rabbit. *Brain*, 1951, **74**, 233–247.

ADEY, W. R., AND MEYER, M. Hippocampal and hypothalamic connections of the temporal lobe in the monkey. *Brain*, 1952, **75**, 358–384.

AKAGI, K., AND POWELL, E. W. Differential projections of habenular nuclei. *Journal of Comparative Neurology*, 1968, **132**, 263–274.

ALLEN, W. F. Degeneration in the dog's mammillary body and ammon's horn following transsection of the fornix. *Journal of Comparative Neurology*, 1944, **80**, 283–292.

ALLEN, W. F. Fiber degeneration in Ammon's horn resulting from extirpations of the piriform and other cortex areas and from transection of the horn at various levels. *Journal of Comparative Neurology*, 1948, **88**, 425–438.

ANDERSON, P., BRULAND, H., AND KAADA, B. R. Activation of the field of CA1 of the hippocampus by septal stimulation. *Acta Physiologica Scandinavica*, 1961, **51**, 29–40.

ANDY, O. J., AND STEPHAN, H. *The septum of the cat*. Springfield, Ill.: Charles C Thomas, 1964a, pp. 1–84.

ANDY, O. J., AND STEPHAN, H. Cytoarchitectonics of the septal nuclei in Old World monkeys (*Ceropithecus* and *Colobus*). *Journal für Hirnforschung*, 1964b, **7**, 1–23.

ANDY, O. J., AND STEPHAN, H. The septum in the human brain. *Journal of Comparative Neurology*, 1968, **133**, 383–410.

BAN, T., AND ZYO, K. Experimental studies on the fiber connections of the rhinencephalon. I. Albino rat. *Medical Journal of Osaka University*, 1962, **12**, 385–424.

CARMAN, J. B., COWAN, W. M., AND POWELL, T. P. S. The organization of cortico-striate connexions in the rabbit. *Brain*, 1963, **86**, 525–562.

CLARK, G., AND CLARK, M. P. *A primer in neurological staining procedures*. Springfield, Ill.: Charles C Thomas, 1971.

CRAGG, B. G. The connections of the habenulae in the rabbit. *Experimental Neurology*, 1961, **3**, 308–409.

CRAGG, B. G. Afferent connexions of the allocortex. *Journal of Anatomy*, 1965, **99**, 339–357.

CRAIGIE, E. H. *An introduction to the finer anatomy of the central nervous system based on that of the albino rat*. Philadelphia: P. Blackiston's Son and Company, 1925.

CROSBY, E. C., DEJONGE, B. R., AND SCHNEIDER, R. C. Evidence for some of the trends in the phylogenetic development of the vertebrate telencephalon. In R. Hassler and H. Stephan (Eds.), *Evolution of the forebrain*. New York: Plenum Press, 1967, pp. 117–135.

DAITZ, H. M., AND POWELL, T. P. S. Studies on the connexions of the fornix system. *Journal of Neurology, Neurosurgery, and Psychiatry*, 1954, **17**, 75–82.

DOMESICK, V. B. Projections from the cingulate cortex in the rat. *Brain Research*, 1969, **12**, 296–320.

DOMESICK, V. B. The fasciculus cinguli in the rat. *Brain Research*, 1970, **20**, 19–32.

DOMESICK, V. B. Thalamic relationships of the medial cortex in the rat. *Brain Behavior and Evolution*, 1972, **6**, 457–483.

ELUL, R. Regional differences in the hippocampus of the cat. II. Projections of the dorsal and ventral hippocampus. *Electroencephalography and Clinical Neurophysiology*, 1964, **16**, 489–502.

EMMERS, R., AND AKERT, K. A Stereotaxic atlas of the brain of the squirrel monkey. Madison: University of Wisconsin Press, 1963, p. 66.

FALCK, B., AND HILLARP, N.-Å., THIEME, G., AND TORP, A. Fluorescence of catecholamines and related compounds condensed with formaldehyde. *Journal of Histochemistry and Cytochemistry*, 1962, **10**, 348–354.

FINK, R. P., AND HEIMER, L. Two methods for selective silver impregnation of degenerating axons and their synaptic endings in the central nervous system. *Brain Research*, 1967, **4**, 369–374.

FOX, C. A. Amygdalo-thalamic connections in *Macaca mulatta*. *Anatomical Record*, 1949, **103**, 537.

GARDNER, W. D., AND FOX, C. A. Degeneration of the cingulum in the monkey. *Anatomical Record*, 1948, **100**, 663–664.

GASTAUT, H., AND LAMMERS, H. J. *Anatomie du rhinencéphale: Les grandes activités du rhinencéphale.* Paris: Masson et Cie, 1961.

GUILLERY, R. W. A quantitative study of the mammillary bodies and their connexions. *Journal of Anatomy,* 1955, **89**, 19–32.

GUILLERY, R. W. Degeneration in the post-commissural fornix and the mammillary peduncle of the rat. *Journal of Anatomy,* 1956, **90**, 350–370.

GUILLERY, R. W. Afferent fibres to the dorso-medial thalamic nucleus in the cat. *Journal of Anatomy,* 1959, **93**, 403–419.

GURDJIAN, E. S. Olfactory connections in the albino rat, with special reference to the stria medullaris and the anterior commissure. *Journal of Comparative Neurology,* 1924/1925, **38**, 127–164.

GURDJIAN, E. S. The diencephalon of the albino rat. *Journal of Comparative Neurology,* 1927, **43**, 1–114.

HINES, M. Studies in the growth and differentiation of the telencephalon in man: The fissura hippocampi. *Journal of Comparative Neurology,* 1922, **34**, 73–171.

HINES, M. The brain of *Ornithorhynchus anatinus. Philosophical Transactions of the Royal Society of London, Series B,* 1929, **217**, 155–287.

HOFFMAN, H. H. The hippocampal and septal formations in anurans. In R. Hassler and H. Stephan (Eds.), *Evolution of the forebrain.* New York: Plenum Press, 1967, pp. 61–72.

JOHNSON, T. N. Studies on the brain of the guinea pig. I. The nuclear pattern of certain basal telencephalic centers. *Journal of Comparative Neurology,* 1957, **107**, 353–378.

JOHNSON, T. N. Studies on the brain of the guinea pig. II. The olfactory tracts and fornix. *Journal of Comparative Neurology,* 1959, **112**, 121–139.

KEMPER, T. L., WRIGHT, S. J., AND LOCKE, S. Relationship between the septum and the cingulate gyrus in *Macaca mulatta. Journal of Comparative Neurology,* 1972, **146**, 465–477.

KNOOK, H. L. *The fibre-connections of the forebrain.* Philadelphia: Davis, 1966, pp. 230–255.

KÖNIG, J. F. R., AND KLIPPEL, R. A. *The rat brain: A stereotaxic atlas of the forebrain and lower parts of the brain stem.* Baltimore: Williams and Wilkins, 1963.

LIBERSON, W. T. Epileptic discharges in hippocampal activity. In P. Passouant (Ed.), *Physiologie de l'hippocampe.* Paris: C.N.R.S., 1962, pp. 351–387.

LOO, Y. T. The forebrain of the opossum. *Didelphis virginiana.* Part II. Histology. *Journal of Comparative Neurology,* 1931, **52**, 1–148.

LORENTE DE NÓ, R. Studies on the structure of the cerebral cortex. II. Continuation of the study of the ammonic system. *Journal of Psychology and Neurology (Leipzig),* 1934, **46**, 113–177.

McLARDY, T. Some cell and fiber peculiarities of uncal hippocampus. In W. Bargmann and J. P. Schadé (Eds.), *The Rhinencephalon and Related Structures.* New York: Elsevier Publishing Company, 1963, pp. 71–88.

MOORE, R. Y., BJORKLUND, A., AND STENEVI, U. Plastic changes in the adrenergic innervation of the rat septal area in response to denervation. *Brain Research,* 1971, **33**, 13–35.

MORGANE, P. J., AND STERN, W. C. The chemistry of the medial forebrain bundle. *Anatomical Record,* 1972, **172**, 369.

MORIN, F. An experimental study of hypothalamic connections in the guinea pig. *Journal of Comparative Neurology,* 1950, **92**, 193–214.

NAUTA, W. J. H. An experimental study of the fornix system in the rat. *Journal of Comparative Neurology,* 1956, **104**, 247–272.

NAUTA, W. J. H. Hippocampal projections and related neural pathways to the mid-brain in the cat. *Brain,* 1958, **81**, 319–340.

NAUTA, W. J. H. Fibre degeneration following lesions of the amygdaloid complex in the monkey. *Journal of Anatomy,* 1961, **95**, 515–531.

NAUTA, W. J. H., AND GYGAX, P. A. Silver impregnation of degenerating axons in the central nervous system (a modified technique). *Stain Technology,* 1954, **29**, 91–93.

NAUTA, W. J. H., AND VALENSTEIN, E. Some projections of the amygdaloid complex in the monkey. *Anatomical Record,* 1958, **130**, 346.

POWELL, E. W. Septal efferents revealed by axonal degeneration in the rat. *Experimental Neurology,* 1963, **8**, 406–422.

POWELL, E. W. Corticolimbic interrelations revealed by evoked potential and degeneration techniques. *Experimental Neurology,* 1964, **10**, 463–474.

POWELL, E. W. Septal efferents in the cat. *Experimental Neurology*, 1966, **14**, 328–337.

POWELL, E. W. Septohabenular connections in the rat, cat and monkey. *Journal of Comparative Neurology*, 1968, **134**, 145–150.

POWELL, E. W. Limbic projections to the thalamus. *Experimental Brain Research*, 1973, **17**, 394–401.

POWELL, E. W., AND HOELLE, D. F. Septotectal projections in the cat. *Experimental Neurology*, 1967, **18**, 177–183.

POWELL, E. W., AND SCHNURR, R. Silver impregnation of degenerating axons; comparisons of postoperative intervals, fixatives and staining methods. *Stain Technology*, 1972, **47**, 95–100.

POWELL, E. W., CLARK, W. M., AND MUKAWA, J. An evoked potential study of limbic projections to nuclei of the cat septum. *Electroencephalography and Clinical Neurophysiology*, 1968, **25**, 266–273.

POWELL, E. W., FURLONG, L. D., AND HATTON, J. B. Influence of the septum and inferior colliculus on medial geniculate body units. *Electroencephalography and Clinical Neurophysiology*, 1970, **29**, 74–82.

POWELL, E. W., WINTER, C. G., KIRBY, M. E., AND AUSTIN, B. Mammillary body fluorescence changes following septal or hippocampal lesions in the rat. *Neurobiology*, 1972, **2**, 149–153.

POWELL, E. W., AKAGI, K., AND HATTON, J. B. Subcortical projections of the cingulate gyrus in cat. *Journal für Hirnforschung*, 1974, **15**, 313–322.

POWELL, T. P. S., GUILLERY, R. W., AND COWAN, W. M. A quantitative study of the fornix-mamillothalamic system. *Journal of Anatomy*, 1957, **91**, 419–437.

POWELL, T. P. S., COWAN, W. M., AND RAISMAN, G. The central olfactory connections. *Journal of Anatomy*, 1965, **99**, 791–813.

RAISMAN, G. The connexions of the septum. *Brain*, 1966, **89**, 317–348.

RAISMAN, G., COWAN, W. M., AND POWELL, T. P. S. The extrinsic afferent, commissural and association fibres of the hippocampus. *Brain*, 1965, **88**, 963–996.

RAISMAN, G., COWAN, W. M., AND POWELL, T. P. S. An experimental analysis of the efferent projection of the hippocampus. *Brain*, 1966, **89**, 83–108.

RAMÓN Y CAJAL, S. *Histologie du systeme nerveux de l'homme et des vertébrés*. Paris: Maloine, 1911, pp. 762–793.

RAMÓN Y CAJAL, S. *The structure of Ammon's horn* (trans. by L. M. Kraft). Springfield, Ill.: Charles C Thomas, 1968.

RIOCH, D. McK. Studies on the diencephalon of carnivora. Part III. Certain myelinated-fiber connections of the dog (*Canis familiaris*), cat (*Felis domestica*), and aevisa (*Crossarchus obscurus*). *Journal of Comparative Neurology*, 1931, **53**, 319–388.

ROSE, M. Der Allocortex bei Tier und Mensch. I. *Journal of Psychology*, 1926, **34**, 1–30.

ROSE, M. Der Allocortex bei Tier und Mensch. II. *Journal of Psychology*, 1927, **35**, 42–76.

SIEGEL, A., AND EDINGER, H. A comparative neuroanatomical analysis of the differential projections of the hippocampus to the septum. In *Proceedings of the Society of Neuroscience*, San Diego, 1973.

SIEGEL, A., AND TASSONI, J. P. Differential efferent projections from the ventral and dorsal hippocampus of the cat. *Brain, Behavior, and Evolution*, 1971*a*, **4**, 185–200.

SIEGEL, A., AND TASSONI, J. P. Differential efferent projections of the lateral and medial septal nuclei to the hippocampus in the cat. *Brain, Behavior, and Evolution*, 1971*b*, **4**, 201–219.

SIEGEL, A., TROIANO, R., ROYCE, A., AND EDINGER, H. Differential efferent projections of the anterior and posterior cingulate gyrus in the cat. *Anatomical Record*, 1972, **172**, 406.

SIMMONS, H. J., AND POWELL, E. W. Septomammillary projections in the squirrel monkey. *Acta Anatomica*, 1972, **82**, 159–178.

SIMPSON, D. A. The efferent fibers of the hippocampus in the monkey. *Journal of Neurology, Neurosurgery, and Psychiatry*, 1952, **15**, 79–92.

SPARKS, D. L., AND POWELL, E. W. Interaction of evoked potentials in the anterior thalamus of the cat. *Electroencephalography and Clinical Neurophysiology*, 1966, **20**, 470–474.

SZENTAGOTHAI, J., FLERKÓ, B., MESS, B., AND HALASZ, B. *Hypothalamic control of the anterior pituitary*. Budapest: Akadémiai Kaidó, 1962.

TREMBLY, B., AND SUTIN, J. Septal projections to the dorsomedial thalamic nucleus in the cat. *Electroencephalography and Clinical Neurophysiology*, 1961, **13**, 880–888.

TSAI, CH. The descending tracts of the thalamus and midbrain of the opossum, *Didelphis virginiana*. *Journal of Comparative Neurology*, 1925, **39**, 217–248.

Valenstein, E. S., and Nauta, W. J. H. A comparison of the distribution of the fornix system in the rat, guinea pig, cat and monkey. *Journal of Comparative Neurology,* 1959, **113,** 337–364.

Valverde, F. Studies on the forebrain of the mouse: Golgi observations. *Journal of Anatomy,* 1963, **97,** 157–180.

Votaw, C. L., and Lauer, E. W. An afferent hippocampal fiber system in the fornix of the monkey. *Journal of Comparative Neurology,* 1963, **121,** 195–206.

Young, M. W. The nuclear pattern and fiber connections of the non cortical centers of the telencephalon of the rabbit (*Lepus cuniculus*). *Journal of Comparative Neurology,* 1936, **65,** 295–402.

Zyo, K., Oki, T., and Ban, T. Experimental studies on the medial forebrain bundle, medial longitudinal fasciculus and supraoptic decussations in the rabbit. *Medical Journal of Osaka University,* 1963, **13,** 193–239.

3

Development of the Hippocampal Region

Jay B. Angevine, Jr.

1. Introduction

A backpacker in the chiseled mountains of the American Southwest searches for panoramas of scorched land or, if such is his pleasure, scours shattered ridges and slopes for abandoned mineshafts. The views illustrate the geomorphology of the Basin and Range country, an interesting and relatively little appreciated part of the face of the land, whereas the mine shafts, if not yielding copper, silver, or gold, offer historical insights—small reward considering the risks of exploration. In any event, the wise tramper watches where he puts his feet.

The prospective explorer of the hippocampal region faces a similar choice. Here may be found perspectives on the design of the cerebral hemispheres or, alternatively, chances to prospect for theoretical percepts leading possibly to better understanding of the forebrain but perhaps only to abandoned approaches to knowledge—conceptual voids with heaps of battered and broken terminology at the bottom.

In writing this chapter, I planned it as a trail guide for those interested in the geography, evolution, and development of the hippocampal region, that sharply delineated but mysteriously ordered expanse of cerebral surface. Unquestionably I should present the terminology, comparative anatomical viewpoints, and embryological concepts that provide horizons for future study of the region; such outlooks are there in abundance. Still, I must admit that I have spent many hours wandering among library stacks—poking around in mineshafts—with certain rewards in terms of evaluating approaches to problems and without more serious mishap than a little wasted time. Therefore, as I guide the reader through the strange landscape of the

Jay B. Angevine, Jr. • Department of Anatomy, College of Medicine, The University of Arizona, Tucson, Arizona.

61

hippocampal region, I remind him of the history of search and discovery here, a history that can be perused with pleasure and scholarly satisfaction (White, 1965; Yakovlev, 1972) and, if security is a concern, without any risk at all.

2.   Terminology and Orientation

The definition of the hippocampal region and the recognition and nomenclature of its components need not present much difficulty. Certainly there are not the thorny problems encountered when classifications of more extensive regions of the brain are attempted under such headings as "rhinencephalon," "limbic lobe," and "limbic system" (Brodal, 1969). Principal subdivisions were surveyed long ago by Ramón y Cajal (1909–1911) and Lorente de Nó (1933, 1934) and later precisely defined by Blackstad (1956) and White (1959). Continued study of hippocampal afferent and efferent systems by modern hodological methods has brought even greater refinement to the resolution of areas and laminae and demonstrated a formal geometry in the region which extends down to the orderly array of diverse terminals at the individual cellular level (Raisman et al., 1965, 1966).

The components of the hippocampal region and their constituent layers are summarized in Table 1 and illustrated in Fig. 2. A review of the principal features of the various parts visible in Nissl preparations (see Fig. 1) of the hippocampal region in the mouse (Angevine, 1965) follows. These characteristics are based on the appearance of the region in horizontal sections, the standard and so-called normal plane (Lorente de Nó, 1934) and the one most desirable for initial study or subsequent detailed analysis. Additional views in horizontal, sagittal, and coronal planes may be had in the *Atlas of the Mouse Brain and Spinal Cord* (Sidman et al., 1971). Study of that book will allow the reader to build a three-dimensional picture of this complex convolution and its challenging topography as seen in certain cuts.

3.   Definition of Areas

The definition of the hippocampal region and the delineation and naming of its parts that I will use follow closely the scheme employed by Blackstad (1956) and White (1959). Their subdivisions of the region are consistent with those of Ramón y Cajal and Lorente de Nó, but also take account of important studies of more recent vintage. The components of the hippocampal region and their respective layers are indicated by arrows or dashed lines in all the reconstructions of autoradiographic findings and summary diagrams presented in this chapter.

The shallow but easily recognized fissura rhinalis superficially marks the lateral boundary between the area entorhinalis (area 28) and the adjacent area perirhinalis (area 35). The distinction between these two areas is even more apparent at a deeper anatomical level and is striking when autoradiographic findings to be described below are considered. Here the cortical plate separates (Figs. 1 and 2) into two broad and sharply defined layers of neurons (laminae principales interna et externa of Maxi-

TABLE 1

Components of the Hippocampal Region and Their Principal Cell Layers,
as Seen in Nissl Preparations (Angevine, 1965)

Hippocampal region	Hippocampal formation	Area dentata[a]	Stratum moleculare / Stratum granulosum[h]	Fascia dentata
			Hilus fasciae dentatae[i]	
		Hippocampus[b] (cornu ammonis)	Stratum moleculare[j] / Stratum lacunosum[j] / Stratum radiatum / Stratum pyramidale / Stratum oriens	
		Subiculum[c]	Stratum moleculare	
			Stratum pyramidale	
	Retrohippocampal formation	Presubiculum[d] (area 27)	Layer I	Lamina zonalis
		Area retrosplenialis e[e] (area 29e)	Layer II / Layer III	Lamina principalis externa (M. Rose, 1927)
		Parasubiculum[f] (area 49)	[Lamina dissecans[k]]	
		Area entorhinalis[g] (area 28)	Layer IV	Lamina principalis interna (M. Rose, 1927)

[Fissura rhinalis]

Area perirhinalis
(area 35)

[a] As defined by Blackstad (1956); emphasizes unity of hilus fasciae dentatae and fascia dentata and eliminates term CA4 of Lorente de Nó (1934).
[b] Subdivided into sectors CA1, CA2, and CA3 of Lorente de Nó (1934) and further subsectors.
[c] Includes prosubiculum of Lorente de Nó (1934).
[d] Recognized by densely packed small pyramidal ("granular") cells in layer II.
[e] "... as devoid of characteristics in thionin sections as it is impressive in silver preparations" (Blackstad, 1956); considered part of parasubiculum.
[f] Boundary with area 28 tentative; lamina dissecans broadens, molecular layer narrows at posterior "tuck" in surface of cerebral hemisphere.
[g] Subdivided into pars medialis (28a) and pars lateralis (28b), separated by a transitional zone, in which large pale cells of 28a yield to scattered, fusiform, vertically arranged cells which still more laterally clump into islands of dark cells.
[h] Subdivided into suprapyramidal and infrapyramidal limbs.
[i] Layer of polymorphic neurons considered part of area dentata but called CA4 by Lorente de Nó (1934).
[j] Considered one layer: stratum lacunosum-moleculare (stratum reticulare of Kupffer, 1859).
[k] Cell-sparse zone, not sharply delimited from adjacent cell layers. Considered part of layer II–III.

milian Rose, 1927); layer IV and layer II–III, as these swaths are called, are separated by a cell-poor zone, the lamina dissecans, and covered by a thick molecular layer, the lamina zonalis. The inner principal layer is composed mainly of small pyramidal cells; the outer principal layer exhibits a mixed population of large and small cells. Islands of darkly stained cells along the superficial border of the outer principal layer terminate abruptly at the rhinal fissure. The division of the cortical ribbon into these two principal layers or broad filaments is evident throughout the entire posterior and posteromedial curvature of the cerebral hemisphere. Perception of these two filaments is an important first act in approaching the troublesome anatomy of this part of the cerebral cortex. The two-layered arrangement of cortical neurons persists and may be seen even more clearly in the cortical areas designated parasubiculum (area 49), area retrosplenialis e (area 29e), and presubiculum (area 27). At the inner edge of the presubiculum, adjacent to the fissura hippocampi, the outer principal cell layer, composed here of numerous, densely packed small pyramidal cells ("granules"), overrides the inner layer (the superpositio lateralis, as Ariëns Kappers called it; where paleopallium and archipallium overlap) and

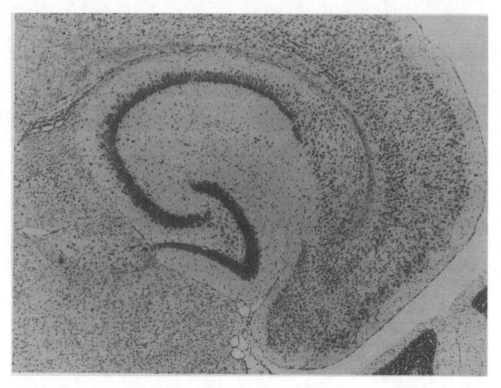

FIG. 1. Components of the hippocampal region and their respective layers. Low-power photomicrograph of region in horizontal section of 33-day-old mouse (same section represented in Fig. 2). Cresyl violet stain. From Angevine (1965); used with permission of the publisher.

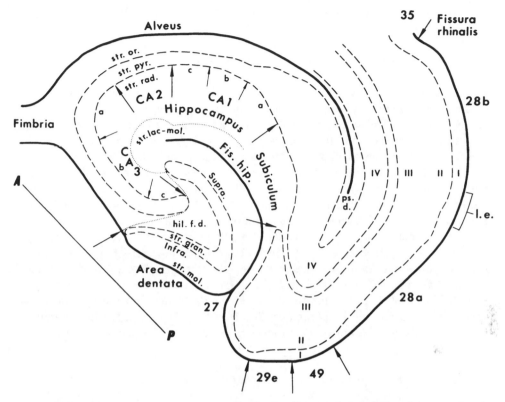

Fig. 2. Diagram drawn from the same section illustrated in Fig. 1 to indicate nomenclature, areal and sector boundaries (long arrows), and subsector boundaries (short arrows). l.e., Transition zone between subareas 28b and 28a; ps.d., psalterium dorsale. For further explanation, see text and Table 1. From Angevine (1965); used with permission of the publisher.

terminates sharply. The cortex is thus reduced in the subiculum to the inner principal layer, which in this area is broad and composed of large- and medium-sized pyramidal cells (stratum pyramidale) surmounted by an exceptionally thick molecular layer (stratum moleculare). The boundary of the subiculum with the hippocampus is also overlapping but still sharp; it occurs at the point where a crowded, densely stained layer of ammonic pyramidal cells suddenly appears along the superficial border of the stratum pyramidale of the subiculum. A progressively narrowing band of distinctly separated pyramidal cells lies beneath this ammonic pyramidal layer; these cells do not appear different from adjacent pyramidal neurons in the subiculum but are considered part of the hippocampus.* With the progressive elimination of these subicular elements, the cortex is reduced to a narrow dense layer of ammonic pyramidal cells (stratum pyramidale), underlain by dispersed polymorphic cells of

* For simplicity, an adjacent area of overlap termed prosubiculum by Lorente de Nó is omitted.

the stratum oriens and covered by scattered cells of the strata radiatum, lacunosum, et moleculare. In the hippocampus, the designations CA1, CA2, and CA3 of Lorente de Nó are recognized; further subdivisions are approximated from his figures.

The area dentata is bounded, according to Blackstad (1956), by an imaginary curved line extending from one tip of stratum granulosum to the end of the ammonic pyramidal layer and thence to the other tip. Within this line is the hilus of the fascia dentata, represented by large scattered polymorphic neurons (termed CA4 by Lorente de Nó). The term "fascia dentata" applies to the outer two of the three layers in the area dentata. In my descriptions, the limb of stratum granulosum adjacent to the hippocampal fissure is termed "suprapyramidal" and the limb next to the deep root of the alveus "infrapyramidal." These terms have precedents in the names for the two components of the mossy fiber system in rodents and have been accepted by at least one authority (Humphrey, 1966b, 1967). Ambiguities arise if other names are employed for these two distinct limbs of the granular layer. For example, one can designate these limbs in rodents with reference to their distance from the hippocampal fissure (McLardy, 1960). Such designations, however, may lead to difficulty if applied to certain other mammals, notably primates and man, where the area dentata has enlarged and all parts of its granular layer are equidistant from the fissure. The stratum moleculare of the area dentata blends into the stratum lacunosum-moleculare of the hippocampus. The stratum radiatum of the hippocampus, on the other hand, fades out beneath the fascia dentata in the region of its hilus.

4. Comparative Neuroanatomical Aspects—Evolution of the Region

A renaissance of comparative neurology (Petras and Noback, 1969) has produced a significant and rapidly widening reappraisal of the homology of neuronal subsystems (Nauta and Karten, 1970). The proven experimental and histochemical methodologies so useful in mammalian neuroanatomical research are now employed with great success on nonmammalian brains and have finally broken the stalemate imposed by the purely descriptive approach of classical comparative neuroanatomy. The recent explorations suggest that nature employs the same neurons, or operational sets of neurons, to construct neuronal aggregates of different design and anatomical organization, as a contractor might use the same bricks, or subassemblies of bricks, to construct buildings of different architecture and distinctive appearance (Karten, 1969; personal communication). The findings imply, as the embryologist Källén (1962) had sensed earlier, a frugal and flexible use of genes. Numerous examples strengthening the impression that a major hypothesis has been formulated are offered by the auditory (Karten, 1967, 1968), the visual (Karten et al., 1973), and now the somesthetic sensory systems (Karten, 1973; Zeigler and Karten, 1973). The most striking contrast in apparent neural equivalencies is in the motor systems, between the large external striatum of reptiles and birds and a major element of the neuronal population of the neocortex in mammals (Karten and Dubbledam, 1973). Accordingly, the direction of comparative neuroanatomical research has shifted toward search for and study of homologies between single neurons or groupings of neurons in specific patterns of connectivity, and away from correlations based on similarities of assembly of neurons in nuclear or cortical components. Such territorial resemblances are now known to be illusory and incorrect in major regions, as, for example, in the basal ganglia of birds and mammals.

The pertinence of the foregoing exposition for a modern comparative outlook on the hippocampal region should be obvious. Even more weight, however, should be given these new trends of thought when the hippocampus and its neighboring structures are considered. This region, especially in mammals where it is beautifully landscaped, presents perhaps more than any other part of the brain many highly formalized and instantly recognizable neuronal assemblages, "so distinctive that cytoarchitectonic boundaries are unmistakable" (Angevine, 1965). The sharpness of areal and sectorial boundaries and the ruling-pen neatness of cortical layers reviewed above are attractive features to almost every kind of neuroscientist imaginable, but allure often hides an element of danger, or at least can be a nuisance. The equivalencies of some of the clearly defined neural components as inferred from similar gross morphological appearance may be more apparent than real when their fine structure and connections are studied. The subdivisions of the hippocampal formation into H fields of pyramidal cells (M. Rose, 1927) are useful to the neuropathologist (Blackwood et al., 1967), but have shortcomings to the comparative neurologist, of which Lorente de Nó (1934) was well aware. The condensations of cells found at the subicular and hilar (end-blade) terminations of the pyramidal cell cordon and the architectonics of the area entorhinalis are places where parcellations of neurons visualized in dye-stained material are either of little help or misleading when comparison of different mammals is essayed. Fortunately, the clear-cut features of internal and external connections mentioned earlier and detailed elsewhere in this volume are continually being refined (Hjorth-Simonsen, 1972) and distinctive biochemical and pharmacological characteristics, such as the presence of zinc in the mossy fiber system and the regional distribution of neural transmitter substances, are receiving much attention (see Haug et al., 1971, for references). The emerging data on the connectivity and neurochemistry of the hippocampal region will provide guidelines for the comparative neurologist that are both attractive and safe.

From what we know so far of its comparative anatomy, the hippocampal region would appear to illustrate the statement (Nauta and Karten, 1970) that the limbic system has had a fairly stable evolutionary history. But the extent of our ignorance of the region in so many vertebrates (see below), coupled with an awareness of the reappraisal of other brain components as just described, mandates caution. One can say that when the enormous task of needed new studies is done the region will likely retain that essential character it has displayed in animals studied to date. It receives diverse indirect inputs from other telencephalic regions. In many but not all vertebrates, these inputs consist principally of influences from the olfactory apparatus which proceed through several interposed regions to the ammonic pyramidal cells. It is clear, however, that olfactory inputs are overshadowed in many mammals, such as the rhesus monkey (Pandya and Kuypers, 1969) and man (Nauta, 1971), by contributions from other sensory systems and from the association areas. In fact, the hippocampal region attains impressive size and degree of differentiation even in the absence of the peripheral olfactory apparatus, as in the cetacean brain (Jacobs *et al.*, 1971). On the efferent side, the region participates in multiple and often reciprocal connections with the septal and preoptic areas, hypothalamus, and medial parts of the mesencephalic tegmentum (Nauta and Haymaker, 1969). The latter structures, although surprisingly flexible throughout the catalogue of vertebrates, also preserve clearly fundamental identities in their connections and functional roles (Crosby and Showers, 1969).

The present lack of experimental data and consideration of space preclude a full description here of the hippocampal region, or its presumed equivalent, in the various vertebrates. The monumental work of Ariëns Kappers *et al.* (1936) is still the obvious and best place to start in acquiring such information. Additional references to a vast literature may be found in the texts by Papez (1929) and Crosby *et al.* (1962), in the writings of Herrick (1948), and most recently in the works of Nieuwenhuys (1969). Nevertheless, it may be useful to review the characteristics of the region in those vertebrates which appear to represent ends of branches on the phylogenetic tree (Romer, 1966, 1968; Hodos, 1970).

4.1. Cyclostomes

The petromyzonts (lampreys) and myxinoids (hagfishes) present striking contrasts in the position and adult appearance of the hippocampal region, immediately and dramatically illustrating the problems outlined above. The lamprey cerebral hemisphere evaginates in the familiar manner, but according to many workers, including Nieuwenhuys (1967), not completely. Thus two upright nonevaginated structures or tag ends are attached to but between the two hemispheres, facing each other and separated by a narrow dorsal part of the third ventricle. Although these structures, and their presumed homologues in other vertebrates, are called primordia hippocampi, the modern comparative neuroanatomist would agree with the late Stanley Cobb (personal communication) that such structures are adult, fully differentiated, legitimate parts of highly diversified and specialized animals and not primordia of anything! Nevertheless, the confinement of neuronal perikarya to a narrow periventricular zone, a modest input of fibers from the olfactory bulb, and scattered fascicles swinging downward from the hippocampal region into an apparent fornix system and medial forebrain bundle (Crosby and Showers, 1969) are a relatively simple arrangement.

The cerebral hemisphere of the hagfish, unlike that of the lamprey, evaginates completely in early development, so that the hippocampal region is found for a time in the lateral wall of the hollow telencephalic anlage, an unusual location for one used to mammalian ontogeny. Subsequently, however, the region invaginates into the large unpaired ventricle, fuses with overlying olfactory bulb rudiments, and ultimately unites with its opposite member in the midline, thus obliterating the telencephalic ventricular cavity. The connections of this single broad medial region, which resembles a diencephalic nucleus in a mouse brain, are similar to those in the lamprey, but the cells are scattered.

4.2. Cartilaginous Fishes

The cerebral hemispheres of elasmobranchs all come about by evagination and have well-developed olfactory bulbs, but present considerable structural differences (Nieuwenhuys, 1967) in the three main groups (chimaeras or rat-fishes, sharks, and the skates and rays) of this class. These differences relate principally to the degree of separation of the lateral ventricles, which communicate broadly across the midline, and to the extent of union between the medial walls of the hemispheres, a feature of sharks, skates, and rays resulting from migration of neuronal elements into the interhemispheric connecting membranes. Within each hemisphere, dorsal pallial and basal subpallial areas blend massively with their counterparts across the midline. In larval sharks and rays a distinct outer sheet of cells, or cortical layer, is recognized in the pallium. The hippocampal region is the most medial field of cells in this sheet and therefore participates in the union just described. In adults, however, the pallium shows only a wide zone of scattered cells, in which the outer part has a somewhat higher cell density than the periventricular region and thus probably is a remnant of the larval cortex. Afferent secondary olfactory and possibly striatal fibers reach the hippocampal region, especially its more cephalic part, but caudal parts, too; earlier workers concentrated on myelinated fibers and probably overlooked many fine unmyelinated secondary olfactory axons. Interhemispheric connections are plentiful and have been debated (Ariëns Kappers *et al.*, 1936) as to identity as decussations or commissures and consequently as to homology. The discussions include the speculation that the connection between the two primordia hippocampi is the forerunner of the corpus callosum.* Efferent fibers lead to the habenula through the medial corticohabenular tract and to the hypothalamus via the tractus pallii and tractus medianus, bundles recognized by Edinger and extensively described in the work just cited. Although the tractus medianus corresponds apparently to the fornix system, experimental analysis of both these tracts is clearly needed.

4.3. Bony Fishes

The bony fishes pose formidable neuroanatomical problems. Despite the expeditions of Johnston, Herrick, and their followers, which have led to magnificent perspectives on the evolution of the nervous system, this class of aquatic vertebrates offers little solid ground for the comparative anatomist searching for homologies of specific mammalian neuronal subsystems, such as those of the hippocampal region. Hence, although an ocean for study and exploration lies here, we must pass by at a respectful distance on the shore.

Early in the Paleozoic Era, the ancestral bony fishes appeared. Early fossil records are sketchy (Romer, 1962; Young, 1962), but evidently in the Devonian period or Age of Fishes these progenitors branched into actinopterygians (ray-finned

---

* Such conjecture had a logical base in bygone days and retains interest even today, but perhaps the reader will feel, as the author did while tramping through the older accounts, that he, like the hiker mentioned earlier, has just narrowly missed stepping into an uncovered shaft.

fishes) and sarcopterygians (fleshy-finned fishes). The latter line led apparently to tetrapods. The former branch ends in the longest and most diversified subclass of modern vertebrates, comprising more than 30,000 species. To make matters worse, the forebrain in these fishes develops in a manner entirely different from the process of evagination seen in most other vertebrates. While the ventral walls of the telencephalon retain the troughlike appearance noted in the primary forebrain vesicle, the dorsal walls do not arch medially (invert) to form a canopy over the medial parts but diverge and recurve laterally (evert) into space. Only a thin telencephalic roof plate stretches over the medial structures and it attaches at the lateral margins of the everted dorsal walls. The degree of eversion differs in various representatives, as does the thickness of the everted parts, which in most bony fishes are so swollen that the lateral ventricles exist only as potential spaces between the dorsal walls and their enveloping membranes.

As might be expected, the everted dorsal area shows marked cytoarchitectonic variation in the four actinopterygian superorders: the primitive, sparsely represented paleoniscoids (African bichir), chondrosteans (sturgeon and paddle fish), and holosteans (bowfin and garpike); and the elaborate, innumerable teleosts which comprise the majority of recent bony fishes. No attempt will be made to describe these variations, which have been detailed by Nieuwenhuys (1962). He notes three phylogenetic parallels in the pallia of evaginated and everted forebrains: the tendency for outward dispersion of neuronal perikarya toward the meningeal surface, the trend toward laminar condensation into a cortical formation, and the evolution of nonolfactory pallial regions and progressive elaboration of pallial connections with other brain components. In his opinion (1967), "the segregation of fields with different cyto-architectonics in the pallial region of bony fish must be regarded as a differentiation process which has taken place entirely within the actinopterygian stock. Consequently these fields cannot be homologized to cellular areas present in the forebrain of other vertebrates." While this appraisal is certainly unassailable, Nieuwenhuys also stresses that the highly differentiated cortex in the everted forebrains of some bony fishes represents refinement of periventricular gray, rather than of superficial migratory elements as in the mammal. If recent advances in neuroembryology and comparative neuroanatomy are considered, however, less emphasis might be placed on position of the cell bodies (periventricular vs. superficial) and degree of laminar organization. From the developmental standpoint, Morest (1970) questions the existence of true ameboid migration of neuroblasts. His Golgi studies show that migration of a prospective neuron really represents displacement of the perikaryon along the preexisting distal process of the bipolar epithelial cell (spongioblast of His), a seemingly standard parenchymal element of the developing central nervous system. Possibly, if such studies are extended to include other vertebrates, the difference between an animal which has a cortex derived from periventricular neurons and one with a cortex of superficial origin may turn out to be only a question of timing. Neurogenetic events are now known to have a wide range of independent variability (Angevine, 1973; Hinds and Hinds, 1974). Developmental neurobiology seems to be making the same point recently established by comparative neuroanatomy: the *connections* and *identity* of the neurons are the crucial matters for the homologist—more important and significant that the position of neuronal cell bodies before, as well as after, attainment of characteristic patterns of arrangement.

In one group of sarcopterygians, the lung fishes (Dipnoi), the forebrain develops not by eversion as in the ray-finned fishes but by evagination, as in most other vertebrates. Here the process takes place predominantly in a rostral direction. In genera with paired lungs, the cerebral hemispheres, which lie forward of a telencephalon medium, consist entirely of nervous tissue and the olfactory bulbs are sessile. In single-lunged representatives, part of the medial and dorsal wall is an ependymal membrane; the olfactory bulbs lie at the ends of short, hollow peduncles. The dorsal pallium has a cortex, clearly separated from well-developed periventricular gray, in which a hippocampal region and other areas have been described by some workers in certain forms. Nieuwenhuys and Hickey (1965), however, consider these fields of neurons results of mechanical effects, not true structural entities. Local inward curvature of the brain wall tends to compress and thicken its histological texture, especially in the periventricular zone, whereas outward curvature, as in the everted forebrains found in primitive actinopterygians, leads to thinning.

In the other group of sarcopterygians, the crossopterygians (lobe-finned fishes, of which the coelacanth, *Latimeria,* is the sole survivor), the forebrain has a structural plan intermediate between those of the other two groups of bony fishes. The dorsal wall is thickened, suggesting eversion in a widened ependymal roof plate as in actinopterygians; the subpallium is clearly evaginated, enclosing an extensive lateral ventricle, as in dipnoans. Rostrally, a layer of small periventricular cells is seen, but elsewhere the neurons of the pallium are dispersed through the entire thickness of the brain wall (Nieuwenhuys, 1965). In this group, as in the other bony fishes, it is evident that it is not possible or wise at this time to specify the hippocampal region.

4.4. Amphibians

The generally accepted tripartite subdivision of the amphibian pallium (see Herrick, 1948, and Nieuwenhuys, 1967, for details and references) places the primordium hippocampi dorsomedially in the wall of the evaginated hemisphere. In urodeles and anurans, but not gymnophionts (the apoda, or limbless wormlike amphibians of the tropics), the hippocampal cell bodies are dispersed outward, almost to the meningeal surface, in contrast to the chiefly periventricular somata seen elsewhere in the pallium. It is difficult (Young, 1962)* to decide whether amphibians are higher animals than fishes, and certainly not possible (Nieuwenhuys, 1967) to homologize specific parts of their forebrains. Nevertheless, the obviously more complex plan of distribution of cell bodies in hippocampus than in other pallial fields is interesting. Similar differences in the distribution of hippocampal perikarya vs. other forebrain cell bodies have been noted in the cartilaginous fishes, at least in larval sharks and rays. Although it has been emphasized that modern comparative neuroanatomists stress homologies based on similarities of connectivity and intrinsic neuronal properties and shy away from spatial patterns of nerve cell bodies, the mat-

* And I think unnecessary.

ter of how a given spatial arrangement meets certain requirements or serves a particular need is still important. In studying the hippocampal region, we are dealing with the forebrain, the most typical function of which according to Nauta and Karten (1970) "appears to lie in the perception of goals and goal priorities, as well as in the patterning of behavioral strategies serving the pursuit of these goals. . . ." We cannot suppress a question, probably somewhat facetious and certainly unanswerable at the moment: is the selective outward dispersion of hippocampal neurons in the forebrains of the first vertebrates to venture on the land and of those carnivores which have ruled the deep since the Devonian of any significance for the ethologist, or is the observed similarity purely fortuitous? All that is certain is that the amphibian hippocampal region receives a few direct projections from the olfactory bulb, but has many connections with other pallial and subpallial fields, including a well-developed fornix system leading caudally through the septum and preoptic region to the hypothalamus.

4.5. Reptiles

The refinement, in terms of outward dispersion of cell bodies from periventricular locations into nuclei or laminae, of the reptilian cerebral hemispheres is so advanced, especially in the pallium, that a true cortex throughout its extent (Nieuwenhuys, 1967) has been generally recognized since the time of the classic studies by Johnston (1915) and Crosby (1917) on the turtle and alligator until the present (Hall and Ebner, 1970). The apparently universal agreement, however, that all the reptiles have a cortex has attracted intense interest down the years (Herrick, 1926; Ariëns Kappers et al., 1936; Goldby and Gamble, 1957; Northcutt, 1967). The results have been a qualified gain; valuable descriptions have been set forth, but attempts to analyze even the dorsal sector alone have engendered "Diversity of opinion . . . almost without parallel in the field of comparative anatomy" (Kruger, 1969). And, concerning efforts to reconcile the considerable differences in forebrain structures within the four remaining reptilian orders (Young, 1962), "Expert opinion . . . has led to contradiction, confusion and discouragement" (Riss et al., 1969).

Three principal cortical fields or longitudinal cytoarchitectonic strips are recognized, variously interpreted and named by different investigators; the dorsomedial sector is generally considered the hippocampal region after Crosby (1917). Her terms and those of several others are helpfully tabulated by Riss et al. (1969) in a manner that clearly presents the greater or lesser extent accorded the hippocampal region by each investigator.* The attention of Riss et al. and other recent workers is

---

* These authors, trying not to add to the number of contradictory claims or make commitments to homologies prior to acquisition of verified experimental data (notably absent in many earlier accounts), leave these fields unnamed in their descriptions and use numerical designations instead. A table provides equivalencies of the numbers and the terms of others. This approach is wise and commendable; however, the nondescript labels for the zones may cost these authors the attention and discussion that their observations deserve. Such injustice has been meted out to others in the past who have also tried this way to lighten the burden of nomenclature. Now that emphasis is shifting to experimental approaches, names and interpretations must stand the test of proof.

focused on turtles, because these forms, unlike the relic *Sphenodon* (the tuatara), modern and highly specialized lizards and snakes, and birdlike crocodiles, seem most closely related to the stem reptiles, and their forebrains are thought to retain features present in the reptilian ancestors of mammals. The connections of the reptilian hippocampal region include a meager input from the medial olfactory tract and an extensive pre- and postcommissural fornix system, as in amphibians. Unlike amphibians, however, reptiles are getting much more attention from neuroanatomists using modern experimental techniques (Ebbesson, 1970), for reasons already stated. A better view of forebrain connections in this class of vertebrates is now emerging (e.g., Butler and Ebner, 1972; Northcutt and Butler, 1974) and will likely provide additional insights into where the hippocampal region fits in the scheme of forebrain organization.

4.6. Birds

Although an avian hippocampal region is generally recognized (Ariëns Kappers *et al.,* 1936) and labeled in atlases (van Tienhoven and Juhász, 1962; Karten and Hodos, 1967), attempts to define it (Craigie, 1928, 1932; Fleischhauer, 1957) have been handicapped by the lack of experimental studies described earlier and by the divergence of avian and mammalian brains from the extinct stem reptiles. It would be outside the scope of this chapter, ill-timed, and manifestly unwise to discuss cortical and corticoid areas of birds here. Such areas, like their counterparts in reptiles, are now under intensive study with modern methods (Karten *et al.,* 1973). New insights into the plan of the vertebrate forebrain are certain to come. The major outgoing pathway from the parahippocampal and hippocampal areas is essentially a fornix system (Crosby and Showers, 1969) but contains additional telencephalic components, including fibers from the septum. Analysis of such complex avian tracts is also going forward (Zeier and Karten, 1973) and previous comments apply here as well.

4.7. Mammals

The size and shape, fiber connections, and topographic relations of the hippocampal region vary greatly in different mammals and the region is much more than, if perhaps it was originally, an analyzer field of neurons indirectly connected to the olfactory structures of the forebrain (see above). The reader should turn to Ariëns Kappers *et al.* (1936) for details and references to the older literature. In acallosal mammals, the monotremes and marsupials, the hippocampal region forms a paramedian crescent—for Elliot Smith, a true limbic lobe—beginning just above the olfactory area, where the major commissures (anterior and hippocampal) lie, and extending superiorly and posteriorly to the amygdala, thus forming most of the rim of the large foramen of Monro. Gastaut and Lammers (1961) and White (1965) present thorough discussions, helpful diagrams, and additional references. Although the dentate gyrus, cornu ammonis, and subiculum are less well differentiated cytoarchitectonically, their unmistakable appearance and straightforward succession in the unfolded acallosal pallial margin were exploited to great advantage by Elliot Smith in his many studies on the evolution of the cerebral commissures and his farsighted interpretations of the supracallosal structures encountered in the placental mammals, where the corpus callosum makes its appearance.

Where the callosum is present, the overlying components of the crescent persist but are attenuated and rudimentary. The fasciola cinerea, indusium griseum, and medial and lateral longitudinal striae correspond in turn to the dentate gyrus, hippocampus, and fimbria-fornix fiber system. The more massive the bulk of the callosum, the more insignificant the rudiments are likely to appear and to be, and *vice versa*. Thus a somewhat indistinct field of neurons, the anterior hippocampus, is preserved precommissurally, adjacent to the septal area, while a tremendously enlarged and elaborated ensemble of neuronal sectors is found in retrocallosal regions, especially in the temporal lobe where that extension of the cerebral hemisphere is present. In this event, concomitant with growth of the neopallium, the hippocampal region is involuted (rolled upon itself) into the lateral ventricle.* Other trends which seem to parallel expansion of the neocortex are a progressive shift of the cordon of hippocampal pyramidal neurons below and underneath the row of dentate granule cells (Lorente de Nó, 1934) and a marked increase in number of granule cells and consequent elaboration of the stratum granulosum. The overlap, or "superpositio medialis" as Ariëns Kappers calls it, of pyramidal and granule cells occurs at the point where a "break" between these cells occurs ontogenetically (see below). The result of these trends is an intimate interlocking of the hippocampus and dentate gyrus, such that a V-shaped sheath of granule cells encloses the enlarged tip of a straight ammonic end-blade, as in the mouse or rat, or that an extensive, elaborate curvilinear hedge of granule cells surrounds an end-blade reflected so much upon itself that it resembles a brush-hook, as in monkeys and man (Lorente de Nó, 1934). The cingulum, which conveys fibers from cingulate and retrosplenial cortical areas to the hippocampal region, attains the dimensions and extent of a cable, while originally modest association fiber tracts between the pyriform cortex and hippocampal region are powerfully augmented to form the temporoammonic or sphenocornual bundle of Ramón y Cajal (1901–1902, see 1955).

The foregoing account is a general overlook on the hippocampal region in representative vertebrates for a wide range of interested readers from many disciplines. The view is neither comprehensive nor detailed; it will not, and should not, satisfy the comparative neuroanatomist, for whom it was not intended and from whom (Karten, personal communications) valued assistance has been required and given. A more intensive and abundantly referenced overview by Crosby *et al.* (1966) is highly recommended as well as other contributions providing important perspectives in the volume edited by Hassler and Stephan (1966). I have paid little attention to variations within the different vertebrate classes; I have not even compared the region in the frog and the newt. I have given short shrift to connections—the pallial

---

* I tried to avoid, wherever possible, descriptions which imply causation; here I made an exception and used a time-honored explanation for sake of clarity but at the risk of misleading the reader. The lack of fidelity of the words selected to what actually occurs in phylogenesis or ontogenesis of the hippocampal region will be sensed from the embryological section which follows.

commissures of nonmammalian vertebrates and the dorsal and ventral psalterium in mammals. Especially, but understandably, I have not dealt in a rich merchandise of specialized neuropil, the marketplace (Nauta) of neuronal interaction: numbers of neurons and packing density; characteristics of cells, axons, or dendrites; structure and arrangement of synapses; arrangements of neurons in space; amounts and distribution of gray and white matter—in a word, in the *fine* structure. Such detail has not yet been obtained in the brains of most adult vertebrates and certainly not in those of the larval stages many vertebrates go through. The tadpole, for example, has a functional and well-developed nervous system with radically different inputs, internal medium, and outputs than that of the frog (Kemali and Braitenberg, 1969).* These authors stress that the adult brain may bear traces of its larval past ("ontogenetic hangovers"). Such structural clues can be significant in the quest for homologies; valuable insights into the origin of vertebrates have been gathered from comparisons of the larvae of echinoderms and hemichordates (Fell, 1948). The long and sometimes illogical search for ontogenetic–phylogenetic parallels should be continued, especially now when many investigators are employing modern and sophisticated methods of neurobiological investigation.† The routes to success in this search may lie not only in selection of the right animal for study by the right method, but also in recognition of the appropriate time in development and maturation of its nervous system for that study.

5. Morphogenetic Aspects—Shaping of the Region

5.1. Classical Studies

The hippocampal region offers a logical and attractive place for study of cortical development. The fascia dentata is the fringe of the cerebral cortex (Elliot Smith, 1896)—the hem of the cortical mantle, almost certainly the first part to be gathered up. Then, too, this neuronal assembly, in section an arrowhead, is so striking, and the adjoining cortical formations so distinctive and sharply bounded, that for a century the region has intrigued the great embryologists as much as the comparative neurologists whose work has just been considered. References to the extensive literature are presented by Hines (1922), Elliot Smith (1923), M. Rose (1926, 1927), Ariëns Kappers *et al.* (1936), Tilney (1938), Macchi (1951), Crosby *et al.* (1962), and Godina and Barasa (1964). Especially useful, well-illustrated, and accessible accounts are the recent papers by J. W. Brown (1966) and Humphrey (1966a) on development of the hippocampal formation in certain insectivorous bats and man, respectively. These two studies (by close associates) complement and largely reinforce one another, but demonstrate obvious differences in the development of the region. In

---

* I am aware of the inadequacy of my own account as I read the introduction by those authors to their atlas of the frog's brain. I urge readers interested in evolution of the hippocampal region or any other brain component to examine this farsighted and illuminating work.

† A glance at the *Contents* on the back covers of the past 2 or 3 years' issues of *The Journal of Comparative Neurology* should impress this point upon a reader who may not already follow the developments in this field.

the human embryo, unlike that of the bat, the first component to be recognized is the dentate gyrus, presaged by the appearance of a cell-free marginal zone in the dorso-medial wall of the telencephalic vesicle. The earliest migrations of neuroblasts (or their perikarya, to restate Humphrey's observations in light of recent findings by Morest and others) from the ventricular zone, however, are into the adjacent prospective hippocampus. Only after small clusters of cells (perikarya) have accumulated in the hippocampus·do scattered cell bodies appear in the rarefied marginal zone of the future dentate gyrus. These cell bodies belong to the neuroblasts derived directly from the underlying ventricular zone and are not marginal elements near the overlying hippocampus which have slipped down, as previously thought (Hines, 1922). The latter cells only rarely contact the layers of the immature dentate gyrus and contribute few, if any, cells to it; instead, they appear to be transient elements which later are incorporated into the stratum lacunosum hippocampi, or else degenerate and disappear.

Subsequently, formation and differentiation of the cell layers of the hippocampus precede these events in the dentate gyrus, despite the precocious indication of the latter by its conspicuous marginal zone. Pyramidal sectors CA1, CA2, and CA3 and the suprapyramidal layers are all identifiable before the initial single cell layer of the incipient dentate gyrus becomes further enlarged and modified (see below). CA3, however, lags somewhat behind the other two sectors in its differentiation, consistent with its protracted period of origin, at least when compared to CA2, disclosed in the autoradiographic study described below.

Humphrey notes ontogenetic–phylogenetic parallels in the development of the human hippocampal formation. At certain stages it resembles, in terms of progressive detachment of neuroblast perikarya from the ventricular zone and their alignment into distinct continuous cell cordons, the region in lungfishes, frogs, many reptiles, and those birds nearer to the main evolutionary line. The superposito medialis of dentate gyrus and hippocampus as seen in some reptiles, however, has no counterpart in human development, in which the ammonic pyramidal layer precedes and hence overlaps *externally* (not internally) the migrating perikarya of the dentate anlage. Only later, when the definitive stratum granulosum of area dentata appears, does a superposition of the type seen in mammals occur.

In her account of the development of the human dentate gyrus, Humphrey calls attention to an intermediate stage, illustrated in studies by others on the fetal brains of the spiny anteater and many but not described for any adult vertebrate. In this stage, the single cell layer of the dentate gyrus thickens into an undifferentiated ball-like mass, along the outer surface of which elements concentrate to form a primordial granular layer. As the definitive stratum granulosum develops, it breaks away from the pyramidal layer of the hippocampus. This feature is also consistent with autoradiographic findings to be detailed shortly. We know now that the oldest neurons (first to originate in terms of final division of antecedents within the ventricular zone) of stratum granulosum are those cells which come to lie along its external boundary. The multitudes of granule cells which originate later and over a long time course* are in the adult animal encountered progressively more deeply in the granule cell

---

* During the last week of gestation and for 3 wk thereafter in the mouse; the human is inaccessible for study at comparable developmental stages.

layer, where neuron origin eventually occurs *in situ* rather than in the ventricular zone, as it does initially for stratum granulosum and throughout the proliferative periods of neurons destined for most other brain components. The deeper regions of the undifferentiated cell mass described by Humphrey for the human fetus probably correspond to the ectopic source (subventricular zone) of late-originating granule cells demonstrated by autoradiographic studies. The deepest part of the ball of cells, according to Humphrey, forms the polymorphic layer of the dentate gyrus, a group of neurons in continuity with the ammonic pyramidal cell layer. As the dentate gyrus enlarges and rotates toward the hippocampus, a shallow hippocampal fissure is consistently identifiable opposite (medial to) the latter structure. Humphrey believed that no comparable fissure exists in reptiles and birds and moreover that this fissure is different from that found by some investigators in other vertebrates and by Hines in man. In a subsequent paper, which was to be the last study of the hippocampal region by this distinguished contributor, Humphrey (1967) presented in full a review of the nearly century-long controversy concerning the origin and development of this fissure in the human brain and the evidence, gathered from study of serial sections of over 60 human embryos and fetuses, for her beliefs. Since she could find no evidence of the fissure in fetuses prior to 10 wk of menstrual age in freshly fixed material without artifacts of any kind, she concluded that the sulci identified earlier in development by many previous workers are artifacts. "The definitive hippocampal fissure first appears when differences in the growth rate of the gyrus dentatus and the cornu ammonis result in a thicker telencephalic wall in the dentate gyrus."

The findings of Brown in the bat also stress that the hippocampus, once it has appeared, develops more rapidly in the early stages than the dentate gyrus, which appears later. Certain differences noted in the bat seem to represent omission of stages seen in man. No initial cell layer formation occurs in the primordial dentate gyrus, a mass of cells which develops in continuity with the hippocampal pyramidal cell layer. The definitive stratum granulosum appears without any preliminary condensation of cells and is immediately separated from the ammonic pyramidal layer. On the other hand, migration of cell bodies into the external part of the dentate gyrus is more marked in the bat than in man.

An especially interesting statement by Brown is that initial differentiation of the granular and polymorphic layers of the dentate gyrus begins near the cornu ammonis and then spreads progressively around the undifferentiated cell mass such that it occurs later in the medial or ventromedial part of the gyrus, "which develops last from the ependymal layer." This gradient of differentiation follows closely the calendar of neuronal birthdates presented by the autoradiographic findings: There is a sequence in time of origin of the granule cells, beginning at the tip of the suprapyramidal limb and proceeding along that limb toward the apex of the fascia dentata and then into the infrapyramidal limb, and also from the outer edge of both limbs toward the deeper reaches of the granule cell layer ("outside-in"). Brown also emphasizes the common embryonic origin of the granule cells of the fascia dentata and the polymorphic cells of its hilus, an observation once again supported by the autoradiographic findings. Finally, he emphasizes that the shallow hippocampal sulcus found above the bulging dentate gyrus remains only a superficial indentation. A pale-staining diffuse zone extending inward from the sulcus, containing a few marginal cells and in most instances blood vessels, resembles at first glance a fissure but is actually an intrinsic feature. Humphrey (1967) rules out the possibility of fusion accounting for the diffuse zone in these embryonic bat brains and calls attention to its subsequent compression with continued development. From consideration of the findings of Brown and other workers on a variety of developing mammalian brains, including her own studies on the human, she feels that "probably a diffuse zone is a transitory feature in the development of the hippocampal fissure for all mammals having a characteristic dentate gyrus, and a compression zone occurs in such adult brains also." Approximation of the walls of the adjoining dentate gyrus and hippocampus obviously takes place in man and certain other mammals and deepens a hippocampal sulcus into a true fissure. Subsequent fusion and obliteration of the walls also occur, to the degree that enlargement of adjacent structures and other cortical regions limits the space available for expansion of the hippocampal formation.

The reader will find detailed descriptions and vivid illustrations of the development of the hippocampal formation in the paper by Godina and Barasa (1964)* on the pig and sheep and in a study by Humphrey (1966b; amplifying material presented in 1966a) on the human. The size of the cerebral hemispheres, thickness of the walls, and clarity of the various cortical strata during and after migration of neuroblasts (or their perikarya) in the large brains of these mammals are advantages for the classical approach to growth and development of the cerebral cortex with dye stains. Space does not permit thorough review of these valuable papers here, but highlights will be selected for comment. Godina and Barasa stress that development of the hippocampal formation begins at a time when the overlying neopallium is already at relatively advanced stages. It must be remembered that such observations relate to recognizability, not formal origin, of cortical components—to attainment, by means of cellular migration and differentiation, of characteristic and identifiable groupings or layerings of perikarya of cells which have originated at earlier times from the ventricular zone, perhaps simultaneously with the neurons of other cortical regions. These cellular events and other chapters in the life history of a nerve cell vary greatly in timing and details, as noted above. Godina and Barasa acknowledge that involution of the hippocampus and development of the hippocampal fissure must involve intrinsic, genetically determined properties, but recognize that mechanical factors—the precocious and predominant growth of the neopallium and the early atrophy of the medial wall of the hemisphere in the region of the prospective choroid plexus—also must play a role. These authors note the early origin and advanced differentiation of deeper neurons in most cortical regions and the colony of undifferentiated cells that remain in the hippocampal primordium (end-blade region) even at late stages of development. These cells continue to divide, forming the fascia dentata through superficial accumulation of their progeny. These observations accord with the autoradiographic findings described below. The paper includes excellent drawings and photomicrographs illustrating the differentiation, as studied by the rapid Golgi method, of pyramidal and other neurons in the hippocampus and of granule cells in the dentate fascia.

Humphrey's account (1966b) reminds us that the human hippocampal forma-

* This paper will be difficult for those not proficient in Italian; nevertheless, perusal of the illustrations and sampling of the text are strongly recommended.

tion is a long structure in anteroposterior (with emergence of the temporal lobe, posteroinferior) extent throughout which development has not progressed to the same degree for any age period or reached a comparable degree for the dentate gyrus and hippocampus at any given point. These observations in man agree with autoradiographic findings in the mouse; there is a caudorostral gradient in neuronal time of origin for the cerebral cortex (Angevine and Sidman, 1962) and a hippocampodentate gradient for the hippocampal formation (Angevine, 1965, 1970a). Humphrey describes the rapid and continued migration of ventricular cell bodies* to form the undifferentiated ball-like mass at the fimbrial end of the dentate gyrus (see above) and gives the subsequent fate of these elements. Condensation, and later perpendicular orientation and separation from the center, of peripheral cells of the mass establishes the granular and polymorphic layers. These events begin where the mass blends with the ammonic pyramidal cell layer and progress around the periphery toward the side into which neuroblasts are still arriving in large numbers from the region near the fimbria.† This sequence of differentiation, which includes acquisition of a minute amount of powdered Nissl substance in the cytoplasm of some of the peripheral cells during their characteristic orientation, suits the observations of Godina and Barasa in sheep fetuses and of Brown in insectivorous bats and the autoradiographic findings of Angevine in the mouse and of Altman and Das in the rat (see below and elsewhere in this volume). In discussing this progression, Humphrey enunciates and elaborates a point emphasized by the present author, that "it is quite possible that the sequences for the cessation of new cell formation, for the beginning of cytologic differentiation and the final maturation of neurons may all differ, particularly from one mammal to another." Moreover, Humphrey finds that by the time that hippocampal sectors CA3, CA2, and CA1 are distinguishable, more cells with fine Nissl granules are found in CA2 than in any other part of the pyramidal layer; this sector is the first (Angevine, 1965) for which the ventricular zone completes its period of neuron formation. "Through the 13.5-week age level, area CA2 remains the best differentiated area of the cornu ammonis." Finally, Humphrey agrees with Brown and Angevine, and other recent workers and many of the classical anatomists, that on the basis of its developmental history as well as other considerations the polymorphic neurons derived from the center of the initial cell mass should be regarded as a deep layer of the dentate gyrus, not part of the hippocampus as Rose, Lorente de Nó, McLardy, and also Godina and Barasa see it.

5.2. Autoradiographic Studies

My monograph on development of the hippocampal region (Angevine, 1965), alluded to above, presented a new approach to study of neurogenesis—the systematic

---

* For consistency and timeliness, the terminology of the Boulder Committee (1970) and the new concepts (Morest, 1970) underlying it are used in this chapter except where I feared changing the author's original meaning. In this instance, Humphrey's words are "cell migration from the ependymal layer. . . ."

† This region, which Humphrey again alludes to as the ependymal layer, is sometimes called the deep root (tief-Wurzel) of the alveus, or the extraventricular alveus. In later stages of development, it is one of several areas of enlargement and high proliferative activity of the subventricular zone defined by the Boulder Committee.

regional analysis of brain development with tritiated thymidine (a DNA precursor) and the autoradiographic technique. For decades, major features of neurogenesis—cell proliferation, migration, differentiation, and degeneration—had been studied and even, for the last few years, approached with these new tools. But the emphasis was on intrinsic peculiarities and overall significance of these four critical cellular events. Surveillance of *particular* neuroblasts, in terms of their time and place of origin, movements, and fate, had been impossible; one could no more hope to identify and follow a particular bat* among those swarming from a cavern at nightfall. First of all, one did not know the time of final mitotic division and imminent departure of a given cell in the suspected germinal region of the ventricular zone, any more than one could see that that bat was leaving now and not just darting to and fro at the cavern mouth. Thereafter, one could not keep track of groups of migratory neuroblasts, let alone a single cell, since there are so many; moreover, the migrations are rapid and extend over long distances in which collisions and intermingling with other groups are inevitable. (The analogy still holds good, as several chiropterists I know have impressed upon me.) Finally, one could not foretell or ascertain the fates of neuroblasts as specific neurons, types of neurons, or components of particular neuronal assemblies, since neuroblasts are relatively undifferentiated and look pretty much alike during proliferative and migratory stages. In fact, many times one cannot even tell neurons from neuroglia, either during development or in the adult, unless he uses methods of metallic impregnation of cell processes to supplement dye stains of their cell bodies in his embryological or cytoarchitectonic investigations. It is not surprising, then, that the resolution of histogenetic studies was poor—at least when compared to the fine detail in studies on development of gross features of the brain, appearance (or disappearance) of landmarks, differentiation of neurons or groups of neurons, elongation and elaboration of axons and dendrites, and mode or sequence of myelination of axons or tracts.

With advent of this new method, now well known and extensively used, tracing the movements of groups of neuroblasts or, alternatively, determining the histories of groups of adult neurons became possible. Sometimes the findings could be narrowed down to the single cell, resolution comparable to that brought to cytology by the electron microscope or that given neuroanatomy by resurgence of the Golgi method. As with banding bats, one could now follow individual elements of teeming populations from beginning to end. A label affixed to the DNA of a neuroblast† is

---

* A hackneyed analogy is to the study of bird migrations, but bats lend themselves better to these issues.

† The term "neuroblast" troubled (and continues to bother) members of the Boulder Committee because the suffix "-blast" commonly connotes a proliferating cell, whereas the vertebrate neuroblast as usually defined is postmitotic. Morest (1970) avoids the issue in his definition: "The *neuroblast* is a cell that will form a neuron in the normal train of events." So defined, its chief ability is emission of axonal and dendritic growth cones, not procreation. As a member of the now defunct committee, I think that "neuroblast" should no longer be used to designate a cell which does not divide. A dignified and good term for the postmitotic migratory element, which is an immature neuron and not a neuron precursor, is "primitive neurocyte" (Snell, 1972). Such elements are the bipolar neuroblasts or "tadpoles" of older accounts, cells which in fact may never be present in the development of some brain regions. But the work of Sechrist (1969) and of Hinds and Hinds (1974) offers hope that "neuroblast" may yet find a proper place and legitimate use—as *the name* for a ventricular cell which will divide one more time but has now signified by its fine structure, just before or immediately after cell division, that at least one or perhaps each daughter cell will ultimately become a neuron. It is in that sense that I have used the term here.

*permanent**; once the cell leaves the ventricular zone, it loses the capacity to divide.†
Furthermore, rigorous schedules of neuron origin permitted *selective* labeling by appropriate timing of radioactive thymidine doses to the embryo or to its mother.‡
Neurons originating (replicating DNA for the last time) at a given developmental
stage could be identified by noting their positions in autoradiograms of the adult
brain. Any neurons arising earlier or later would be nonradioactive and hence unlabeled¶ in the overlying emulsion.

The results and significance of many studies in which these principles were exploited have been discussed by Angevine (1970*a,b,* 1973) and Sidman (1970*a,b*). But
although important regional studies of time of neuron origin followed and clarified
events in other brain regions (e.g., Hinds, 1968*a,b*; Shimada and Nakamura, 1973;
Creps, 1974*a,b*); the hippocampal region provided the ideal setting in which to view,
for the first time with complete clarity, the results of workings of a general and unquestionably important developmental mechanism: the patterned cell proliferation
for specific brain components. The place was right, because of the striking neuronal
assemblies and razor-sharp boundaries between them found in this brain locale.

The salient findings of this study are reviewed below. To appreciate these
results in full measure, however, the reader must study and compare many autoradiographic reconstructions (Figs. 3, 4, and 5)—maps which show the position and
identity in the adult hippocampal region of neurons labeled at successive developmental stages. The maps reproduced here are about half the number presented in the
original work; only maps showing positions of heavily labeled neurons are included.
Such neurons are considered the definitive population arising on the day of development represented by each map. The findings concern numerous components and
extend over a long time (1 month—a week before birth and 3 wk after); they must be
considered as a whole and read like a book.

In the mouse, which is born 19 days after conception, the period of neuron
formation for the hippocampal region begins on the tenth day of gestation. Few
neurons of the isocortex originate that early, although proliferation for those regions
is well under way by the following day. Deep cells of the adult pyriform cortex,
however, arise at this time, as well as cells in various subcortical centers and here
and there in the brain stem. Some motor neurons of the cranial nerves even originate
on day 9. The simultaneity of onset of cell proliferation for so many neural
components, whose subsequent histories lead to degrees of neglect or elaboration all
the way from rudiments to grandeur, should give us pause when we use such terms
as "neocortex" and "old brain." From the standpoint of the ventricular zone,
construction for all these buildings—edifices or shacks (Nauta)—begins at about the

---

* Unless the cell becomes a trephocyte, which in this context serves not some nutrient function for its
neighbors but gives up its DNA in death.

† Unless it becomes a component of the subventricular zone, as it does if it is a late-forming granule cell
originating in or near the fascia dentata.

‡ [^{3}H]Thymidine is available in the circulation to cells (whether or not they incorporate it into DNA) for
only a short time (from ½ to 1 h, approximately) and hence serves as a pulse label in studies utilizing
mammals; continuous labeling techniques involving injections into the yolk sac are used in avian experiments. See Sidman (1970*a*) for thorough discussion of these technological considerations.

¶ But see interpretation of lightly labeled cells in Angevine (1970*b*). It is hoped that the number of
footnotes for this paragraph will not undermine confidence in a proven methodology.

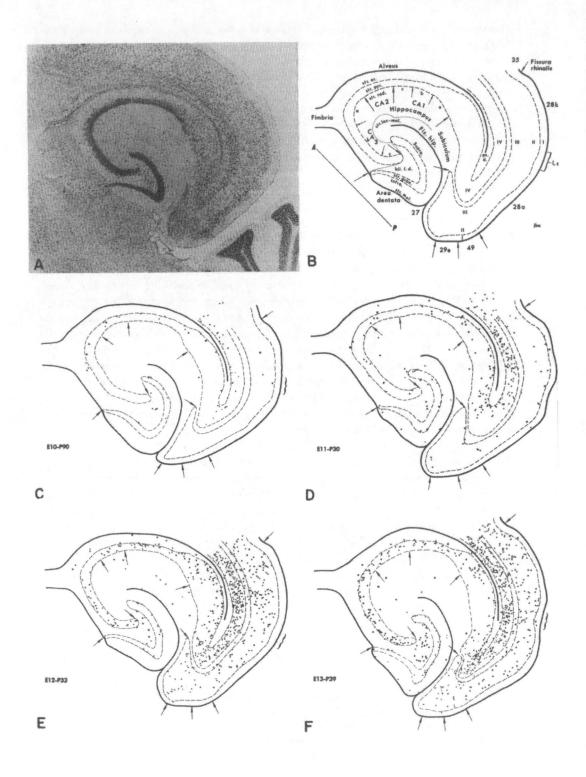

same time. Comparative neuroanatomy is hard at work making the same point (see above).

The first neurons to form are usually those of the deepest layers of the various components of the region, reminiscent of earlier observations in the isocortex on progressions in time of neuron origin (Angevine and Sidman, 1961) and sequences in cell recognizability and cytodifferentiation (Godina and Barasa, 1964). Simultaneous origin of neurons in outer layers occurs in some places (entorhinal and dentate areas), but on the whole the hippocampal region displays the now well-publicized "inside-out" sequence of neuronal time of origin. The pyramidal or other neurons originate in the ventricular zone and migrate into the cortex, more or less directly if the findings of Godina and Barasa, of Brown, and of Humphrey are recalled. In the adult, the positions of neurons generated later are usually superficial to neurons originating earlier. The inference was that the younger neurons somehow migrated past older neurons already in place. Subsequent studies (Morest, 1970) have clarified these matters and shown, as stated above, that in many brain regions examined with the Golgi method, it is really the perikarya that are transposed.

Neurons of the retrohippocampal formation continue to arise through the fifteenth day of gestation, when many cells of Rose's outer principal layer are formed. The small pyramidal cells of the presubiculum (at the superpositio lateralis of Ariëns Kappers) arise the next day and cell proliferation for the retrohippocampal formation is over by the seventeenth day. Meanwhile, the granule cells of the adult isocortex (lamina ii) beneath the molecular layer are still arising for the regions on the convexity of the cerebral hemisphere. Because of this longer period of neuron origin for the isocortex, striking contrasts are seen in those parts of the maps where the rhinal fissure separates allocortical and isocortical regions. In the adult brain, neurons incorporating label on the seventeenth and eighteenth days of gestation are encountered immediately lateral to the fissure, but not medial to it. In this area and other areas studied, regional differences in the pattern of radioactive neurons coincided so well with cytoarchitectonic parcellations of the cortex that some plots were repeated to rule out the possibility of error or bias.

FIG. 3. Horizontal cresyl violet stained section of hippocampal region in adult mouse brain (A) and matching diagram of regional components (B), as presented in Figs. 1 and 2 but reduced to aid the reader in interpreting the maps. Autoradiographic reconstructions (C–F) show the inside-out sequence and characteristic calendars of neuronal birth dates for each part of the region (Angevine, 1965). In C–F, arrows indicate sector boundaries as in B; code at lower left gives the day of development on which the mouse received a single injection of [^{3}H]thymidine and the day it was killed. Thus E10-P90 is an autoradiogram from an animal injected on the 10th day of embryonic development and killed on the 90th postnatal day. The dots represent radioactive nerve cell bodies; each dot, however, is an exact outline of the cell *nucleus,* not its perikaryon. The size of the dots in the original maps affords a clue to the size and identity of neurons when compared with a photomicrograph (A), but the degree of reduction here makes such correlations unrealistic. The ages of the mice when killed varied, but all neurons had reached final destinations. Note the early origin (C) and contrasting outside-in sequence of cells in the stratum granulosum of the area dentata (D–F) and also the gradient from suprapyramidal to infrapyramidal limb. From Angevine (1965); used with permission of the publisher.

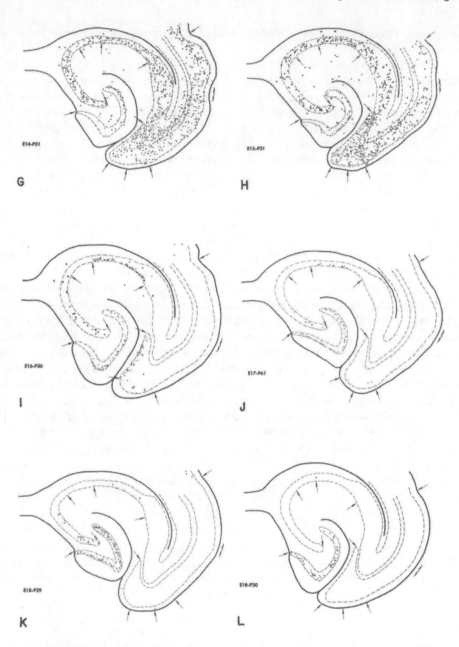

Fig. 4. Series of autoradiographic reconstructions continued, to follow the inside-out sequence and the calendars of neuron origin. Note the abrupt cessation of neuron origin for hippocampal sector CA2 (I), with continued origin for CA1 and CA3. Note also the late-forming isocortical neurons in the area just lateral to the rhinal fissure on days 16 (I) and 18 (K) when neurons of entorhinal area have all arisen. Outside-in and supra-infrapyramidal gradients seen in the stratum granulosum of the area dentata are associated with generation of deeper cells in both limbs, as will be noted in the adult injected later on day 18 (L), just before birth. From Angevine (1965); used with permission of the publisher.

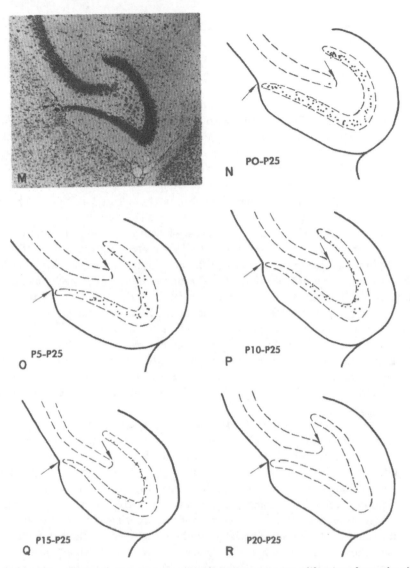

FIG. 5. Autoradiographic reconstructions continued, to show persistent proliferation of granule cells for the fascia dentata, as well as the outside-in and supra-infrapyramidal limb gradients. These five animals received triple injections of [³H]thymidine at 4-h intervals to label a greater percentage of cells preparing for final division; the mice were also younger when killed (25 days old). In the animals injected on the 15th (Q) and 20th (R) postnatal days, the late-forming cells had not completed differentiation. Granule cells originating prenatally or perinatally migrate to their layer from the ventricular zone, but postnatally arise by proliferation of subventricular cells beneath or in the stratum itself. Note the solitary labeled pyramidal cell at the tip of the hippocampal end-blade in the mouse injected on the day of birth, P0 (N). Granule cells arise as late as P20 (R) but also as early as E10 (Fig. 3C). Hence their origin is not triggered at or by birth, which occurs during but appears not to alter the sequence of cell proliferation. (Birth may, however, accelerate the rate.) Observe that in the series of maps shown in Figs. 3–5 neurons representing all components of the adult hippocampal region have been accounted for by label. From Angevine (1965); used with permission of the publisher.

In the hippocampal formation, the subiculum completes its period of neuron origin earlier than adjacent retrohippocampal structures. Its outermost neurons form late on the fifteenth day. Virtually no labeled cells are found in the adult following injections on the sixteenth day, whereas the overhanging presubicular outer layer is peppered with such cells. The hippocampal sector CA2 displays a similar time course, as does the pyriform cortex. But sectors CA1 and CA3 have longer periods of neuron origin, extending into the early part of the eighteenth day, just before birth. Although the difference of a day or so may not seem like much at first, it must be recalled that the time span of murine brain development is compressed and the proliferative activity of the ventricular zone vigorous. Many cells can be formed in a day. The last neurons to form are generally, but not exclusively, the outermost cells of the pyramidal layer. An occasional labeled neuron is found in CA1 or CA3 when [^{3}H]thymidine is injected perinatally.

The longest, and a very interesting, period of neuron origin is in area dentata. Neurons of the molecular and hilar layers are all generated by the end of the fifteenth day, but the innumerable dwarf neurons (Ramón y Cajal) of the granular layer arise from the very beginning of this story, the tenth day of embryonic development, until 3 wk after birth. The sequence of neuron formation in this one layer of this single area of the cerebral cortex breaks the rule, as far as the pattern noted for other laminar structures is concerned. It is "outside-in." The first neurons to incorporate label lie in the mature brain along the outer edge of stratum granulosum, while the last are found along the inner edge. These findings, not available to Humphrey at the time her book chapter (1966a) was prepared, were accepted readily by her and correlated meaningfully with her own findings in her subsequent amplified account (1966b).

Independent autoradiographic studies in the rat brain (Altman, 1966; Altman and Das, 1965, 1966) present a different approach to analysis of the development of the hippocampal region, but offer no conflict with the foregoing observations. On the contrary, the respective findings are complementary. These authors emphasize kinetic and numerical parameters of cell proliferation, rather than its pattern. Where time of origin of specific brain components is concerned, they stress the prolongation of neurogenesis into postnatal development and the possible anatomical and functional significance of this event, especially for the enrichment of the brain that must and does somehow keep up with the increased environmental stimuli present after birth. Postnatal neurogenesis is not ubiquitous, but it has now been described by several workers in several brain areas in several mammals, including man. The cerebellum, hippocampus, and olfactory bulb are well-known examples where it can occur, depending on the animal in question. Das and Altman (1970) have described it in the caudate nucleus and nucleus accumbens septi in the rat, which like the mouse studied above is a precocial mammal. A comprehensive series of autoradiographic studies of regional neurogenesis in sufficient types of mammals to represent the major orders has not yet been made, nor is likely to come soon. Such a survey might show that postnatal neurogenesis, as Hinds (1968a) put it, "is not a necessary feature of mammalian development," but instead, as Das and Altman acknowledge, "the tail-end of neuroembryogenesis." The process occurs vigorously, however, in the right places at the right time, as far as completion of proliferation of numerically large populations of neurons and strategic position with regard to afferent systems are concerned. It deserves description by the neuroembryologist and attention by the behaviorist, and probably also by the comparative neuroanatomist, who has much to learn about changes in the brain during metamorphosis (see above). Recent insights into neonatal neuroendocrine reciprocities exemplify the wisdom of viewing events in relation to their settings, which in that case are environmental influences within the body rather than outside. The next chapter will show the value of this viewpoint and present the effects of surgical interruption of afferentation on the development of the hippocampal region.

The central message of my findings is that there is a *pattern* in the early cellular events in neurogenesis. Nineteenth-century brain scientists and probably earlier thinkers listened for this message—now every neurobiologist hears it. Getting a look at it, however, is what everybody wants. The maps presented in this chapter, although not easy to read, provide a good look. There have been others and soon will be many more, as every neuroscientist today well knows.

5.3. Other Studies and Approaches for the Future

The results of this study of hippocampal neurogenesis have facilitated *in vitro* study of postnatal brain development in hippocampal explants obtained at birth (Kim, 1973; LaVail and Wolf, 1973). The characteristics of the region stimulate cell dissociation–reaggregation experiments on fundamental histogenetic properties: cell migration, aggregation, and capacity for histotypical organization (DeLong, 1970). The simple and precise, if rigid, cytoarchitecture, clear and deliberate manner of construction by patterned cell proliferation, and careful wiring of this structure, from the outside and on the inside, are irresistible features for tissue culturists and also for those seeking structural–behavioral correlations (Wimer *et al.*, 1971). Functional plasticity, demonstrated in this region by snipping wires in the immature brain (Lynch *et al.*, 1973*a,b*), can also be explored by surgical intervention with earlier details of fabrication, which are accessible in the events of postnatal neurogenesis.

The regional analysis narrated above follows the trail of autoradiographic studies on general dynamics of neurogenesis (the "to-and-fro" movements of intermitotic ventricular cells) blazed in the 1950s by Sidman (see 1970*a*). Turnabout, the recent exploration (Caviness and Sidman, 1973) of malpositioned neurons in the hippocampal region of neurological mutant mice (which reveal the extent and suffer the consequences of nature's neurogenetic experiments) departs from places discovered by me. Moreover, a subsequent study (Caviness, 1973) on time of neuron origin in the hippocampal formation of the reeler mutant (whose brain shows many malpositioned neurons, but whose hippocampus and dentate gyrus, although imperfect, have explicit laminar patterns) drew heavily upon my guidebook and used its azimuths of final cell positions for each day of cell generation. The results of this study are fascinating: The hippocampus of normal mice and that of reeler mice contain the same

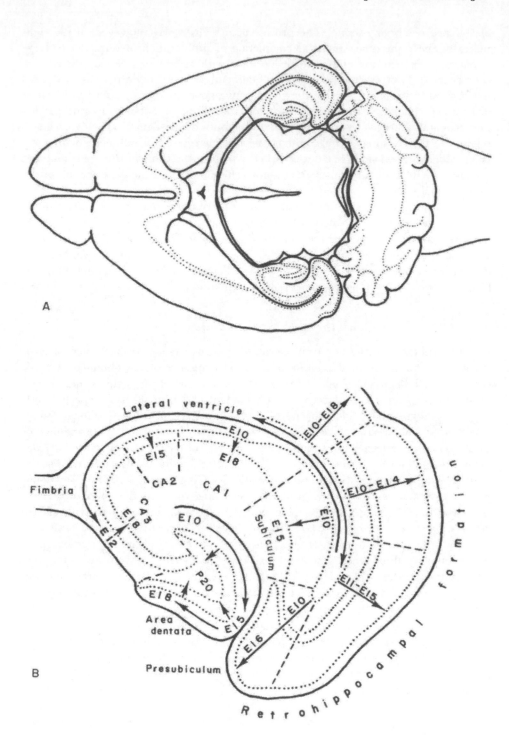

A

B

cells, which have the same birth dates. Most of the cells show the same systematic relationship between relative final position and relative time of origin. But in reeler the nature and direction of this relationship are different, deranged far beyond the mildly abnormal cytoarchitecture described in the first paper on this mutant. Laminar constraints may be absent or delayed, or established later in ineffective and misleading ways, but, whatever the cause, many more of the earlier neurons are generated over a longer early period and end up *scattered* in suprapyramidal layers rather than *aggregated* in the compact stratum pyramidale. And, extraordinary to tell, the later-forming cells that do find or heed constraints take up final positions in "*outside-in*" sequence, "just the reverse" from that expected. (Fittingly, I chose the words that Angevine and Sidman, 1961, used for the "inside-out" sequence they found and that characterizes most of the normal cerebral cortex.) One eagerly looks to the dentate gyrus, to see what happens there. The earlier-formed cells again tend to be nonlaminated and the first granule cells are scattered within the hilus instead of outlining the arrowhead. But the study did not extend beyond E18, and before that time a program of cell proliferation for stratum granulosum could not be discerned. Nevertheless, the findings in both studies indicate that positions of many cortical neurons "are systematically regulated, yet controlled according to some rule that differs in normal and mutant" (Caviness and Sidman, 1973).

These findings permit another good look at a message of pattern, here a different pattern but pattern all the same. Such observations redress the balance of the constant, understandable, but possibly overweighted emphasis placed on the *parallelism* and *interdependence*, rather than *independent variability*, of early cellular events in neurogenesis (not to mention the forgivable sins of framing most of our thoughts in postmitotic cells and four events). Sequences of neuronal degeneration, differentiation, migration, and time and place of origin may indeed reflect each previous event, and so on, all the way back to the genome. But the fidelity is variable, as

FIG. 6. Horizontal section of mouse brain (A) showing location (rectangle) of high-power diagram (B) of the hippocampal region to summarize the inside-out sequence and calendars of neuronal birth dates for each part. To recapitulate, a series of mice received a single (or in a few cases triple) injection of [³H]thymidine on a day of development from E10 to P20; all animals were killed after birth when radioactive neurons had reached their destinations. Arrows show the sequence and duration of neuron origin (not the boundaries between the parts as in the previous figures); dashed lines delineate areas and sectors of the region. The outer dotted line represents the boundary of the molecular layer of the cerebral cortex. The inner dotted line marks the boundary between cortex and white matter (the corpus callosum or the alveus, depending on location). The intermediate dotted line or lines delimit the outer and inner principal cell layers of the allocortex. The outer layer stops short in the presubiculum (superpositio lateralis); its tiny cells originate later than the cells nearby. The inner layer runs from the retrohippocampal formation through the subiculum into hippocampal sectors CA1, CA2, and CA3. Note the early cessation (E15) of neuron origin for CA2 and the early origin, long course, and outside-in sequence shown by the granule cells of the fascia dentata (E10-E15-E18-P20). Observe the many gradients: from supra- to infrapyramidal limb of stratum granulosum, from hippocampus into area dentata (but with an early end for CA2), from lateral to medial regions in the retrohippocampal formation, and from the rhinal fissure rostrally into the isocortex. From Angevine (1970*a*); used with the permission of the publisher.

are the expressions and timing of the events themselves (Hinds and Hinds, 1974), qualities evident and ready for evaluation perhaps earlier than most have cared to look (Sechrist, 1969).

ACKNOWLEDGMENTS

Warm thanks are given to the following persons who provided direct support for my efforts in preparing this chapter: Emeline Angevine, Marilyn Berg, Elaine Puckett, Jack Sechrist, and Zeke Vaughn.

The kind permission of Academic Press, Inc., New York, is gratefully acknowledged for the use of the illustrations in Fig. 1–5, taken and in some cases modified slightly from my article in *Experimental Neurology,* 1965, Supplement **2,** 1–70.

The courtesy of The Rockefeller University Press, New York, is much appreciated in granting permission to use Fig. 6, which first appeared in my chapter in *The Neurosciences: Second Study Program,* F. O. Schmitt, Editor-in-Chief, published in 1970.

6. References

ALTMAN, J. Autoradiographic and histological studies of postnatal neurogenesis. II. A longitudinal investigation of the kinetics, migration and transformation of cells incorporating tritiated thymidine in infant rats, with special reference to postnatal neurogenesis in some brain regions. *Journal of Comparative Neurology,* 1966, **128,** 431–474.

ALTMAN, J., AND DAS, G. D. Autoradiographic and histological evidence of postnatal hippocampal neurogenesis in rats. *Journal of Comparative Neurology,* 1965, **124,** 319–336.

ALTMAN, J., AND DAS, G. D. Autoradiographic and histological studies of postnatal neurogenesis. I. A longitudinal investigation of the kinetics, migration and transformation of cells incorporating tritiated thymidine in neonate rats, with special reference to postnatal neurogenesis in some brain regions. *Journal of Comparative Neurology,* 1966, **126,** 337–390.

ANGEVINE, J. B., JR. Time of neuron origin in the hippocampal region: An autoradiographic study in the mouse. *Experimental Neurology,* 1965, Supplement **2,** 1–70.

ANGEVINE, J. B., JR. Critical cellular events in the shaping of neural centers. In F. O. Schmitt (Ed.), *The neurosciences: Second study program.* New York: Rockefeller University Press, 1970a.

ANGEVINE, J. B., JR. Time of neuron origin in the diencephalon of the mouse: An autoradiographic study. *Journal of Comparative Neurology,* 1970b, **139,** 129–188.

ANGEVINE, J. B., JR. Clinically relevant embryology of the vertebral column and spinal cord. *Clinical Neurosurgery,* 1973, **20,** 95–113.

ANGEVINE, J. B., JR., AND SIDMAN, R. L. Autoradiographic study of cell migration during histogenesis of cerebral cortex in the mouse. *Nature (London),* 1961, **192,** 766–768.

ANGEVINE, J. B., JR., AND SIDMAN, R. L. Autoradiographic study of histogenesis in the cerebral cortex of the mouse. *Anatomical Record,* 1962, **142,** 210.

ARIËNS KAPPERS, C. U., HUBER, G. C., AND CROSBY, E. C. *The comparative anatomy of the nervous system of vertebrates, including man.* 2 vols. New York: Macmillan, 1936 (reissued 3 vols. New York: Hafner, 1960).

BLACKSTAD, T. W. Commissural connections of the hippocampal region in the rat, with special reference to their mode of termination. *Journal of Comparative Neurology,* 1956, **105,** 417–538.

BLACKWOOD, W., McMENEMY, W. H., Meyer, A., Norman, R. M., and Russell, D.S. (Eds.). *Greenfield's neuropathology*. Baltimore: Williams and Wilkins, 1967.

BOULDER COMMITTEE. Embryonic vertebrate central nervous system: Revised terminology. *Anatomical Record*, 1970, **166,** 257–262.

BRODAL, A. *Neurological anatomy in relation to clinical medicine*. 2nd ed. New York: Oxford University Press, 1969.

BROWN, J. W. Some aspects of the early development of the hippocampal formation in certain insectivorous bats. In R. Hassler and H. Stephan (Eds.), *Evolution of the forebrain: Phylogenesis and ontogenesis of the forebrain*. Stuttgart: Georg Thieme Verlag, 1966.

BUTLER, A. B., AND EBNER, F. F. Thalamotelencephalic projections in the lizard *Iguana iguana*. *Anatomical Record*, 1972, **172,** 282.

CAVINESS, V. S., JR. Time of neuron origin in the hippocampus and dentate gyrus of normal and reeler mutant mice: An autoradiographic analysis. *Journal of Comparative Neurology*, 1973, **151,** 113–120.

CAVINESS, V. S., JR., AND SIDMAN, R. L. Retrohippocampal, hippocampal and related structures of the forebrain in the reeler mutant mouse. *Journal of Comparative Neurology*, 1973, **147,** 235–254.

CRAIGIE, E. H. Observations on the brain of the humming bird (*Chrysolampis mosquitus* Linn. and *Chlorostilbon caribeaus* Lawr.). *Journal of Comparative Neurology*, 1928, **45,** 377–481.

CRAIGIE, E. H. The cell structure of the cerebral hemisphere of the humming bird. *Journal of Comparative Neurology*, 1932, **56,** 135–168.

CREPS, E. S. Time of neuron origin in the anterior olfactory nucleus and nucleus of the lateral olfactory tract of the mouse: An autoradiographic study. *Journal of Comparative Neurology*, 1974a, **157,** 139–160.

CREPS, E. S. Time of neuron origin in preoptic and septal areas of the mouse: An autoradiographic study. *Journal of Comparative Neurology*, 1974b, **157,** 161–244.

CROSBY, E. C. The forebrain of *Alligator mississippiensis*. *Journal of Comparative Neurology*, 1917, **27,** 325–402.

CROSBY, E. C., AND SHOWERS, M. J. C. Comparative anatomy of the preoptic and hypothalamic areas. In W. Haymaker, E. Anderson, and W. J. H. Nauta (Eds.), *The hypothalamus*. Springfield, Ill.: Charles C Thomas, 1969.

CROSBY, E. C., HUMPHREY, T., AND LAUER, E. W. *Correlative anatomy of the nervous system*. New York: Macmillan, 1962.

CROSBY, E. C., DE JONGE, B. R., AND SCHNEIDER, R. C. Evidence for some of the trends in the phylogenetic development of the vertebrate telencephalon. In R. Hassler and H. Stephan (Eds.), *Evolution of the forebrain: Phylogenesis and ontogenesis of the forebrain*. Stuttgart: Georg Thieme Verlag, 1966.

DAS, G. D., AND ALTMAN, J. Postnatal neurogenesis in the caudate nucleus and nucleus accumbens septi in the rat. *Brain Research*, 1970, **21,** 122–127.

DELONG, G. R. Histogenesis of fetal mouse isocortex and hippocampus in reaggregating cell cultures. *Developmental Biology*, 1970, **22,** 563–583.

EBBESSON, S. O. E. On the organization of central visual pathways in vertebrates. *Brain, Behavior and Evolution*, 1970, **3,** 178–194.

ELLIOT SMITH, G. The fascia dentata. *Anatomische Anzeiger*, 1896, **12,** 119–126.

ELLIOT SMITH, G. I. The central nervous system. In A. Robinson (Ed.), *D. J. Cunningham's text-book of anatomy*. 5th ed. London: Oxford University Press, 1923.

FELL, H. B. Echinoderm embryology and the origin of the chordates. *Biological Reviews*, 1948, **23,** 81–107.

FLEISCHHAUER, K. Untersuchungen am Ependym des Zwischen- und Mittelhirns der Landschildkröte (*Testudo graeca*). *Zeitschrift für Zellforschung und mikroskopische Anatomie*, 1957, **46,** 729–767.

GASTAUT, H., AND LAMMERS, H. J. Anatomie du rhinencéphale. In Th. Alajouanine (Ed.), *Les grandes activités du rhinencéphale*. Vol. I. Paris: Masson et Cie, 1961.

GODINA, G., AND BARASA, A. Morfogenesi ed istogenesi della formazione ammonica. *Zeitschrist für Zellforschung und mikroskopiche Anatomie*, 1964, **63,** 327–355.

GOLDBY, F., AND GAMBLE, H. J. The reptilian cerebral hemispheres. *Biological Reviews*, 1957, **32,** 383–420.

HALL, W. C., AND EBNER, F. F. Thalamotelencephalic projections in the turtle (*Pseudemys scripta*). *Journal of Comparative Neurology*, 1970, **140**, 101–122.

HASSLER, R., AND STEPHAN, H. (Eds.) *Evolution of the forebrain: Phylogenesis and ontogenesis of the forebrain*. Stuttgart: Georg Thieme Verlag, 1966.

HAUG, F.-M. Š., BLACKSTAD, T. W., SIMONSEN, A. H., AND ZIMMER, J. Timm's sulfide silver reaction for zinc during experimental anterograde degeneration of hippocampal mossy fibers. *Journal of Comparative Neurology*, 1971, **142**, 23–32.

HERRICK, C. J. *Brains of rats and men*. Chicago: University of Chicago Press, 1926.

HERRICK, C. J. *The brain of the tiger salamander*. Chicago: University of Chicago Press, 1948.

HINDS, J. W. Autoradiographic study of histogenesis in the mouse olfactory bulb. I. Time of origin of neurons and neuroglia. *Journal of Comparative Neurology*, 1968a, **134**, 287–304.

HINDS, J. W. Autoradiographic study of histogenesis in the mouse olfactory bulb. II. Cell proliferation and migration. *Journal of Comparative Neurology*, 1968b, **134**, 305–322.

HINDS, J. W., and HINDS, P. L. Early ganglion cell differentiation in the mouse retina: An electron microscopic analysis utilizing serial sections. *Developmental Biology*, 1974, **37**, 381–416.

HINES, M. Studies in the growth and differentiation of the telencephalon in man: The fissura hippocampi. *Journal of Comparative Neurology*, 1922, **34**, 73–171.

HJORTH-SIMONSEN, A. Projection of the lateral part of the entorhinal area to the hippocampus and fascia dentata. *Journal of Comparative Neurology*, 1972, **146**, 219–232.

HODOS, W. Evolutionary interpretation of neural and behavioral studies of living vertebrates. In F. O. Schmitt (Ed.), *The neurosciences: Second study program*. New York: Rockefeller University Press, 1970.

HUMPHREY, T. The development of the human hippocampal formation correlated with some aspects of its phylogenetic history. In R. Hassler and H. Stephan (Eds.), *Evolution of the forebrain: Phylogenesis and ontogenesis of the forebrain*. Stuttgart: Georg Thieme Verlag, 1966a.

HUMPHREY, T. Correlations between the development of the hippocampal formation and the differentiation of the olfactory bulbs. *Alabama Journal of Medical Sciences*, 1966b, **3**, 235–269.

HUMPHREY, T. The development of the human hippocampal fissure. *Journal of Anatomy*, 1967, **101**, 655–676.

JACOBS, M. S., MORGANE, P. J., AND McFARLAND, W. L. The anatomy of the brain of the bottlenose dolphin (*Tursiops truncatus*): Rhinic lobe (rhinencephalon). *Journal of Comparative Neurology*, 1971, **141**, 205–272.

JOHNSTON, J. B. The cell masses in the forebrain of the turtle, *Cistudo carolina*. *Journal of Comparative Neurology*, 1915, **25**, 393–468.

KÄLLÉN, B. Embryogenesis of brain nuclei in the chick telencephalon. *Ergebnisse der Anatomie und Entwicklungsgeschichte*, 1962, **36**, 62–82.

KARTEN, H. J. The organization of the ascending auditory pathway in the pigeon (*Columba livia*). I. Diencephalic projections of the inferior colliculus (nucleus mesencephali lateralis, pars dorsalis). *Brain Research*, 1967, **6**, 409–427.

KARTEN, H. J. The ascending auditory pathway in the pigeon (*Columba livia*). II. Telencephalic projections of the nucleus ovoidalis thalami. *Brain Research*, 1968, **11**, 134–153.

KARTEN, H. J. The organization of the avian telencephalon and some speculations on the phylogeny of the amniote telencephalon. *New York Academy of Sciences, Annals*, 1969, **167**, 164–179.

KARTEN, H. J. Specificity and evolution of the thalamus of amniotes. *University of California at Los Angeles, Bulletin. Brain Information Service*, 1973, Conference Report No. 31, pp. 9–12.

KARTEN, H. J., AND DUBBLEDAM, J. L. The organization and projections of the paleostriatal complex in the pigeon (*Columba livia*). *Journal of Comparative Neurology*, 1973, **148**, 61–90.

KARTEN, H. J., AND HODOS, W. *A stereotaxic atlas of the brain of the pigeon (Columba livia)*. Baltimore: Johns Hopkins University Press, 1967.

KARTEN, H. J., HODOS, W., NAUTA, W. J. H., AND REVZIN, A. M. Neural connections of the "visual wulst" of the avian telencephalon: Experimental studies in the pigeon (*Columba livia*) and owl (*Speotyto cunicularia*). *Journal of Comparative Neurology*, 1973, **150**, 253–278.

KEMALI, M., AND BRAITENBERG, V. *Atlas of the frog's brain*. Berlin: Springer-Verlag, 1969.

KIM, S. U. Morphological development of neonatal mouse hippocampus cultured *in vitro*. *Experimental Neurology*, 1973, **41**, 150–162.

KRUGER, L. Experimental analyses of the reptilian nervous system. *New York Academy of Sciences, Annals,* 1969, **167,** 102–117.

KUPFFER, G. De cornus Ammonis textura disquisitones praecipue in cuniculus institutae. Dorpat: Schünmann and Mattiesen, 1859.

LaVAIL, J. H., AND WOLF, M. K. Postnatal development of the mouse dentate gyrus in organotypic cultures of the hippocampal formation. *American Journal of Anatomy,* 1973, **137,** 47–66.

LORENTE DE NÓ, R. Studies on the structure of the cerebral cortex. I. The area entorhinalis. *Journal für Psychologie und Neurologie,* 1933, **45,** 26–438.

LORENTE DE NÓ, R. Studies on the structure of the cerebral cortex. II. Continuation of the study of the ammonic system. *Journal für Psychologie und Neurologie,* 1934, **46,** 113–177.

LYNCH, G., DEADWYLER, S., AND COTMAN, C. Postlesion axonal growth produces permanent functional connections. *Science,* 1973a, **180,** 1364–1366.

LYNCH, G., STANFIELD, B., AND COTMAN, C. W. Developmental differences in post-lesion axonal growth in the hippocampus. *Brain Research,* 1973b, **59,** 155–168.

MACCHI, G. The ontogenetic development of the olfactory telencephalon in man. *Journal of Comparative Neurology,* 1951, **95,** 245–305.

McLARDY, T. Neurosyncytial aspects of the hippocampal mossy fibre system. *Confinia Neurologica,* 1960, **20,** 1–17.

MOREST, D. K. A study of neurogenesis in the forebrain of opossum pouch young. *Zeitschrift fur Anatomie und Entwicklungsgeschichte,* 1970, **130,** 265–305.

NAUTA, W. J. H. The problem of the frontal lobe: A reinterpretation. *Journal of Psychiatric Research,* 1971, **8,** 167–187.

NAUTA, W. H. J., AND HAYMAKER, W. Hypothalamic nuclei and fiber connections. In W. Haymaker, E. Anderson, and W. J. H. Nauta (Eds.), *The hypothalamus.* Springfield, Ill.: Charles C Thomas, 1969.

NAUTA, W. J. H., AND KARTEN, H. J. A general profile of the vertebrate brain, with sidelights on the ancestry of cerebral cortex. In F. O. Schmitt (Ed.), *The neurosciences: Second study program.* New York: Rockefeller University Press, 1970.

NIEUWENHUYS, R. Trends in the evolution of the actinopterygian forebrain. *Journal of Morphology,* 1962, **111,** 69–88.

NIEUWENHUYS, R. The forebrain of the crossopterygian *Latimeria chalumnae* Smith. *Journal of Morphology,* 1965, **117,** 1–24.

NIEUWENHUYS, R. Comparative anatomy of olfactory centres and tracts. *Progress in Brain Research,* 1967, **23,** 1–64.

NIEUWENHUYS, R. A survey of the structure of the forebrain in higher bony fishes (Osteichthyes). *New York Academy of Sciences, Annals,* 1969, **167,** 31–64.

NIEUWENHUYS, R., AND HICKEY, M. A survey of the forebrain of the Australian lungfish *Neoceratodus forsteri. Journal für Hirnforschung,* 1965, **7,** 433–452.

NORTHCUTT, R. G. Architectonic studies of the telencephalon of *Iguana iguana. Journal of Comparative Neurology,* 1967, **130,** 109–147.

NORTHCUTT, R. G., AND BUTLER, A. B. Retinal projections in the northern water snake *Natrix sipedon sipedon* (L.). *Journal of Morphology,* 1974, **142,** 117–136.

PANDYA, D. N., AND KUYPERS, H. G. J. M. Cortico-cortical connections in the Rhesus monkey. *Brain Research,* 1969, **13,** 13–36.

PAPEZ, J. W. *Comparative neurology.* New York: Thomas Y. Crowell Company, 1929.

PETRAS, J. M., AND NOBACK, C. R. (Eds.) Comparative and evolutionary aspects of the vertebrate central nervous system. *New York Academy of Sciences, Annals,* 1969, **167,** 1–513.

RAISMAN, G., COWAN, W. M., AND POWELL, T. P. S. The extrinsic afferent, commissural and association fibres of the hippocampus. *Brain,* 1965, **88,** 963–996.

RAISMAN, G., COWAN, W. M., AND POWELL, T. P. S. An experimental analysis of the efferent projection of the hippocampus. *Brain,* 1966, **89,** 83–108.

RAMÓN Y CAJAL, S. *Histologie du système nerveux de l'homme et des vertébrés,* (trans. by L. Azoulay). 2 vols. Paris: Maloine, 1909/1911 (reissued Madrid: Consejo Superior de Investigaciones Cientificas, 1955).

94 JAY B. ANGEVINE, JR.

RAMÓN Y CAJAL, S. *Studies on the cerebral cortex* (trans. by L. M. Kraft). Chicago: Year Book Publishers, 1955.

RISS, W., HALPERN, M., AND SCALIA, F. The quest for clues to forebrain evolution—The study of reptiles. *Brain, Behavior and Evolution,* 1969, **2,** 1–50.

ROMER, A. S. *The vertebrate body.* 3rd ed. Philadelphia: W. B. Saunders Company, 1962.

ROMER, A. S. *Vertebrate paleontology.* 3rd ed. Chicago: University of Chicago Press, 1966.

ROMER, A. S. *The procession of life.* Cleveland: World, 1968.

ROSE, M. Der Allocortex bei Tier und Mensch. I. Teil. *Journal für Psychologie und Neurologie,* 1926, **34,** 1–111.

ROSE, M. Die sog. Riechrinde beim Menschen und beim Affen. II. Teil. Des "Allocortex bei Tier und Mensch." *Journal für Psychologie und Neurologie,* 1927, **34,** 261–401.

SECHRIST, J. W. Neurocytogenesis. I. Neurofibrils, neurofilaments, and the terminal mitotic cycle. *American Journal of Anatomy,* 1969, **124,** 117–134.

SHIMADA, M., AND NAKAMURA, T. Time of neuron origin in mouse hypothalamic nuclei. *Experimental Neurology,* 1973, **41,** 163–173.

SIDMAN, R. L. Autoradiographic methods and principles for study of the nervous system with thymidine-H^3. In W. J. H. Nauta and S. O. E. Ebbesson (Eds.), *Contemporary research methods in neuroanatomy.* New York: Springer-Verlag, 1970a.

SIDMAN, R. L. Cell proliferation, migration, and interaction in the developing mammalian central nervous system. In F. O. Schmitt (Ed.), *The neurosciences: Second study program.* New York: Rockefeller University Press, 1970b.

SIDMAN, R. L., ANGEVINE, J. B., JR., AND TABER PIERCE, E. *Atlas of the mouse brain and spinal cord.* Cambridge, Mass.: Harvard University Press, 1971.

SNELL, R. S. *Clinical embryology for medical students.* Boston: Little, Brown and Company, 1972.

TILNEY, F. The hippocampus and its relations to the corpus callosum. *New York Neurological Institute Bulletin,* 1938, 7, 1–77.

VAN TIENHOVEN, A., AND JUHÁSZ, L. P. The chicken telencephalon, diencephalon and mesencephalon in stereotaxic coordinates. *Journal of Comparative Neurology,* 1962, **118,** 185–197.

WHITE, L. E., JR. Ipsilateral afferents to the hippocampal formation in the albino rat. *Journal of Comparative Neurology,* 1959, **113,** 1–41.

WHITE, L. E., JR. A morphologic concept of the limbic lobe. *International Review of Neurobiology,* 1965, **8,** 1–34.

WIMER, C. C., WIMER, R. E., AND RODERICK, T. H. Some behavioral differences associated with relative size of hippocampus in the mouse. *Journal of Comparative and Physiological Psychology,* 1971, **76,** 57–65.

YAKOVLEV, P. I. A proposed definition of the limbic system. In C. H. Hockman (Ed.), *Limbic system mechanisms and autonomic function.* Springfield, Ill.: Charles C Thomas, 1972.

YOUNG, J. Z. *The life of vertebrates.* 2nd ed. New York: Oxford University Press, 1962.

ZEIER, H. J., AND KARTEN, H. J. Connections of the anterior commissure in the pigeon (*Columba livia*). *Journal of Comparative Neurology,* 1973, **150,** 201–216.

ZEIGLER, H. P., AND KARTEN, H. J. Trigeminal structures and feeding behavior in rat and pigeon. *Program and Abstracts, Society for Neuroscience,* San Diego, 1973, p. 409.

4

Postnatal Development of the Hippocampal Dentate Gyrus Under Normal and Experimental Conditions

Joseph Altman and Shirley Bayer

1. Introduction

Recent studies using [³H]thymidine autoradiography produced convincing evidence that in the development of a particular brain region the small, short-axoned cells come into existence after the larger long-axoned cells. Indeed, in an altricial rodent, the rat, the granular nerve cells of the olfactory bulb, hippocampus, cerebellum, and cochlear nucleus are formed exclusively or predominantly after birth (Altman and Das, 1965a). There are indications that these short-axoned neurons (microneurons) arise from late-forming secondary germinal matrices (like the subependymal layer of the forebrain ventricles and the external germinal layer of the cerebellar cortex), in contrast to the long-axoned neurons (macroneurons) which originate from the periventricular primary matrix, the neuroepithelium (Altman, 1969). We do not as yet have an adequate explanation of the delayed formation of microneurons but the importance of these elements in the maturation of brain functions is indicated by behavioral studies. For instance, interference with the postnatal acquisition of cerebellar granule cells by experimental means produces behavioral deficits comparable to those seen after decerebellation (Wallace and Altman, 1969a,b; Altman et al., 1971; Brunner and Altman, 1973).

Joseph Altman and Shirley Bayer • Laboratory of Developmental Neurobiology, Department of Biological Sciences, Purdue University, West Lafayette, Indiana.

In this chapter, we will attempt to review available information on the postnatal maturation of the dentate gyrus. Autoradiographic studies in mice (Angevine, 1965) and rats (Altman and Das, 1965*b*) have established that a considerable proportion of the dominant neuronal elements, the granule cells, of this hippocampal substructure are of postnatal origin. As yet, we lack an understanding of the role played by the dentate gyrus in hippocampal functioning but it must be a crucial one. This was indicated by recent studies which showed that when the acquisition of the postnatally forming hippocampal granule cells is prevented experimentally, the animals display behavioral symptoms comparable to those seen after surgical destruction of the hippocampus as a whole (Bayer *et al.*, 1973; Haggbloom *et al.*, 1974).

2. Normal Development of the Dentate Gyrus

A recent evaluation (Altman, 1975) of the postnatal neurogensis of the cerebellar cortex suggested that this process can be meaningfully and conveniently subdivided into five distinct, essentially sequential phases. These were referred to as (neuro)cytogenesis, (neuro)morphogenesis, synaptogenesis, gliogenesis, and myelogenesis. In this chapter, we will attempt to describe our knowledge of dentate neurogenesis in these terms.

Cytogenesis is the phase of cell production or acquisition, and its consideration includes such problems as the locus and kinetics of cell proliferation, and the time of origin of neurons of different types. The latter event is usually equated with the period when the precursor cells cease to divide and begin to differentiate. Differentiation consists of several steps. *Morphogenesis* refers to the phase when the postmitotic cells begin to develop their axons and dendrites in accordance with regional characteristics. This is the sculpturing phase of neurogenesis. For some types of cells, this involves migration of the soma and the extrusion of trailing processes, while in the case of others the stationary soma sends out processes that invade close or distant brain regions over simple or tortuous paths. *Synaptogenesis* is the next step, when the cell processes begin to establish contacts with one another. This is the connectivity phase when cell adhesion, membrane specializations, and the maturation of functional synapses result in the establishment of the characteristic regional wiring circuitry. The subsequent phase of *gliogenesis* has many functions, including the segregation of contiguous but independent neuronal elements, the ensheathing of related elements, such as pre- and postsynaptic regions of contact, and the production of materials for the myelination of axons. The last process represents the final phase of regional development, the phase of *myelogenesis*.

It is a reflection of the uneven advances made in the study of the development of different parts of the nervous system that although such a stepwise analysis has proven profitable for cerebellar neurogenesis, in the case of the hippocampus such an attempt merely calls attention to the fragmentary nature of the available information.

2.1. Cytogenesis

The study of neurocytogenesis has been greatly aided in recent years by the use of [^{3}H]thymidine autoradiography. Proliferating cells, but not postmitotic cells, selectively incorporate labeled thymidine into their duplicating chromosome strands (Taylor *et al.*, 1957). Because of the metabolic stability of DNA, the dividing cells become permanently tagged, unless the labeling of DNA molecules becomes diluted by further subdivisions. The visualization of the labeled cells is accomplished with the photographic technique of autoradiography (Messier and Leblond, 1957). If animals are killed shortly after injection, the sites of cell proliferation can be located. If animals are killed following a schedule of graduated survival times, the migratory movements of the originally labeled cells may be reconstructed. If animals are killed after the lapse of several weeks or months, the ultimate fate of the labeled cells can be established with some certainty. Finally, the technique also allows the dating of the time of origin (or "birth date") of different nerve cells, although this task poses several technical and methodological difficulties.

Because of some discrepancies in reports of the time of origin of hippocampal nerve cells, an explanatory technical note is in order here. The usual procedure for dating the time of origin of cells is to count all heavily labeled cells in a selected brain region and plot them as a function of age at injection. For example, if an animal is injected with [^{3}H]thymidine on day 15 of gestation, the heavily labeled cells seen in any particular brain region are assumed to have differentiated (or been "born") soon thereafter. This assumption is justified by the argument that had the cells in question continued to divide repeatedly after the injection they would be lightly labeled due to dilution of the radiochemical. But there are two major difficulties with this interpretation. One is the technical uncertainty as to what constitutes a heavily labeled cell—the size of the nucleus, the specific activity of the radiochemical, aspects of the photographic procedure, and other factors will have an influence on how much label will be visualized over a cell nucleus. The other difficulty is that high concentrations of isotope need not reflect immediate cessation of cell proliferation. It is conceivable that the cells in question entered a dormant period after the injection and resumed a few divisions at some later date. This procedure evidently would result in the premature dating of the time of origin of the neurons in question. This can be remedied by employing another procedure. In this, the question is asked as to how long injections can be delayed and still label all or a specified proportion of the cells considered. Only when this can no longer be done is it safe to assume that the examined cells have stopped dividing and entered their phase of life as nondividing or differentiated neurons.

The time of origin of pyramidal cells of Ammon's horn was examined in a pioneering study by Angevine (1965). Heavily labeled pyramidal cells appeared as early as 10 days of gestation. They were seen in larger numbers in animals labeled at 12 days, and the highest concentration occurred when [^{3}H]thymidine was injected on day 14 or 15. By gestational day 17, only a few heavily labeled cells were seen in the late-maturing CA1 and CA3 regions; none was labeled after day 15 in CA2. With regard to the dentate gyrus, intensely labeled granule cells were seen in mice injected from gestational day 13 onward; they were at that time essentially restricted to the superficial zone bordering the molecular layer. The highest concentration of heavily labeled granule cells was present in animals injected on day 18, these cells being distributed in the upper two-thirds of the layer. Because a few heavily labeled granule

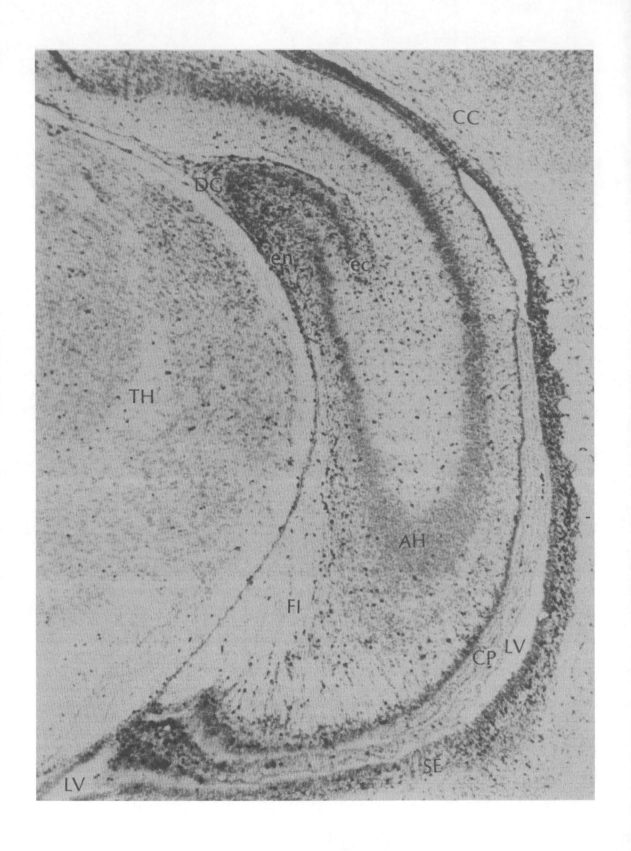

cells could be seen at the base of the granular layer in animals injected up to 20 days postnatally, it was concluded that granule cells continue to arise up to that date. Angevine also examined the other layers of dentate gyrus and concluded that "the molecular and hilar layers of area dentata complete their periods of neurons formation by embryonic day 15" (Angevine, 1965, p. 34).

Similar results were obtained recently by Hine and Das (1974) in the rat, except that in this more slowly maturing species heavily labeled pyramidal cells began to appear in Ammon's horn in animals injected on day 16 and they could be seen in appreciable numbers up to day 19. In the dentate gyrus, heavily labeled granule cells began to appear in a superficial position as early as day 15 where they greatly increased in numbers in rats injected on days 20, 21, and 22 prenatally. In the molecular layer of the dentate gyrus, heavily labeled cells were seen as early as day 15 of gestation.

These prenatal studies established that in mice and rats the acquisition of the pyramidal cells of Ammon's horn is completed before birth. However, in view of the previously described limitations for determining the onset of the acquisition of a class of cells by counting heavily labeled cells, the dates set for the commencent of neuron formation in the hippocampus cannot be considered to be adequately established. This reservation is reinforced by other studies (to be described below) in which rats were injected with [³H]thymidine postnatally. These studies confirmed that the pyramidal cells are of prenatal origin (as none of them was ever labeled), but they also showed that some of the cell types that were assumed to be formed prenatally, as the cells of the molecular layer of the dentate gyrus, are really of postnatal origin. Moreover, these studies revealed that a much larger proportion of granule cells are formed postnatally than would be predicted on the basis of observations of prenatally injected animals.

In the newborn rat, Ammon's horn is clearly delineated by the maturing pyramidal cells, but the dentate gyrus is less distinct because few of its granule cells have started to differentiate in its ectal arm (the arm facing the cerebral hemispheres) and few are recognizable as mature granule cells (defined as round-to-ovoid cells with large, pale nuclei) in the endal arm (Fig. 1). A count made in animals ranging in age from newborn to 300 days of age (Altman and Das, 1965b) indicated that less than 20% of all the granule cells present at 60–90 days of age could be identified during the first week of life. The scarcity of identifiable granule cells in the infant does not necessarily imply that the cells are absent, it may merely indicate a lack of differentiation of cells already formed. In infant rats, the hilus of the dentate gyrus and the base of the granular layer (where it was identifiable) are packed with small, darkly staining cells with occasional mitotic figures. The nature and fate of these cells were clarified with short-survival and long-survival autoradiography.

When young rats were injected with [³H]thymidine and killed several hours af-

FIG. 1. Low-power photomicrograph of a gallocyanin–chromalum stained autoradiogram of a coronal brain section from a rat injected with [³H]thymidine at the age of 6 h and killed 24 h later. AH, Ammon's horn; CC, cerebral cortex; CP, choroid plexus; DG, dentate gyrus; ec, ectal arm of DG; en, endal arm of DG; FI, fimbria; LV, lateral ventricle; SE, subependymal layer; TH, thalamus. Slightly modified from Altman and Das (1965a).

terward, a large proportion of the small darkly staining cells in both the hilus and the base of the granular layer became labeled (Altman and Das, 1966). This can be dramatically demonstrated with cumulative labeling as shown in Fig. 2. The concentration of these labeled cells was high during the first week, began to decline by the end of the second week (the exact time course of this process has yet to be determined), and in adolescent and young adult rats only a few labeled cells could be seen. In the animals killed at spaced intervals after injection, there was (Fig. 3A,B,C) for up to 6 days an increase in the concentration of labeled cells in the hilus but then the number of these cells declined and there was an accumulation of labeled undifferentiated and differentiated granule cells in the basal aspect of the granular layer by the twelfth day after injection (Fig. 3D). Finally, when animals ranging in age from newborn to 8 months of age were injected with [³H]thymidine and killed several months later (Altman and Das, 1965b; Altman, 1966) it became apparent, in agreement with Angevine's (1965) observations, that the granule cells are acquired in a sequence. The superficially situated cells were formed first, while the later acquired cells were progressively added to the base (Fig. 4).

In a more recent study (Bayer and Altman, 1974), we estimated the proportion of granule cells that is formed prenatally and the proportion that was added subsequently over blocks of 4 days postnatally. The procedure of dated, sequential

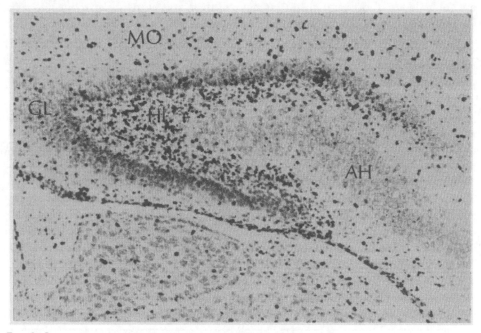

FIG. 2. Low-power photomicrograph of an autoradiogram of the dentate gyrus from a rat injected cumulatively with [³H]thymidine on days 5, 6, 7, and 8 after birth and killed on day 9. Note the high concentration of labeled cells in the hilus (HI) of the dentate gyrus and also in its molecular layer (MO). The concentration is low in the pyramidal layer of Ammon's horn (AH), and relatively few cells are seen in the granular layer. Slightly modified from Altman (1966).

comprehensive (or cumulative) labeling was used (Fig. 5). The evidence indicated (Fig. 6) that 15% of the granule cells are acquired prenatally, 72% are added between 0 and 16 days, and 13% are formed thereafter. Earlier studies established that granule cells are continued to be formed in adult rats (Altman, 1963) at least up to 8 months of age (Altman and Das, 1965b).

2.2. Morphogenesis

Cell proliferation in the dentate gyrus, according to the previously considered evidence, is most pronounced in the hilus and at the base of the granular layer. These are the direct proliferative sites of this system. However, not all the primitive cells at the base of the granular layer, where they form a more or less distinct zone, become labeled. The properties of this zone of the granular layer, which could be the site of the onset of neuronal differentiation, have never been adequately examined. It is interesting to note that this subgranular zone is more conspicuous in the precocial guinea pig (Fig. 7) and the slowly maturing kitten (Fig. 8) than in mice and rats. A previous study (Altman and Das, 1967) showed that in the guinea pig a proportion of these primitive cells become labeled with [^{3}H]thymidine and differentiate subsequently as granule cells. But there is no information about their proliferative properties in the kitten.

To shed some light on this problem, we have recently tested the sensitivity of these primitive cells to low-level X-irradiation. The rationale of this approach is based on some results on the effects of X-irradiation in the developing cerebellar cortex. The postnatally acquired granule, stellate, and basket cells of the cerebellum arise from the multiplying cells of the superficial zone of the external germinal layer. In the deeper part of this germinal layer, the cells that ceased to divide (and cannot be labeled with [^{3}H]thymidine) assume a horizontal bipolar shape in the coronal plane. This signals the first step in the differentiation of these cells, the onset of the outgrowth of the horizontal portion of the parallel fibers, the axons of granule cells (Altman, 1972a). When the developing cerebellum is exposed to low-level X-ray (100–200 r), most of the cells in the proliferative zone of the external germinal layer are killed within 3–12 h. This is recognized by the drastic condensation of the affected cells and their rounding up into dark spherules (cell pyknosis). But unlike the multiplying cells of this layer, the bipolar cells are largely spared by the irradiation (Altman and Nicholson, 1971). This suggested that the great radiosensitivity of primitive cells disappears as soon as they begin to differentiate. Applying this technique to the hippocampus (Bayer and Altman, 1974), we found that many of the darkly staining cells at the base of the granular layer became pyknotic (Fig. 9) while others, as well as the maturing or mature granule cells, remained visibly unaffected by the radiation.

On analogy with results in the cerebellar cortex, it is assumed that the radioresistant primitive cells at the base of the granular layer constitute the group that has just started to differentiate, perhaps by sending out axons, the mossy fibers. But tentatively we also have to postulate that the mossy fibers at this stage of their development have not assumed the characteristic biochemical properties of mature

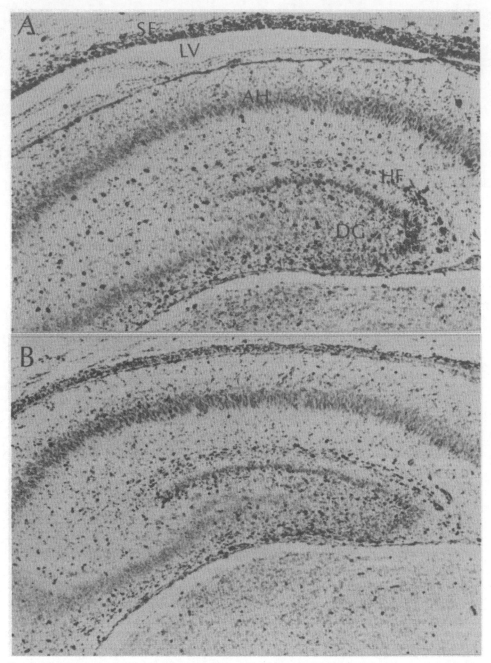

FIG. 3. Photomicrographs of autoradiograms of the dorsal hippocampus illustrating the sequence of events in the proliferation and migration of the precursors of dentate granule cells. All animals were injected with [^{3}H]thymidine 6 h after birth but were killed at different intervals thereafter. A: From a rat killed 6 h after injection to show the sites of cell proliferation in the hippocampus. Heavily labeled cells abound in the hilus of the dentate gyrus (DG), at the base of the granular layer, and in the region of the unformed extension of the endal arm. Scattered labeled cells are also seen in the various layers of Ammon's horn (AH) and the hippocampal fissure (HF). B: From a rat killed 24 h after injection. Apart from an increase in the number of labeled cells and with an associated label dilution within cells (not discernible at this magnification), there are no major changes at this period.

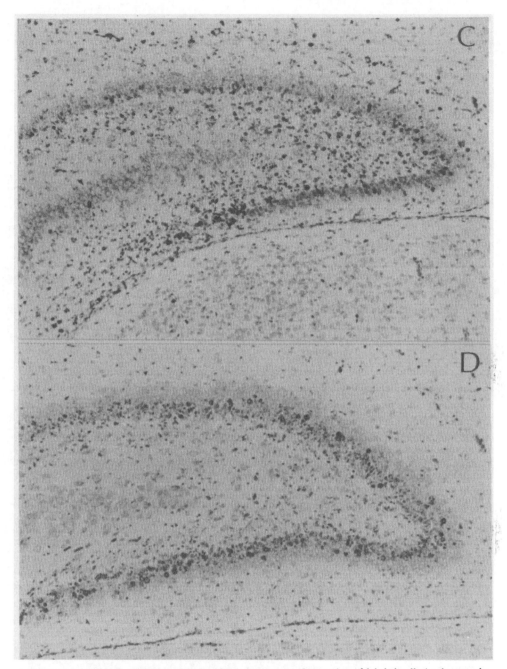

C: From a rat killed 6 days after injection. Note increase in the number of labeled cells in the greatly expanded endal and ectal arms of the granular layer. But the concentration of labeled cells in the hilus is still high. D: From a rat killed 12 days after injection. Note the reduction of labeled cells in the hilus and the high proportion of labeled granule cells. The heavily labeled granule cells are located in the superficial aspect of the layer, the lightly labeled cells at its base. Apparently a large proportion of the originally labeled precursor cells and their descendants have migrated into the granular layer and have become differentiated. Slightly modified from Altman and Das (1966).

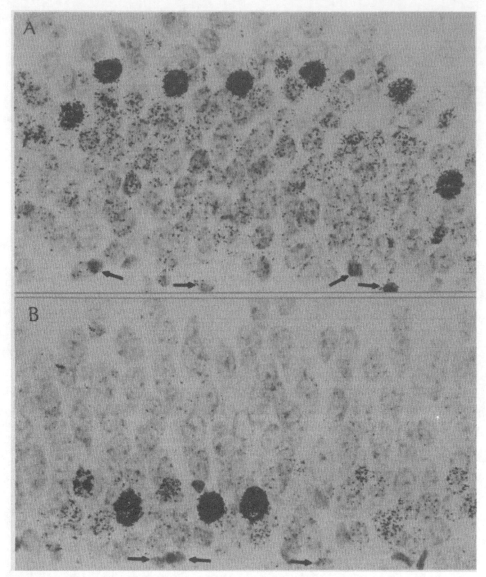

FIG. 4. High-power photomicrographs of autoradiograms of granule cells in the granular layer of rats injected with [³H]thymidine when 2 days old (A) or 13 days old (B). Both animals were killed 60 days after injection. Note that in the animal injected at 2 days of age the heavily labeled granule cells (those that differentiated first) are located in a superficial position while the lightly labeled cells (whose precursors multiplied many times before they began to differentiate) are located more basally. The basally situated cells were heavily labeled in the animal injected at 13 days and all the cells above them are unlabeled because they were formed before the injection. Arrows point to undifferentiated primitive cells. From Altman (1966).

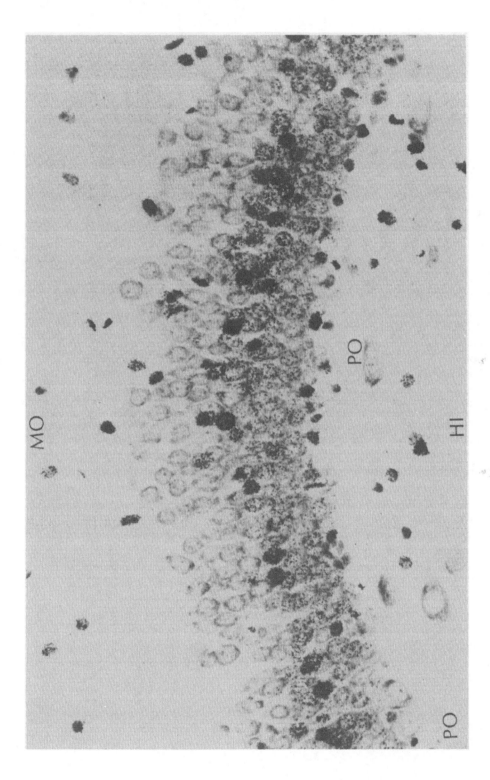

FIG. 5. Photomicrograph of an autoradiogram of the dentate gyrus from a rat injected repeatedly on days 4, 5, 6, and 7 after birth and killed when 60 days old. With this cumulative labeling technique, the great majority of cells formed after day 4 become reliably labeled. In addition to granule cells, many of the cells in the lower half of the molecular layer (MO) and in the hilus (HI) become labeled. The polymorph cells (PO), which are prenatally formed, remain unlabeled. From Bayer and Altman (unpublished data).

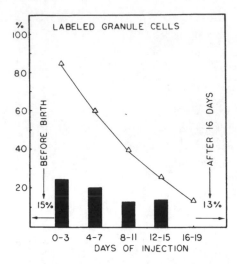

FIG. 6. Proportion of labeled granule cells in groups of 60-day-old rats that were injected daily with [³H]thymidine on days 0–3, 4–7, 8–11, 12–15, 16–19. Since only 85% could be labeled with injections on days 0–3, it was concluded that 15% of the cells were formed prenatally. The proportion of differentiated cells formed during the blocks of days, as indicated in the histograms, was arrived at by deducting the proportion of cells that could be relabeled during successive periods. Since 13% of the cells were labeled with injections made after 16 days, it is concluded that this proportion of cells is formed after that date. This technique does not permit us to specify either the exact prenatal onset or postnatal cessation of granule cell acquisition. Modified from Bayer and Altman (1974).

axons. This is suggested by the recent report that the high concentration of zinc in the hippocampus associated with mossy fiber terminals does not become apparent in the rat until about 18–22 days postnatally (Crawford and Connor, 1972).

Probably in most central nervous structures the outgrowth of the axon antedates the development of the cell's dendritic system. In the granular layer of the dentate gyrus, situated between the subgranular zone of darkly staining small cells and the superficial zone of mature granule cells, cells may be seen which are intermediate in both size and staining intensity. These typically have a vertically oriented elongated shape. Often these cells have a recognizable apical extension or a thick dendritic shaft. In the development of cerebellar Purkinje cells, it was observed that the outgrowth of the richly arborizing dendritic system is preceded by the formation of an apical cone filled with mitochondria (Altman, 1972b). These mitochondria are produced in an apical position in the vicinity of the nucleus and then, judged by the transient high concentration and shape of the mitochondria, they stream upward and become distributed in the rapidly expanding dendritic branches. In analogy with this event, it is assumed that the pear-shaped cells with apical cones or shafts in the subgranular zone are in a comparable stage of development.

This assumption is supported by histochemical observations on the concentration and distribution of various oxidative enzymes in the developing dentate gyrus. Biochemical studies have established that the majority of oxidative enzymes in brain tissue homogenates are concentrated in mitochondria. In the newborn rat (Meyer et al., 1972), oxidative enzyme (succinic dehydrogenase and cytochrome oxidase) activity was low and restricted to the perikarya of hippocampal granule cells. Staining intensity increased subsequently, but by day 10 perikaryal activity began to decline and the stronger staining shifted to the molecular layer. The adult pattern of

FIG. 7. The granular layer (GL) of the dentate gyrus in guinea pigs aged 6 h (A), 6 days (B), 18 days (C), and 36 days (D). Note the subgranular zone of primitive cells (arrows) at the base of the granular layer and the hilus (HI) of the dentate gyrus. From Altman and Das (unpublished data).

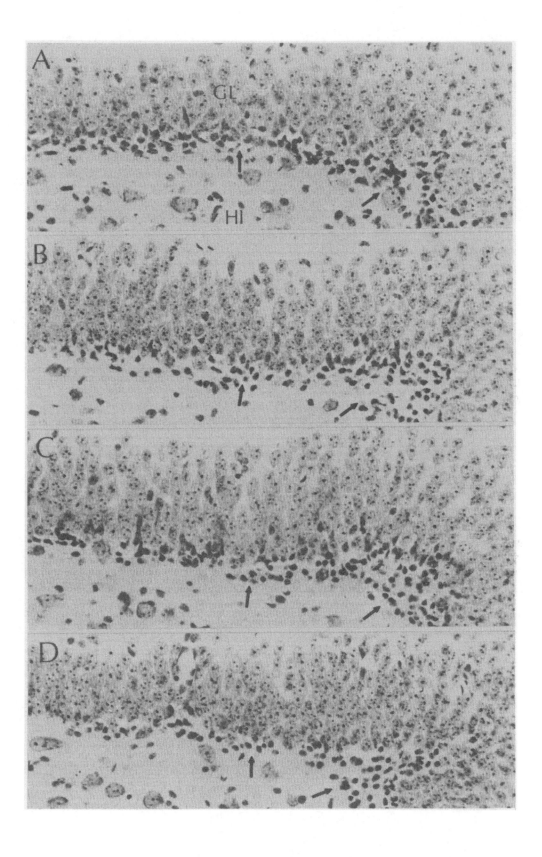

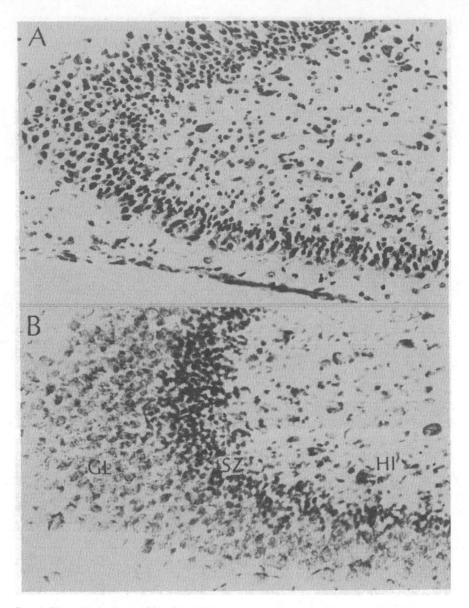

Fig. 8. The granular layer (GL) of the dentate gyrus in kittens aged 9 days (A), 21 days (B), 30 days (C), and 60 days (D). Primitive cells are seen in the hilus (HI) but are most numerous in the subgranular zone (SZ). The presence of pronounced subgranular zone in 60-day-old (postnatal) kittens suggests that hippocampal neurogenesis may be more protracted in the cat than in the rat. Unpublished photomicrographs.

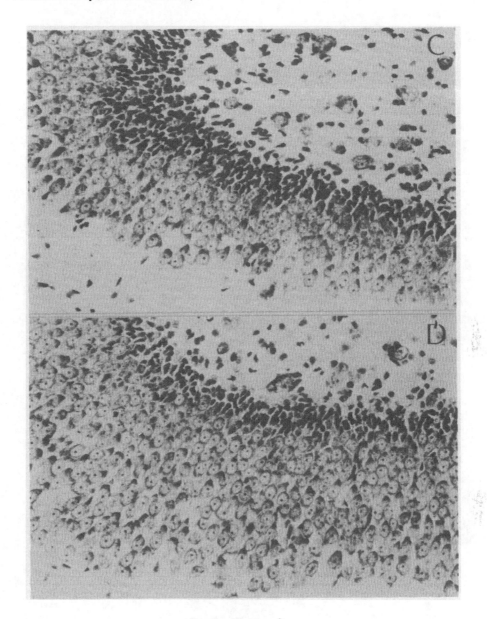

FIG. 8. (*Continued*)

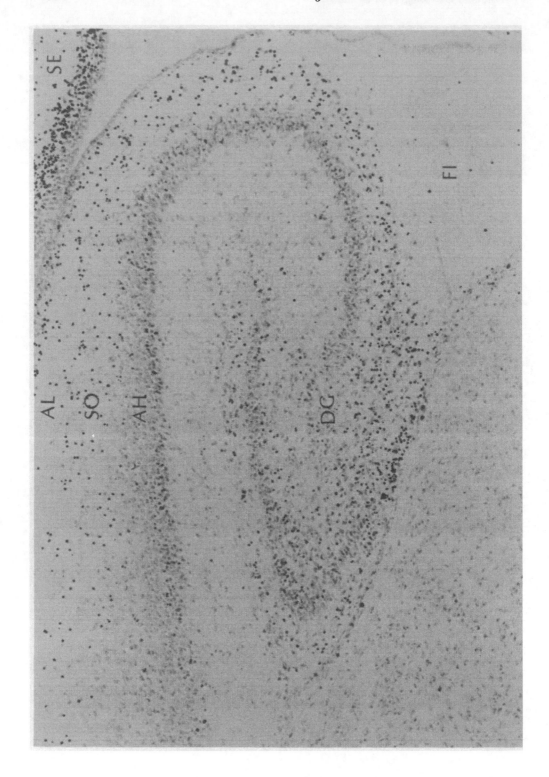

oxidative enzyme activity was reached by the fourth week of life. These observations agreed with earlier results of Das and Kreutzberg (1967) and were more recently confirmed by Mellgren (1973). They noted during early development a stronger staining of differentiating granule cell perikarya in a superficial position than in the less-differentiated deeper cells, and the small primitive cells of the subgranular zone tended to remain unstained.

In the absence of electron microscopic studies, little more can be said about the morphogenetic phase of dentate development. The fragmentary information that is available suggests that the "outside-in" pattern of cytogenesis is paralleled by a similar gradient of differentiation. The radiosensitive, multiplying cells are situated basally mixed with cells that have become radioresistant and presumably have commenced to differentiate by starting to grow axons. The cells lying above them have started to grow dendrites. As the dendritic system is developing, the cell acquires adult appearance, first in shape and then in size. The developmental course of the differentiation of mature granule cells is summarized in Fig. 10.

2.3. Synaptogenesis

The appropriate examination of synaptogenesis requires ultrastructural methods, but unfortunately very few published electron microscopic studies are available that deal with the development of the dentate gyrus. In the pioneering studies of Schwartz et al. (1968) and in the associated report by Purpura and Pappas (1968), relatively little attention was paid to the development of the dentate gyrus, and the impression these investigators gained was that the dentate gyrus is quite mature in newborn kittens. Their conclusion on the basis of Golgi observations was that "Nonpyramidal neurons of the neonatal kitten hippocampus, like pyramidal neurons, have a remarkably mature appearance and exhibit little change in overt characteristics in the postnatal period" (Purpura and Pappas, 1968, p. 389). Similarly, they interpreted their electron microscopic observations to indicate that "complex axon terminals resembling typical mossy fiber endings described in adult animals are also well developed in neonatal kitten" (Schwartz et al., 1968, p. 394). In light of other evidence of the late and protracted acquisition of dentate granule cells in the kitten, it must be assumed that these investigators observed the prenatally formed complement of granule cells.

In a more recent study (Crain et al., 1973), synapse counts were made in the molecular layer of the dentate gyrus of rats aged 4, 11, 25, and about 90 days. The number of synapses was very low at 4 days, constituting less than 1% of those seen in adults. Between 4 and 11 days, the extrapolated number of synapses nearly doubled every day. But even by 11 days less than 5% of the adult concentration of synapses

Fig. 9. Photomicrograph of the hippocampus of a 1-day-old rat whose head was irradiated with 200 r X-ray and killed 6 h later. The radiosensitive pyknotic cells appear as opaque dots. They are abundant in the hilus and molecular layer of the dentate gyrus (DG), the subependymal layer (SE) of the cerebral cortex and the alveus (AL) and stratum oriens (SO) of Ammon's horn (AH). Only a few pyknotic cells are seen in the fimbria (FI), where gliogenesis has barely started. From Bayer and Altman (1974).

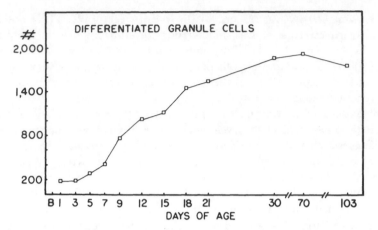

Fig. 10. Histological assessment of postnatal cell acquisition in the granular layer from birth to 103 days of age. Means are based on counts made in six animals at all ages indicated in matched coronal sections of the dorsal hippocampus. Modified from Bayer and Altman (1974).

was obtained in the endal arm of the dentate gyrus. The more than hundredfold increase in synapses was reached by 25 days of age, with no apparent increases thereafter.

Several histochemical and biochemical studies dealing with the maturation of transmitter-related enzymatic activity are relevant to the discussion of synaptogenesis and the development of the circuitry of the dentate gyrus. Evidence is available (Storm-Mathisen and Blackstad, 1964; Shute and Lewis, 1966; Lewis and Shute, 1967; Mellgren and Srebro, 1973) that the histochemically demonstrable acetylcholinesterase activity of the hippocampus is associated with septal afferents. These septohippocampal fibers are distributed in a laminar pattern—in the dentate gyrus in two separate bands, one infragranular and the other supragranular. Ritter et al. (1972) reported that acetylcholinesterase reaction is not present in the rat hippocampus until day 3 (however, faint monoamine oxidase reaction was noticeable at birth). Between days 10 and 20, there was a large increase in acetylcholinesterase activity and the adult pattern of distribution and intensity was reached by day 35.

In a more detailed study, Matthews et al. (1974) reported that acetylcholinesterase activity was not pronounced in the rat hippocampus at 4 days of age. Staining became more distinct by day 6 in the ectal (lateral) arm of the dentate gyrus; by day 8, the hilus was staining heavily and thereafter there was an increase in staining intensity in all regions of the hippocampus. Staining of the commissural zone was not detectable until day 16 and was not obvious until day 25. Because the staining reaction appeared earlier in the septal (anterior) end of the hippocampus than its more temporal (posterior) portions, it was postulated that the progression of staining marked the growth of septal afferents. Matthews et al. (1974) suggested that septal afferents first reach the dentate gyrus by day 4 and are distributed throughout the hippocampus by day 11. In a correlated quantitative histochemical study (Nadler et al., 1974), the attempt was made to obtain some indirect evidence about "cholinergic synaptogenesis" by determining choline acetyltransferase activity (together with acetylcholinesterase activity) in discrete layers of the dentate gyrus. Choline acetyltransferase activity was low at 11 days, but there was a sharp increase around 16–17 days. It was suggested that the latter dates may mark the onset of rapid synaptogenesis in the dentate gyrus. In view of the uncertain correlations between "cholinergic" afferents and synapses on the one hand and acetylcholinesterase and choline acetyltransferase activity on the other, we must await further information gained with other techniques, especially electron microscopy, for a more definitive assessment of synaptogenesis in the dentate gyrus.

2.4. Gliogenesis and Myelogenesis

The fimbria and fornix are recognizable in the rat by day 18 of gestation (Bayer and Altman, 1974). Insofar as a large proportion of the fornix is made up of efferents of the hippocampus, the outgrowth of axons of pyramidal cells is presumably in progress by this time. But there are few cells in the fimbria and fornix until the end of the first week after birth, suggesting that the glia cells (oligodendroglia) that are responsible for myelination do not appear in appreciable numbers for some time after the formation of axons. Myelination is heralded by a spurt in the proliferation of glia cells, the event is sometimes referred to as "myelination gliosis" (Roback and Scherer, 1935). This was examined in a recent study with the dated cumulative autoradiographic labeling technique. The results (Bayer and Altman, 1975) showed that about 73% of the cells of the fimbria are formed postnatally (Fig. 11). Similar results were obtained with the method of X-ray-produced cell pyknosis (see Fig. 9). There may be a delay of several days between the peak of myelination gliosis and the onset of myelination (Bensted *et al.,* 1957; DeRobertis *et al.,* 1958). The results of Jacobson (1963) indicated that the stainability of the fornix for myelin (Weigert's technique) is still faint on postnatal day 12 and moderate by day 21. By day 25, the fornix as well as the perforant bundle stain

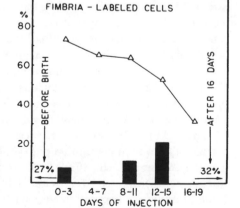

Fig. 11. Proportion of labeled cells in the fimbria in groups of 60-day-old rats that were injected daily with [³H]thymidine on days 0–3, 4–7, 8–11, 12–15, and 16–19. Method of evaluation the same as in Fig. 6. Modified from Bayer and Altman (1974).

strongly. The late myelination of the fornix was also reported for man. Yakovlev and Lecours (1967) found that the fornix remains unmyelinated until the end of the fourth postnatal month and that the process is a protracted one and may not be completed until the third postnatal year. In none of these studies has the gliogenesis and myelogenesis of the hippocampus been examined in great detail, and on this subject as on so many other aspects of hippocampal neurogenesis there is a great need for further investigations.

3. Development of the Dentate Gyrus Under Experimental Conditions

Many lines of evidence suggest that the immature brain has greater powers of recovery after insult than the mature brain. However, in many respects it is the developing nervous system which is the more vulnerable. An example already discussed is the extreme radiosensitivity of multiplying cells. Other treatments that selectively damage the developing brain include under- or malnutrition, drug administration, and hormonal manipulations. Because small neurons and the glia cells of the hippocampus are to a large extent formed postnatally and because exposure to environmental hazards increases greatly after birth, the susceptibility of the developing hippocampus has been the subject of several investigations.

3.1. Effects of Interference with Cytogenesis

Since irradiation of the hippocampus leads to the death of a large proportion of the multiplying precursors of granule cells (Fig. 9), it was logical to inquire whether or not this treatment will result in a permanent loss in differentiated granule cells. Earlier studies with the cerebellum showed that the long-term effects of X-irradiation are complicated by the capacity of the subtotally eradicated external germinal layer to regenerate (Altman *et al.*, 1969). To prevent this recovery, the cerebellum was irradiated repeatedly at daily or longer intervals. This approach showed that with the delivery of up to five successive daily doses of 200 r, the speed of regeneration of the external germinal layer was inversely related to the number of exposures. Because the external germinal layer disappears naturally at 21 days of age, the progressive delay in recovery of the germinal matrix resulted in a graded reduction of the postnatally formed cells of the cerebellum (Altman and Anderson, 1971). We have as yet no information about the time course and nature of recovery following irradiation of the hippocampus. However, it has been recently established (Bayer *et al.*, 1973) that irradiation with eight doses of 150–200 r between 2 and 15 days reduced the granule cell population of dentate gyrus to 15–18% of its normal concentration (Figs. 12 and 13A). The treatment presumably prevented the formation of all the postnatally acquired population of granule cells. The prenatally formed granule cells, like the pyramidal cells (Fig. 13B), were spared. A subsequent study (Bayer and Altman, 1975) showed that with two doses of 200 r delivered on days 2 and 3 the granule cell population was halved and the effect of four and six successive doses was similar to that of eight doses, resulting in an asymptotic reduction of

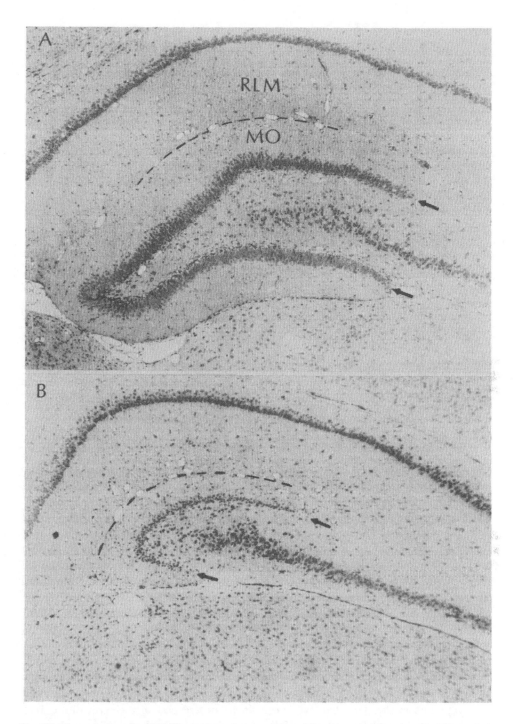

FIG. 12. Photomicrographs of the hippocampus in a control rat (A) and a rat irradiated with eight doses of 150–200 r between days 2 and 15 (B). Broken lines delineate the hippocampal fissure separating the dentate gyrus from Ammon's horn. Note the drastic reduction in the cell thickness of the granular layer (arrows) in the irradiated animal and also in the width of the molecular layer of dentate gyrus (MO). The width of strata radiatum, lacunosum, and moleculare of Ammon's horn (RLM) is not obviously affected. Slightly modified from Bayer *et al.* (1973).

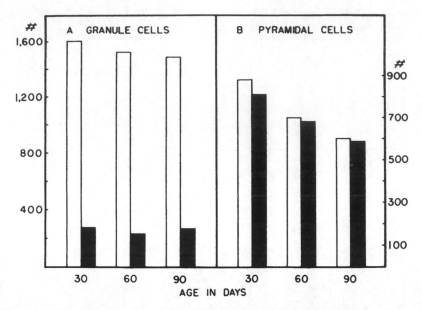

FIG. 13. A: Number of dentate granule cells in matched sections of the dorsal hippocampus in control rats (white bars) and X-irradiated rats (black bars). Each bar represents the mean of data from five animals. The reduction is highly significant. B: Number of pyramidal cells in the same group of animals. The age-dependent reduction in granule cells in control animals and in pyramidal cells in both groups is attributed to the volumetric expansion of the hippocampus between days 30 and 90. Modified from Bayer *et al.* (1973).

granule cells (Fig. 14B). The effects on cells of the molecular layer, which presumably consist of a mixed population of neurons and glia, was somewhat different (Fig. 14C). The delivery of two doses had no obvious effect, and with progressively more doses (four, six, and eight) there was a proportional reduction in these cells, with the magnitude never reaching that seen in granule cells. In view of the fact that these cells are largely of postnatal origin, we have tentatively assumed, in agreement with earlier observations (Altman *et al.*, 1968a), that the precursors of glia cells are more radioresistant than the precursors of neurons. None of the irradiation schedules had an effect on the number of pyramidal cells (Fig. 14A). It is relevant in this context to refer to behavioral studies (Bayer *et al.*, 1973; Haggbloom *et al.*, 1974) which showed that interference with the acquisition of the postnatally forming granule cells and other cellular elements of the hippocampus results in deficits similar to those observed when the hippocampus is destroyed *in toto*.

Do manipulations of milder kind than X-irradiation interfere with cell acquisition in the hippocampus? To answer this question, it will be necessary to carry out investigations of the type that have been done with respect to the cerebellum. It is now well established (*cf.* Altman, 1975) that treatments such as undernutrition or hypo- and hyperthyroidism interfere with cell acquisition in the cerebellum. In one study in which the food intake of infant rats was restricted by increasing litter size, biochemical assessment of DNA concentration indicated a reduction in cell number in

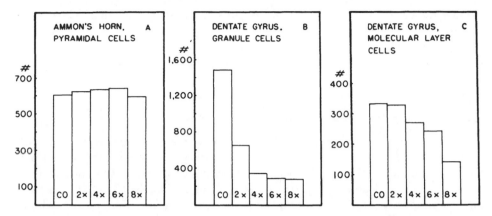

FIG. 14. The effects of two, four, six, and eight successive doses of 150–200 r delivered from day 2 onward on different cell types of the hippocampus. CO, Control. There was no significant effect with any of the radiation schedules used on the pyramidal cells of Ammon's horn (A). There was substantial reduction in granule cells with two doses and near asymptotic reduction was obtained with four doses (B). However, two doses had no effect on cell concentration in the dentate molecular layer and graded reduction was obtained with four, six, and eight doses (C). From Bayer and Altman (unpublished data).

the hippocampus at 17 days of age (Fish and Winick, 1969). Our own results have so far been somewhat ambiguous on this point.

Another approach that requires further investigation is the possible effect of behavioral manipulations in young animals on cell acquisition in the hippocampus. In an exploratory study (Altman *et al.*, 1968b), rats were "handled" daily from day 2 to day 11 after birth. On day 11, these animals and unhandled controls were injected with [³H]thymidine and killed 6 h or 3 or 30 days thereafter, at which time the brains were processed for autoradiography. The results showed (Fig. 15) a higher concentration of labeled granule cells at all the three ages in the handled animals.

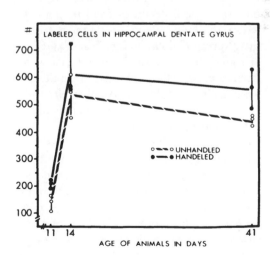

FIG. 15. Number of labeled cells in matched sections of the dorsal hippocampus in five 200-μm-wide strips of the granular layer. From Altman *et al.* (1968b).

Neither the nature of handling nor the meaning of the higher rate of cell labeling is adequately understood at present. For instance, a higher concentration of labeled hippocampal granule cells in the handled animals could be interpreted as a higher rate of cell production or just as a delay in hippocampal maturation, a hypothetical phenomenon that we referred to as "infantilization" (Altman *et al.*, 1968*b*).

3.2. Effects of Interference with Morphogenesis

The concept of neuromorphogenesis includes, on the cytological level, the acquisition by differentiating nerve cells of specific perikaryal, dendritic, and axonal patterns and, on the histological level, the aggregation of cell bodies and of their proximal and distal processes in specific ways. This developmental phase antedates and is a prerequisite to the establishment of the gross and fine circuitry in any brain region.

In one study (Das, 1971), the anterior portion of the dorsal hippocampus was surgically severed in 8- and 15-day-old rabbits so that the precursors of granule cells that migrate to the dentate gyrus could not reach their destination. The morphological organization of the hippocampus was examined when the animals were 40 days old. Das found that the dorsal dentate gyrus remained underdeveloped as a result of the cut but that the proliferating granule cell precursors formed a hypertrophied accessory dentate gyrus, rudiments of which are present in normal rabbits.

A different approach was taken by Lynch, Cotman, and their associates (Lynch *et al.*, 1973*a*). It is well established that the commissural projection to the molecular layer of the dentate gyrus is confined to a narrow zone above the granular layer. In a group of 11-day-old rats, large lesions were made in the entorhinal cortex to eliminate the entorhinal projection to the hippocampus. When these animals became mature, a second lesion was made to sever the commissural projection; similar lesions were also made in a control group. In the rats with previous entorhinal lesions, the commissural afferents extended over 90% of the width of the molecular layer. The investigators concluded that the developing commissural system will spread out along the granule cell dendrites from its normally restricted domain if the entorhinal afferents that normally innervate the outer dendritic field are eliminated. In addition to this "spreading" effect, Lynch *et al.* (1973*b*) also observed an increase in the density of commissural terminals in animals whose entorhinal cortex was removed in infancy. Similar effects of lesser magnitude were noted when the entorhinal lesions were made in adulthood. Another effect of entorhinal lesions was the intensification of cholinesterase staining in the outer part of the molecular layer, which could be attributed to the spreading of septal afferents (Cotman *et al.*, 1973). Electron microscopic investigations suggested that this augmentation could be partly attributed to an increase in the number of acetylcholinesterase-rich synaptic endings.

However, a caution is in order against interpreting the latter type of findings as a reflection of the greater "plasticity" of the infant than adult hippocampus and, by implication, a greater potential for "recovery of function." Morphological findings of "spreading" or "sprouting" of axons and axon terminals, even when coupled with electron microscopic demonstration of synaptic junctions formed, cannot be taken as evidence of the establishment of coordinated functional contacts. This cautionary attitude is the outcome of some recent results regarding the effects of X-irradiation of the cerebellum in infant rats, where we found that numerous indications of structural reorganization were not paralleled by functional recovery (Brunner and Altman, 1974). For instance, when the acquisition of cerebellar granule cells is prevented by X-irradiation during infancy, the mossy fibers, which primarily synapse with granule cells in the granular layer, invade the molecular layer. This was first suggested in a histochemical study dealing with the distribution of acetylcholinesterase (Altman and Das, 1970) and was subsequently confirmed with electron microscopy (Altman and Anderson, 1972). These mossy fibers apparently form synapses with the soma and dendrites of Purkinje cells, thus displaying considerable potential for "reorganization." However, when these animals are tested as adults severe motor deficits are observed (Wallace and Altman, 1969b; Altman et al., 1971) which resemble the effects of decerebellation. Moreover, our studies seem to indicate that schedules of X-irradiation which produce massive reduction in the number of cells lead to less behavioral deficits than schedules which have less effect on the cell population but which result in drastic reorganization of the circuitry of the cerebellar cortex (Altman, 1975).

As yet, we have little information about the nature of the structural changes produced in the hippocampus by the prevention of the recruitment of the postnatally forming granule cells. No gross malformation has been noted in hippocampal morphogenesis, nor is there an evident effect on the number, size, and appearance of pyramidal cells. Tentatively it may be assumed that neither the extrinsic input to Ammon's horn nor its output is seriously interfered with. Nevertheless, as referred to earlier, the irradiated animals display fully all symptoms of surgical destruction of the hippocampus as a whole (Bayer et al., 1973; Haggbloom et al., 1974). The conclusion seems inescapable that, as in the case of the cerebellar cortex, the granule cells play a vital role in the integrated functions of the hippocampus.

ACKNOWLEDGMENTS

Some of the work reviewed here was done in collaboration with William J. Anderson, Robert L. Brunner, Gopal D. Das, and R. B. Wallace. We are grateful for the technical assistance of Sharon Evander and Zeynep Kurgun. This research program is supported by the National Institute of Mental Health and the Atomic Energy Commission.

4. References

ALTMAN, J. Autoradiographic investigation of cell proliferation in the brains of rats and cats. *Anatomical Record*, 1963, **145**, 573–591.
ALTMAN, J. Autoradiographic and histological studies of postnatal neurogenesis. II. A longitudinal investigation of the kinetics, migration and transformation of cells incorporating tritiated thymidine in infant

rats, with special reference to postnatal neurogenesis in some brain regions. *Journal of Comparative Neurology,* 1966, **128,** 431–474.

ALTMAN, J. Autoradiographic and histological studies of postnatal neurogenesis. IV. Cell proliferation and migration in the anterior forebrain, with special reference to persisting neurogenesis in the olfactory bulb. *Journal of Comparative Neurology,* 1969, **137,** 433–458.

ALTMAN, J. Postnatal development of the cerebellar cortex in the rat. I. The external germinal layer and the transitional molecular layer. *Journal of Comparative Neurology,* 1972a, **145,** 353–398.

ALTMAN, J. Postnatal development of the cerebellar cortex in the rat. II. Phases in the maturation of Purkinje cells and of the molecular layer. *Journal of Comparative Neurology,* 1972b, **145,** 399–464.

ALTMAN, J. Effects of interference with cerebellar maturation on the development of locomotion: An experimental model of neurobehavioral retardation. In N. Buchwald (Ed.), *Brain mechanisms in mental retardation.* New York: Academic Press, 1975, pp. 41–91.

ALTMAN, J., AND ANDERSON, W. J. Irradiation of the cerebellum in infant rats with low-level X-ray: Histological and cytological effects during infancy and adulthood. *Experimental Neurology,* 1971, **30,** 492–509.

ALTMAN, J., AND ANDERSON, W. J. Experimental reorganization of the cerebellar cortex. I. Morphological effects of elimination of all microneurons with prolonged X-irradiation started at birth. *Journal of Comparative Neurology,* 1972, **146,** 355–406.

ALTMAN, J., AND DAS, G. D. Post-natal origin of microneurones in the rat brain. *Nature (London),* 1965a, **207,** 953–956.

ALTMAN, J., AND DAS, G. D. Autoradiographic and histological evidence of postnatal hippocampal neurogenesis in rats. *Journal of Comparative Neurology,* 1965b, **124,** 319–335.

ALTMAN, J., AND DAS, G. D. Autoradiographic and histological studies of postnatal neurogenesis. I. A longitudinal investigation of the kinetics, migration and transformation of cells incorporating tritiated thymidine in neonate rats, with special reference to postnatal neurogenesis in some brain regions. *Journal of Comparative Neurology,* 1966, **126,** 337–390.

ALTMAN, J., AND DAS, G. D. Postnatal neurogenesis in the guinea-pig. *Nature (London)* 1967, **214,** 1098–1101.

ALTMAN, J., AND DAS, G. D. Postnatal changes in the concentration and distribution of cholinesterase in the cerebellar cortex of rats. *Experimental Neurology,* 1970, **28,** 11–34.

ALTMAN, J., AND NICHOLSON, J. L. Cell pyknosis in the cerebellar cortex of infant rats following low-level X-irradiation. *Radiation Research,* 1971, **46,** 476–489.

ALTMAN, J., ANDERSON, W. J., AND WRIGHT, K. A. Differential radiosensitivity of stationary and migratory primitive cells in the brains of infant rats. *Experimental Neurology,* 1968a, **22,** 52–74.

ALTMAN, J., DAS, G. D., AND ANDERSON, W. J. Effects of infantile handling on morphological development of the rat brain: An exploratory study. *Developmental Psychobiology,* 1968b, **1,** 10–20.

ALTMAN, J., ANDERSON, W. J., AND WRIGHT, K. A. Early effects of X-irradiation of the cerebellum in infant rats: Decimation and reconstitution of the external granular layer. *Experimental Neurology,* 1969, **24,** 196–216.

ALTMAN, J., ANDERSON, W. J., AND STROP, M. Retardation of cerebellar and motor development by focal X-irradiation during infancy. *Physiology and Behavior,* 1971, **7,** 143–150.

ANGEVINE, J. B. Time of neuron origin in the hippocampal region. *Experimental Neurology, Supplement,* 1965, **2,** 1–70.

BAYER, S. A., AND ALTMAN, J. Hippocampal development in the rat. Cytogenesis and morphogenesis examined with autoradiography and low-level X-irradiation. *Journal of Comparative Neurology,* 1974, **158,** 55–80.

BAYER, S. A., AND ALTMAN, J. Radiation-induced interference with postnatal hippocampal cytogenesis in rats and its long-term effects on the acquisition of neurons and glia. *Journal of Comparative Neurology,* 1975, in press.

BAYER, S. A., BRUNNER, R. L., HINE, R., AND ALTMAN, J. Behavioural effects of interference with the postnatal acquisition of hippocampal granule cells. *Nature New Biology,* 1973, **242,** 222–224.

BENSTED, J. P. M., DOBBING, J., MORGAN, R. S., REID, R. T. W., AND WRIGHT, G. P. Neuroglial development and myelination in the spinal cord of the chick embryo. *Journal of Embryology and Experimental Morphology,* 1957, **5,** 428–437.

BRUNNER, R. L., AND ALTMAN, J. Locomotor deficits in adult rats with moderate to massive retardation of cerebellar development during infancy. *Behavioral Biology*, 1973, **9**, 169–188.

BRUNNER, R. L., AND ALTMAN, J. The effects of interference with the maturation of the cerebellum and hippocampus on the development of adult behavior. In D. G. Stein *et al.* (Eds.), *Recovery of function in the central nervous system.* New York: Academic Press, 1974, pp. 129–148.

COTMAN, C. W., MATTHEWS, D. A., TAYLOR, D., AND LYNCH, G. Synaptic rearrangement in the dentate gyrus: Histochemical evidence of adjustments after lesions in immature and adult rats. *Proceedings of the National Academy of Sciences U.S.A.* 1973, **70**, 3473–3477.

CRAIN, B., COTMAN, C., TAYLOR, D., AND LYNCH, G. A quantitative electron microscopic study of synaptogenesis in the dentate gyrus of the rat. *Brain Research*, 1973, **63**, 195–204.

CRAWFORD, I. L., AND CONNOR, J. D., Zinc in maturing rat brain: Hippocampal concentration and localization. *Journal of Neurochemistry*, 1972, **19**, 1451–1458.

DAS, G. D. Experimental studies on the postnatal development of the brain. I. Cytogenesis and morphogenesis of the accessory fascia dentata following hippocampal lesions. *Brain Research*, 1971, **28**, 263–282.

DAS, G. D., AND KREUTZBERG, G. W. Postnatal differentiation of the granule cells in the hippocampus and cerebellum: A histochemical study. *Histochemie*, 1967, **10**, 246–260.

DEROBERTIS, E., GERSCHENFELD, H., AND WALD, F. Cellular mechanisms of myelination in the central nervous system. *Journal of Biophysical and Biochemical Cytology*, 1958, **4**, 651–658.

FISH, I., AND WINICK, M. Effect of malnutrition on regional growth of the developing rat brain. *Experimental Neurology*, 1969, **25**, 534–540.

GODINA, G., AND BARASA, A. Morfogenesi istogenesi della formazione ammonica. *Zeitschrift für Zellforschung und Mikroskopische Anatomie*, 1964, **63**, 327–355.

HAGGBLOOM, S. J., BRUNNER, R. L., AND BAYER, S. A. Effects of hippocampal granule cell agenesis on acquisition of escape from fear and one-way active-avoidance responses. *Journal of Comparative and Physiological Psychology*, 1974, **86**, 447–457.

HINE, R. J., AND DAS, G. D. Neuroembryogenesis in the hippocampal formation of the rat: An autoradiographic study. *Zeitschrift für Anatomie und Entwicklungsgeschichte*, 1974, **144**, 173–186.

JACOBSON, S. Sequence of myelinization in the brain of the albino rat. A. Cerebral cortex, thalamus and related structures. *Journal of Comparative Neurology*, 1963, **121**, 5–29.

LEWIS, P. R., AND SHUTE, C. C. D. The cholinergic limbic system: Projections to hippocampal formation, medial cortex, nuclei of ascending cholinergic reticular system, and the subfornical organ and the supra-optic crest. *Brain*, 1967, **90**, 521–540.

LYNCH, G. S., MOSKO, S., PARKS, T., AND COTMAN, C. W. Relocation and hyperdevelopment of the dentate gyrus commissural system after entorhinal lesions in immature rats. *Brain Research*, 1973a, **50**, 174–178.

LYNCH, G., STANFIELD, B., AND COTMAN, C. W. Developmental differences in post-lesion axonal growth in the hippocampus. *Brain Research*, 1973b, **59**, 155–168.

MATTHEWS, D. A., NADLER, J. V., LYNCH, G. S., AND COTMAN, C. W. Development of cholinergic innervation in the hippocampal formation of the rat. I. Histochemical demonstration of acetylcholinesterase activity. *Developmental Biology*, 1974, **36**, 130–141.

MELLGREN, S. I. Distribution of acetylcholinesterase in the hippocampal region of the rat during postnatal development. *Zeitschrift für Zellforschung und Mikroskopische Anatomie*, 1973, **141**, 375–400.

MELLGREN, S. I., AND SREBRO, B. Changes in acetylcholinesterase and distribution of degenerating fibres in the hippocampal region after septal lesions in the rat. *Brain Research*, 1973, **52**, 19–36.

MESSIER, B., AND LEBLOND, C. P. Preparation of coated radioautographs by dipping sections in fluid emulsion. *Proceedings of the Society for Experimental Biology and Medicine*, 1957, **96**, 7–10.

MEYER, U., RITTER, J., AND WENK, H. Zur Chemodifferenzierung der Hippocampusformation in der postnatalen Entwicklung der Albinoratte. I. Oxydoreductasen. *Journal fuer Hirnforschung*, 1972, **13**, 235–253.

NADLER, J. W., MATTHEWS, D. A., COTMAN, C. W., AND LYNCH, G. S. Development of cholinergic innervation in the hippocampal formation of the rat. II. Quantitative changes in choline acetyltransferase and acetylcholinesterase activities. *Developmental Biology*, 1974, **36**, 142–154.

PURPURA, D. P., AND PAPPAS, G. D. Structural characteristics of neurons in the feline hippocampus during postnatal ontogenesis. *Experimental Neurology*, 1968, **22**, 379–393.

RITTER, J., MEYER, U., AND WENK, H. Zur Chemodifferenzierung der Hippocampusformation in der postnatalen Entwicklung der Albinoratte. II. Transmitterenzyme. *Journal fuer Hirnforschung*, 1972, **13**, 255–278.

ROBACK, H. N., AND SCHERER, H. J. Ueber die feinere Morphologie der frühkindlichen Hirnes unter besonderer Berücksichtigung der Gliaentwicklung. *Virchows Archiv, Abteilung A, Pathologische Anatomie*, 1935, **294**, 365–413.

SCHWARTZ, I. R., PAPPAS, G. D., AND PURPURA, D. P. Fine structure of neurons and synapses in the feline hippocampus during postnatal ontogenesis. *Experimental Neurology*, 1968, **22**, 394–407.

SHIMADA, M. Oxidative enzymes in maturing hippocampus and cerebellum of hamsters. *Experientia*, 1970, **26**, 181–182.

SHUTE, C. C. D., AND LEWIS, P. R. Electron microscopy of cholinergic terminals and acetyl-choninesterase-containing neurones in the hippocampal formation of the rat. *Zeitschrift für Zellforschung und Mikroskopische Anatomie*, 1966, **69**, 334–343.

STORM-MATHISEN, J., AND BLACKSTAD, T. W. Cholinesterase in the hippocampal region. *Acta Anatomica*, 1964, **56**, 216–253.

TAYLOR, J. H., WOODS, P. S., AND HUGHES, W. L. The organization and duplication of chromosomes as revealed by autoradiographic studies using trititium-labeled thymidine. *Proceedings of the National Academy of Sciences, U.S.A.*, 1957, **43**, 122–128.

WALLACE, R. B., AND ALTMAN, J. Behavioral effects of neonatal irradiation of the cerebellum. I. Qualitative observations in infant and adolescent rats. *Developmental Psychobiology*, 1969a, **2**, 257–265.

WALLACE, R. B., AND ALTMAN, J. Behavioral effects of neonatal irradiation of the cerebellum. II. Quantitative studies in young-adult and adult rats. *Developmental Psychobiology*, 1969b, **2**, 266–272.

WENDER, M., AND KOZIK, M. Zur Chemoarchitektonik des Ammonshorngebietes während der Entwicklung des Mausegehirns. *Acta Histochemica (Jena)*, 1968, **31**, 166–181.

WENDER, M., AND KOZIK, M. Studies of the histoenzymatic architecture of the Ammon's horn region in the developing rabbit brain. *Acta Anatomica (Basel)*, 1970, **75**, 248–262.

YAKOVLEV, P. I., AND LECOURS, A.-R. The myelogenetic cycles of regional maturation of the brain. In A. Minkowski (Ed.), *Regional development of the brain in early life*. Oxford: Blackwell, 1967, pp. 3–65.

5

The Hippocampus as a Model for Studying Anatomical Plasticity in the Adult Brain

GARY LYNCH AND CARL W. COTMAN

1. Introduction

The suggestion that intact axons might grow new branches ("sprout") in response to damage of their neighbors appears to have been made at various times throughout the century-long argument that revolved around the issue of peripheral nerve regeneration. One of the earliest references was made by Haighton, who in 1795 reported a series of "physiological" studies on regeneration (or "reproduction") which included controls for "a difficulty which naturally presents itself here, and this is, the possibility of the stomach and vocal organs having received an additional supply of nervous energy from another source" (p. 198). Exner (1885), according to Edds (1953), provided a clear description of sprouting at the neuromuscular junction, and Kennedy in his 1897 review listed several authors who mentioned the need to control for growth by undamaged nerves into deafferented sites in evaluating studies on regeneration.

Ramón y Cajal (1928) and Lugaro (1906), in the course of their classic experiments on regeneration, performed several studies to directly examine possible growth by intact fibers. These included placing a graft of degenerating sciatic nerve on the uncut sciatic (to subject it to the hypothetical "neurotropic" substance) and partial cuts of the nerve. In neither of these situations did they observe any evidence of growth by the intact axons, and so concluded that the phenomenon did not occur.

GARY LYNCH AND CARL W. COTMAN • Department of Psychobiology, University of California, Irvine, California.

It was not until the 1940s that definitive evidence of collateral sprouting was obtained and appreciated as a demonstration of an important phenomenon. Van Harreveld (1945), Edds (1950), and Hoffman (1950), working separately, found that the intact fibers of partially cut peripheral nerves would grow sprouts which invested the Schwann cells of the degenerating axons and finally reinnervated the muscle which had lost its inputs. Shortly after the description of collateral sprouting at the neuromuscular junction, Murray and Thompson (1957) found that intact roots of the preganglionic chain would sprout when their neighbors were transected, and thereby would reestablish innervation of the superior cervical ganglion (see also Guth and Bernstein, 1961).

An even more impressive and ultimately more influential demonstration came from the experiments of Liu and Chambers (1958). These workers used the then recently devised Nauta technique to show that if the dorsal roots to the rostral spinal cord were cut the intact caudal roots would sprout for considerable distances to invade the sites deafferented by the transections. They found similar effects after removal of descending tracts from supraspinal centers. These important experiments, replicated and extended by Goldberger (1974), were a vital catalyst to the study of sprouting.

The experiments of Liu and Chambers indicated that sprouting was not restricted to the peripheral nervous system, and research during the intervening years has suggested that "sprouting-like" effects may be obtained after lesions throughout the central nervous system. These reports have aroused a great deal of interest, since it is apparent that the occurrence of sprouting in the brain would profoundly influence the interpretation of neurobehavioral studies using lesion techniques and a variety of clinical phenomena, including recovery from brain damage. Furthermore, the observation that undamaged afferents in the adult brain possess the capacity for various forms of growth has important implications for theories of brain plasticity. Specifically, the types of reactive changes observed after lesions may be a part of the brain's operation under normal conditions.

However, in order to understand the significance and interpret the functional consequences of growth by intact neurons after removal of their neighbors, a number of questions must be answered.

First, it will be necessary to establish the types of axonal and terminal changes that take place following deafferentation. The term "sprouting" has a recent years acquired a perhaps undesirably broad denotation. In studies done in the peripheral nervous system it was used to indicate growth of collateral axons, and this is probably the usage intended by Liu and his coworkers in their spinal cord work. It is now commonly used to indicate the reoccupancy of a vacated synaptic site by a new afferent, a process which does not necessarily require axonal growth. It is conceivable that in partially deafferented brain sites intact terminals might expand to reach sites which have lost their inputs or dendrites might extend themselves to contact undamaged presynaptic elements. These distinctions are of some importance in any effort to accurately describe sprouting in the brain, particularly if we hope to relate it to well-established phenomena which occur after lesions in the peripheral nervous system.

Second, we need to consider the generality of the sprouting process. This question takes two forms: does some form of sprouting occur at every deafferented site and does every afferent remaining to the partially denervated site show reactive changes, or is the effect restricted to certain classes of intact fibers? Evidence on these two points would provide important information regarding the mechanisms underlying sprouting as well as point to an interpretation of its physiological and behavioral significance.

Third, it is critical to establish whether reactive axonal and terminal growth produces functional synapses. There is some evidence that aberrant contacts in the peripheral nervous system are in certain cases nonfunctional (Marotte and Mark, 1970a).

Fourth, understanding of the functional significance of sprouting will be greatly increased by measuring the time course of the effect. Establishing a correlation between the time course of postlesion axonal changes and the physiological/behavioral effects of the lesion is a necessary first step in searching for causal links.

Finally, we need to begin an analysis of the mechanisms which govern "sprouting"-like effects in the brain.

In the following review, we will assess the evidence that sprouting occurs in the adult brain, first in areas outside the hippocampus and then in the hippocampus. As this analysis will show, the literature clearly indicates that in several brain regions some form of axonal or terminal change transpires in afferents remaining to partially denervated sites, but for reasons described it is not certain that these are necessarily reflective of collateral sprouting. In fact, only a very few studies clearly suggest axonal collateral sprouting of the type observed at the neuromuscular junction occurs in brain. In the section on the hippocampus, we will consider work from our laboratories in which the hippocampus was used as a model system and try to gain answers to the questions concerning existence, generality, functionality, and timing raised above.

2. Studies on Sprouting Outside the Hippocampus

Research on postlesion growth in brain areas other than the hippocampus has been approached almost exclusively by anatomical methods (degeneration—employing silver stains, histochemistry, and electron microscopy) and the degree of certainty of the nature of the growth depends on the method used. Thus it is appropriate for the purposes of a critical analysis to organize the data on the basis of methodology.

2.1. Light Microscopic Studies

The first report of sprouting in the brain was that of Goodman and Horel (1966) (however, see Rose et al., 1960), who studied the distribution of the optic tract projections to the thalamus after lesions of the visual cortex. Goodman and Horel used the same experimental design and anatomical techniques as did Liu and

Chambers (1958). A lesion of the visual cortex was used to deafferent the lateral geniculate nucleus. This was followed by an extended survival period to allow for the removal of the degeneration products produced by this lesion, and then a second lesion was performed on the intact afferents (optic tracts) which were suspected of sprouting. Within days of the second lesion, the brain was processed for axonal degeneration by the Nauta technique and the course of the second afferent was plotted and compared with that for control (no primary lesion) animals; the long interoperative interval allowed the investigators to distinguish degeneration products caused by the second lesion from those remaining from the first.

Using this paradigm, Goodman and Horel found two sites at which the optic tract increased its density of innervation after the cortical ablation; they also found several deafferented sites at which sprouting did not occur, even though the optic tract passed near them. This selective failure of sprouting was also reported by Liu and Chambers in their spinal cord studies and, as will be argued below, may provide some important clues about the mechanisms controlling the effect.

More recently, Goodman et al. (1973) have returned to this system using essentially the same paradigm to study possible sprouting by the ipsilateral optic projections after removal of the contralateral eye. They found increased density of degeneration in each region of normal innervation which was partially deafferented by the loss of the crossed projections and, more importantly, obtained evidence that the ipsilateral projections expanded their field of termination into the caudal superior colliculus, a region in which these afferents are not normally found.

The histofluorescent method for the demonstration of catecholamines has also produced data suggestive that "sprouting" occurs in the brain. Moore et al. (1971) have found that the number of fluorescent terminal varicosities in the septum increases after lesions of the hippocampus and attributed this to sprouting by norepinephrine-containing fibers traveling in the medial forebrain bundle. Moore et al. found that the increase in the number of terminals was detectable within a week of the hippocampal lesion, a time period for sprouting which is comparable with that given for this effect at the neuromuscular junctions (see above). Stenevi et al. (1972) observed an increase in the number of fluorescent norepinephrine (NE) terminals in the lateral geniculate nucleus after lesions of the visual cortex, but *not* after enucleation. This indicates a selectivity in sprouting, since removal of one input produces the effect while destruction of another equally important input does not. There was no evidence for proliferation of the NE terminals in other sites shared by the degenerating cortical fibers and the NE system (e.g., superior colliculus).

It can be seen from the above necessarily brief review that almost all of the light microscopic evidence for sprouting in the adult brain outside the hippocampus comes from studies reporting an increased density of terminals generated by an intact afferent in a partially denervated site.

A central problem in the interpretation of increases in terminal density is the difficulty of controlling for the contribution of shrinkage to apparent increases in density. While there have been no parametric studies, subtantial shrinkage has been observed at several deafferented brain sites, including the septum (Raisman, 1969), the hippocampus (see below), and the optic thalamus (Guillery, 1972). An increase of density of the afferents remaining to a partially deafferented site would be expected on this basis alone. Goodman and Horel (1966) were aware of this problem and attempted to control for it by counting the number of cell bodies contained within a fixed area in the regions in which terminal density increased. They found no increase in cell density and concluded that no shrinkage had occurred; they argued, therefore, that the apparent increase in terminal density must be due to the proliferation of existing population of endings rather than a compression of that population. However, the fact that they detected no tissue loss whatsoever must cause us to question the sensitivity of their measurement technique, since shrinkage is so commonly observed at deafferented sites elsewhere in the brain.

The same criticism can be applied to studies demonstrating sprouting using the histofluorescent technique. Raisman and Field (1973) report that the septum shows a 20% shrinkage ipsilateral to a transection of the fimbria, and there is every reason to assume that most of this occurs in the areas innervated by hippocampal and medial forebrain bundle (MFB) fibers. On this basis alone, a considerable increase in the density of catecholaminergic terminals would be expected in the septum after hippocampal lesions. It is necessary, then, to ask what percentage of the density increase observed by Moore and his associates was due to this shrinkage factor and what portion actually represented terminal proliferation.

Shrinkage necessarily complicates the interpretation of any increase in the density of elements in a deafferented area, and for this reason studies showing that a given afferent *expands* its area of innervation into deafferented zones are particularly important. As discussed, Liu and Chambers and Goldberger obtained strong evidence for expanded zones of innervation in the spinal cord, but the only report of this in the adult brain outside of the hippocampus is that of Goodman et al. (1973), in which they found that the ipsilateral optic projections expanded into the caudal superior colliculus after removal of the contralateral inputs to that zone.

Another criticism pertinent to the findings of increased density of innervation must be considered in evaluating the expansion of terminal fields. The possibility exists that increased density of innervation or expanded zones of termination are not due to the growth of nerve endings but rather the "uncovering" of existing ones. It is widely suspected that silver methods leave unstained certain populations of degenerating endings, and even in the same animal the optimal parameters for these procedures vary from one brain system to the next. If the "stainability" of a given population of axons were altered by the chemical changes it experiences as a result of neighboring degeneration, then it is conceivable that a greater percentage of this population might be detected by the silver methods. This would then be interpreted as an increased density of innervation, whereas in fact an increased probability of successful staining was the true basis for the density increase.

Analogous problems exist with regard to the interpretation of histochemical evidence of sprouting. If the deafferentation were to increase the concentration of norepinephrine per terminal, then it is quite possible that more terminals would be detected in a partially deafferented region; rather than actually increasing the number of terminals in a given area, one would have increased the likelihood of detecting a population that was already present. Moore *et al.* (1971) were aware of this problem and presented circumstantial evidence against it; however, definitive data are still lacking.

2.2. Electron Microscopic Studies

The material reviewed immediately above pertaining to light microscopic experiments indicates that some form of axonal or paraterminal growth takes place in intact afferents of partially deafferented regions but that a number of problems make this a qualified conclusion. The electron microscope could resolve certain of these difficulties, particularly those related to endings not normally detected by the light microscopic methods.

Shortly after the appearance of the results of Goodman and Horel, Raisman (1969) provided the first electron microscopic description of sprouting. The medial septal nucleus receives afferents from both the MFB and the hippocampus, but the ultrastructural appearance and location of the terminals of these systems are different. The inputs from the hippocampus terminate on dendrites only, while those from the MFB innervate both dendrites and cell bodies. Raisman removed one of these projections and then, after waiting a sufficient time for the degenerating debris to be removed, made a lesion of the second input and studied its distribution by plotting the location of degenerating terminals. After either lesion sequence, he found that the number of synaptic contacts per terminal of the remaining afferent increased; it appeared, then, that the surviving endings occupied the sites made available by removing a population of terminals.

Raisman and Field (1973) have reexamined these effects in a quantitative electron microscopic study and found that in the first few postlesion days many sites are apposed by degenerating terminals, but this number drops rapidly, and by 30 days after lesion essentially all postsynaptic sites are occupied by intact endings, so the total number of normal synapses is restored. They also found that the number of terminals with multiple contacts increased during the "reoccupancy" period, again suggesting that a given intact ending was making new connections with sites left vacant after the lesion.

Westrum and Black (1971) and Lund *et al.* (1973; Lund and Lund, 1971) have also reported electron microscopic evidence which indicates that an intact population of terminals will reoccupy sites vacated by a lesion.

As noted by Raisman and Field, the reoccupation of synaptic sites by intact presynaptic elements could be due to factors other than proliferation of an existing set of terminals. It is also possible that the number of endings generated by the remaining afferents remains constant but that these form synaptic contacts with postsynaptic specializations which have lost their input. In support of this suggestion is the observation that the number of multiple contacts made by a single terminal increases dramatically during the postlesion period (Raisman, 1969; Raisman and Field, 1973).

The central conclusion from the electron microscopic investigations is that new synaptic junctions in fact are formed. It is not clear, however, whether axon collateral sprouting *per se* is involved. Minor extensions of terminals or simple restructuring of dendrites are alternate explanations.

2.3. Negative Results

In contrast to the positive studies just cited, there have also been reports in which no evidence of sprouting whatsoever was obtained in brain regions where it might have been expected to occur (Kerr, 1972; Guillery, 1972; Lund *et al.*, 1973).

Guillery (1972) studied the distribution of ipsilateral optic projections to the lateral geniculate nucleus after prior removal of the contralateral eye. Recognizing the problems associated with shrinkage, Guillery made no effort to assess increased density of ipsilateral projections in the deafferented sites. Instead, he investigated the possibility of translaminar growth of retinal fibers to the lateral geniculate of adult cats. Following unilateral enucleation and a survival time of 4 months to 4 years, the remaining eye was removed and its terminal fields were analyzed using the Fink–Heimer and Nauta techniques. Guillery focused his analysis on lamina A of the lateral geniculate ipsilateral to the second enucleation. This lamina normally receives contralateral retinal input only and is separated from an ipsilateral retinal terminal field, lamina A1, by a small intralaminar fiber plexus (partially of contralateral retinal origin). Following the second enucleation, it was found that the ipsilateral retinal fibers remained confined to lamina A1, showing no abberant extension into lamina A or the intralaminal plexus (however, some aberrant growth was noted in young kittens).

Kerr (1972), in a carefully controlled study using the double lesion procedure, could not detect any evidence of sprouting by dorsal root projections after lesions of the descending spinal root of the fifth cranial nerve.

3. Studies on Sprouting in the Hippocampus

During the past several years, we have been conducting a multidisciplinary study on the consequences of partial deafferentation to the organization and operation of the hippocampus. During the course of this work, we have obtained evidence that axonal sprouting takes place in this structure after the elimination of its primary afferent and that this results in the formation of new, functional circuitry. We have also found that the remaining afferents show chemical or anatomical adjustments after both neonatal and adult lesions, but the extent and topography of the reorganization are different.

We have centered our studies on the dentate gyrus of the hippocampal formation because its normal synaptic organization is simple and well defined, and because the functional significance of structural changes can be assessed. The afferents (both intrinsic and extrinsic) to any given subfield of the hippocampus occupy discrete dendritic layers, and with few exceptions these do not overlap; this makes it possible to correlate neuroanatomical and electrophysiological techniques to a degree unparalleled in the mammalian central nervous system. The unusual anatomical simplicity of the hippocampus, shown in Fig. 1, has long been appreciated by neuroanatomists. The hippocampus contains only two major cell populations, the pyramidal cells of the hippocampus proper and the granule cells of the dentate gyrus. As is evident from the drawing, the dendritic fields of pyramidal and granule cells are completely separate. Interneurons are present, but these are greatly outnumbered by both the pyramidal or granule cells.

The granule cells' dendrites ramify in the zone called the molecular layer, and the major afferents of these cells contact the dendrites in a precise laminated arrangement. As shown schematically in Fig. 1, the outer three-fourths of the molecular layer receives the massive projection from the ipsilateral entorhinal cortex (Blackstad, 1956; Raisman *et al.*, 1966; Hjorth-Simonsen, 1972; Hjorth-Simonsen and Jeune, 1972), while the inner one-fourth of the molecular layer is occupied by commissural fibers from the contralateral hippocampus (Blackstad, 1956; Raisman *et al.*, 1966; Mosko *et al.*, 1973) and associational fibers from the ipsilateral hippocampus (Zim-

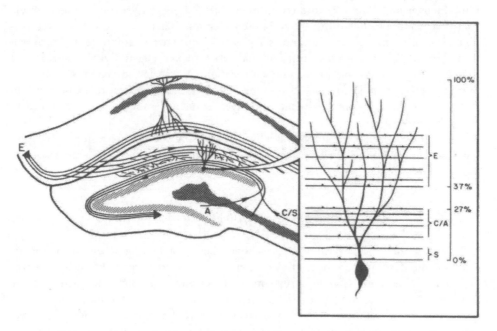

FIG. 1. Schematic representation of the organization of the hippocampus and the distribution of some of its fiber tracts. Shown are the course of fibers from the entorhinal cortex (E), septum (S), and the commissural (C) and associational (A) pathways. Also indicated by drawings of cells are the pyramidal cell layer (larger cell at top of figure) and the granule cell layer of the dentate gyrus. The drawing to the right provides further details of the distribution of the fiber systems entering the dentate gyrus. The location of the various afferents is indicated as a percentage of the distance from the granule cell bodies to the tops of their dendritic trees; the manner in which these values were arrived at is described in the text (see also Figs. 4 and 5). The zone between commissural and associational and entorhinal fibers is occupied by terminals and paraterminal elements from the entorhinal projection. Not shown in the schematic are the small septal and crossed entorhinal projections to the outer dendritic fields.

mer, 1970; Gottlieb and Cowan, 1973; Segal and Landis, 1974). There is virtually no overlap between the entorhinal and commissural/associational systems. A third, less dense afferent fiber system originates in the septum and terminates in a narrow layer between the granule cell bodies and the zone of termination of commissural/associational afferents, as well as in a zone beneath the granule cell layer; a sparser field of septal fibers in the outer two-thirds of the molecular layer has also been suggested (see Mosko *et al.*, 1973).

In addition to its well-defined structural organization, the dentate gyrus contains two identified transmitter systems. Several lines of evidence suggest that the septal input is cholinergic. Destruction of the septum or transection of the fimbria causes a nearly complete disappearance of acetylcholinesterase (AChE) and choline acetyltransferase (ChAc) throughout the hippocampus (Lewis *et al.*, 1967), whereas lesions of other extrinsic afferents produce no such loss (Lewis and Shute, 1967; Storm-Mathisen, 1970, 1972). In addition, both AChE and ChAc are organized in laminae (Fonnum, 1970) which closely correspond to patterns of septal terminal degeneration observed by silver degeneration methods (see Mosko *et al.*, 1973). From these data, it is reasonable to conclude that the hippocampus receives a significant cholinergic input which originates in the septum.

It has also been suggested that GABA may serve as a neurotransmitter for the interneurons. Glutamic acid decarboxylase activity is not reduced by lesions of hippocampal extrinsic afferents (Storm-Mathisen, 1972), so the GABAnergic system appears to originate from interneurons. The presence of an extrinsic cholinergic system and an intrinsic GABAnergic population allows an analysis of chemical changes induced by denervation through the use of both microchemical and histochemical techniques.

3.1. Studies with Adult Rats

In order to analyze axonal changes produced by lesions, we have removed the ipsilateral entorhinal cortex in adult rats and examined the consequence of the denervation on each of the remaining synaptic populations. Ablation of the entorhinal cortex deprives the outer three-fourths of the molecular layer of about 80% of its synapses (unpublished observations). The operation is simple and reproducible and because the projection to the granule cells is essentially unilateral the opposite side serves as a control.

Our work on possible sprouting in the dentate gyrus began with a histochemical study. As described above, the septal projection to the hippocampus contains AChE. We examined the distribution of this enzyme after lesions of the entorhinal cortex. The results were quite dramatic. A very dense band of AChE activity appeared in the outer molecular layer of the dentate gyrus, the target of the entorhinal projections (Fig. 2). That this band represented a change in the septal inputs was indicated by the fact that it was completely eliminated by a secondary septal lesion (Lynch *et al.*, 1972). The increased staining appeared within 4–5 days of the entorhinal lesion (Cotman *et al.*, 1973), a time course which is compatible for that reported for sprouting at the neuromuscular junction (see above) and for changes in the NE system in the septum after hippocampal lesions (Moore *et al.*, 1971). More recently, the intensification of AChE staining after an entorhinal lesion has been confirmed (Storm-Mathisen, 1974).

Direct microchemical analyses are a particularly critical adjunct to histochemical studies, alone, histochemical methods remain inconclusive because they

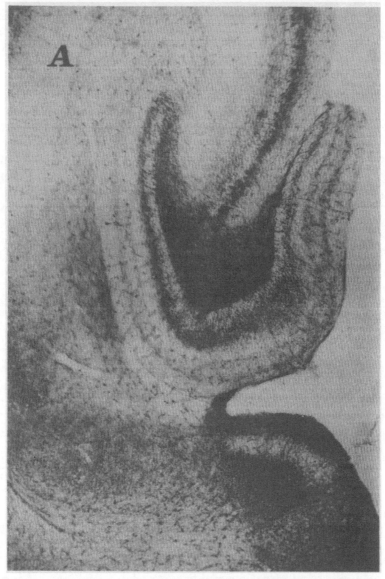

FIG. 2. Effects of a unilateral lesion of the entorhinal cortex on the acetylcholinesterase (AChE) staining of the dentate gyrus. The photomicrographs are of horizontal sections ipsilateral (panel B) and contralateral (panel A) to the lesion. Note the dense band of AChE staining (arrow) present in the dentate gyrus ipsilateral to the lesion which is missing from the contralateral side.

are subject to many potential artifacts possibly highlighting particular subfields (see Nadler *et al.*, 1973). The histochemical results are indicative of some change in the state of AChE, but the exact basis of this change requires extensive microchemical analysis.

Recently, microchemical analyses on discrete zones of the molecular layer have shown an increase in the specific activity of AChE and ChAc following deafferenta-

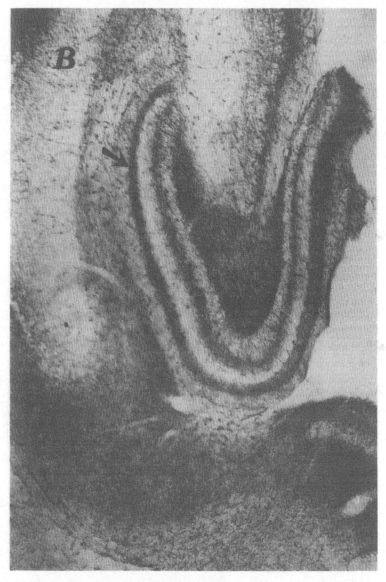

Fig. 2. (*Continued*)

tion (Storm-Mathisen, 1974). However, measurements of total activity have not been reported and these are required to exclude the possibility the specific activity increase results from a protein loss.

A reasonable explanation of these results is that a small population of AChE-containing septal fibers in the outer molecular layer sprout when that zone loses its primary afferent. It might be argued that some portion of the effect could be due to shrinkage of the deafferented zone, concentrating the AChE elements to a smaller area, and thereby producing a denser appearance. However, the effect is evident within 4–5 days after entorhinal lesion (Lynch *et al.*, 1975c) and at this time there is very little shrinkage of the molecular layer. A more serious consideration, similar to the one raised with regard to the histofluorescent evidence for sprouting, is that the activity of enzyme per terminal or axon is increasing rather than that the AChE-containing endings proliferate. To resolve these issues, an electron microscopic count must be made of the number of degenerating terminals caused by a septal lesion in the normal outer molecular layer and in that zone some months after an ipsilateral entorhinal lesion. An experiment of this type is currently in progress.

In our next study we examined the effects of the entorhinal lesion on the distribution of the commissural projections to the dentate gyrus. As described above, this system normally occupies the inner one-fourth of the molecular layer, adjacent to the zone of entorhinal innervation. In our studies we used the serial lesion paradigm followed by Liu and his associates; that is, we first placed a lesion in the entorhinal cortex, waited a sufficient time for the terminal degeneration products to be removed, and then plotted the distribution of the degeneration 4 days after lesions of the commissural projections. With this approach, we found that the terminal field of the commissural system expanded into the deafferented outer molecular layer, with the degree of growth being greater in young rats than in mature ones (Lynch *et al.*, 1973c; Zimmer, 1974). In adult rats, the portion of the molecular layer occupied by the commissural terminals increased 35–50 μm on the average, an expansion of about 40% (Lynch *et al.*, 1973b,c). Zimmer (1974) has recently replicated and extended the results obtained after lesions in immature rats.

Certain aspects of these findings should be emphasized. Silver degeneration (Blackstad, 1956; Raisman *et al.*, 1966; Lynch *et al.*, 1973b; Mosko *et al.*, 1973) and autoradiographic (Gottlieb and Cowan, 1973) methods, as well as experimental electron microscopy (Alksne *et al.*, 1966), are in agreement that the commissural termination field is restricted to the inner portion of the granule cell dendritic layer. In addition, preliminary electron microscopic analysis indicates that the 35-μm zone adjacent to the normal commissural field is deprived of most of its terminals by a complete entorhinal lesion, and with time is totally repopulated (Lynch *et al.*, 1975a). Therefore, it is hardly likely that the expansion is due to an "uncovering" of terminals which normally escape analysis. It should also be emphasized that these fibers are moving into "new" dendritic territory rather than increasing their density in a zone that they normally occupy. We feel that the increased size of the terminal field of this system represents a genuine case of collateral sprouting.

A critical question raised by these results is whether the postlesion growth results in the formation of permanent functional synapses. The genesis of new functional synapses by a foreign input may be difficult if the transmitters are dissimilar; receptors would have to be induced or modified. It is possible that functional contacts are not formed or that these are abortive, as is sometimes the case in the peripheral nervous system (Marotte and Mark, 1970a,b). The commissural growth is well suited to the analysis of its functional properties. In normal animals, stimulation of the opposite CA3 field produces a monosynaptic short-latency (1.5–2.5 ms) response whose polarity and magnitude depend on the depth of the electrode in the molecular layer (Lynch et al., 1973a; Deadwyler et al., 1975). At this site of active synaptic contact, the extracellular field potentials are negative, while outside this zone these field potentials are reversed and positive. This negativity represents the depolarization of the dendrite at the site of synaptic excitation and the formation of a current sink. Outside the active zone, the potentials are positive and represent currents leaving the activated cells (see Deadwyler et al., 1975, for a discussion). Thus the precise distribution of commissural afferents results in a comparable distribution of extracellular field potentials to stimulation of the CA3 field. These physiological properties of the commissural projections allow a test of the functional state of the new synapses. If the commissural axons establish functional synapses in the outer molecular layer, the distribution of extracellular responses to commissural stimulation would be expanded accordingly.

After an entorhinal lesion, analysis of the extracellular field potentials at various levels in the molecular layer in response to contralateral CA3 stimulation clearly illustrates that commissural stimulation activates a larger section of the granule cell dendrite than it does in the normal animals. In twelve control rats the negativity was never found to extend beyond 150 microns, whereas in all the lesioned rats tested the negativity could be recorded more than 200 microns above the granule cells. These changes appeared permanent. The latency to onset and waveform of the commissural potential appeared indistinguishable from normal animals. Thus, the combined anatomical and electrophysiological data indicate a growth of new functional commissural synapses after an entorhinal lesion (West et al., 1975). We have also measured the time required for onset of these changes and have found the spread negativity is first observed nine days after an entorhinal lesion (West et al., 1975).

The commissural fibers share the inner molecular layer with the recently discovered associational projections originating somewhere in the regio inferior of the hippocampus (Zimmer, 1970; Gottlieb and Cowan, 1973). We have completed a study of the distribution of this system in normal rats and in animals with entorhinal lesions using the horseradish peroxidase (HRP) histochemical method for tracing anterograde projections (Lynch et al., 1975a). Enzyme was injected by pressure through very fine glass pipettes into the dentate hilus region of the hippocampus and the rats were given 18–72 h of survival before sacrifice. The brains were then removed, postfixed, and stained for HRP by the methods described in Lynch et al. (1974a). In agreement with others (Zimmer, 1970), we found that the associational system originates deep in the hilus and projects at a very acute angle across the septotemporal axis of the hippocampus. More pertinent to the present discussion, our experiments demonstrated that the associational system occupies the same proportion of the inner molecular layer as the commissural fibers, and that it shows the same degree of sprouting as that projection. These systems together probably account for the repopulation of the 30-μm zone overlying the normal commissural/associational fields that was observed in the electron microscopic analysis. (see above) Zimmer (1973), using the Fink–Heimer technique, found much greater growth of this associational system after lesions in neonatal rats.

Thus, after removal of the major extrinsic input to the granule cells, the extrinsic afferent from the septum undergoes a change in its transmitter-related enzymes which may signify growth, and the two intrinsic hippocampal afferents, the commissural/associational systems, expand their terminal fields and invade the adjacent denervated entorhinal zone.

In addition to these systems, there is another fiber input to the hippocampus, originating in the contralateral entorhinal cortex. This afferent, the crossed temporoammonic system (CTA), has its major termination in the apical dendrites of the regio superior and extends the length of the hippocampus. In addition, a very minor and previously undescribed projection appears to exist to the outer molecular layer of the dentate gyrus exclusively at the septal end of the hippocampus (Goldowitz et al., 1975).

We tested the possibility that after a unilateral entorhinal lesion the CTA system would invade portions of the denervated dentate gyrus where it is normally undetectable or proliferate in areas where it is present as a very minor projection (Steward et al., 1973, 1974). We use the autoradiographic method for analyzing potential reorganization of the projection instead of a secondary lesion or Fink–Heimer analysis in order to avoid complications due to residual, long-lasting degeneration from the primary lesion. Radioactive leucine or proline was injected into the remaining entorhinal cortex 20–40 days after a unilateral entorhinal lesion. The brain was removed and processed for autoradiography 3–6 days after the injection (Steward et al., 1974). We found that the CTA invades or proliferates within the denervated zone. Our finding that a minor, previously undetected projection exists to the molecular layer necessitates that we reconsider our initial interpretation that the CTA fibers must grow from the regio superior to reach the denervated outer molecular layer. At present, we cannot distinguish between a shrinkage effect, terminal proliferation, and/or a partial invasion by intact fibers. Electron microscopic analysis of the relative number of CTA synapses present before and after a lesion and light microscopic tracing of the fibers of origin are required; these studies are currently in progress.

With the exception of the interneurons, we have now analyzed every major afferent to the dentate gyrus and its response to an entorhinal lesion. The molecular layer also contains a population of GABAnergic terminals. Their response can be assessed in part by determining the levels of glutamic acid decarboxylase (GAD), the enzyme responsible for synthesis of GABA. GAD activity is confined mainly to the synaptic bouton portion of GABAnergic neurons (Nadler et al., 1975a), and in regional studies is closely related with that of the transmitter (Baxter, 1970). Thus changes in GAD activity there may signify changes in the number or size of GABAnergic terminals.

GAD activity, measured by microchemical methods, increases in the deafferented outer molecular layer relative to the activity measured in the contralateral dentate gyrus. The specific activity increases 63% and the total activity 45%, so that although a loss of protein affects the activity the change does not appear due solely to protein loss. At 90 days or more after the operation, the difference in specific activity is still significant, although reduced. We suggest that additional GABAnergic boutons may be formed or that an increase in GAD within existing connections may be transynaptically induced.

In summary, we have found marked changes in each of the afferents to the molecular layer of the dentate gyrus when its major input is removed in adult rats:

1. There is an increase in AChE-staining intensity in the afferents originating in the septum and terminating in the outer molecular layer.
2. The commissural fibers invade the aspects of the deafferented territory closest to their normal terminal field, and thereby enlarge that field to about 140% of normal.
3. The associational projections also expand their terminal field in a manner identical to that seen in the commissural system.
4. The crossed temporoammonic tract appears to increase the density of its innervation of the deafferented outer molecular layer.
5. The GABAnergic interneurons show an increase in glutamic acid decarboxylase activity.

We feel that the commissural/associational system shows collateral sprouting and that this is also a plausible but not an unequivocal explanation for the septal and CTA effects. In the commissural system, we have obtained evidence that the growth results in formation of functional terminals and that these begin to operate at about 9 days after lesion.

3.2. Studies with Immature Rats

There have been several studies showing that considerable sprouting takes place in the dentate gyrus following lesions of the entorhinal cortex in immature rats. Histochemical studies have shown that AChE activity is greatly increased in a narrow band at the top of the molecular layer (Cotman et al., 1973; Zimmer, 1973), and electron microscopic work has shown a marked increase in AChE-positive terminals in this area (Cotman et al., 1973). Microchemical analyses have confirmed the increase in AChE activity and shown that it is accompanied by a rise in ChAc (Nadler et al., 1973). The terminal zone of the hippocampal commissural and associational systems expands far beyond its normal field in the inner molecular layer and occupies almost the entire height of the granule cells' dendritic trees (Lynch et al., 1973c; Zimmer, 1973). The growth of these afferents appears to stop at about the level of the intensified septal projection so that, as in normal animals and in animals lesioned as adults, these two systems occupy adjacent layers. It is important to note that the commissural stimulation elicits a response across a much greater proportion of the molecular layer after lesions in neonatal rats than after similar damage in adults (Lynch et al., 1973a). Thus the degree of physiological change is related to the degree of anatomical growth. This correlation provides evidence that the anatomical growth is causing the spread of the commissural potentials. The projection from the contralateral entorhinal cortex becomes hyperdeveloped in the molecular layer and seems largely excluded from the commissural/associational zone, although some overlap exists (Steward *et al.*, 1973).

3.3. *Negative Results*

In striking contrast to the plasticity observed after a unilateral entorhinal lesion in young and adult rats, deafferentation of the inner molecular layer by transection of the commissural and associational projections does not cause any obvious changes in the distribution of the immediately adjacent ipsilateral entorhinal projections. Specifically, we eliminated the commissural (by contralateral hippocampectomy) or commissural and associational (by longitudinal transection of the hippocampus) projections to the granule cells in 11-day-old rats and analyzed the effects of this on the distribution of the entorhinal fibers to the dentate gyrus. In this way, we tested the hypothesis that the entorhinal fibers which innervate the outer portion of the granule cell dendrites would sprout into the deafferented inner molecular layer. Somewhat to our surprise, we found that the entorhinal projections gave no indication of growth, despite the fact that they were immediately adjacent to dendritic fields which had lost a large part of their afferent innervation (Lynch *et al.*, 1974*b*). We have recently repeated these studies using longitudinal transections of the hippocampus in adult rats and the autoradiographic technique to trace the entorhinal projections and again found no evidence for a strong invasion by this fiber system (Goldowitz *et al.*, 1975).

4. *Status of Sprouting in the Brain*

In the above sections, we have tried to give an overview of the efforts to detect collateral sprouting in the adult brain, with particular emphasis on the hippocampus. It can be seen that many techniques have been used, several brain areas studied, and a diversity of results obtained. Table 1 summarizes some of these points; before discussing these, we would hasten to note that the table is not complete, and in particular does not contain reference to developmental studies or experiments in which regeneration was involved (e.g., Bernstein and Bernstein, 1973).

The table catagorizes the studies according to method employed and type of result obtained. From the point of view of technique, electron microscopy would seem to be the strongest approach to the study of sprouting. It is disappointing, then, to find that ultrastructural studies have demonstrated alterations within a terminal field but that little evidence for terminal proliferation or axon growth into new territory has been reported.

The light microscopic techniques used to study sprouting all possess certain drawbacks. In studies showing increased density of innervation, it is difficult to control satisfactorily for the contributions of shrinkage. Even more difficult to evaluate is

TABLE 1

Technique	Result		
	Reoccupancy of sites	Increased density of terminals	Expansion of terminal field
Degeneration (silver)		Horel and Goodman	Liu et al. Goldberger Zimmer Lynch et al. Goodman et al.
Transport (autoradiography) (horseradish peroxidase)			Goldberger et al. Lynch et al. Lynch et al.
Histochemistry		Moore et al. Stenevi et al. Lynch et al.	
Electron microscopy	Lund and Lund Raisman Raisman and Field Westrum and Black		

the possibility that an already existing population of terminals normally not detected is revealed to the neuroanatomical or histochemical method.

For these reasons, light microscopic experiments showing an expansion of a terminal field into areas in which it is not normally found or is very sparse are more convincing demonstrations of some form of sprouting. There is good evidence arrived at in different laboratories using different techniques (Table 1) that this type of growth takes place in the spinal cord. In the brain, two types of transport methods, as well as degeneration techniques, indicate that expansion of terminal fields occurs in the partially deafferented dentate gyrus. Experiments using the Fink–Heimer method suggest that this effect may also occur in the superior colliculus. While not shown in the table, there are also several well-documented cases in which an intact afferent does not invade a neighboring region which has been deprived of its dominant input.

Taken together, these results suggest the following conclusions:

1. Both light and electron microscopic studies indicate that intact afferents do react to removal of their neighbors, and ultrastructural evidence suggests that vacated postsynaptic sites may be reoccupied.
2. The evidence that intact fibers sprout new branches is much less compelling. While collateral growth seems to be the most parsimonious explanation of the available data, it is only in the hippocampus and spinal cord that several lines of evidence point to this conclusion.

5. Hypotheses Regarding Mechanisms Controlling Sprouting in the Hippocampus

From the above discussion, we feel that it is apparent that some form of growth often takes place in the intact afferents of a partially denervated brain structure and that in the hippocampus this includes the occurrence of axonal growth. It is appropriate now to turn to a discussion of the possible mechanisms responsible for these effects. Logically this analysis might be broken into two parts: (1) what events initiate or release the growth, and (2) what factors are responsible for the regulation and distribution of the growing elements. This analytical framework is hardly new; something akin to it was used by Edds (1953) in his farsighted review of collateral sprouting at the neuromuscular junction. It is also inherent in Ramón y Cajal's classical analysis of regeneration of peripheral nerves.

In the following section, we would first like to consider how the growth process might be initiated and then conclude with a discussion of the regulation of the spatial organization of sprouted afferents in the hippocampus. We will examine and compare similar phenomena as they operate in the peripheral nervous system and argue that certain analogies exist which help explain the mechanisms of axon sprouting in the hippocampus.

5.1. Initiation

Three positions have been advanced with regard to initiation of sprouting. Edds (1953) and Hoffman (1951) proposed that the axonal degeneration, or more likely the Schwann cell activity associated with it, released a trophic substance which acted on the intact fibers, causing them to emit new branches. Hoffman succeeded in extracting with ether a substance from normal or denervated rabbit muscle, ox brain white matter, or egg yolk which produced a 30–50% increase in ultraterminal branches 3–4 days after injection. This stimulating agent, which he labeled "neurocletin," was characterized as a moderately unsaturated fatty acid. It has been hypothesized that neurocletin is released from disintigrating myelin or perhaps Schwann cells, diffuses to adjacent intact axons, and initiates axon sprouting (Hoffman, 1951; Hoffman and Springell, 1951).

The idea that degenerating matter releases neurotrophic substances dates back to the beginning of this century and was the subject of much research by Ramón y Cajal, Lugaro, and others (see Ramón y Cajal, 1928, for a review of the early literature). Ramón y Cajal believed that degeneration caused a division of the Schwann cells and that these mitotic elements released a diffusible factor which directed or, more accurately, facilitated axon regeneration. It might be mentioned parenthetically that Ramón y Cajal felt that the absence of Schwann cells in the brain accounted for the failure of central axons to regenerate. However, as noted in the Introduction, neither Ramón y Cajal nor Lugaro felt that the neurotropic substance was effective in influencing intact fibers. Hoffman, of course, reached a different conclusion.

A second mechanism postulated to be responsible for the initiation of sprouting is the removal of substances which inhibit collateral growth. This could be the elimination of a kind of contact inhibition or the loss of a diffusible substance normally emitted by intact fibers. Aguilar *et al.* (1973) have very recently shown that blockade of axoplasmic flow in one nerve will produce electrophysiological evidence of sprouting in surrounding fibers. This certainly suggests that intact fibers are releasing compounds which suppress growth in their neighbors. Lugaro performed experiments on the regenerating dorsal roots and felt that they were actively repelled from entering the spinal cord, and used this as an argument for the existence of negative neurotrophic factors; Ramón y Cajal, however, did not share this interpretation of Lugaro's experiments.

Related to both these ideas is the hypothesis that intact axons are constantly sprouting within their environments and this process is simply accelerated by removal of neighboring inputs. Burgess *et al.* (1973, 1974) have discovered that the sural nerve, which innervates the hind leg skin of the adult cat, changes the position of its terminal elements over an extended period of time. This might indicate continuous turnover of elements in the nerve's receptor field. In the central nervous system, Sotelo and Palay (1971) have proposed that synaptic turnover is part of a continuous remodeling process. If intact axons are constantly sprouting into their environments, this process might be simply accelerated by removal of neighboring inputs. The nature of the accelerating agent could, of course, be either the release of stimulants by the degeneration or the loss of inhibitors, but the emphasis in this explanation is placed on the idea that sprouting is an exaggeration of a normal process (e.g., Goodman *et al.*, 1973).

There are no experimental data which might cause us to favor any of the above hypotheses as an explanation for the anatomical changes we have observed in the dentate gyrus. However, it should be emphasized that denervation in the hippocampus initiates a variety of drastic morphological changes and that several of these are not implausible sources for triggering compounds. Within hours of the lesion, the astroglia cells in the zone of degenerating terminals undergo hyperdevelopment and begin the process of phagocytosis (Westrum and Black, 1971; unpublished observations) (Fig. 3). We have recently observed that 3–4 days after lesion the outer molecular layer becomes occupied by a dense population of cells corresponding to del Rio Hortega's "microglia" (Lynch *et al.*, 1975c). These peculiar cells are found only infrequently in the normal dentate gyrus and their sudden appearance in the deafferented zone precedes the development of sprouting by the commissural and associational systems. It is tempting to hypothesize that growth by intact systems represents a late step in a chain of events and is initiated by the link which precedes it, in this case perhaps the "microglial" cell. As we have noted above, many earlier workers assigned the origins of neurotrophic substances in the peripheral nervous system to mitotic and phagocytic glial elements.

5.2. Principles for the Reorganization of Terminal Fields

Once the sprouting process has been initiated, it becomes pertinent to ask what mechanisms regulate or direct the growing elements. Several suggestions have been made in the literature which bear on this question. It seems quite likely that if more

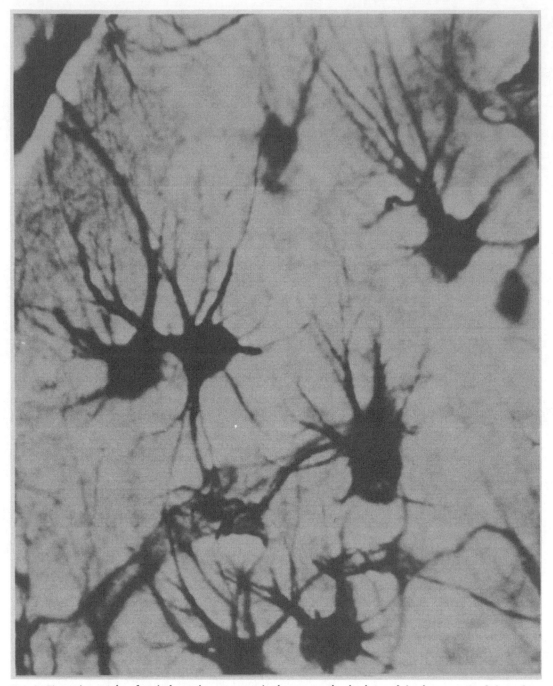

FIG. 3. Photomicrographs of typical reactive astrocytes in the outer molecular layer of the dentate gyrus 5 days after a lesion of the ipsilateral entorhinal cortex. The thick, numerous processes radiating from the cell bodies are not seen in this area contralateral to the lesion. Note the relationship of the glial processes to the blood vessel in the lower left hand corner of the figure. del Rio Hortega's method; initial magnification ×308.

than one afferent were to sprout into a denervated site, these different systems would interact with each other to produce a new organization. For example, sprouting afferents might compete with each other for synaptic sites such that their relative success determines the final organization of the area. It is also possible that the postsynaptic cell plays a critical role in regulating growth, accepting innervation by some afferents but not from others. For example, the formation of functional synapses could depend on the biochemical match of pre- and postsynaptic membranes. Both of these ideas have also been used as explanations for the failure of certain intact systems to gain territory made available by the removal of normal inputs.

In the section that follows, we will present new findings which indicate, first, that sprouting is a regulated process and, second, that the key element in this regulation is not an interaction between growing elements or a property of the postsynaptic cell but rather the orientation of intact fibers to the degenerating elements.

As described above, removal of the primary input to the dentate gyrus results in marked changes in its four remaining afferents, and this probably includes collateral sprouting. It appears that in this situation growth is a nonspecific phenomenon, occurring in every input located in or adjacent to the deafferented territory. However, these changes do not eliminate the stereotyped laminated pattern of dentate gyrus afferents, but instead simply change the location of the borders between the newly expanded inputs.

In order to develop the arguments for the above conclusion, it is necessary that we first provide further data on the size and location of the terminal fields generated by dentate gyrus afferents in the normal rat. Since several techniques have been used to study these projections and because these cause different amounts of tissue shrinkage, it is inappropriate to compare absolute values of the size of terminal fields obtained with them. To obviate this problem, we have chosen to compare the percentage of the molecular layer occupied by each afferent rather than their absolute values. Measurements were made at three sites on three standard coronal planes through the dorsal hippocampus (Fig. 4). Figure 5 summarizes our calculations for the four major dentate gyrus afferents. It can be seen that the commissural and associational projections occupy the inner 27% of the molecular layer, while the ipsilateral entorhinal inputs innervate the remaining 63%. Note the close agreement between the results obtained from studies using the autoradiographic, Fink–Heimer,

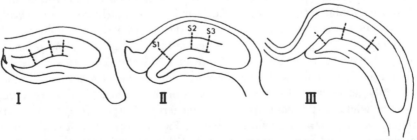

FIG. 4. Sites at which the distribution of the afferents to the dentate gyrus were measured. Note that three points (S1, S2, S3) on three coronal planes were used.

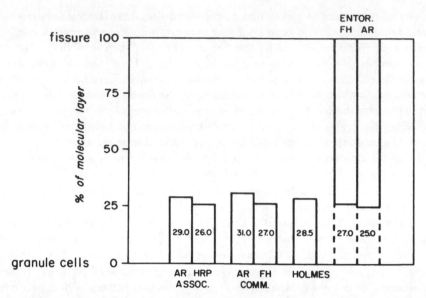

FIG. 5. Percentage of the molecular layer occupied by the various dentate gyrus afferents as measured at the sites illustrated in the preceding figure. Three projections were assessed: (1) associational (assoc.), (2) commissural (comm.), and (3) entorhinal (entor.). Several techniques were employed: autoradiography (AR), horseradish peroxidase (HRP), the Fink–Heimer method (FH), and the Holmes technique.

and horseradish peroxidase methods for tracing connections. Also shown in the figure are measurements of the extent of the axonal plexus (revealed by the Holmes silver stain) generated by the commissural and associational projections. This will be discussed presently, but for the moment note that it agrees well with the values obtained for the associational and commissural terminal fields with the use of the other techniques.

From this study, it is apparent that the afferents of the normal dentate gyrus occupy one of two zones, separated by a line running at approximately 27% of the distance from the granule cells to the hippocampal fissure.

In order to measure the distribution of the afferents which invade the outer zone after removal of its dominant input (i.e., the entorhinal cortex), it was necessary to deal with the shrinkage problem discussed above. A bilateral comparison of the width of the two molecular layers 20 days after a unilateral entorhinal lesion reveals that the deafferented side has shrunk to 80% of the contralateral, nondenervated side. Measurements of the percent of the molecular layer occupied by the remaining or invading afferents to the deafferented side would be distorted by this reduced width. To control for this, we measured the absolute width of the terminal fields of inputs to the partially denervated dentate gyrus but calculated their percentage occupancy against the width of the contralateral, nondenervated side. Using this control for shrinkage, we arrived at the values shown in Fig. 6. From this, it can be seen that the associational and commissural terminal fields expand outward to 37% of the distance from the granule cells to the hippocampal fissure (44% uncorrected for shrinkage); in other words, they expand their terminal fields to about 140% of the normal values (a distance of about 40 μm). Note also that the inner plexus of axons revealed by the Holmes stain expands by about the same degree.

The observation that the outward expansion of the commissural and associational projections stops after a gain of only 30–50 μm could be explained by assuming an interaction or competition between the sprouting afferents. Specifically, it is possible that the commissural and associational fibers sprout outward until they encounter the invading or expanding population of crossed entorhinal inputs. However, both commissural and associational systems show exactly the same degree of growth after bilateral removal of the entorhinal cortex (which eliminates the crossed temporoammonic tract) as they do after unilateral lesions of the entorhinal cortex (Lynch *et al.*, 1975*a*). Clearly, then, the crossed temporoammonic fibers are not suppressing the growth of the fibers from the inner molecular layer into the outer molecular layer.

Another test of the interaction hypothesis was performed on the septal projection to the dentate gyrus, which, as described above, undergoes considerable changes after entorhinal lesion in both adult and immature animals. However, the hyperdeveloped AChE band is in no way influenced by any of several combinations of lesions (unilateral entorhinal and commissural, bilateral entorhinal, or bilateral entorhinal and commissural) which remove afferents that might compete for the space made available by the initial ipsilateral entorhinal lesion (Nadler *et al.*, 1975*a*). Again, it seems unlikely that a simple version of the interaction hypothesis will serve to explain the laminated reorganization seen in the dentate gyrus after the entorhinal lesion.

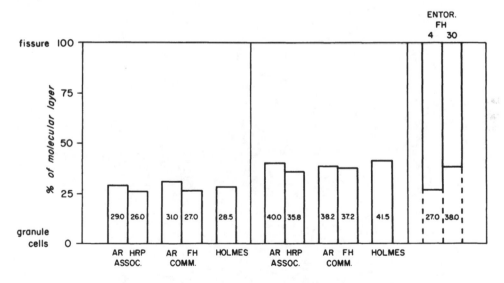

FIG. 6. Effect of a unilateral lesion of the entorhinal cortex on the distribution of afferents of the dentate gyrus molecular layer. Abbreviations are as in the previous figure. The left-hand panel illustrates the percentage occupancy of commissural and associational projections in the normal rat, while the middle panel gives these values following the lesion. The small panel to the far right shows the distribution of degeneration products produced by the lesion 4 and 30 days following the lesion.

What then does serve as the limiting barrier for axonal growth in the dentate gyrus? In the following section, we will describe neuroanatomical data which lead us to offer the following hypothesis: *the extent of collateral sprouting is greatly influenced by the relationship of the growing fibers to those degenerating as a result of the lesions.* Specifically, growth by the commissural and associational systems seems to be blocked precisely at the level of the degenerating entorhinal fibers.

In the molecular layer of the normal rat, fibers are organized into two layers separated by a zone of terminal and paraterminal elements. Both Holmes and unsuppressed Nauta techniques show that the molecular layer contains two dense axonal plexuses, one of which occupies the inner one-fourth of the layer and an outer plexus which begins half way up the layer and continues to the fissure. The location of the inner plexus corresponds exactly to the terminal fields of the commissural and associational projections and therefore is probably generated by their axons: the outer plexus disappears after entorhinal lesions and must therefore be composed of the axons of the temporoammonic projection. Between the two plexuses is a "clear" zone which contains only very fine-caliber axons. A comparison of the distribution of entorhinal terminal field with the location of the fiber plexuses shown by these silver stains indicates that this clear zone is occupied by the terminals of the entorhinal projection. Figure 7 summarizes this interpretation of the organization of the molecular layer.

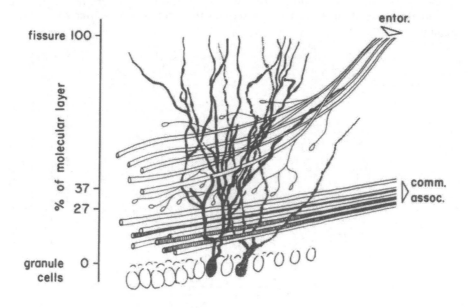

FIG. 7. Semischematic drawing of an interpretation of the distribution of three fiber projections to the dentate gyrus molecular layer. Note that terminal and paraterminal elements from the entorhinal (entor.) fibers occupy the zone between the two fiber plexuses; the terminals of the commissural (comm.) and associational (assoc.) axons are not shown. The two cells are reconstructions from material stained with the Golgi technique.

It is widely held that the elements impregnated by the Fink–Heimer method shortly after a lesion are predominantly degenerating terminals, while degenerating axons are the primary elements detected weeks after a lesion. This distinction suggested a means of checking the conclusions drawn from the Holmes method about the location of the entorhinal axons and terminals in the molecular layer. Accordingly, the distribution and character of the post-entorhinal-lesion degeneration were analyzed in the molecular layer of rats with different survival times. The results of this experiment accord well with our analysis of the Holmes material. At 4 days after lesion, fine, randomly distributed degeneration products occupy the molecular layer from the fissure to a line approximately 27% of the distance from the granule cells to the fissure. By 20 days after lesion, the degeneration has "retreated" up the molecular layer to occupy only the outermost 62% of this zone (see Fig. 6). Furthermore, the character of the degeneration products is different at this time in that it is coarse and arrayed into rows; it appears axonal.

These results point to the same conclusions as did the study of the Holmes material. The entorhinal axons occupy a plexus occupying the outer 60% of the granule cell dendrites but generate a terminal field which includes the outer 73%. The discrepancy of the two figures presumably accounts for the "clear" zone seen in the Holmes stain.

Measurements of the commissural/associational system suggest that these projections expand to occupy the "clear" zone but stop their outward movement at the boundary of the degenerating entorhinal axons. That is, the location of the outer plexus seen in sections treated with the Holmes method and the inner boundary of the persistent axonal degeneration detected with the Fink–Heimer method correspond with the outward border of the expanded commissural/associational systems. This point can be best appreciated by examination of Fig. 6, which summarizes our measurements.

Taken together, these observations suggest that the axons of the commissural and associational system are able to sprout quite readily through the zone of rapidly eliminated entorhinal terminal and paraterminal elements (the "clear" zone in silver stains) but that their outward movement is stopped when they encounter the fascicles of degenerating myelinated entorhinal axons.

The hypothesis that degenerating axons serve as barriers to sprouting might also explain some of the cases in which an intact afferent failed to invade a neighboring deafferented territory. That is, this hypothesis predicts that expansion of a terminal field will occur only when the remaining projection is bordered by terminal fields of a degenerating input; if it were to lie adjacent to the axons of the lost afferent, successful sprouting would not occur.

As an example, consider the failure of the entorhinal projections to expand into the inner molecular layer after that zone was deafferented by a transection of the commissural and associational projections. Figure 7 indicated that the fibers of the commissural/associational plexus bordered a terminal field generated by the entorhinal projections; conversely, the entorhinal endings are immediately adjacent to the axons of the commissural/associational system. If degenerating axons do serve as barriers to sprouting, then we would expect sprouting in one direction (into the terminal field of the entorhinal projection) but not in the other (across the axons of the commissural/associational plexus); this result was in fact obtained.

Whether this argument applies to other experiments in which no sprouting was obtained can be tested only by establishing the distributions of the fibers and terminals of the degenerating and intact afferents. Even if the hypothesis possesses some validity, it is presumably applicable only in situations in which afferents laminate themselves; in other circumstances, other factors presumably serve as regulatory agents.

A legitimate question at this point concerns the relationship of the growth of the crossed temporoammonic tract in the deafferented outer molecular layer to the hypothesis of degenerating axons as barriers to sprouting. The answer to this can be achieved only with a conclusive resolution to the problem of the extent to which the crossed entorhinal projections normally make contact with the dentate gyrus. As described above, we have data which indicate the existence of a very slight, previously undetected projection from the crossed system to the rostralmost part of the dentate gyrus. If this projection is coextensive in distribution with the much denser crossed entorhinal connection found after removal of the ipsilateral entorhinal cortex, then we would postulate that a small collection of terminals simply proliferates in a zone it already occupies. That is, the degenerating axons of the ipsilateral projections do not serve as barriers in this case because the system exhibiting growth is already present in the dendritic territory deafferented by the lesion. If, however, the crossed projections are found to have expanded their septotemporal distribution, a modification of the axon-barrier hypothesis will clearly be required.

In summary, in the above sections we have (1) suggested that sprouting in the hippocampus is a process which is triggered possibly by an expanding population of glial, perhaps "microglial," cells and (2) presented evidence that the extent of growth is regulated by some element associated with fascicles of degenerating axons.

These are admittedly tentative hypotheses based purely on correlative evidence. The involvement of glia is based on analogous functions they are believed to serve in the peripheral nervous system and on their proliferation shortly before the onset of sprouting in the hippocampus. The idea that factors associated with degenerating axons play a regulatory role is indicated by the observation that these elements are located at a time and place to provide a boundary to the expanded axonal and terminal fields.

We hardly need to emphasize, first, that further work will be required to buttress these anatomical studies, particularly with the electron microscope, and, second, that alternative explanations can be offered for the results so far obtained. With regard to this second point, several authors have suggested that sprouting may be the uncovering of developmental processes still active in the adult brain (e.g., Guillery, 1972; Goodman *et al.*, 1973; Sotelo and Palay, 1971). Axonal systems might be continuously growing and therefore require no triggering event; sprouting would then be the removal of obstacles to growth. Similarly, the process might well be regulated by the very mechanisms which were responsible for the segregation of afferents in the developing dentate gyrus. Growing fibers might establish connections with dendritic fields in which appropriate "matching" of recognition molecules of terminals and spines were possible; in other areas, sprouts would be rejected.

Two lines of reasoning have led us to the types of hypotheses we have proposed in this chapter rather than to developmental models. First, it must be recognized that deafferented dendritic territories do contain elements and chemistries which are unlikely to exist in normal developing or adult brain regions. We find it difficult to accept the idea that the changes in astroglia and oligodendroglia which accompany degenerating afferents, to say nothing of the exploding population of microglia, do not interact with the fiber systems that remain connected to the partially deafferented cell. Deafferentation creates a new milieu. Second, we have been greatly influenced in our hypothesizing by the careful descriptions and analyses of collateral sprouting in the peripheral nervous system. It must be admitted that the effects seen there need not be comparable to those found in the central nervous system, but it would seem unwise to ignore the results obtained in these more easily studied situations. Several authors have argued that trophic compounds are involved in sprouting at the neuromuscular junction and that these are derived from the Schwann cells. While the evidence on this point is hardly conclusive, it is certainly suggestive. It is more certain that the Schwann cell plays a crucial role in directing the growing collateral branches, a process which certainly is not the resurrection of a developmental mechanism. It appears then that the formation of aberrant connections at the neuromuscular junction and in the superior cervical ganglion is dominated by the abnormal situation created by degeneration, and we propose that this generality will hold true for the brain as well.

6. Conclusion and Summary

In conclusion, we have analyzed the types of synaptic changes in intact neurons induced by lesions. In the Introduction, we raised a series of questions on the existence, generality, functionality, and timing of sprouting in the brain.

It is abundantly clear that intact neurons are often, but not always, affected if they reside in or near fields of deafferentation (see Table 1 and Section 4). Essentially all responses in neighboring neurons observed so far are consistent with axon collateral or paraterminal sprouting. We find, however, that the evidence is sufficient to demonstrate sprouting in only a few cases. At present, the clearest evidence that sprouting exists in the brain is the demonstration by electron microscopy that new synaptic boutons repopulate a previously denervated area. As described in Section 3, we have observed the loss and partial reappearance of synapses in the molecular layer of the dentate gyrus. Raisman and Field had earlier noted a drastic loss of synapses in the septal nucleus and their reappearance. In both these situations, sprouting must have occurred since synapses are lost and replaced.

At present, there is really very little evidence which identifies the *type* of sprouting response. Collateral sprouting refers to a phenomenon in which existing axons grow new branches, so that it is necessary to identify axonal growth in order to establish this effect. In the dentate gyrus, it is likely that axon collaterals of the commissural/associational system have grown and invaded the denervated zone, but further evidence on this is desirable.

Other data on changes in denervated zones based on histofluorescence, histochemical, microchemical, or autoradiographic methods is at present compatible with the idea that axon sprouting has occurred, but these techniques cannot discriminate between changes in preexisting systems and the growth of new ones. Metabolic changes in preexisting fibers or terminals might give rise to the same appearance as would the addition of new fibers or terminals. Results with any method are misleading and uninterpretable if account is not taken of shrinkage in the field of deafferentation, which, by condensing a terminal field, gives the impression of terminal proliferation. In addition, it is essential to be certain that the synapses or fibers under study do not preexist and are not uncovered by the treatments. Similarly, neurophysiological results which indicate a postlesion change in the remaining circuitry require careful interpretation. Independent of anatomical evidence, it is difficult if not impossible to distinguish among sprouting, supersensitivity, or increases in synaptic efficacy in preexisting synaptic boutons.

While it is clear that some form of sprouting occurs in several systems in the brain, it is also clear that this is not a general response to denervation. The effect is absent or very limited in the dorsal root after removal of the spinal root of the fifth nerve, in the lateral geniculate after unilateral eye enucleation, and in the projections from the entorhinal cortex after commissural/associational lesions. Interestingly, the absence of a sprouting response is seen in some systems in both developing and adult animals. Systems which sprout in developing animals may or may not do so in adults, but the reciprocal effect, presence of sprouting in adults and absence in developing animals, has not been shown. Studies purporting to demonstrate the absence of sprouting can be criticized on the basis of the possibility that some sprouting may have occurred but it was too small to detect or of only transient existence so that the experimental design failed to measure it.

An understanding of the mechanisms of the neuronal response to lesions is at a very early stage and the mechanisms are particularly difficult to analyze because the exact nature of the response is not always clear. In the case of presumed axon collateral sprouting, we examined the mechanisms believed to operate and have relied on analogous mechanisms involved in the peripheral nervous system. It is at least plausible to hypothesize that similar mechanisms exist in the central nervous system. The growth response in the CNS was divided into two stages: (1) an initiation state and (2) a stage of control of fiber growth. We have suggested the possibility that glial cells form the first and essential link in a series of events which ultimately produce sprouting in the brain; specifically, we hypothesized that degeneration attracts and stimulates glial activity, which in turn initiates neuronal growth. New axon branches are then guided by their relationship to those which are degenerating.

The analysis of neuronal changes in deafferented brain sites has provided new clues about the anatomical events underlying recovery from brain damage; in the future, it should also generate clues about the nature of plasticity in the normal nervous system. The finding that, under pathological conditions, adult brains are capable of forming new synaptic connections and can probably grow new axonal branches as well leads naturally to questions about the occurrence of effects of these sorts during normal brain operation. It is hoped that further analysis of the mechanisms of postlesion "sprouting" in model systems such as the hippocampus will ultimately provide means of experimentally investigating suggestions of this type.

7. References

AGUILAR, C. E., BISBY, M. A., COOPER, E., AND DIAMOND, J. Evidence that axoplasmic transport of trophic factors is involved in the regulation of peripheral nerve fields in salmanders. *Journal of Physiology (London)*, 1973, **234,** 449–464.

ALKSNE, J. F., BLACKSTAD, T. W., WALBERG, F., AND WHITE, L. E., JR. Electron microscopy of axon degeneration: A valuable tool in experimental neuroanatomy. *Ergebnisse der Anatomie und Entwicklungsgeschichte*, 1966, **39,** 1–31.

BAXTER, C. The nature of aminobutyric acid. In A. Lajtha (Ed.), *Handbook of neurochemistry*. Vol. 3. New York: Plenum Press, 1970, pp. 289–353.

BERNSTEIN, J. J., AND BERNSTEIN, M. E. Neuronal alteration ad reinnervation following axonal regeneration and sprouting in mammalian spinal cord. *Brain, Behavior and Evolution*, 1973, **8,** 135–161.

BLACKSTAD, T. W. Commissural connections of the hippocampal region in the rat with special reference to their mode of termination. *Journal of Comparative Neurology*, 1956, **105,** 417–537.

BURGESS, P. R., 1973. Specific regeneration of cutaneous fibers in the cat. *Journal of Neurophysiology 36,* 101–114.

COTMAN, C. W., MATTHEWS, D. A., TAYLOR, D., AND LYNCH, G. Synaptic rearrangement in the dentate gyrus: Histochemical evidence of adjustments after lesions in immature and adult rats. *Proceedings of the National Academy of Sciences U.S.A.*, 1973, **70,** 3473–3477.

DEADWYLER, S., WEST, J., LYNCH, G., AND COTMAN, C. W. A neurophysiological analysis of the commissural projection to the dentate gyrus of the rat. *Journal of Neurophysiology*, 1975, in press.

EDDS, M. V., JR. Collateral regeneration of residual motor axons in partially denervated muscles. *Journal of Experimental Zoology*, 1950, **113,** 517–552.

EDDS, M. V., JR. Collateral nerve regeneration. *Quarterly Review of Biology*, 1953, **28,** 260–276.

EXNER, S. Notiz zu der Frage von der Faservertheilung mehrer Nerven in Muskel. *Pfluegers Archiv für die Gesamte Physiologie*, 1885, **36,** 572–576.

FONNUM, F. Topographical and subcellular localization of choline acetyltransferase in rat hippocampal region. *Journal of Neurochemistry*, 1970, **17,** 1029–1037.

GOLDBERGER, M. E. Recovery of movement after CNS lesions in monkeys. In D. Stein, J. Rosen, and N. Butters (Eds.), *Plasticity and recovery of function in the CNS*. New York: Academic Press, 1974, pp. 265–339.

GOLDOWITZ, D., WHITE, W. F., STEWARD, O., COTMAN, C., AND LYNCH, G. Anatomical evidence for a projection from the entorhinal cortex to the contralateral dentate gyrus of the rat. *Experimental Neurology*, 1975, in press.

GOODMAN, D. C., AND HOREL, J. A. Sprouting of optic tract projections in the brain stem of the rat. *Journal of Comparative Neurology*, 1966, **127,** 71–88.

GOODMAN, D. C.; BAGDASARIAN, R. S., AND HOREL, J. A. Axonal sprouting of ipsilateral optic tract following opposite eye removal. *Brain, Behavior and Evolution*, 1973, **8,** 27–50.

GOTTLIEB, D. I., AND COWAN, W. M. Autoradiographic studies of the commissural and ipsilateral association connections of the hippocampus and dentate gyrus of the rat. I. The commissural connections. *Journal of Comparative Neurology*, 1973, **149,** 393–422.

GUILLERY, R. W. Experiments to determine whether retinogeniculate axons can form translaminar collateral sprouts in the dorsal lateral geniculate nucleus of the cat. *Journal of Comparative Neurology*, 1972, **146,** 407–420.

GUTH, L., AND BERNSTEIN, J. J. Selectivity in the re-establishment of synapses in the superior cervical sympathetic ganglion of the cat. *Experimental Neurology*, 1961, **4**, 59–69.

HAIGHTON, J. An experimental inquiry concerning the reproduction of nerves. *Philosophical Transactions of the Royal Society (London)*, 1795, **1**, 190–200.

HARRIS, A. J. Inductive functions of the nervous system. *Annual Review of Physics*, 1974.

HJORTH-SIMONSEN, A. Projection of the lateral part of the entorhinal area to the hippocampus and fascia dentate. *Journal of Comparative Neurology*, 1972, **146**, 219–232.

HJORTH-SIMONSEN, A., AND JEUNE, B. Origin and termination of the hippocampal perforant path in the rat studied by silver impregnation. *Journal of Comparative Neurology*, 1972, **144**, 215–232.

HOFFMAN, H. Local re-innervation in partially denervated muscle: A histo-physiological study. *Australian Journal of Experimental Biology and Medical Science*, 1950, **28**, 383–397.

HOFFMAN, H. A study of the factors influencing innervation of muscles by implanted nerves. *Australian Journal of Experimental Biology and Medical Science*, 1951, **29**, 289–308.

HOFFMAN, H., AND SPRINGELL, P. H. An attempt at the chemical identification of "neurocletin" (The substance evoking axon sprouting). *Australian Journal of Experimental Biology and Medical Science*, 1951, **29**, 417–424.

KENNEDY, R. The regeneration of nerves. *Philosophical Transactions of the Royal Society (London)*, 1897, **188**, 256–305.

KERR, F. W. L. The potential of cervical primary afferents to sprout in the spinal nucleus of V following long term trigeminal denervation. *Brain Research*, 1972, **43**, 547–560.

LEWIS, P. R., AND SHUTE, C. C. D. The cholinergic limbic system: Projections to hippocampal formation, medial cortex, nuclei of the ascending cholinergic reticular system, and the subfornical organ and supra-optic crest. *Brain*, 1967, **90**, 520–540.

LEWIS, P. R., SHUTE, C. C. D., AND SILVER, A. Confirmation from choline acetylase analyses of a massive cholinergic innervation of the rat hippocampus. *Journal of Physiology (London)*, 1967, **191**, 215–224.

LIU, C.-N., AND CHAMBERS, W. W. Intraspinal sprouting of dorsal root axons. *Archives of Neurology and Psychiatry*, 1958, **79**, 48–61.

LUGARO, E. Osservazioni sui "gomitoli" nervosi nell rigererazione dei nervi. *Rivista di Patologia Nervosa e Mentale*, 1906, **2**, 337–348.

LUND, R. D., AND LUND, J. S. Synaptic adjustment after deafferentation of the superior colliculus of the rat. *Science*, 1971, **171**, 804–807.

LUND, R. D., CUNNINGHAM, T. S., AND LUND, J. S. Modified optic projections after unilateral eye removal in young rats. *Brain, Behavior and Evolution*, 1973, **8**, 51–72.

LYNCH, G., MATTHEWS, D. A., MOSKO, S., PARKS, T., AND COTMAN, C. W. Induced acetylcholinesterase-rich layer in rat dentate gyrus following entorhinal lesions. *Brain Research*, 1972, **42**, 311–318.

LYNCH, G., DEADWYLER, S., AND COTMAN, C. W. Postlesion axonal growth produces permanent functional connections. *Science*, 1973a, **180**, 1364–1366.

LYNCH, G., MOSKO, S., PARKS, T., AND COTMAN, C. Relocation and hyperdevelopment of the dentate gyrus commissural system after entorhinal lesions in immature rats. *Brain Research*, 1973b, **50**, 174–178.

LYNCH, G., STANFIELD, B., AND COTMAN, C. W. Developmental differences in post-lesion axonal growth in the hippocampus. *Brain Research*, 1973c, **59**, 155–168.

LYNCH, G., GALL, C., MENSAH, P., AND COTMAN, C. W. Horseradish peroxidase histochemistry: A new method for tracing efferent projections in the central nervous system. *Brain Research*, 1974a, **65**, 373–380.

LYNCH, G., STANFIELD, B., PARKS, T., AND COTMAN, C. W. Evidence for selective post-lesion axonal growth in the dentate gyrus of the rat. *Brain Research*, 1974b, **69**, 1–11.

LYNCH, G., GALL, C., ROSE, G., WEST, B., AND COTMAN, C. W. Changes in the distribution of the dentate gyrus associational and commissural systems after unilateral and bilateral entorhinal lesions in adult rats. 1975a, submitted.

LYNCH, G., LEE, K., STANFORD, E., MCWILLIAMS, J. R., AND COTMAN, C. W. Ultrastructural evidence for axon sprouting in the adult hippocampus, 1975b, in preparation.

LYNCH, G., ROSE, G., GALL, C., AND COTMAN, C. W. The response of the dentate gyrus to partial deaf-ferentation. In R. Santini (Ed.), *The Golgi centennial symposium.* New York: Raven Press, 1975c, in press.

MAROTTE, L. R., AND MARK, R. F. The mechanism of selective reinnervation of fish eye muscle. I. Evidence from muscle function during recovery *Brain Research,* 1970a, **19,** 41–51.

MAROTTE, L. R., AND MARK, R. F. The mechanism of selective reinnervation of fish eye muscle. II. Electron microscopy of nerve endings. *Brain Research,* 1970b, **19,** 53–62.

MATTHEWS, D. A., COTMAN, C. W., AND LYNCH, G. Ultrastructural alterations in the molecular layer of the dentate gyrus following lesions of the ipsilateral entorhinal cortex. 1975, in preparation.

MOORE, R. Y., BJORKLUND, A., AND STENEVI, U. Plastic changes in the adrenergic innervation of the rat septal area in response to denervation. *Brain Research,* 1971, **33,** 13–35.

MOSKO, S., LYNCH, G., AND COTMAN, C. W. Distribution of the septal projection to the hippocampal formation of the rat. *Journal of Comparative Neurology,* 1973, **152,** 163–174.

MURRAY, M., AND THOMPSON, J. W. The occurrence and function of collateral sprouting in the sympathetic nervous system of the cat. *Journal of Physiology (London),* 1957, **135,** 133–162.

NADLER, J. V., COTMAN, C. W., AND LYNCH, G. S. Altered distribution of choline acetyltransferase and acetylcholinesterase activities in the developing rat dentate gyrus following entorhinal lesion. *Brain Research,* 1973, **63,** 215–230.

NADLER, J. V., COTMAN, C. W., AND LYNCH, G. S. Biochemical plasticity of short-axon interneurons: Increased glutamate decarboxylase activity in the denervated area of rat dentate gyrus following entorhinal lesions. *Experimental Neurology,* 1975a, **45,** 404–413.

NADLER, J. V., PAOLETTI, C., COTMAN, C. W., AND LYNCH, G. Histochemical evidence of altered development of cholinergic fibers in the rat dentate gyrus following lesions. II. Effects of partial entorhinal and simultaneous multiple lesions. 1975b, in preparation.

RAISMAN, G. Neuronal plasticity in the septal nuclei of the adult rat. *Brain Research,* 1969, **14,** 25–48.

RAISMAN, G., AND FIELD, P. A quantitative investigation of the development of collateral regeneration after partial deafferentation of the septal nuclei. *Brain Research,* 1973, **50,** 241–264.

RAISMAN, G., COWAN, W. M., AND POWELL, T. P. S. The connexions of the septum. *Brain,* 1966, **89,** 317–348.

RAMÓN Y CAJAL, S. *Degeneration and regeneration in the nervous system.* New York: Hafner, 1928 (reprinted 1968).

ROSE, J. E., MALIS, L. K., KRUGER, L., AND BAKER, C. D. Effects of heavy ionizing monoenergic particles on the cerebral cortex. *Journal of Comparative Neurology,* 1960, **115,** 243–296.

SEGAL, M., AND LANDIS, S. Afferents to the hippocampus of the rat studied with the method of retrograde transport of horseradish peroxidase. *Brain Research,* 1974, **78,** 1–15.

SOTELO, C., AND PALAY, S. L. Altered axons and axon terminals in the lateral vestibular nucleus of the rat: Possible example of axon remodeling. *Laboratory Investigation,* 1971, **25,** 653–671.

STENEVI, U., BJORKLUND, A., AND MOORE, R. Y. Growth of intact central adrenergic axons in the denervated lateral geniculate body. *Experimental Neurology,* 1972, **35,** 290–299

STEWARD, O., COTMAN, C. W., AND LYNCH, G. Re-establishment of electrophysiologically functional entorhinal cortical input to the dentate gyrus deafferented by ipsilateral entorhinal lesions: Innervation by the contralateral entorhinal cortex. *Experimental Brain Research,* 1973, **18,** 396–414.

STEWARD, O., COTMAN, C. W., AND LYNCH, G. Growth of a new fiber projection in the brain of adult rats: Reinnervation of the dentate gyrus by the contralateral entorhinal cortex following ipsilateral entorhinal lesions. *Experimental Brain Research,* 1974, **20,** 45–66.

STORM-MATHISEN, J. Quantitative histochemistry of acetylcholinesterase in the rat hippocampal region correlated to histochemical stain. *Journal of Neurochemistry,* 1970, **17,** 739–750.

STORM-MATHISEN, J. Glutamate decarboxylase in the rat hippocampal region after lesions of the afferent fiber systems: Evidence that the enzyme is localized in intrinsic neurones. *Brain Research,* 1972, **40,** 215–235.

STORM-MATHISEN, J. Increase of choline acetylase (ChAc) and acetylcholinesterase (AChE) in stratum moleculare fascia dentate following degeneration of the perforant path. *Acta Physiologica Scandinavica Supplement,* 1974, **396,** 33.

VAN HARRELVELD, A. Re-innervation of denervated muscle fibers by adjacent functioning motor units. *American Journal of Physiology*, 1945, **144**, 477–493.

WEST, J. R., DEADWYLER, S., COTMAN, C. W., AND LYNCH, G. Time dependent changes in commissural field potentials in the dentate gyrus following lesions of the entorhinal cortex in adult rats: A neurophysiological study of axon sprouting. *Brain Research*, 1975, submitted.

WESTRUM, L. E., AND BLACK, R. G. Fine structural aspects of the synaptic organization of the spinal trigeminal nucleus (pars interpolaris) of the cat. *Brain Research*, 1971, **25**, 265–288.

ZIMMER, J. Ipsilateral afferents to the commissural zone of the fascia dentata demonstrated in decommissurated rats by silver impregnation. *Journal of Comparative Neurology*, 1970, **142**, 393–416.

ZIMMER, J. Extended commissural and ipsilatural projections in postnatally de-entorhinated hippocampus and fascia dentata demonstrated in rats by silver impregnation. *Brain Research*, 1973, **64**, 293–311.

ZIMMER, J. Proximity as a factor in the regulation of aberrant growth in postnatally deafferented fascia dentata. *Brain Research*, 1974, **72**, 137–142.

6

Organization of Hippocampal Neurons and Their Interconnections

PER ANDERSEN

1. *Extrinsic Connections of the Hippocampus*

In his famous work, Ramón y Cajal (1911) has a drawing of the main connections of the hippocampal formation (Fig. 1). This is indeed a remarkable figure. With few exceptions, further investigations with different techniques have shown Ramón y Cajal to be correct, despite the fact that he had to infer the direction of impulse traffic from histological data only. Physiological and biochemical techniques have, however, added details in the scheme devised by Ramón y Cajal, and changed it on some points. The purpose of this chapter is to give a review of what these studies have added to our knowledge about hippocampal neuronal organization.

1.1. *The Perforant Path*

As was emphasized by Ramón y Cajal and later corroborated by Lorente de Nó (1934), the main input to the hippocampal formation is from the entorhinal area by way of a fiber system called the perforant path. When the perforant path fibers or the entorhinal area itself is stimulated, large potentials can be recorded from the hippocampal formation due to the close packing of cells and their arrangement parallel to each other. the size of these potentials is so large that an electrode can pick up activity although situated several millimeters away from the cells that generate the activity. In

---

PER ANDERSEN • Institute of Neurophysiology, University of Oslo, Oslo, Norway.

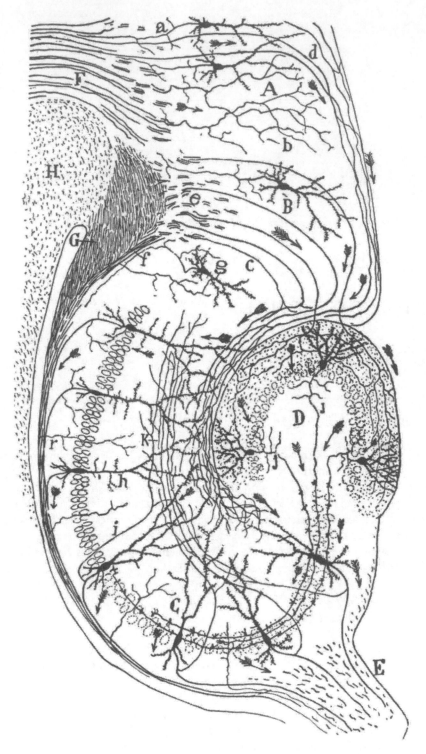

Fig. 1. Schematic drawing by Ramón y Cajal (1911) of the main cells, connections, and flow of impulse traffic in the hippocampal formation.

order to determine the specific termination area and therefore the cells which are excited by the perforant path, a laminar profile analysis has to be performed based on a series of tracks through the hippocampal formation. Little happens with the initial part of such records as long as the recording electrode traverses the field CA1 or CA3 overlying the dentate area. In contrast, a dramatic change takes place when the electrode approaches and traverses the hippocampal fissure (Fig. 2). The potential recorded is a large negative wave with a maximum in the middle part of the dentate molecular layer. When the electrode is moved toward the granular layer, this potential reverses and superimposed on it there appears a large negative population spike. Based on information with fine electrodes, we know that this population spike is composed of many individual unitary discharges. Since these discharges can be found in a narrow band of tissue, $80-100\,\mu$m thick, which histologically corresponds to the upper blade of the dentate fascia, these units are most likely due to granule cell discharges. Consequently, the large negative population spike recorded in this layer is due to the near-

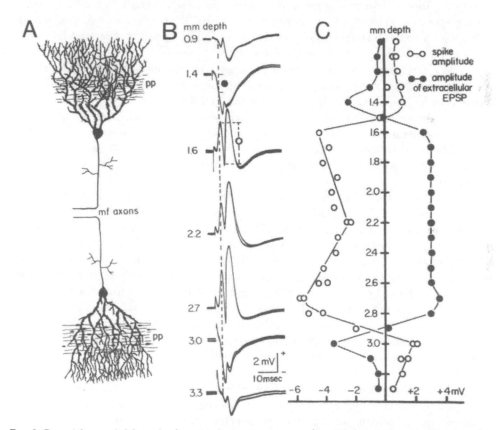

FIG. 2. Potentials recorded from the dentate area in response to perforant path activation. A: Diagram of two granule cells drawn to scale and in a position appropriate to the potentials and scale in B and C. B: Sequence of potentials recorded at the indicated depth in response to perforant path stimulation. C: Plotting of the size of the population spike (open circles) and extracellular EPSP (filled circles) in relation to the recording depth. From Lømo (1971a).

synchronous activation of many granule cells. The potential may measure several millivolts, probably indicating discharge of several thousand cells.

The negative wave in the molecular layer is causally related to the population spike, since the spike occurs only when the negative wave in the molecular layer has reached a certain magnitude. Intracellular recording from granule cells excited by the perforant path shows depolarization with characteristics of an excitatory postsynaptic potential (EPSP) which coincides in time with the negative wave in the molecular layer. The latter is therefore interpreted as the extracellular sign of the perforant path EPSP. The extracellular EPSP has its maximum in the middle of the molecular layer (Andersen *et al.*, 1966b; Lømo, 1971a), corresponding to the localization of the perforant path synapses emerging from the medial part of the entorhinal area (Blackstad, 1958; Nafstad, 1967; Hjorth-Simonsen and Jeune, 1972).

After the population spike, an electrode located in the cell body layer records a large positive wave. This corresponds to an intracellularly recorded hyperpolarization and inhibition of unitary discharges of granule cells as well as reduction of the granule cell population spike due to a subsequent test volley. This granule layer positivity is therefore accepted as the extracellular sign of an IPSP. As will be described later, this is due to a recurrent inhibitory mechanism.

The description above refers to the perforant path arising in the medial entorhinal area only. No detailed physiological analysis has been made of the part of the perforant path (Hjorth-Simonsen, 1972) coming from the lateral enthorinal area.

When the recording electrode is advanced beyond the upper blade of the granule cells, it penetrates the hilus, in which the granule cell population spike is smaller or even absent; thereafter, it enters the lower blade of the granule cells, where the same type of potentials as were found in the upper blade is recorded, although in the opposite sequence.

The input from the perforant path to the dentate area is organized in parallel lines (Lømo, 1971a). Stimulation of a discrete part of the perforant path leads to excitation of granule cells in a thin band of tissue, roughly transverse to the longitudinal axis of the hippocampal formation. From recording of the population spike at various distances along this band, it appears that the size remains remarkably constant from the caudal to the anterior end. Since the density of granule cells is similar throughout the dentate area, the number of perforant path excited granule cells is nearly the same per unit volume of tissue inside this band. This fits the histological observation that the perforant path fibers run for the whole length from the caudal to the rostral part of the structure. Thus an afferent volley excites a strip of evenly spaced granule cells, stretching from the caudal to the anterior part of the dentate fascia and crossing the longitudinal axis of the hippocampal formation.

The arrangement of the perforant path differs slightly in different animals, but the principle is the same, namely that successive regions of the entorhinal area are connected with a series of parallel strips or lamellae (see Fig. 7). The orientation of the dentate lamellae differs somewhat in different animals from roughly transverse to the longitudinal axis of the hippocampal formation in rats and guinea pigs to a more oblique orientation (anterior end pointing more septally) in rabbits and even more so in cats. This is related to the relatively more temporal position of the entorhinal area in the latter two animals.

1.2. Other Afferents to the Dentate Area

Another main afferent source to the dentate area is the commissural pathway. In spite of many attempts, there is still no clear account of the accurate origin of these fibers. However, it is possible that stimulation of the fimbria near the midline excites the commissural pathway as well. Such a procedure may synaptically excite the dentate granule cells (Cragg and Hamlyn, 1957).

There are also several reports on septal activation of dentate cells. However, it is difficult to be sure that the stimulation selectively excites septal cells or their axons and not also fibers coursing through this complicated area. In my opinion, there is as yet no conclusive physiological report on a pure septodentate projection.

Similarly, although suggestive, there is still no firm physiological evidence to say whether the catecholamine-containing terminals in the dentate area have any synaptic function.

1.3. Afferents to the CA3 Region

A major afferent fiber system to the CA3 pyramidal cells is the mossy fibers, the axons of the dentate granule cells. With Golgi staining, the thin (about 0.2 μm) mossy fibers have large synaptic swellings at intervals of about 250–450 μm and course in a suprapyramidal situation along the whole of field CA3 (Blackstad and Kjaerheim, 1961; Hamlyn, 1961). The histological structure of the coupling between these fibers and the CA3 pyramids is unique. Branched dendritic spines, situated on the proximal part of the CA3 apical dendrites, are completely invested in one of the synaptic swellings mentioned above. This bouton may be 3–5 μm in diameter and 3–6 μm long and may contain several hundreds or thousands of vesicles with asymmetrical (type I) synaptic contacts on several of the digitlike processes of the dendritic spine. The mossy fiber synapses are typically of the *en passage* type, so that a single volley influences a number of CA3 pyramids.

The stimulation of a set of granule cells produces in the CA3 a large extracellular negative wave with a maximum corresponding to the level of the mossy fiber synapses (Fig. 3). This negative wave coincides with intracellularly recorded EPSP, and is therefore regarded as the extracellular sign of the excitatory synaptic action. Provided that a sufficient number of granule cells have been excited, the EPSP is large enough to give rise to an extracellular population spike. Judged from the occurrence of unitary discharges correlated with the population spike, the latter is regarded as the sign of CA3 cell discharges synchronously driven by the synaptic action. Both histological evidence (Hjorth-Simonsen and Jeune, 1972) and physiological evidence (Lømo, 1971b; Danscher *et al.*, 1974) indicate that mossy fibers are arranged in parallel lamellae (see below).

As for the perforant path–granule cell synapse, the mossy fiber–CA3 pyramidal cell synapse shows remarkable frequency potentiation, with an optimal frequency around 8–10/s. Of importance for the interpretation of hippocampal networks, not least the data on θ activity, is the fact that a considerable degree of convergence seems necessary in order to discharge the CA3 pyramids. Usually, the population spike does not occur until the extracellular EPSP has reached a size of several millivolts. Similarly, with single-unit recording it is often necessary to stimulate the afferent

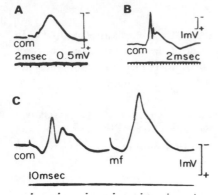

FIG. 3. Responses of CA3 neurons to stimulation of commissural and mossy fiber pathways. A, B: Extracellular response from the surface of CA3 in response to weak (A) and stronger (B) commissural activation. C: Commissural (com) followed by mossy fiber (mf) activation of CA3 neurons recorded from the layer of the mossy fiber synapses. Note the inverted commissural EPSP recorded from this depth.

system strongly or even use repetitive stimulation around 10/s to bring forward a discharge.

The CA3 commissural pathway arises from the homotopic point on the contralateral side. Stimulation of this area or the fimbria closer to the midline produces an initial diphasic deflection which corresponds in part to an afferent volley in the fimbrial fibers and in part to antidromic invasion of CA3 pyramids (Fig. 3A). Following this initial deflection, there is a negative wave of 5–10 ms duration, which under favorable conditions can be seen to correlate with an intracellulary recorded EPSP (Andersen and Lømo, 1966). The intracellular records show little of an EPSP unless the cell has a high membrane potential or is artificially hyperpolarized. Nevertheless, since the negative wave seems causally related to a large negative population spike, and since its maximum is found among the basal dendrites in the stratum oriens where histologically the commissural termination is found (Blackstad, 1956), its interpretation as the extracellular sign of an EPSP seems reasonable. As with other afferent systems, unit recording shows the population spike to be composed of the discharge of many individual units. Following the discharge of the cells, there is a large positive wave with a maximum at the cell body layer. This is invariably associated with a large IPSP recorded intracellularly, and with cessation of spontaneous or induced discharges (Andersen et al., 1964a). Also, the commissural pathway shows prominent frequency potentiation at around 10/s.

No sign has been seen of any connection from CA1 to CA3 (Hjorth-Simonsen, 1973; Andersen et al., 1973a). One apparent contradiction to this statement can be seen in certain physiological experiments in which the Schaffer collaterals are excited. Because the Schaffer collaterals of CA3 pyramidal axons (see below) make synaptic connection with other CA3 cells, a synaptic effect on these will be seen also when the fibers conduct antidromically. Unless this interpretation is taken into account, one might falsely conclude that there is a connection from CA1 to CA3.

As with other areas, a proper septofugal connection is difficult to demonstrate physiologically due to the difficulty with selective stimulation of septal cells or septofugal fibers.

1.4.   Afferents to the CA1 Region

Two main afferents to the CA1 pyramidal cells are the Schaffer collaterals of the CA3 pyramids and the commissural pathway. In addition, there is a possibility that CA1 cells may be excited by their neighbors through a local circuit. The possibility of direct excitation of CA1 cells from the subicular or entorhinal area and neighboring regions needs to be further investigated.

The most prominent synaptic excitation of CA1 cells is seen following activation of CA3 cells or their Schaffer collaterals. The activation takes the form of a very large negative wave with a maximum in the apical dendritic layer (Fig. 4D). The depth distribution of this wave varies slightly with the animal used. The most selective localization is found in rabbits, in which the Schaffer collaterals form a very narrow band (less than 100 μm) at about 0.8 mm from the alvear surface. In this animal, there is a sharp maximum of the negative wave at this layer. With greater recording depths, this wave slowly subsides and is replaced by activity of the granule cells. With recording closer to the pyramidal layer, the synaptic wave from the Schaffer collateral activity is reduced and eventually reversed when the electrode is withdrawn through the pyramidal cell layer. With sufficient strength, the Schaffer collateral wave carries a population spike near its top. The shape and size of this spike vary considerably with the recording depth. When recorded from the synaptic layer itself (i.e., 0.7–0.8 mm in rabbit), the

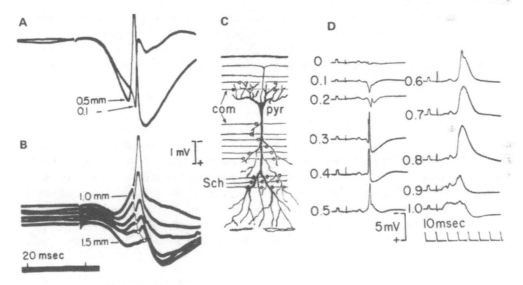

FIG. 4. Synaptic activation of CA1 pyramidal cells. A: Superimposed records taken extracellularly near the pyramidal layer (0.5 mm) and near the surface (0.1 mm) in order to show conduction of the spike. B: Montage of records taken at each 0.1-mm interval from 1.0 to 1.5 mm to show somatofugal conduction of the spike and its decrease in amplitude. C: Diagram of CA1 pyramidal cell with the territory for the commissural and Schaffer collateral synapses. D: Extracellular responses due to Schaffer collateral activation recorded at the indicated depths. Note the different depth distribution of the spike and the synaptic wave, the latter having a maximum at 0.8 mm. From Andersen (1960).

spike is often seen reversed as a small positive or diphasic wave of relatively small amplitude. However, when the electrode tip is moved toward the pyramidal layer the spike increases greatly in size in parallel with a reduction of the Schaffer synaptic wave. The population spike may be single, but two or, more rarely, three spikes may appear, having their maximum amplitude at the cell body layer or the proximal part of the apical dendrite. On further removal of the electrode toward the alveus, the spike rapidly decreases in size and attains a diphasic shape. Individual units discharge corresponding to the negative population spikes, supporting the idea that these are composed of synchronous discharges of many neurons. As with other afferents, the discharges are followed by a large wave which is positive in the cell body layer and is associated with an intracellularly recorded IPSP.

In the CA1 the commissural pathway also terminates on the apical dendrites, but in addition on the basal dendrites (Fig. 4A,B). However, since the former location is the largest, the usual appearance of a CA1 commissural response is a large negative wave among the apical dendrites which reverses to a positive wave between the cell body and the surface (Cragg and Hamlyn, 1957). The commissural excitatory synaptic wave is usually somewhat smaller than a Schaffer collateral wave and has less probability of discharging the cell. Otherwise, the two inputs show remarkably similar properties, including the need for a relatively large synaptic convergence in order to discharge the cells, and a prominent frequency potentiation at the appropriate frequency of stimulation.

The fact that stimulation of the CA1 may induce a synaptic wave in nearby parts of the same region makes it possible that there exist CA1 → CA1 synaptic connections in a plane nearly transverse to the longitudinal axis of the hippocampus. However, this statement must be qualified since intracellular recording on such a postulated connection is not available.

1.5. Afferents to the Subiculum

Little information exists on the afferent connection to the subicular region as studied with physiological techniques. In the study of the output of hippocampal neurons, Andersen *et al.* (1973a) found that CA1 neurons send their fibers to the subicular region, supporting the independent anatomical study by Hjorth-Simonsen (1973). Unfortunately, scant precise physiological information exists on the connection to the entorhinal area and neighboring regions. Apart from large field potential studies, usually without laminar analysis, there exists no study with unit or intracellular recording.

2.   Intrinsic Connections: Basket Cell Inhibition

A remarkable finding of all investigators using intracellular recording in the hippocampal formation is the ubiquitos hyperpolarization associated with inhibition of cell discharge which follows excitation of the cell from all afferent sources studied so far. The inhibition typically has a slightly longer latency than that of the excitation. An important feature is that it is seen in nearly all cells even without previous excitation. Spencer and Kandel (1961) found that after chronic deafferentation antidromic stimulation of the CA3 axons resulted in large IPSPs or antidromic invasion followed by an IPSP in the surviving neurons, showing that the inhibition was due to collaterals of the efferent axons. The IPSP is associated with an extracellular positive wave which has its maximum at the cell body layer (Andersen *et al.*, 1964a). The fact that the amplitude of the extracellular IPSP showed a sharp maximum at the cell body layer and that even a single IPSP might give an increase of the membrane potential to the equilibrium potential for the IPSP (around −85 mV) suggests that inhibition took place at the cell body itself. Coupled with the information that the inhibition was very much more widespread than synaptic excitation and the onset of the IPSP was typically 1.2–1.5 ms longer than the antidromic invasion, this led to the postulate that the inhibition was produced by an interneuron with its synapses terminating on the cell bodies of the pyramidal cells. Since the only known interneuron to have this property were the basket cells (Ramón y Cajal, 1911; Lorente de Nó, 1934), these neurons were supposed to be inhibitory interneurons in the hippocampal formation (Andersen *et al.*, 1964a). Thus, the unique characteristics of the hippocampal inhibition and the special histology of the region made it possible to identify the first inhibitory interneuron in the mammalian brain (Fig. 5).

As indicated in the drawings of Ramón y Cajal (1911) and Lorente de Nó (1934), the basket cell is located in a position where it can be easily activated by collaterals of the pyramidal cell axons. This applies to both the CA3 and the CA1 region. Since the basket cells typically discharge at the same threshold as the antidromic population spike, although 0.8–1.2 ms later, they are postulated to be synaptically excited by collaterals of the CA1 or CA3 pyramidal axons. This pathway then is an example of recurrent inhibition. This postulate is supported by experiments with the conditioning-test stimulation technique, in which the population spike produced by the test stimulus

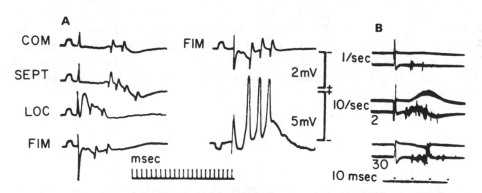

FIG. 5. Properties of basket cells. A: Extracellular records in response to stimulation of various pathways. COM, commissural, SEPT, septal, LOC, local, and FIM, fimbrial, afferents. In the middle are shown extracellular records to fimbrial stimulation just before penetration (above) and an intracellular record (below) from a basket cell. Note the high discharge frequency. B: Each pair gives the extracellular field potential (above) and basket cell discharge (below) in response to 1/s stimulation (upper pair) and 10/s stimulation (lower two pairs after 2 and 30 s, respectively).

does not start to show inhibition until the strength of the conditioning response passes the threshold of spike production.

Using extracellular recording from stratum oriens, presumed basket cells were recorded from and shown to be activated by a large variety of afferent inputs (Andersen *et al.*, 1964*b*). Typically, the basket cell interneurons discharged repetitively. The number of spikes and the initial frequency of the burst both increased with increasing afferent bombardment. Occasionally, the pyramidal IPSP showed increments on the rising phase at a frequency corresponding to the typical discharge frequency of the basket cell, which is around 5–700/s. The basket cell followed the stimulating frequency very well, as was also the case for the IPSPs. Basket cells show great afferent convergence, indicated by reduced latency and increased spike number and frequency to convergent stimulation of pathways. Some basket cells discharge at such a weak strength of the afferent volley to suggest the presence of forward inhibition—i.e., afferent fibers going directly to the basket cells.

As for the pyramidal cells, intracellular records from dentate granule cells have shown an initial excitation followed by a prominent IPSP. Because of the slightly longer latency of the IPSP compared to the EPSP, the wide distribution of the IPSP, and its association with an extracellular positive wave which is maximal at the dentate cell body layer, Andersen *et al.* (1966*b*) suggested that the inhibition of the granule cells also is caused by interneurons with branching axons and terminals at the granule cell body itself. The most obvious candidates for these interneurons are the basket cells, having their cell bodies below the granule cells with widely branching axonal plexuses terminating on the granule cell bodies. Since a similar inhibition follows antidromic activation of granule cells, the basket cells are probably excited by collaterals of the mossy fibers. This inhibition is, therefore, also of the recurrent type (Lømo, 1971*a*).

Although the evidence on excitatory interneurons is much less direct, there are some data to suggest the presence of such cells. For example, prolonged facilitation of the response to the second of the identical volleys is shown by most afferent pathways impinging on pyramidal or granule cells. Furthermore, long-lasting changes have been seen following repeated periods of tetanic stimulation at around 10/s (Lømo, 1971*b*; Bliss and Lømo, 1973). Although a part of this long-standing excitation increase may be due to changes at the synapses of the afferent pathways, an alternative explanation is the activation of excitatory interneurons, which may increase the responses to subsequent stimuli in a train.

3. Output from the Hippocampal Formation

The study of the output from various portions of the hippocampal formation is a classical topic of experimental neuroanatomy. Recently, several physiological studies have added to the data in this field. The earlier studies on spread of afterdischarges and various studies on field potentials due to synaptic activation (Cragg and Hamlyn, 1957; Andersen, 1960; Gloor *et al.*, 1964) were informative, although accurate localization was difficult with these techniques. However, by use of antidromic stimulation more precise information is possible. Employing the fact that an antidromic population spike signals the near-synchronous activation of many pyramidal cells (Andersen *et al.*, 1971*a*), Andersen, Bland, and Dudar (1973*a*) made a study on the projection of CA3 and CA1 pyramidal cells. Stimulation of the fimbria gave antidromic invasion of CA3 pyramidal cells only (Fig. 6). In contrast, no CA3 cell could be antidromically invaded from the alveus overlying the CA1. On the other hand, the CA1 pyramidal cells were antidromically invaded by stimulation of the alveus or the subiculum in a particularly oriented strip of tissue (see Section 4). These studies confirmed the independent histological studies of Hjorth-Simonsen (1973).

By studying the distribution of the antidromic field potential of dentate granule cells, Lømo (1971*a*) showed that the mossy fibers could be antidromically invaded from the CA3 region but not from areas beyond the CA3/CA1 border. This is in full accord with the histological studies that show that the mossy fibers do not go beyond this limit (Ramón y Cajal, 1911; Blackstad *et al.*, 1970).

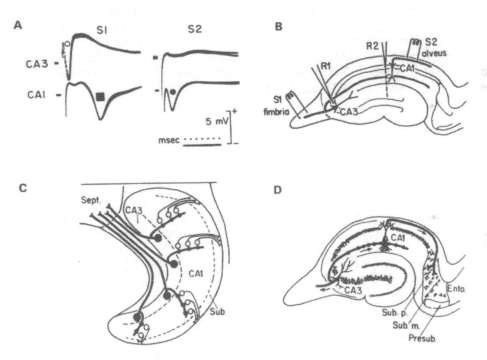

Fig. 6. Output from the CA3 and CA1 neurons. A: Responses from the pyramidal layer of CA3 (upper records) and CA1 (lower records) in response to stimulation of the fimbria (S1) and the alveus (S2). Fimbrial stimulation gives antidromic population spike of the CA3 neurons (open circle), whereas alveus stimulation gives antidromic activation of CA1 pyramids only (filled circle). The CA1 deflection given by the filled square is Schaffer collateral synaptic activation following fimbria stimulation. C: Diagram showing the distribution of the axons of CA3 pyramids (filled circles and thick lines) and of the CA1 pyramids (open circles and thin lines). D: Diagram of a transverse section through the hippocampus to show the distribution of the CA3 and CA1 axons (heavily outlined).

4. Lamellar Organization of the Hippocampal Formation

Schematically, the hippocampus can be regarded as a four-membered neuronal loop, consisting of (1) perforant path, (2) mossy fibers, (3) fimbrial axons as well as Schaffer collaterals, and (4) CA1 axons in the alveus. By plotting the pathway taken by the perforant path fibers in activating the dentate granule cells, Lømo (1971a) showed that stimulation of a bundle of perforant path fibers excited the granule cells along a narrow strip. Stimulation of a neighboring bundle excited a strip of tissue parallel to the first. The orientation of these strips was roughly sagittal in the dorsal part of the hippocampus. The region next to the midline (2–3 mm) deviated from this general pattern since the dentate fascia here is curved back on itself.

Studying the distribution of the other members of the four-neuron loop, Andersen et al. (1971b) found that both orthodromic and antidromic activation of the mossy fibers showed these also to be arranged in a parallel fashion, with the direction of the fibers like the teeth of a comb and parallel to the perforant path fibers (Fig. 7). Similarly, the Schaffer collaterals and the CA1 alvear fibers were found to be arranged in a striplike fashion, somewhat obliquely to the longitudinal axis of the hippocampus. Thus the axons of all major neuronal members inside the hippocampal formation are arranged in bands parallel to each other, so that a narrow, nearly transverse slice of the hippocampal formation should contain a complete neuronal loop from the perforant path through the mossy fibers and Schaffer collaterals to the CA1 axons ending in the

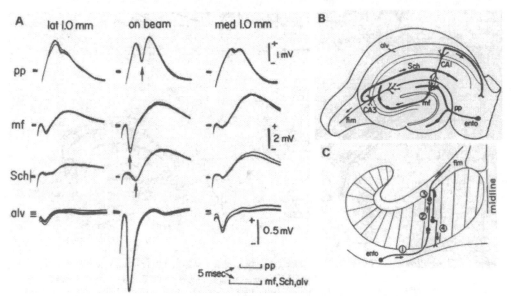

FIG. 7. Hippocampal lamellae. A: Records taken on beam (middle column) along a hippocampal lamella, and 1 mm lateral and medial to it, in response to stimulation of the perforant path (upper row), the mossy fibers (second row), the Schaffer collaterals (third row), and the alveus (lower row). B: Diagram of a lamella with the four-membered neuronal loop heavily outlined. C: Diagram of the hippocampal formation seen from above. The transverse thin lines indicate the orientation of the various lamellae. The encircled numbers give the members of the neuronal loop as they would appear from above.

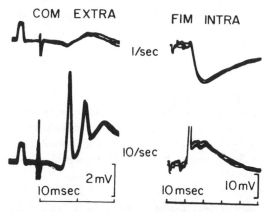

FIG. 8. Frequency potentiation of hippocampal responses. The two left-hand records show extracellular responses of CA1 cells to commissural stimulation at 1/s (above) and after 2 s of 10/s stimulation (below). The two right-hand traces are intracellular records from a CA3 neuron activated by fimbrial stimulation at 1/s (above) and after 12 s of 10/s stimulation (below).

subiculum. In their very first part, the CA3 axons coursing out toward the fimbria may also follow this pattern.

Dissection of the arteries of the brain after they had been filled with a fast-setting colored plastic material showed that the arteries in the hippocampal fissure are parallel and course in the same direction as the fibers of the four pathways.

The ultimate proof that the hippocampal intrinsic pathway is in a plane nearly transverse to the longitudinal axis is given by experiments with isolated hippocampal slices 3–400 μ m thick. If the slices are taken out with the appropriate orientation and placed on a net in a chamber perfused with artificial cerebrospinal fluid and overstreamed with moist oxygen with 5% CO_2, the presence of all four pathways having excitatory function can be shown (Skrede and Westgaard, 1971).

5. Frequency Characteristics of Hippocampal and Dentate Synapses

A remarkable feature of all excitatory synapses in the hippocampal formation so far studied is their ability to increase their synaptic effect at increasing rates of stimulation. Thus, most pathways give relatively weak responses when the rate is as low as 0.5–1 stimulus per second, but the size of the field potentials (both synaptic potentials and population spikes) increases when the stimulus rate is increased (Fig. 8). Usually, the optimal frequency—i.e., the frequency giving the largest size of the potentials—is around 10–15/s. Typically, the second and the third shock in such a series show reduced responses, most likely due to the recurrent inhibition induced by the first stimulus. However, after a few shocks in the train the potentials show a steady growth until maximal responses may be obtained after 10–30 stimuli at about 10/s. When the optimal stimulus rate is maintained, the duration of the period in which large responses occur does vary with the excitability of the hippocampus, being cut short in situations leading to electrical afterdischarges. With high rates of stimulation, the duration of the enhanced potentials is shorter, giving way to a period of depression. With rates as high as 40–50/s, increased responses are seen only to a few stimuli, whereupon a severe depression occurs which lasts a few minutes after the cessation of the tetanic stimulation. Even after very high rates of tetanic stimulation, the depressed state usually does not last longer than 2 min, and full recovery is seen after 5 min, provided that electrical afterdischarges do not occur. If the latter appears, restitution may take as long as half an hour, probably due to the much larger number of neurons involved.

The mossy fibers–CA3 pathway and the Schaffer collateral–CA1 pyramidal cell pathways seem more sensitive to high-frequency stimulation than the perforant path–granule cell pathway.

Interestingly, repeated tetani give rise to increased responses not only to subsequent tetanic stimulation but also to single-shock stimuli delivered at a low rate. Such increased responsiveness is seen as an increase of both the amplitude and the rate of rise of the extracellular EPSP of granule and pyramidal cells, an increase of the number and amplitude of the population spikes, and a decrease of their latency, and as an increase of the speed with which frequency potentiation manifests itself (Fig. 9). The process shows accumulation in that the effect on the size and latency of the responses improves with the number of tetani given. This phenomenon requires that the tetani be not too long and not given too frequently. The effect can be impressively large (100–300%), occurs after four or five periods of 10/s tetani each lasting for 10–15 s, and has a duration of several hours (Bliss and Lømo, 1973). The facilitatory effect is not associated with an increase of the antidromic invasion of pyramidal cells and is therefore probably not of a postsynaptic nature. It is more likely explained by a changed ability of the presynaptic terminals to deliver increased amounts of transmitter.

A final point is that tetanic stimulation of the perforant path leads to varying probability of discharge along the four-neuronal group inside the hippocampal formation. Thus at low rate of perforant path stimulation (up to 5/s), the frequency potentiation effect is usually mild, leading to granule cell discharge only. With a slightly higher rate of stimulation (5–8/s), the frequency potentiation is evident on both granule cells and CA3 neurons, but the effect may still be too weak to drive the CA1 cells. Finally, when the stimulus rate is increased to about 10–12/s this is sufficient to drive impulses all around the loop. In this way, varying the input frequency may determine to what degree the impulses traverse the multineuronal loop. However, in order to know the physiological importance of such a mechanism similar studies must be performed in unanesthetized animals.

6. Localization of Synapses on Hippocampal Neurons and Axons

As already stated, there is reason to believe that the inhibitory synapses causing the large IPSPs are produced by the basket cell neurons. Thus the best-known inhibition of the hippocampal pyramids is due to synapses at the cell body itself. Indirect evidence for this conclusion is found from the fact that the degeneration of

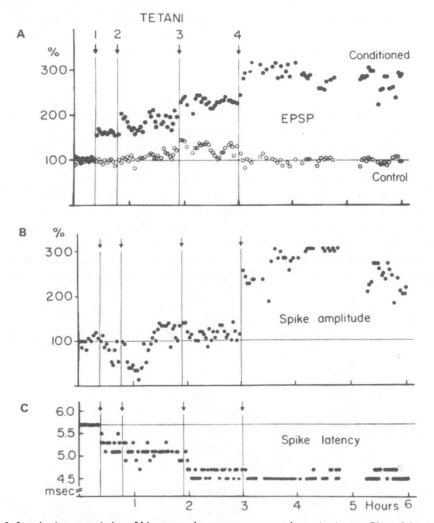

FIG. 9. Lost-lasting potentiation of hippocampal responses to repeated tetanization. A: Plots of the size of extracellular EPSP recorded from the right side (open circles) and on the left side (filled circles) in response to one stimulus every 3 s before and after four tetani at 15/s each lasting for 15 s. The amplitude of the conditioned population spike (B) and its latency (C) are plotted for the same stimulation sequence. (Lømo, unpublished observations).

boutons following section of all known afferent pathways is found in the dendritic territory but not in the pyramidal layer itself. In contrast, localization of excitatory synapses seems confined to the dendritic territory. Studying four different excitatory pathways, Andersen et al. (1966a) found the degenerating boutons always associated with dendrites, mostly on dendritic spines. Thus, as a general rule, excitatory synapses of the hippocampal formation are found mainly on dendrites and most often on dendritic spines, whereas inhibitory synapses are found on the cell body.

Whether there are exceptions to this rule is not easy to decide. Dendritic location of inhibitory synapses is difficult to establish with histological techniques until unequivocal morphological criteria for inhibitory synapses are found. Likewise, the very large IPSP seen by an electrode which penetrates the soma masks any presence of dendritic inhibition. However, the recently developed transverse hippocampal slice may offer an opportunity to study possible dendritic inhibition by recording either from inside dendrites with hyperfine electrodes or using direct iontophoretic application of putative inhibitory transmitters to known dendritic locations. The possibility of excitatory synapses on the cell bodies or proximal dendrites seems more remote since clear histological evidence points to only one type of synapse on the cell bodies of granule and pyramidal cells, namely those belonging to the basket cell system.

There may exist several types of hippocampal excitatory synapses. Andersen *et al.* (1973*b*) have made three-dimensional reconstructions of the Schaffer collateral–CA1 synapse from serial electron micrographs (Fig. 10). This synapse has a remarkable structure in being based on frequent Ranvier nodes (60–70 μm apart) from which one to three unmyelinated axons emerge to bend and course parallel to the parent fiber. Each branch has numerous synaptic swellings 3–5 μm apart, each of which makes contact with one or two spines of neighboring pyramidal apical dendrites. Because of

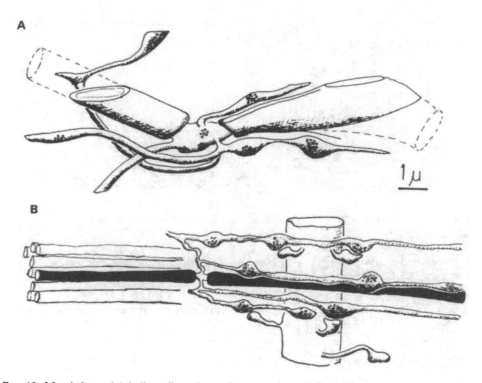

Fig. 10. Morphology of Schaffer collaterals. A: Drawing of a Schaffer collateral node with emerging unmyelinated branches, reconstructed from 30 serial electron micrographs. B: Diagram drawn on the basis of 25 series of the type shown in A to show the pattern of the nodal branch synapses on the CA1 pyramidal dendritic spines.

the very large number of side branches and associated synaptic swellings belonging to a single axon, an impulse in a single Schaffer collateral may give rise to synaptic excitation through several hundred or perhaps several thousand synaptic knobs, exciting a very large number of target CA1 cells. This situation is clearly different from that prevailing in lower parts of the central nervous system, where the axons usually terminate in a few branches, each with a terminal knob.

Andersen and Vaaland (1974) have also found evidence for synapses on Ranvier nodes within the hippocampal formation. The nodes receiving synaptic connections are all found within the stratum oriens of CA3, and at least some belong to Schaffer collaterals, sometimes at the bifurcation of the CA3 parent axon. The synapses on these Ranvier nodes may be of two types, either asymmetrical synapses with clear postsynaptic density and round vesicles but more often of another type which shows no great postsynaptic specialization and contains flattened or pleomorphic vesicles embedded in a more electron-dense background material (Fig. 11). Although no physiological evidence exists on their function, the histology suggests that these structures may be engaged in presynaptic control of afferent and possibly efferent connections of the hippocampus.

7. Localization of Putative Transmitters in the Hippocampal Formation

Iontophoretic ejection techniques have shown that several compounds excite various cells of the hippocampal formation (Fig. 12). For example, Biscoe and Straughan (1966) showed that a good many pyramidal cells were excited by acetylcholine. However, as often is the case in the central nervous system, the current needed to excite cells was relatively high and the latency of the effect was long. Therefore, this approach has not finally solved the problem of whether acetylcholine plays a transmitter role in the hippocampal formation.

The distribution of acetylcholinesterase shows a remarkably stratified pattern (Storm-Mathisen and Blackstad, 1964). Furthermore, acetylcholine transferase is found in appreciable quantities in the hippocampal formation. Both acetylcholine transferase and acetylcholinesterase are drastically reduced following section of the pathways from the septal region to the hippocampus. Furthermore, acetylcholinesterase piles up on the septal side of such a lesion (Shute and Lewis, 1961, 1963). Finally, injection of eserine in relatively moderate doses produces a remarkable enhancement of the θ activity, which probably is dependent on a septohippocampal pathway (Brücke et al., 1958), possibly the pathway associated with acetylcholinesterase. From these data, it is probable that the septohippocampal pathway associated with θ activity may operate by releasing acetylcholine.

Of the naturally occurring acidic amino acids, glutamic acid is the most important excitant of hippocampal cells (Biscoe and Straughan, 1966). Due to the lack of a specific blocker, it is, however, difficult to decide with certainty whether glutamic acid may be functioning as an excitatory transmitter in this region (see Curtis and Crawford, 1969). However, circumstantial evidence obtained by ejection of glutamic

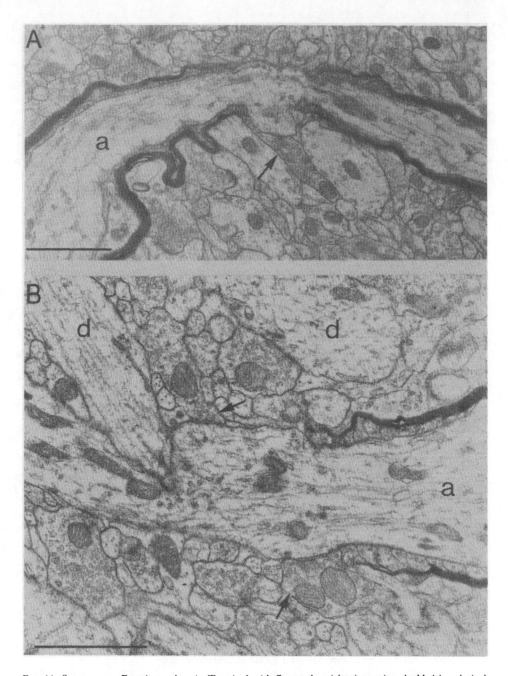

FIG. 11. Synapses on Ranvier nodes. A: Terminal with flattened vesicles (arrow) embedded in relatively dense background making contact with the nodal membrane of an axon in the stratum oriens of CA3 (rabbit). B: Terminals with flattened vesicles (arrow) terminating on the unmyelinated portion of a presumed afferent axon found in the stratum oriens of CA3 in rabbit. a, Axon; d, dendrite.

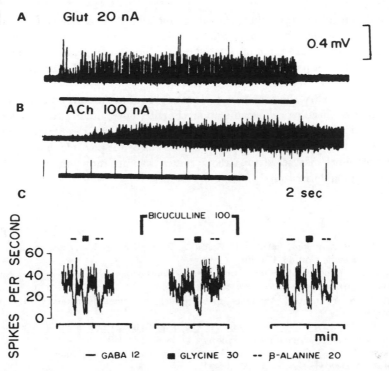

FIG. 12. Responses of hippocampal cells to iontophoretic ejected drugs. A: Short-latency response of CA3 neuron to small dose of glutamic acid. B: Typical long-latency response of CA3 cells to ejection of acetylcholine with relatively large current. The horizontal bars under the records show the time of ejection. C: Inhibitory effect of GABA, glycine, and β-alanine shown to the left. Simultaneous ejection of bicuculline (middle) prevents most of the effect of GABA and β-alanine, whereas the glycine depression remains. The right-hand record shows recovery after 2 min. From Curtis *et al.* (1971).

acid in an isolated transverse slice of the hippocampus (Dudar, 1972) suggests that glutamic acid may indeed have such a role. Dudar found that ejection of the glutamic acid was effective only in the dendritic tree of the cell recorded from. This would fit the interpretation that the synaptic sites were excited by the glutamic acid. Schwartzkroin and Anderson (1975) have found that the sensitivity to glutamic acid is remarkably high in the dendritic field. This suggests that the glutamic acid is acting on those areas containing synapses and not on the naked soma or proximal dendritic regions.

The most likely inhibitory transmitter in the hippocampal formation is γ-aminobutyric acid (GABA). The evidence for this is that when GABA is ejected toward the soma it inhibits both spontaneous and induced hippocampal pyramidal activity (Stefanis 1964; Curtis *et al.*, 1971). Furthermore, Curtis *et al.* (1971) found that the specific inhibitor of GABA receptors, bicuculline, when given in small doses, prevented the hippocampal inhibition, whereas strychnine had no effect. The enzyme producing GABA from glutamic acid, glutamic acid carboxylase (GAD), has its highest concentration just at the soma, where the recurrent inhibition takes place (Fonnum and Storm Mathisen, 1969).

In addition to the transmitter candidates mentioned, it is relevant that the hippocampal formation contains considerable quantities of serotonin (5-hydroxytryptamine, 5-HT) and of catecholamines (norepinephrine and epinephrine). As is the case in other parts of the brain, a proportion of the catecholamine is probably located in dense-core vesicles. However, in contrast to the enzymes associated with acetylcholine or GABA, the distribution of both 5-HT and the catecholamines does not follow a known histological stratification. Furthermore, in electron micrographs dense-core vesicles are found in all identified fiber systems, both excitatory and inhibitory, as well as in the postsynaptic dendrites. Therefore, it is possible that the serotonin and catecholamines may serve a metabolic rather than an immediate transmitter role in the hippocampal formation.

8. References

Andersen, P. Interhippocampal impulses. II. Apical dendritic activation of CA1 neurones. *Acta Physiologica Scandinavica*, 1960, **48**, 178–208.

Andersen, P., and Lømo, T. Mode of activation of hippocampal pyramidal cells by excitatory synapses on dendrites. *Experimental Brain Research*, 1966, **2**, 247–260.

Andersen, P., and Vaaland, J. Axo-axonal contact in the hippocampal formation. 1974, unpublished observations.

Andersen, P., Eccles, J. C., and Løyning, Y. Location of postsynaptic inhibitory synapses of hippocampal pyramids. *Journal of Neurophysiology*, 1964a, **27**, 592–607.

Andersen, P., Eccles, J. C., and Løyning, Y. Pathway of postsynaptic inhibition in the hippocampus. *Journal of Neurophysiology*, 1964b, **27**, 608–619.

Andersen, P., Blackstad, T. W., and Lømo, T. Location and identification of excitatory synapses on hippocampal pyramidal cells. *Experimental Brain Research*, 1966a, **1**, 236–248.

Andersen, P., Holmqvist, B., and Voorhoeve, P. E. Entorhinal activation of dentate granule cells. *Acta Physiologica Scandinavica*, 1966b, **66**, 448–460.

Andersen, P., Bliss, T. V. P., and Skrede, K. K. Unit analysis of hippocampal population spikes. *Experimental Brain Research*, 1971a, **13**, 208–221.

Andersen, P., Bliss, T. V. P., and Skrede, K. K. Lamellar organization of hippocampal excitatory pathways. *Experimental Brain Research*, 1971b, **13**, 222–238.

Andersen, P., Bland, B. H., and Dudar, J. D. Organization of the hippocampal output. *Experimental Brain Research*, 1973a, **17**, 152–168.

Andersen, P., Teyler, T., and Vaaland, J. Morphological specialization of a hippocampal synapse. 1973b, unpublished observations.

Biscoe, T. J., and Straughan, D. W. Micro-electrophoretic studies of neurones in the cat hippocampus. *Journal of Physiology (London)*, 1966, **183**, 341–359.

Blackstad, T. W. Commissural connections of the hippocampal region in the rat, with special reference to the mode of termination. *Journal of Comparative Neurology*, 1956, **105**, 417–537.

Blackstad, T. W. On the termination of some afferents to the hippocampus and fascia dentata. *Acta Anatomica (Basel)*, 1958, **35**, 202–214.

Blackstad, T. W., and Kjaerheim, Å. Special axo-dendritic synapses in the hippocampal cortex: Electron and light microscopic studies on the layer of mossy fibers. *Journal of Comparative Neurology*, 1961, **117**, 133–159.

Blackstad, T. W., Fuxe, K., and Høkfelt, T. Noradrenalin nerve terminals in the hippocampal region of the rat and the guinea pig. *Zeitschrift für Zellforschung*, 1967, **78**, 463–473.

Blackstad, T. W., Brink, K., Hem, J., and Jeune, B. Distribution of hippocampal mossy fibers in the rat: An experimental study with silver impregnation methods. *Journal of Comparative Neurology*, 1970, **138**, 433–449.

BLISS, T. V. P., AND LØMO. T. Long-lasting potentiation of synaptic transmission in the dentate area of the anaesthetized rabbit following stimulation of the perforant path. *Journal of Physiology (London)*, 1973, **232**, 331–356.

BRÜCKE, F., SAILER, S., AND STUMPF, C. Wechselwirkungen zwischen Physostigmin einerseits und Evipan, Procain, Largactil und Scopolamin andererseits auf die rhinencephale Tätigkeit des Kaninchens. *Archiv für Experimentelle Pathologie und Pharmakologie*, 1958, **232**, 433–441.

CRAGG, B. G., AND HAMLYN, L. H. Some commissural and septal connections of the hippocampus in the rabbit: A combined histological and electrical study. *Journal of Physiology (London)*, 1957, **135**, 460–485.

CURTIS, D. R., AND CRAWFORD, J. M. Central synaptic transmission—Microelectrophoretic studies. *Annual Review of Pharmacology*, 1969, **9**, 209–240.

CURTIS, D. R., DUGGAN, A. W., FELIX, D., JOHNSTON, G. A. R., AND McLENNAN, H. Antagonism between bicuculline and GABA in the cat. *Brain Research*, 1971, **33**, 57–73.

DANSCHER, G., SHIPLEY, M. T., AND ANDERSEN, P. Persistent funciton of mossy fibre synapses after metal chelation with DEDTC (Antabuse). *Brain Research*, 1975, **85**, 522–526.

DUDAR, J. D. Glutamic acid sensitivity of hippocampal pyramidal cell dendrites. *Acta Physiolica Scandinavica*, 1972, **84**, 28A.

FONNUM, F., AND STORM-MATHISEN, J. GABA synthesis in rat hippocampus correlated to the distribution of inhibitory neurones. *Acta Physiologica Scandinavica*, 1969, **76**, 35–37A.

GLOOR, P., VERA, C. L., AND SPERTI, L. Electrophysiological studies of hippocampal neurons. III. Responses of hippocampal neurons to repetitive perforant path volleys. *Electroencephalography and Clinical Neurophysiology*, 1964, **17**, 353–370.

HAMLYN, L. H. Electron microscopy of mossy fibre endings in Ammon's horn. *Nature (London)*, 1961, **190**, 645–646.

HJORTH-SIMONSEN, A. Projection of the lateral part of the entorhinal area to the hippocampus and fascia dentata. *Journal of Comparative Neurology*, 1972, **146**, 219–232.

HJORTH-SIMONSEN, A. Some intrinsic connections of the hippocampus in the rat: An experimental analysis. *Journal of Comparative Neurology*, 1973, **147**, 145–161.

HJORTH-SIMONSEN, A., AND JEUNE, B. Origin and termination of the hippocampal perforant path in the rat studied by silver impregnation. *Journal of Comparative Neurology*, 1972, **144**, 215–232.

LØMO, T. Patterns of activation in a monosynaptic cortical pathway: The perforant path input to the dentate area of the hippocampal formation. *Experimental Brain Research*, 1971a, **12**, 18–45.

LØMO, T. Potentiation of monosynaptic EPSPs in the perforant path–dentate granule cell synapse. *Experimental Brain Research*, 1971b, **12**, 46–63.

LORENTE DE NÓ, R. Studies on the structure of the cerebral cortex. II. Continuation of the study on the ammonic system. *Journal für Psychologie und Neurologie (Leipzig)*, 1934, **46**, 113–117.

NAFSTAD, P. H. J. An electron microscope study on the termination of the perforant path fibers in the hippocampus and the fascia dentata. *Zeitschrift für Zellforschung*, 1967, **76**, 532–542.

RAMÓN Y CAJAL, *Histologie du système nerveux de l'Homme et des Vertébrés*. Paris: A. Maloine, 1911.

SCHWARTZKROIN, P. A., AND ANDERSEN, P. Glutamic acid sensitivity of dendrites in hippocampal slices *in vitro*. In: G. Kreuzberg (Ed.), *Properties of dendrites*. New York: Raven Press, 1975.

SHUTE, C. C. D., AND LEWIS, P. R. The use of cholinesterase techniques combined with operative procedures to follow nervous pathways in the brain. *Bibliotheca Anatomica*, 1961, **2**, 34–49.

SHUTE, C. C. D., AND LEWIS, P. R. Cholinesterase-containing systems of the brain of the rat. *Nature (London)*, 1963, **199**, 1160–1164.

SKREDE, K. K., AND WESTGAARD, R. H. The transverse hippocampal slice: A well-defined cortical structure maintained *in vitro*. *Brain Research*, 1971, **35**, 589–593.

SPENCER, W. A., AND KANDEL, E. R. Hippocampal neuron responses to selective activation of recurrent collaterals of hippocampofugal axons. *Experimental Neurology*, 1961, **4**, 149–161.

STEFANIS, C. Hippocampal neurons: Their responsiveness to microelectrophoretically administered endogenous amines. *Pharmacologist*, 1964, **6**, 171.

STORM-MATHISEN, J., AND BLACKSTAD, T. W. Cholinesterase in the hippocampal region. *Acta Anatomica (Basel)*, 1964, **56**, 216–253.

7

An Ongoing Analysis of Hippocampal Inputs and Outputs: Microelectrode and Neuroanatomical Findings in Squirrel Monkeys

1. Introduction

In this chapter, I shall deal principally with a series of microelectrode studies centering around the hippocampal formation, which, as will be explained, occupies a pivotal position in the evolution of the paleomammalian-type brain. First, I shall summarize the results of our investigations concerned with the nature and mechanism of action of intero- and exteroceptive inputs. Then I shall conclude by describing findings on the influence of the hippocampus on unit activity in structures of the brain stem recognized to be involved in somatovisceral and neuroendocrine aspects of "prosematic" behavior. "Prosematic," meaning rudimentary signaling, refers to any kind of nonverbal signal—vocal, bodily, chemical—used by animals or human beings in communication (MacLean, 1974c).

The work to be reviewed involved testing the responsiveness of more than 12,000 units, of which nearly 40% were located in various limbic cortical and subcortical areas. All experiments were performed on squirrel monkeys, most of

PAUL D. MacLEAN • Laboratory of Brain Evolution and Behavior, National Institute of Mental Health, Bethesda, Maryland.

which were prepared for exploration of the brain under awake, sitting conditions. Findings will also be presented from neuroanatomical studies in which improved silver techniques were used in an attempt to corroborate the electrophysiological observations.

In the rest of the Introduction, I shall explain the conceptual framework of this research, dealing first with some evolutionary considerations and then citing antecedent studies.

1.1. Evolutionary Considerations

In its evolution, the primate forebrain expands along the lines of three basic patterns which may be characterized both anatomically and biochemically as reptilian, paleomammalian, and neomammalian. Radically different in chemistry and structure and in an evolutionary sense countless generations apart, the three basic formations constitute, so to speak, three brains in one, a *triune* brain (MacLean, 1970, 1973b).

From an evolutionary standpoint, the reptilian forebrain is of particular interest because it allows one to visualize how developments at a critical locus in the so-called hypopallium (see a, Fig. 1) described by Elliot Smith (1918/1919) may have tipped the scales so that some animals evolved in the direction of birds while others went the mammalian way. The critical area lies near the ventrolateral base of what J. B. Johnston (1916) called the dorsal ventricular ridge, presumably because it reminded him of a mountain ridge. In an extension of Johnston's geological analogy, the proliferating hypopallial area might be imagined as comparable to a turbulent volcanic zone. In birds its continued eruption resulted in a piling up of ganglia on ganglia, whereas its explosion in mammals was responsible for the mushrooming of cortex forming the dorsolateral part of the brain.

At the base of the dorsal ventricular ridge are structures that have remained firmly embedded in the brains of reptiles, birds, and mammals. Evidence in support of this conclusion has been strengthened by recent comparative histochemical studies—particularly those demonstrating the presence and localization of cholinesterase (Parent and Olivier, 1970) and dopamine (Juorio and Vogt, 1967). In mammals the structures corresponding to the ganglia in this part of the reptilian forebrain include the olfactostriatum, the corpus striatum (caudate nucleus and putamen), the globus pallidus, and satellite collections of gray matter. Since there is no name that applies to all of these structures, they may be simply referred to as the striatal complex. Recent experimental studies in monkeys indicate that the striatal complex plays a fundamental role in such genetically constituted, species-typical, prosematic forms of behavior as displays in aggression, courtship, and greeting (MacLean, 1972a, 1973a, 1974a).

On the dorsomedial surface of the cerebral hemispheres of reptiles (well removed from the mentioned ventrolateral site of turbulence) is a rudimentary cortex, which on the basis of its anatomical structure (Humphrey, 1966) and histochemistry (see Baker-Cohen, 1969) appears to correspond to the archicortex of the hippocampal formation. This area failed to develop further in avian evolution, but in mammals became a principal site of unfolding of the limbic cortex.

Since it is generally not explained in textbooks, it is worth mentioning the origin of the term "archipallium," for which the word "archicortex" is now often substi-

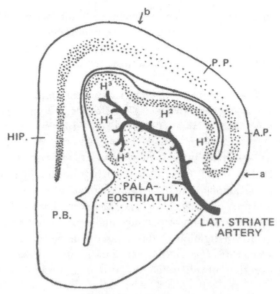

Fig. 1. Reproduction of first figure of Elliot Smith's paper of 1918/1919, illustrating a frontal section through the forebrain of the tuatara (*Sphenodon punctatum*). The diagram is useful for suggesting how a proliferating area represented by the U-shaped collection of cells (arrow at *a*) may have been influential in determining the divergent evolution of birds and mammals. It was as though a proliferation of cells in the left limb of the U had led to a piling up of ganglia on ganglia in birds, whereas activity in the right limb resulted in a burgeoning of cortex in mammals (see text). H^1–H^5 designate areas to which Elliot Smith gave the name "hypopallium," and which J. B. Johnston had previously referred to as the "dorsal ventricular ridge." The lateral striate artery marks the boundary between the ridge and the underlying striatal complex, which is found as a constant feature in the brains of reptiles, birds, and mammals. HIP. identifies the hippocampal formation, which becomes a principal site of unfolding of the limbic cortex. Other abbreviations: A.P., area pyriformis [*sic*]; P.B., paraterminal body; P.P., parahippocampal pallium.

tuted. It was originally attributed to Elliot Smith. Incensed by being credited for the "invention" of a term of which he did not approve, Smith published "A Disclaimer" in 1910, explaining that his original expression "old pallium" had been inappropriately translated into the German by Edinger as "archipallium." The word amounts to a redundancy, Smith argued, "if we apply [it] exclusively to the hippocampal formation, as many writers are doing." (Smith himself, 1901, had included part of the piriform lobe as "old pallium".) The hippocampal formation is usually understood to include the dentate gyrus, hippocampus, and parahippocampal gyrus.

In the lost transitional forms between reptiles and mammals—the so-called mammal-like reptiles—the archipallium is presumed to have undergone further development and differentiation. In 1878, Broca published his comparative studies showing that the large convolution which he called the "great limbic lobe" is found as a common denominator in the brains of all mammals. The word "limbic" was used descriptively to indicate that this lobe surrounds (literally, forms a border around) the brain stem. Broca's fundamental observations provided evidence that the cortex of the limbic lobe represents a paleomammalian inheritance.

The question has been raised as to whether or not Broca included the hippocampus in the limbic lobe (Stephan, 1964), but he could hardly have meant otherwise because the archipallium, stretching from the septum to the amygdala, constitutes part of the innermost ring of cortex surrounding the brain stem.

In the latter part of the nineteenth century, the entire limbic lobe including the hippocampus came to be referred to as the "rhinencephalon." As stated in 1900 in such an authoritative source as Schäfer's *Text-Book of Physiology*: "Broca was of opinion that the whole of the limbic lobe . . . is related to the sense of smell. . . ." A footnote cites other authorities, including Turner, who adhered to this view. It should be pointed out, however, that Turner, who had reintroduced the term "rhinencephalon" in 1890 (*cf.* Smith, 1901), did not include the cingulate gyrus or hippocampus in the rhinencephalon. Ramón y Cajal, writing in the early 1900s, stated: "Since the memorable works of Broca the general opinion has been that the limbic convolutions are the station for the transmission of primary and secondary fibres" (translation by Kraft, 1955). His own investigation of this matter will be mentioned later. In their *Comprehensive Textbook of Human Anatomy* published in 1962, Crosby *et al.* have continued in the tradition of referring to the entire limbic lobe as part of the rhinencephalon.

In 1937, Papez published his now famous paper in which he stated that "there is no clinical or other evidence to support [the] view that the hippocampal and cingulate parts of the limbic lobe are concerned with olfactory function." Rather, he proposed that these structures are concerned with the experience and expression of emotion.

The classical study of Klüver and Bucy (1939) on the effects of bilateral temporal lobectomy in monkeys provided support for the Papez thesis, as did other developments to be mentioned shortly. By the early 1950s it was becoming evident that the limbic lobe and related subcortical structures were functionally, as well as anatomically, an integrated system. In 1952, MacLean reverted to the use of Broca's descriptive term "limbic" and referred to the counterpart of the paleomammalian brain as the *limbic system*. The system was defined as the limbic cortex (and perilimbic cortex of the temporal pole) and structures of the brain stem with which it has primary connections. The term "limbic," as Broca (1878) originally pointed out, has the advantage that it carries no implications as regards function.

Developments in clinical electroencephalography, beginning with the identification by Gibbs *et al.* (1948) of a condition that they called "psychomotor epilepsy," have provided the most convincing evidence that the limbic system is involved in the elaboration of emotional states. *During the initial epileptic discharge, patients may experience one or more of a wide variety of vivid emotional feelings.* Gibbs *et al.*, concluded that the epileptogenic focus was usually in the temporal polar region. In basal lead studies, however, MacLean and Arellano (1950) uncovered cases in which there was evidence of foci in medial basal temporal structures. In their surgically treated patients, Penfield and Jasper (1954) found that electrical stimulation in or near the limbic cortex elicited affective and other subjective states similar to those experienced during the aura.

The pathological studies of Sano and Malamud (1953) and of Margerison and Corsellis (1966) have revealed that Ammon's horn (hippocampal) sclerosis is, in

Malamud's words (1966), the "common denominator in cases of psychomotor epilepsy." The sclerosis, however, frequently extends into other medial temporal structures, and for this reason Falconer *et al.* (1964) prefer the term "medial temporal sclerosis." Their analysis is based on two series of 100 cases in which the temporal lobe on one side was resected *en bloc,* a procedure which, contrary to the usual neurosurgical practice of aspiration, affords a thorough gross and microscopic examination.

The aura experienced in psychomotor epilepsy is believed to be a subjective manifestation of the initial discharge. In some cases, it may be associated with sensations or distortions of perception involving one or more of the interoceptive and exteroceptive systems. In 1949, I elaborated upon the Papez theory of emotion by suggesting that impulses from *all* of the intero- and exteroceptive systems find their way to the hippocampus via the hippocampal gyrus. The hypothetical pathways were schematized in Fig. 3 of that paper. It was suggested that the hippocampal formation combines information of internal and external origin into affective feelings that find further elaboration and expression through connections with the amygdala, septum, striatum, and hypothalamus, as well as through the reentry pathway to the limbic lobe via the mammillothalamic tract and thalamic projections to the cingulate gyrus (the so-called Papez circuit).

1.2. The Problem of "Sensory" Inputs

Aside from the stated hypotheses, the question as to the nature of limbic inputs presents an important problem in its own right. Even if the original claims of anatomists about the olfactory functions of the limbic lobe were correct, there would remain the question of how olfactory information is integrated with that of other sensory systems.

In reasoning from analogy, it might be expected that the limbic cortex, like the neocortex, would receive information from more than one sensory system, and might also have special areas for their representation. One might imagine the neomammalian formation as constituting an assembly of first-order, second-order, etc., "biocones" related to each of its special sensory systems. The projecting nucleus would represent the apex of the cone and the recipient cortex the base of the cone. (Corticocortical associations between cones, whether ipsilateral or commissural, would be comparable to hourglass cones.) It is known from clinical observations that destruction of any part of the cone formed by the lateral geniculate nucleus and its projections to the striate cortex results in a scotoma within the part of the visual field involved. What accounts for the subjective awareness of the picture generated by the cone remains a mystery, but it may be presumed that, remotely analogous to a television tube, it requires both the cortical screen and nuclear workings in the cone. Compounding the mystery is the question of how there results a subjective integration of information derived from a whole assembly of cones with their apices interlocking centrally.

A question we shall return to following the presentation of recent findings is: Does any such collection of analyzing cones (and associating cones) exist in the limbic system?

1.3. Antecedent Findings

Except for olfaction (e.g., Ramón y Cajal, 1955; Clark and Meyer, 1947; Fox *et al.*, 1944; Meyer and Allison, 1949; Penfield and Erickson, 1941), there existed prior to 1950 hardly any experimental neuroanatomical or electrophysiological information about the relationship of other sensory systems to the limbic lobe. On the basis of normal material, Herrick (1921) concluded that the anterior olfactory nucleus receives various ascending non-olfactory systems from the diencephalon. Johnston (1923) referred to the amygdala as "a complex in which olfactory, gustatory, and general somatic sense impressions are brought into correlation." In 1938, Bailey and Bremer reported that vagal stimulation evoked synchronized activity in the orbital gyrus of the cat—a region that would correspond to the limbic cortex overlying the claustrum. In 1951, Dell and Olson reported that in *encephale isolé* preparations of cats, vagal volleys evoked slow potentials in the buried cortex of the anterior rhinal fissure and in the amygdala. The following year, MacLean *et al.* (1952) reported that gustatory and noxious stimulation resulted in rhythmically recurring olfactory-like potentials in the piriform area of the rabbit. Sometimes rhythmic potentials appeared in the hippocampus following olfactory and gustatory stimulation. In pursuing this lead, Green and Arduini (1954) found in experiments on *unanesthetized,* curarized animals that different forms of sensory stimulation evoked rhythmic θ activity in the hippocampus. Such changes appeared to be characteristic of macrosmatic animals, but not of primates, and were regarded as nonspecific in nature. In 1957, MacLean (1957*b*) described some ancillary experiments on rats in which recordings from the dorsal hippocampus revealed that rhythmic activity of about 6/s "characteristically appeared when an animal seemed to focus its attention or was exploring and disappeared when it was eating or drinking." In the Volume 2 of this series, Winson analyzes various conditions under which hippocampal θ activity appears.

Since 1952, it has been shown anatomically (Daitz and Powell, 1954) and physiologically (Green and Adey, 1956; Green and Arduini, 1954) that the septum acts as a connecting link between the hypothalamus and the hippocampal formation. This input presumably provides a source of interoceptions.

In the paper of 1949 mentioned above, I suggested that somatic, auditory, and visual information might be channeled to the hippocampal gyrus by transcortical connections from the primary receiving areas. Subsequently, Pribram and MacLean reported strychnine neuronographic findings in the cat (MacLean and Pribram, 1953) and monkey (Pribram and MacLean, 1953) that were compatible with this hypothesis. Four years ago, Jones and Powell (1970) described experimental anatomical observations in the macaque that would afford stepwise cortical connections from the primary receiving areas to the hippocampal gyrus. Additional support for this inference has been provided by an anatomical study by Van Hoesen *et al.* (1972).

Various authors have reported that visual, auditory, and somatic stimulation evokes slow-wave potentials in the hippocampus (Brazier, 1964; Robinson and Lennox, 1951) or other areas of limbic cortex (Desmedt and Mechelse, 1959; Harman and Berry, 1956; Hughes, 1959; Ingvar and Hunter, 1955; O'Leary and Bishop,

1938), but in such experiments there may be doubt about the origin of a response because of the possibility of volume-conducted potentials from neighboring structures.

2.   Microelectrode and Anatomical Findings on Limbic Inputs

Because of the mentioned limitations of the evoked slow potential technique for
purposes of localization, we developed methods for microelectrode recording of
evoked unit responses in chronically prepared, awake, sitting squirrel monkeys. Such
experimentation not only makes it possible to be sure of the locus of a response but
also avoids the depressant effects of anesthesia on neural transmission. Monkeys were
chosen as subjects because it was desirable to use animals which, like man, have a
well-developed visual system. Thus far, we have explored virtually all the cortex of
the limbic lobe except the posterior orbital and piriform areas. I shall first review the
results of stimulating exteroceptive systems and then describe progress on a study
testing the effects of vagal stimulation.

It must be emphasized that the primary purpose of these investigations was to
learn whether or not there are areas of limbic cortex in which cells, like those of neocortical sensory areas, are specifically and regularly activated by stimulation of
particular sensory systems. In view of the findings discussed earlier on rhythmic
slow-wave responses, a convergence of more than one sensory system on some
neurons was anticipated, but proved to be unlikely. In relying primarily on extracellular recording for detecting unit excitation or inhibition, we were not unmind-
ful of the importance of inputs that have a partial depolarizing effect on neurons. For
example, it will be seen that with intracellular recording, stimulation of the olfactory
bulb produced excitatory postsynaptic potentials in hippocampal and entorhinal
neurons, but no spike discharge. Such changes reflect modulatory effects that might
be significant with respect to the conditioning of neurons and to mechanisms of learning and memory. But as regards the influence of a particular sensory system on
limbic cortical areas, stimulus time-locked activation or inhibition of units would
provide stronger evidence of direct, effective connections than partial depolarizing
responses.

2.1.   Methods

The stereotaxic device used for exploring the brain in the frontal plane of the
stereotaxic atlas (Gergen and MacLean, 1962) consists of a platform with electrode
guides which is fixed above the scalp on four screws cemented into the skull (Mac-
Lean, 1967) (see Fig. 9). The device provides a closed system for exploration with
either metal or glass microelectrodes and among other advantages makes it possible
to obtain serial histological sections in the same plane as all of the electrode tracks.
During an experiment, the monkey sits in a special chair which prevents flexion of
the neck and resulting movement of the brain. Gently swaddled, it sits quietly and is
periodically given its favorite forms of liquid nourishment. After exploration of a
track, the electrode is fixed at a desired locus for histological localization of the tip

and reconstruction of the track. About eight explorations are conducted on each animal over a period of 6 wk. A brief description of the techniques for sensory stimulation will be given under the appropriate heading.

2.2. Visual Input

Because of the progressive importance of vision in the evolution of primates, we initially focused on the question of a visual input to the hippocampus and other limbic areas. Photic stimulation was performed with a stroboscope, a tungsten light, and moving patterns. The conditions of the experiment made it impractical to plot receptive fields. In pilot studies conducted under chloralose anesthesia, photic stimulation elicited in about half the animals high-amplitude slow potentials in the hippocampus, with maximum negativity in the radiate layer (Gergen and MacLean, 1964). In the awake, sitting animal, none of the 596 units recorded from the hippocampus responded to visual stimulation (MacLean et al., 1968). As topographically shown in Fig. 2, the only limbic areas found to contain consistently responding units were the posterior hippocampal gyrus (H), the parahippocampal portion of the lingual gyrus (L), and the retrosplenial cortex (R). Units in the perilimbic cortex of the fusiform gyrus were also responsive. As will deserve further comment (see below), the responsive parahippocampal area did not extend forward into the entorhinal cortex.

A majority of the responsive cells in the posterior hippocampal gyrus were of particular interest because they gave a sustained on-response during illumination of the eye (see Figs. 2 and 3). No such responses were observed in testing numerous units in the striate and peristriate cortex and in the superior colliculus and subdivisions of the pulvinar. The only other structures to show sustained on-responses were the lateral geniculate body and the lateral tegmental process of the pons.

Some cells in the retrosplenial cortex were characteristic insofar as they responded only to stimulation of the contralateral eye, suggesting the possibility that impulses originated in the primitive temporal monocular crescent.

The photically activated cells appeared to be modality specific insofar as they were unresponsive to auditory stimulation and to sample testing with somatic stimuli.

The shortest latencies for the responsive units in the limbic areas (about 41 ms) were more than 2 times those observed for cells in the striate cortex (20 ms). Nevertheless, the latency values and the regularity and character of the responses were suggestive of a subcortical rather than transcortical pathway. Evidence in support of this inference was obtained in a neuroanatomical study in which improved silver techniques for demonstrating fine fibers were used to trace degeneration from lesions in the lateral geniculate body and the pulvinar (MacLean and Creswell, 1970). A continuous band of degeneration extended from the lateral geniculate body into the core of the hippocampal gyrus (Fig. 4). Some fibers entered the cortex of the posterior hippocampal gyrus as well as contiguous areas in the lingual gyrus and fusiform cortex. Traced caudally, the medial degenerating fibers in the optic radiation appeared to intermingle with those of the cingulum. In agreement with the electrophysiological findings, degeneration did not extend forward into the entorhinal

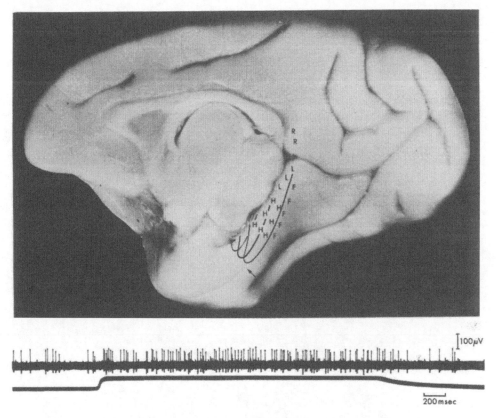

Fig. 2. Medial view of squirrel monkey's brain showing topographical location of limbic (H, L, R) and perilimbic (F) areas with units activated by photic stimuli. The curved black lines indicate the temporal detour of optic fibers traced anatomically to these areas. The arrow points to the termination of the rhinal fissure and the caudal boundary of the entorhinal cortex. Oscillographic record underneath illustrates a photically sustained on-response characteristic of units found in the posterior hippocampal gyrus (H). Lower tracing shows response of photocell signaling ocular illumination for 2.5 s. Negativity up in this and subsequent records. Other abbreviations: F, perilimbic cortex of fusiform gyrus; L, parahippocampal portion of lingual gyrus; R, retrosplenial cortex. Adapted from MacLean *et al.* (1968).

cortex. Lesions of the inferior pulvinar resulted in a coarser type of degeneration involving the same limbic and perilimbic areas. The pulvinar projections were contained in a band of fibers just lateral to the optic radiations. These findings will perhaps prove to correlate with those of Mishkin and Pribram, who found retrograde degeneration in the inferior pulvinar following ventral temporal lesions in the baboon (Mishkin and Pribram, 1954) and macaque (Mishkin, 1954).

2.3. Comment

In view of the recent findings that the superior colliculus projects to the inferior pulvinar (Mishkin, 1972; Myers, 1963; Snyder and Diamond, 1968), it might be in-

A

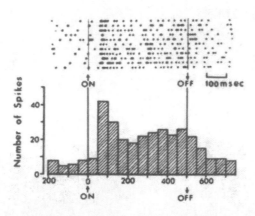

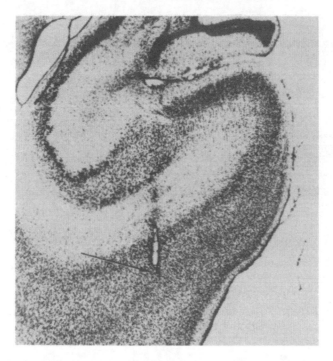

FIG. 3. Arrows in the three histological sections show marks of microelectrode tips at loci in posterior hippocampal gyrus where units responded to photic stimulation. The superimposed oscillographic dot displays with pre- and poststimulus time histograms show series of photic responses of units at these respective loci. Adapted from MacLean *et al.* (1968).

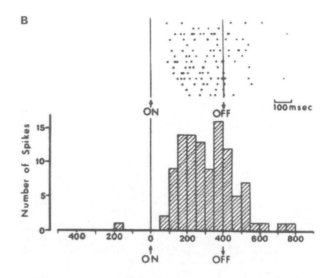

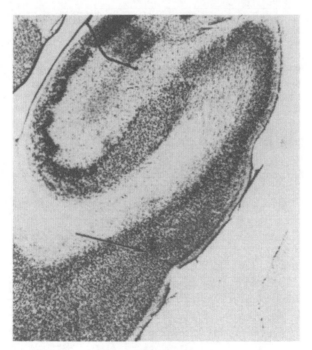

FIG. 3. (*Continued*)

C

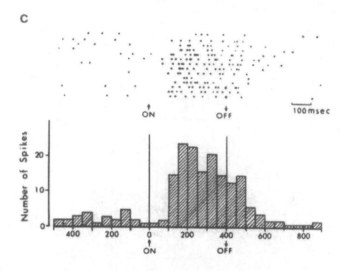

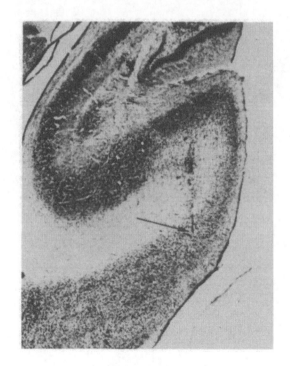

FIG. 3. (*Continued*)

ferred that the pulvinar rather than the sparse lateral geniculate connections accounted for the limbic photic responses. Against this supposition, however, were the
observations that no units in either the superior colliculus or pulvinar gave a sustained on-response to illumination of the eye.

In the human brain, the temporal detour of the optic radiations around the
lateral ventricle is referred to as Flechsig's "knee" and Meyer's "loop." Polyak's
(1957) description of this part of the radiations would apply to the picture of
degeneration seen in Fig. 4: "The bent-in lower edge . . . thins out into a narrow
fiber sheet, which slips medially underneath the ventricle and becomes the core of the
hippocampal gyrus." The reason for this temporal detour has always been puzzling.
The present findings indicate that in the course of phylogeny the optic radiation was
connected with the limbic cortex before the great ballooning out of the neopallium.
Putnam's reconstructions of the optic radiations show the presence of a "temporal
knee" in the rabbit, cat, monkey, and ape (1926).

In agreement with the microelectrode findings, it is significant that the
degeneration does not extend forward into the entorhinal cortex (area 28). The entorhinal cortex (so named because its lateral border is bounded by the rhinal sulcus)
has a distinctive architecture, and in macrosmatic mammals occupies a triangular
area in the posterior part of the hippocampal gyrus. Ramón y Cajal referred to it as
the angular ganglion (Kraft's translation, 1955). He wished to discover whether or
not it was a part of the limbic lobe that should be included in the so-called rhinencephalon. He pointed out that it had three similarities to visual cortex. "The importance of this ganglion," he emphasized, "is based on its very strong relationships
with Ammon's horn and the fascia dentata. . . . Thus it can be said without fear of error that if the angular ganglion is olfactory, then so should Ammon's horn be olfactory.
If the former is optic, so is the latter, etc." He concluded his study by favoring the
hypothesis that it is "a special olfactory centre" (see Chapter 1 in regard to recent
findings).

Caudally, the entorhinal cortex ends abruptly with the termination of the rhinal
sulcus. (See Fig. 1 in Cuénod et al., 1965, for the histological picture in monkey and
Fig. 8.5 in MacLean, 1972b, for man.) As Crosby et al. (1962) point out, the entorhinal cortex becomes greatly expanded in the human brain. It accounts partly for
what Spatz (1964) called the great "promination" of the human brain in the
frontotemporal region. As will be discussed later, it is in a position to integrate
sensory information transmitted to it from several limbic areas, including the visually
responsive area under consideration. In primates, the latter occupies the posterior
part of the hippocampal gyrus, (thi of Beck; area 36 of Brodmann), as well as the
parahippocampal part of the lingual (area 26 of Beck) and retrosplenial (area 29 of
Brodmann) cortex (see MacLean, 1966, for further discussion). Sanides and
Vitzthum (1965) include areas 36, 26, and 29 under the designation prostriata, a
term signifying a primitive form of visual cortex (see Fig. 2 in Vitzthum and Sanides,
1966).

In the past, the emphasis on the close relationship of the hippocampal formation
to the rhinal fissure seems to have diverted attention from its close relationship to the
calcarine sulcus (MacLean, 1966, 1972b). Early in this century, Elliot Smith (1902)

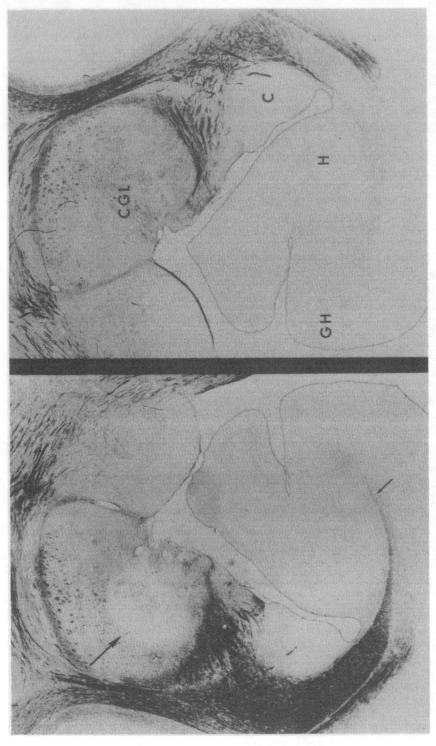

Fig. 4. Photomicrograph on left shows degeneration in the optic radiations leading from a lesion in the ventrolateral part of the lateral geniculate body (large arrow) to the core of the posterior hippocampal gyrus (small arrow). Some degenerating fibers were traced to cells in the posterior hippocampal gyrus and adjacent areas labeled in Fig. 2. Photomicrograph on right is from the same section showing the corresponding structures on the opposite side where only normal fibers are stained. Abbreviations: C, nucleus caudatus; CGL, corpus geniculatum laterale; GH, gyrus hippocampi; H, hippocampus. From MacLean and Creswell (1970).

pointed out that the calcarine sulcus is one of the three oldest furrows in the brain and that its counterpart, the stem of the splenial sulcus in carnivores and ungulates, marks the boundary between the striate and limbic cortex. In his cytoarchitectural studies, Campbell (1905) noted that in the pig it is difficult to identify the point of transition between the striate and limbic cortex. Rose and Malis (1965) have found in the rabbit that the limbic cortex medial to the splenial sulcus receives projections from the "dorsal lateral geniculate body." Thompson *et al.* (1950) evoked potentials in this same area by stimulation of the contralateral eye (*cf.* microelectrode findings above).

In the original paper describing the tonic on-responses of units in the posterior hippocampal gyrus (MacLean and Creswell, 1970), it was suggested that these units are possibly involved in mechanisms by which light contributes to a state of wakefulness and changes in light induce alerting and attention. It was also suggested that they may play some role in neuroendocrine functions affected by diurnal or seasonal changes in light. These considerations raise the question of a possible input to the posterior hippocampal gyrus from the nucleus opticus tegmenti, which is innervated by the accessory optic tract and is known to contain on-units (Marg, 1964). Although comparative neuroanatomy offers no evidence of such connections, the nucleus falls within the ventral tegmental area of Tsai, which besides being a recipient of fornix projections (Poletti *et al.*, 1973; Valenstein and Nauta, 1959) has been found to contain dopamine cells (Dahlström and Fuxe, 1965) which innervate the olfactostriatum (Ungerstedt, 1971) and possibly the limbic cortex (Hökfelt *et al.*, 1974).

In macrosmatic animals, the hippocampus shows rhythmic θ activity during the rapid eye movement (REM) phase of sleep (Jouvet *et al.*, 1960). Since the posterior parahippocampal cortex is a source of afferents to the hippocampus, it is possible that the archicortex may be implicated in the visual aspects of dreaming and in autonomic manifestations of REM sleep.

In clinical observations on the effects of brain stimulation of deep structures, Horowitz *et al.* (1968) reported that "the posterior hippocampus was the site of greatest interest in terms of visual events." Thirteen of 26 reports of fully formed imagery occurred upon stimulation with electrodes believed to be in the posterior hippocampus. In a number of instances, "the visual imagery occurred in conjunction with an electrically evoked aura or seizure." In 1922, Cushing had commented, "One would naturally expect hallucinations to be a feature of occipital rather than temporal lobe tumors, but the former are less common in my series." Years ago, Penfield and Jasper (1954) had demonstrated at operation that stimulation of the hippocampal gyrus of patients with epilepsy might evoke feelings of *déjà vu* or visual *recollections* (Penfield and Perot, 1963).

2.4. *"Approach-Type" Units*

The only other limbic area of interest with respect to visual stimulation was the cortex overlying the claustrum (Sudakov *et al.*, 1971). The cortex of this area is by definition limbic because it forms part of the phylogenetically old cortex bordering the brain stem. We identified a few units in this area that responded with a brisk discharge to an approaching object (Fig. 5). This finding recalls the observation by

Penfield and Jasper (1954) that discharges in the parainsular cortex may result in macropsia, a condition in which objects appear to become larger.

2.5. Auditory Input

As was mentioned earlier, no hippocampal units responded to auditory stimulation. The same proved true in regard to cells of various areas of the parahippocampal gyrus and the supracallosal and precallosal cingulate cortex. The only responsive limbic area was found in the limbic cortex overlying the caudal part of the claustrum (Sudakov *et al.*, 1971) (*cf.* Fig. 6). Auditory stimulation consisted of clicks and pure tones applied in the form of 5-ms triangular pulses at frequencies ranging from 500 Hz to 20 kHz. Eighteen hundred and twenty-eight units were tested while exploring the insula itself and surrounding areas. Of nearly 500 units in the claustrocortex, approximately 11% responded to auditory stimulation. The "auditory" units were unaffected by visual and somatic stimulation. There were two main types of units. One type responded with the discharge of one to six spikes at short latencies ranging from 7 to 15 ms and could follow a stimulus frequency up to 10/s. The other type responded at longer latencies with a discharge persisting as long as 250 ms; there was a low probability of firing at stimulus rates of more than 1–2/s. A few units were encountered which responded to tones only in the range between 500 Hz and 1.5 kHz.

The finding that some units responded with latencies as short as 7 ms suggests a direct orthodromic pathway. This inference is supported by anatomical studies in which both anterograde and retrograde degeneration has indicated that the insular cortex overlying the claustrum receives projections from the medial geniculate body (for references, see Sudakov *et al.*, 1971).

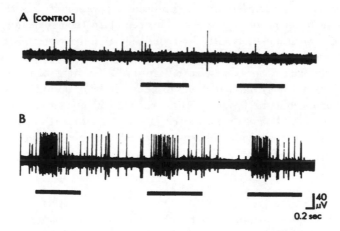

FIG. 5. Lower tracing (B) shows responses of an "approach-type" unit recorded in insular cortex overlying the claustrum. The control recording shown in the upper tracing (A) was made during approach movements in darkness, ruling out the possibility of adventitious stimulation of tactile receptors by currents of air. From Sudakov *et al.* (1971).

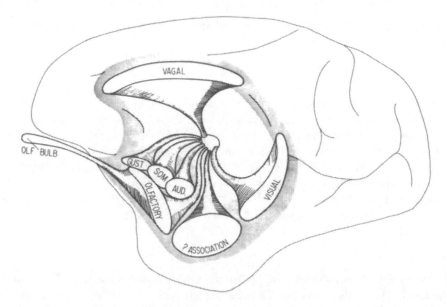

Fig. 6. Diagram of medial aspect of squirrel monkey brain indicating location of limbic "sensory biocones." The buried insular cortex overlying the claustrum is externalized for showing the partially overlapping areas with units specifically activated by gustatory, somatic, or auditory stimulation. The areas marked OLFACTORY, VISUAL, and VAGAL overlie, respectively, the piriform, posterior parahippocampal, and cingulate cortex. The entorhinal area (association), in conjunction with the hippocampus, to which it projects, possibly serves to integrate information received from the various "sensory" areas.

2.6. Somatic Input

The claustrocortex of the insula was also the only limbic area with cells responding to somatic stimulation (Sudakov *et al.*, 1971). Most of the units were located rostral to the auditory zone (see Fig. 6). Somatic stimulation was performed by light touch with a camel's hair brush or exertion of blunt and sharp pressure by means of a probe attached to a strain guage for signaling the onset, duration, and strength of the stimulus. Approximately 9% of the 497 tested units responded to somatic stimulation, with latencies ranging from 30 to 300 ms. There were only three units that responded solely to noxious stimulation. The rest were also activated by light touch. The receptive fields were usually large and bilateral. Fields confined to one side of the body were invariably contralateral to the hemisphere being explored. As in the case of "auditory" units, the "somatic" units appeared to be modality specific.

2.7. Gustatory Stimulation

Of 497 units isolated in the claustrocortex, 437 were tested during gustatory stimulation (Sudakov *et al.*, 1971). Under the conditions of the experiment, there was also the possibility of simultaneously stimulating tactile, thermal, and proprioceptive receptors. Consequently, we attempted no more than to distinguish units responding during gustatory stimulation from those that showed a change in firing rate only with mechanical stimulation of the oral cavity and associated masticatory movements. Given these conditions, there were 14 (3.2%) of 437 units that responded to gustatory stimulation (Fig. 7). The majority were located in the rostral part of the insula (Fig. 6) within an area from which Benjamin and Burton (1968) recorded evoked slow potentials with stimulation of the chorda tympani in the squirrel monkey. In their analysis of gustatory deficits following cortical ablations in the monkey, Bagshaw and Pribram (1953) concluded that the focal cortical area for gustation in the monkey is located at the junction of the frontal operculum and anterior insula. Their study of retrograde degeneration indicated that the "nucleus medialis ventralis" (nucleus reuniens) of the thalamus projects to the rostral part of the insula.

2.8. Comment

Since the claustrocortex projects to the hippocampal formation (Pribram and MacLean, 1953), the foregoing findings suggest pathways by which gustatory, somatic, and auditory information could be transmitted to the brain stem via the hippocampus.

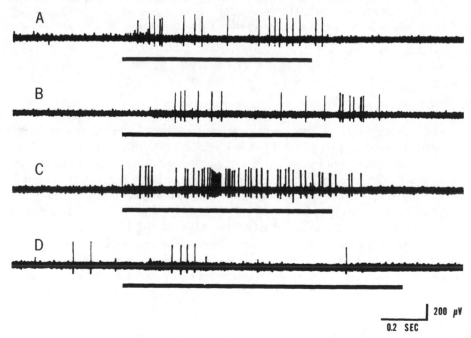

Fig. 7. Examples of responses of unit in cortex of anterior insula following lingual application of cow's milk (A, B), 10% sucrose (C), and distilled water (D). The response to saline was similar to that of glucose. Bars underneath records coincide with signal from foot pedal at moment of application and are not reliable for measurements of response latencies. From Sudakov *et al.* (1971).

2.9. Stimulation of Olfactory Bulb

We have made no systematic study of the effects of natural olfactory stimulation on units of the limbic cortex. We have, however, investigated the differential effects of electrical stimulation of the olfactory bulb and the septum on intracellular and extracellular potentials of hippocampal neurons, in the awake, sitting squirrel monkey (Yokota *et al.*, 1970). Observations were also made on a few neurons in the entorhinal cortex. With the glass microelectrodes in an extracellular position, no spike discharges were observed following application of double or triple shocks to the olfactory bulb (Fig. 8). Such stimuli, however, elicited excitatory postsynaptic potentials (EPSPs) in 14 (26%) of 54 hippocampal cells that were successfully penetrated. The response latencies ranged from 15 to 17.5 ms. Similar EPSPs without spikes were elicited in entorhinal cells. The latencies for the entorhinal units were shorter by 2–2.5, a finding consistent with other evidence that the entorhinal area may be a site of transmission of olfactory impulses to the hippocampus (Cragg, 1960; Gergen and MacLean, 1962).

2.10. Vagal Input

One of the present challenges in neurophysiology is to obtain further information about the influence of interoceptive systems on the functions of the forebrain. We have recently attempted to learn whether or not vagal volleys affect units in specific parts of the diencephalon and telencephalon. The first investigation, which focused on the cingulate cortex, has just been completed (Bachman *et al.*, in preparation). Units of this cortex had been found in an earlier study (Bachman and MacLean, 1971) to be virtually unresponsive to exteroceptive stimulation. The vagal experiments were performed on awake, sitting squirrel monkeys prepared with electrodes chronically implanted on the cervical vagus or inside the jugular foramen next to the vagus (Bachman *et al.*, 1972a). The accessory and hypoglossal nerves were sectioned below the foramen. Of 518 cingulate units, 100 (19.3%) were responsive to vagal volleys, with 50 showing initial excitation and 51 showing initial inhibition. In the supracingulate cortex, 110 (22%) of 498 units were responsive, with 82 initially excited and 28 initially inhibited. Analysis of latencies revealed that cingulate units responded with significantly shorter latencies (12–20 ms for the majority) than those of the supracingulate cortex.

Control tests were performed in an attempt to show that the unit responses did not result from adventitious somatic stimulation. In this respect, it is significant that in a preceding study (Bachman and MacLean, 1971) somatic stimulation failed to excite cingulate units, but activated 14% of those tested in the supracingulate cortex. As a further control, we injected microamounts of serotonin through a catheter in the superior vena cava as a means of exciting pulmonary receptors (Bachman *et al.*, 1972b). Of 80 cingulate units tested with repeated injections, 18 showed excitatory or inhibitory effects, with the ratio of excited to inhibited units being 3:1.

The short-latency responses of the cingulate units suggest a fairly direct oligosynaptic pathway. What are some of the possibilities? Morest (1967) showed

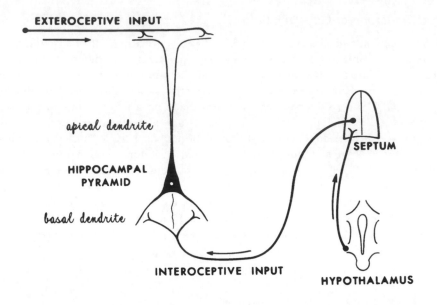

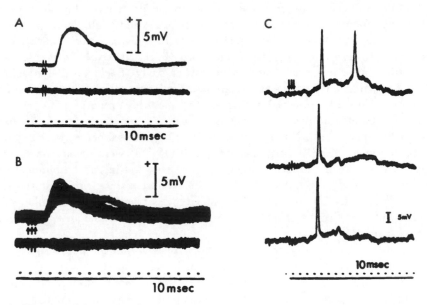

FIG. 8. Diagram of hippocampal circuitry described in text with respect to differential action of interoceptive and exteroceptive inputs. As shown in A and B, olfactory volleys generated only EPSPs without spikes, whereas septal volleys (C) were highly effective in eliciting EPSPs and neuronal discharge. Lower recordings in A and B are recordings just outside cell. In B, several sweeps are superimposed; C shows three successive responses. Resting membrane potentials in each case were stable and measured between −40 and −50 mV. Drawing from MacLean (1969) and records from Yokota *et al.* (1970).

that the nucleus solitarius projects to the dorsal tegmental nucleus of Gudden, suggesting the possibility that vagal impulses may be transmitted via the mammillopeduncular system to the mammillary bodies, anterior thalamic nuclei, and cingulate gyrus. Since there are norepinephrine-containing cells in the nucleus solitarius (Dahlström and Fuxe, 1964) and fine NE terminals in the anterior ventral nucleus (Fuxe, 1965), it is also conceivable that there may be some ascending noradrenergic fibers, possibly involving inhibitory mechanisms. In a current microelectrode study, we found that vagal volleys activated 50% of a small population of units tested in the anterior ventral nucleus (Hallowitz and MacLean, 1974). An unexpected finding was a large percentage of responsive units in the paracentral nucleus. There is evidence that this nucleus projects to the cingulate and supracingulate cortex (Murray, 1966). Significantly, one of Morest's diagrams shows degenerating fibers reaching this nucleus after experimental lesions of the dorsal tegmental nucleus (1961).

2.11.   The Question of Limbic Sensory Integration

In our initial studies on squirrel monkeys anesthetized with α-chloralose, we found evidence of convergence of sensory inputs on some limbic cortical units. It turned out, however, in the awake, sitting monkey that all responsive units appeared to be modality specific, indicating a high degree of selectivity. Moreover, it was found with extracellular recording that there was no limbic cortical meeting place for intero- and exteroceptive inputs. Rather, there was evidence of separate sensory cortical areas that hypothetically represent parts of "biocones" (see Introduction) diagrammatically portrayed in Fig. 6. These considerations raise the question of how information reaching the limbic cortex from intero- and exteroceptive systems is integrated and processed. Metaphorically stated, where does the viewer (or viewers) reside in the limbic system? One likely candidate for this role is the entorhinal cortex, which receives connections from all the responsive areas, namely, the piriform (olfactory), anterior insula (gustatory, somatic, and auditory), posterior parahippocampal cortex (visual), and cingulate cortex (vagal).* As mentioned earlier, the entorhinal cortex forms an extensive area in the human brain. It projects to the hippocampus via the perforant and alveolar pathways.

Located farther "downstream" than the entorhinal area, the hippocampus itself might be imagined in computer terms as serving as "coordinator of requests for various services." It receives information not only from the entorhinal cortex but also from some of the same cortical areas that feed into the entorhinal cortex. The anterior hippocampus receives connections from the frontotemporal cortex (Pribram and MacLean, 1953), and there are afferent connections to the posterior hip-

---

* **Note Added in Proof:** Since this article went to press, we have begun to explore the hippocampus and amygdala. Vagal volleys have elicited unit and slow-wave responses in the hippocampus and central nucleus of the amygdala (Radna and MacLean, in preparation). It is the first time in the awake monkey that we have seen hippocampal unit responses to stimulation of the peripheral nerve.

pocampus from the posterior parahippocampal cortex and from the cingulate cortex via the cingulum and lamina medullaris superficialis. Through such corticocortical connections (representing what were referred to in the Introduction as "hourglass biocones") olfactory, gustatory, somatic, auditory, visual, and visceral information could be processed and selectively scheduled for transmission to the hypothalamus and other structures of the brain stem considered in the next section.

The septum, which receives connections from the hypothalamus (Fig. 8) (Green and Adey, 1956; Green and Arduini, 1954), is probably another source of interoceptive information reaching the hippocampus. The septal projections are believed to terminate largely in the stratum oriens, possibly on the basal dendrites of the hippocampal pyramids. Lewis and Shute have provided evidence that this is a cholinergic system (1967). The perforant pathway, on the contrary, which is hypothesized as the main source of exteroceptive information, terminates on the apical dendrites of hippocampal pyramids. In the intracellular study described above, we found that septal stimuli elicited excitatory postsynaptic potentials associated with a spike discharge, whereas the stimulation of the olfactory bulb (Fig. 8) generated EPSPs but never spikes (Yokota et al., 1970). In terms of classical conditioning, the impulses traveling via these respective intero- and exteroceptive inputs would be comparable to unconditional and conditional stimuli (Gergen and MacLean, 1964).

The intracellular study offers a possible explanation of why impulses conducted via the perforant pathway from exteroceptive and other systems might fail to elicit hippocampal unit responses detectable by extracellular recording. If, for example, visual impulses transmitted from the parahippocampal cortex acted in a manner similar to those arising from olfactory volleys, it would explain the failure to record spikes with extracellular electrodes.

At present, there is no explanation of the function of the norepinephrine-containing terminals in the radiate layer of the hippocampus that were originally observed by Fuxe (1965) in the rat. They presumably arise from cells in the midbrain or pontomedullary region (Dahlström and Fuxe, 1965) and possibly exert an inhibitory or some modulatory action. In a comparative histofluorescence study including observations on the pygmy marmoset and squirrel monkey, Jacobowitz and I (in preparation) have found that the pattern of organization of the three recognized aminergic systems of the brain has been preserved with remarkable consistency in the evolution of primates.

2.12.   Inputs to Hypothalamus

In concluding this account of sensory inputs, it is relevant to point out that, contrary to what has been reported for other species, it was found in awake, sitting unmedicated monkeys that no hypothalamic units responded to photic or somatic stimulation (including noxious stimuli) and only a few were affected by auditory stimulation (Poletti et al., 1973). The results are summarized in Table 1. These findings indicate that exteroceptive information affecting the hypothalamus is first integrated and processed in related structures, of which the limbic cortex would be one

TABLE 1

Hypothalamic Units Tested by Visual, Auditory, and Somatic Stimulation

	Basal forebrain	Preoptic region	Hypothalamus
Visual, tested/responsive	110/0	78/0	321/2*
Auditory, tested/responsive	109/0	77/0	315/8
Somatic, tested/responsive	43/0	27/0	168/0

* The two responsive units were activated by an approaching object.

example. Because of technical difficulties, we were unable to obtain satisfactory intracellular recordings from hypothalamic units, and it is possible that, as in the case of the hippocampus and entorhinal cortex, exteroceptive stimuli might have elicited postsynaptic potentials that would not have been apparent with extracellular recording of single units.

3. Hippocampal Influence on the Brain Stem

Having considered the question of inputs to the hippocampus, we turn now to the converse question of how the hippocampus influences unit activity of structures of the brain stem. In one of the two studies to be summarized, we tested the effects of hippocampal volleys on units of the basal forebrain, preoptic region, and hypothalamus (Poletti et al., 1973). The term "basal forebrain" is used for designating a group of neighboring structures including nuclei of the septum, diagonal band, and stria terminalis, the nucleus accumbens, olfactory tubercle, and caudal portion of the gyrus rectus. Because of anatomical (Simpson, 1952), electrophysiological (MacLean, 1957b), and behavior (MacLean, 1957b, 1968; Siegel and Flynn, 1968) evidence of differences in function of the anterior and posterior hippocampus, we compared the effects of applying stimuli through electrodes implanted on each side at anterior and posterior loci (Fig. 9). In addition to testing with hippocampal volleys, we observed at every half-millimeter interval along a track the effects of hippocampal afterdischarges on unit and slow-wave activity. Because of the prolonged effect of afterdischarges on excitability (Gergen and MacLean, 1961), we did not resume testing with hippocampal volleys until more than 5 min had elapsed.

As shown in Table 2, more than 30% of the units in the basal forebrain and preoptic areas responded to hippocampal volleys, whereas only 14% of hypothalamic units were affected. Specific hypothalamic areas such as the perifornical area, mammillary region, and posterior hypothalamus, however, contained a large percentage of responsive units. There were only three hypothalamic areas (the supraoptic, arcuate, and premammillary nuclei) with no responsive units, but in each case there was but a small sample.

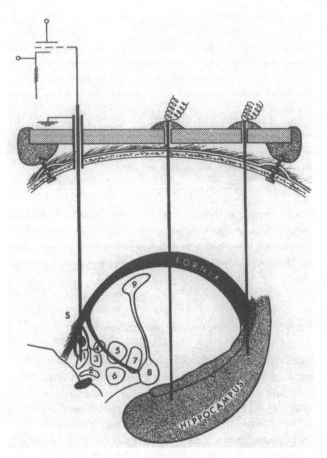

FIG. 9. Diagram of closed-system, stereotaxic device and electrode array used in testing effects of hippocampal volleys on unit activity of hypothalamus, preoptic region, and basal forebrain. Coaxial, bipolar, stimulating electrodes are chronically implanted in the anterior and posterior hippocampus of each side. Diagram shows exploring microelectrode in preoptic region. S, Septal area; 1, preoptic region; 2, supraoptic nucleus; 3, anterior hypothalamus; 4, paraventricular nucleus; 5, dorsomedial nucleus; 6, ventromedial nucleus; 7, posterior hypothalamus; 8, mammillary body; 9, anterior thalamic nucleus. From Poletti *et al.* (1973).

As will be discussed later, Table 2 lists the significant finding that more than 83% of the responsive units in each of the three regions showed initial excitation, while the rest were initially inhibited. An analysis showed that this result could not be attributed to a bias favoring the detection of excitation rather than inhibition because of the slow firing rate typical of most units in the hypothalamus. It should also be noted that the intervals between tests were sufficiently long to avoid potentiation effects. Figure 10A is an example of an oscillographic record and dot display of a unit in the medial preoptic area showing initial excitation, while the lower record in (B) is that of another unit in the same area in which the response is one of inhibition.

TABLE 2

Units Responsive to Hippocampal Volleys

	Basal forebrain	Preoptic region	Hypothalamus
Number tested	177	99	390
Number and percent responding	60 (34%)	30 (30%)	56 (14%)
Initially excited (%)	87	83	86
Initially inhibited (%)	13	17	14

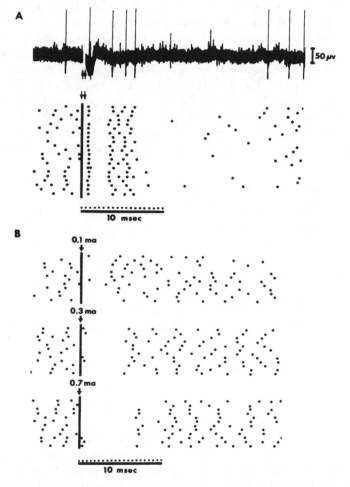

FIG. 10. A: Oscillographic record and dot display of responses of unit in medial preoptic area evoked by hippocampal volleys. B: Dot displays for another medial preoptic unit, showing prolongation of an inhibitory effect by increase of the shock intensity.

Approximately two-thirds of all responsive units showed *only* excitatory changes, manifested by either single or multiple spikes and sometimes recurring bursts of spikes. Six percent showed only inhibitory effects while the rest had complex firing patterns combining alternating phases of excitation and inhibition. Since complex patterns are not seen in spontaneously firing units, they must be presumed to reflect a combination of excitatory changes in the hippocampus and brain stem induced by the unnatural type of stimulus.

As in the case of hippocampal volleys, hippocampal afterdischarges more commonly elicited unit excitation than inhibition, with a ratio of 3:1 for the three regions combined. As shown in Fig. 11, some units showed excitation during seizure burst activity of the hippocampus while others were inhibited. Units that were excited during an afterdischarge usually became silent upon its termination, whereas those that were inhibited showed an increased firing rate. These postafterdischarge changes were manifested for periods ranging from 50 s to 11 min.

For all regions combined, stimulation of the ipsilateral anterior hippocampus activated a significantly larger percentage of units than the posterior hippocampus ($p < 0.001$). The discrepancy could not be explained by differences in the location of the stimulating electrodes. Contralateral stimulation was usually ineffective. When units responded to stimulation at more than one hippocampal locus, there might be a considerable difference in latency, but the pattern of response was usually similar. The medial mammillary nucleus was the only structure in which unit and slow-wave responses reflected differences in the organization of hippocampal projections.

3.1. Neuroanatomical Correlations

Units responding to hippocampal volleys with short, constant latencies of 9.5–12.5 ms were found in the nucleus accumbens, nucleus of the stria terminalis, nucleus of the diagonal band, lateral septum, medial and lateral preoptic areas, anterior hypothalamus, dorsal hypothalamus, perifornical area, medial mammillary nucleus, and posterolateral hypothalamus. The response characteristics of these units were suggestive of a direct orthodromic pathway. In a companion neuroanatomical study using improved techniques for showing degeneration of fine fibers, it was found that unilateral section of the fornix resulted in degeneration leading to all these areas (Poletti *et al.,* 1973).

3.2. Dorsal Thalamus

In another microelectrode study, we compared the effects of fornix and fifth-nerve volleys on activity of intralaminar and nonintralaminar units of the dorsal thalamus (Yokota and MacLean, 1968). It was found that hippocampal or fornix volleys inhibited, but did not augment, responses of caudal intralaminar units to fifth-nerve stimulation. Hippocampal afterdischarges had a similar action, except that the inhibitory effect endured for many seconds. Stimulation of the fifth nerve had no effect on 609 units recorded in the medial dorsal thalamus, whereas fornix volleys activated 51 and inhibited 2. The latencies ranged from 6 to 91 ms.

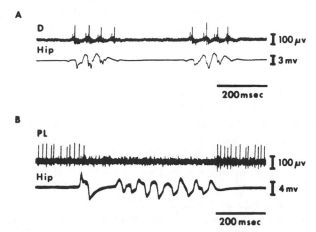

Fig. 11. The two typical effects on unit activity seen during hippocampal afterdischarges (Hip). A: Activation during the seizure burst and silence during the isoelectric period. B: The reverse situation. Abbreviations: D, nucleus of the diagonal band; PL, lateral preoptic area. From Poletti *et al.* (1973).

3.3. Comment

The high percentage of initially excited units in the hypothalamus, preoptic region, and basal forebrain requires consideration in the light of speculation based on behavioral studies that the hippocampus operates primarily through mechanisms of inhibition (e.g., Douglas, 1967; Grastyán, 1959; Isaacson and Wickelgren, 1962; Pribram, 1967). There are a few physiological studies that would provide support for such an inference. It has been reported, for example, that hippocampal stimulation inhibits cortically induced extensor reflexes (Vanegas and Flynn, 1968) and results in a suppression of the release of adrenocorticotropic hormone (ACTH) (Endröczi and Lissák, 1959; Mason, 1958; and Porter, 1954). As mentioned in the microelectrode study described above (Yokota and MacLean, 1968), it was found that hippocampal or fornix volleys inhibited, but did not augment, responses of caudal intralaminar units to fifth-nerve stimulation. Other studies suggest that facilitatory or inhibitory effects with respect to ACTH release (Kawakami *et al.*, 1968), to cardiovascular reflexes (Hockman *et al.*, 1969), or to visceral responsiveness (MacLean, 1957*b*) may depend on the physiological state of the animal at the time of stimulation. The microelectrode findings in the awake, sitting monkey indicate that if the hippocampus exerted a predominantly inhibitory influence on these and other functions mentioned below, it would induce such an effect largely through the agency of neurons that are initially excited.

The findings on hippocampal afterdischarges provide additional evidence that the hippocampal formation is more likely to induce excitation than inhibition of units. The prolonged aftereffects of such discharges on unit activity (up to 11 min) are of interest in light of the correspondingly long "rebound" behavioral and autonomic changes that may occur subsequent to hippocampal seizures. "Rebound" changes seen in different species include prolonged eating and drinking, scratching and grooming of the body, agitation, and vocalization, pleasure reactions, quietude, "taming," cardiac slowing and irregularities, waxing and waning of penile erection, and somnolence (for references, see MacLean, 1968). In the cat, for example, the termination of an afterdischarge is frequently signaled by pupillary dilatation and plaintive meows suggestive of distress, after which there may be prolonged grooming and signs of pleasure. Ethologists refer to such reactions to apparent stress as displacement behavior, a nonspecific kind of activity for which some psychologists would prefer the term "adjunctive behavior." If so interpreted, the accompanying "rebound" autonomic changes such as penile erection and cardiac slowing and irregularities should also be regarded as displacement or adjunctive behavior. The occurrence of "rebound" manifestations lends itself to another, more physiological interpretation. Sherrington's work on spinal reflexes suggested that rebound is the result of an excitatory state outlasting that of inhibition following the simultaneous stimulation of excitatory and inhibitory fibers (1906). As I have suggested elsewhere, there may exist in the brain a reciprocal innervation of opposing behavioral and feeling states that compares to a reciprocal innervation of muscles (MacLean, 1958).

The anatomical findings on the preoptic and perifornical regions are of particular interest because it appears to be the first time that direct hippocampal projections have been traced to these structures in a primate. The microelectrode findings show that the hippocampus exerts primarily excitatory effects on cells in these areas. Long ago, Hess and Brügger (1943) concluded that the perifornical region plays an important role in the expression of angry behavior. The medial preoptic area has been implicated in the control of body temperature, cardiovascular function, water balance, food intake, sexual functions, and mechanisms of sleep (for review, see Haymaker et al., 1969). In recent years, the medial preoptic area has been the subject of intensive research because of the finding in rodents that circulating testosterone during the first 2 wk of life has the capacity to determine sexual differentiation by its action on this area and adjoining parts of the septum and anterior hypothalamus (for review, see Gorski, 1971). Autoradiographic studies have provided evidence that this region has a special affinity for L-testosterone (Pfaff, 1968a) and estradiol (Michael, 1965; Pfaff, 1968b; Stumpf, 1968). Electrical stimulation of the medial preoptic area and adjoining structures results in ovulation (Everett, 1965) and penile erection (MacLean and Ploog, 1962), whereas the direct application of estrogen brings about changes in sexual receptivity (Lisk, 1962). It is believed that in the male the preoptic area exerts a tonic effect on the secretion of gonadotropin whereas in the female the influence is cyclical.

Heretofore, it has been the belief that limbic influences on the medial preoptic area are mediated primarily by the amygdala (e.g., Nauta, 1962). The microelectrode and neuroanatomical findings reviewed here demonstrate that, whatever may be the functions of the hippocampus, it also has the capacity to exert a direct influence on the preoptic area. If the hippocampus should prove to be involved in emotional behavior, the new findings would suggest neural mechanisms by which either the agreeable or disagreeable aspects of affective experience could influence genital and gonadal function.

4. Conclusion and Summary

The introductory section of this chapter deals with evolutionary and functional considerations pertaining to the hippocampal formation and its pivotal position within the limbic system. It is essential for the understanding of hippocampal function to learn the nature and mechanism of action of its inputs and outputs. The major part of the chapter has been devoted to an ongoing analysis of inputs and outputs of the hippocampal formation based on experiments in which improved microelectrode techniques were used for extra- and intracellular recording in awake, sitting, squirrel monkeys. The work has involved testing of more than 12,000 units, of which nearly 40% were located in various limbic cortical and subcortical areas. Parallel neuroanatomical studies employing improved silver techniques have provided corroborative information.

With extracellular recording of unit activity, it was not possible to demonstrate hippocampal unit responses to natural and/or electrical stimulation of exteroceptive systems. Such stimulation, however, activated units in limbic areas known to project to the hippocampus. Units responding to photic stimulation were found in the posterior part of the parahippocampal gyrus and the contiguous perilimbic cortex. A neuroanatomical study revealed separate projections from the lateral geniculate body and inferior pulvinar, respectively, to these areas. Units responding specifically to auditory and somatic stimulation were found only in the limbic cortex of the insula overlying the claustrum. Some units in the most anterior part of the insula were activated by gustatory stimulation. The cingulate gyrus proved unresponsive to exteroceptive stimulation, but in another series of experiments vagal volleys evoked discharges or induced inhibition of nearly 20% of the recorded units. Intravenous microinjections of serotonin (an exciter of pulmonary receptors) also evoked unit responses in the cingulate gyrus. Significantly, in regard to afferent pathways it was found that vagal volleys evoked short-latency responses in the anterior ventral and paracentral nuclei, which are known to project to the cingulate gyrus.

Since all the above limbic responsive sensory areas project to the hippocampal formation, the findings suggest mechanisms by which information of olfactory, gustatory, somatic, auditory, visual, and interoceptive origin might be integrated and transmitted to the hypothalamus and other structures of the brain stem involved in the regulation of somatovisceral and neuroendocrine functions.

A study involving intracellular recording of hippocampal neurons in awake, sitting monkeys suggested a possible explanation of why impulses from exteroceptive and other systems conducted via the perforant pathway might fail to evoke responses detectable by extracellular recording. It was found that electrical stimuli applied to the olfactory bulb evoked excitatory postsynaptic potentials in the hippocampus and entorhinal area, but never spike discharges. Other aspects of the intracellular study are discussed with respect to a differential action of exteroceptive and interoceptive systems on hippocampal neurons.

In regard to the question of outputs, it was found that hippocampal volleys elicited responses in a large proportion of units in certain structures of the hypothalamus, preoptic region, and basal forebrain. In each of the three regions, more than 80% of the responsive units showed initial excitation. Hippocampal afterdischarges also more commonly excited than inhibited units. If, as some studies suggest, the hippocampus influences various forms of behavior through mechanisms of inhibition, the microelectrode findings indicate that it would do so largely through the agency of neurons that are initially excited.

Upon termination of hippocampal afterdischarges, units of the brain stem showed changes in their firing patterns that persisted for as long as 11 min. These findings may help to explain the prolonged "rebound" behavior and autonomic changes seen in various species following hippocampal afterdischarges.

Bearing out microelectrode findings, a parallel neuroanatomical study showed for the first time in a primate that the fornix projects to the medial preoptic area and to the perifornical region. These new findings have been discussed with respect to hippocampal influences on emotional behavior and on genital and gonadal aspects of sexual function.

Finally, recent developments in histochemistry call for comment on the possible modulatory effects of ascending aminergic systems on the function of limbic structures. A current histofluorescence study, including comparative observations on the pygmy marmoset and squirrel monkey, has revealed that the organizational pattern of recognized aminergic systems has been preserved with remarkable consistency in the evolution of primates.

5. References

BACHMAN, D. S., AND MacLEAN, P. D. Unit analysis of inputs to cingulate cortex in awake, sitting squirrel monkeys. I. Exteroceptive systems. *International Journal of Neuroscience*, 1971, **2**, 109–113.

BACHMAN, D. S., KATZ, H. M., AND MacLEAN, P. D. Effect of intravenous injections of 5-hydroxytryptamine (serotonin) on unit activity of cingulate cortex of awake squirrel monkeys. *Federation Proceedings*, 1972a, **31**, 303.

BACHMAN, D. S., KATZ, H. M., AND MacLEAN, P. D. Vagal influence on units of cingulate cortex in the awake, sitting squirrel monkey. *Electroencephalography and Clinical Neurophysiology*, 1972b, **33**, 350–351.

BAGSHAW, M. H., AND PRIBRAM, K. H. Cortical organization in gustation (*Macaca mulatta*). *Journal of Neurophysiology*, 1953, **16**, 499–508.

BAILEY, P., AND BREMER, F. A sensory cortical representation of the vagus nerve. *Journal of Neurophysiology*, 1938, **1**, 405–412.

BAKER-COHEN, K. F. Comparative enzyme histochemical observations on submammalian brains. III. Hippocampal formation in reptiles. *Brain Research*, 1969, **16**, 215–225.

BENJAMIN, R. M., AND BURTON, H. Projection of taste nerve afferents to anterior opercular-insular cortex in squirrel monkey (*Saimiri sciureus*). *Brain Research*, 1968, **7**, 221–231.

BRAZIER, M. A. B. Evoked responses recorded from the depths of the human brain. *Annals of the New York Academy of Sciences*, 1964, **112**, 33–59.

BRÖCA, P. Anatomie comparée des circonvolutions cérébrales: Le grand lobe limbique et la scissure limbique dans la série des mammifères. *Revue d'Anthropologie*, 1878, **1**, 385–498.

CAMPBELL, A. W. *Histological studies on the localization of cerebral functions.* Cambridge: University Press, 1905, 360 pp.

CLARK, W. E. L., AND MEYER, M. The terminal connexions of the olfactory tract in the rabbit. *Brain*, 1947, **70**, 304–328.

CRAGG, B. G. Responses of the hippocampus to stimulation of the olfactory bulb and of various afferent nerves in five mammals. *Experimental Neurology*, 1960, **2**, 547–571.

CROSBY, E. C., HUMPHREY, T., AND LAUER, E. W. *Correlative anatomy of the nervous system*. New York: Macmillan, 1962, 731 pp.

CUÉNOD, M., CASEY, K. L., AND MacLEAN, P. D. Unit analysis of visual input to posterior limbic cortex. I. Photic stimulation. *Journal of Neurophysiology*, 1965, **28**, 1101–1117.

CUSHING, H. The field defects produced by temporal lobe lesions. *Brain*, 1922, **44**, 341–396.

DAHLSTRÖM, A., AND FUXE, K. Evidence for the existence of monoamine neurons in the central nervous system. I. Demonstration of monoamines in the cell bodies of brainstem neurons. *Acta Physiologica Scandinavica*, 1962, **62**: Supplement 232, 1–80.

DAITZ, H. M., AND POWELL, T. P. S. Studies of the connexions of the fornix system. *Journal of Neurology, Neurosurgery, and Psychiatry*, 1954, **17**, 75–82.

DELL, P., AND OLSON, R. Projections "secondaires" mésencéphaliques, diencéphaliques et amygdaliennes des afférences viscérales vagales. *Compte Rendu des Seances de la Societe de Biologie (Paris)*, 1951, **145**, 1088–1091.

DESMEDT, J. E., AND MECHELSE, K. Mise en évidence d'une quatrième aire de projection acoustique dans l'éncorce cérébrale du Chat. *Journal de Physiologie (Paris)*, 1959, **51**, 448–449.

DOUGLAS, R. J. The hippocampus and behavior. *Psychological Bulletin*, 1967, **67**, 416–442.

ELLIOT SMITH, G. Notes upon the natural subdivision of the cerebral hemisphere. *Journal of Anatomy and Physiology*, 1901, **35**, 431–454.

ELLIOT SMITH, G. On the homologies of the cerebral sulci. *Journal of Anatomy*, 1902, **36**, 309–319.

ELLIOT SMITH, G. The term "archipallium," a disclaimer. *Anatomischer Anzeiger Jena*, 1910, **35**, 429.

ELLIOT SMITH, G. A preliminary note on the morphology of the corpus striatum and the origin of the neopallium. *Journal of Anatomy, London*, 1918/1919, **53**, 271–291.

ENDRÖCZI, E., AND LISSÁK, K. The role of the mesencephalon and archicortex in the activation and inhibition of the pituitary–adrenocortical system. *Acta Physiologica Hungaricae*, 1959, **15**, 25.

EVERETT, J. W. Ovulation in rats from preoptic stimulation through platinum electrodes: Importance of duration and spread of stimulus. *Endocrinology*, 1965, **76**, 1195–1201.

FALCONER, M. A., SERAFETINIDES, E. A., AND CORSELLIS, J. A. N. Etiology and pathogenesis of temporal lobe epilepsy. *Archives of Neurology*, 1964, **10**, 233–248.

FOX, C. A., McKINLEY, W. A., AND MAGOUN, H. W. An oscillographic study of olfactory system of cats. *Journal of Neurophysiology*, 1944, **7**, 1–16.

FUXE, K. Evidence for the existence of monoamine neurons in the central nervous system. IV. Distribution of monoamine nerve terminals in the central nervous system. *Acta Physiologica Scandinavica*, 1965, Supplement 247, **64**, 37–84.

GERGEN, J. A., AND MacLEAN, P. D. Hippocampal seizures in squirrel monkeys. *Electroencephalography and Clinical Neurophysiology*, 1961, **13**, 316–317.

GERGEN, J. A., AND MacLEAN, P. D. *A stereotaxic atlas of the squirrel monkey's brain (Saimiri sciureus)*. Washington, D.C.: U.S. Government Printing Office, 1962, 91 pp. (Public Health Service Publication No. 933, 1962).

GERGEN, J. A., AND MacLEAN, P. D. The limbic system: Photic activation of limbic cortical areas in the squirrel monkey. *Annals of the New York Academy of Sciences*, 1964, **117**, 69–87.

GIBBS, E. L., GIBBS, F. A., AND FUSTER, B. Psychomotor epilepsy. *Archives of Neurology and Psychiatry (Chicago)*, 1948, **60**, 331–339.

GORSKI, R. A. Sexual differentiation of the hypothalamus. In H. C. Mack (Ed.), *The neuroendocrinology of human reproduction*. Springfield, Ill.: Charles C Thomas, 1971.

GRASTYÁN, E. The hippocampus and higher nervous activity. In *Second Conference on the Central Nervous System and Behavior, Transactions*. New York: Josiah Macy, Jr. Foundation, 1959, pp. 119–205.

GREEN, J. D., AND ADEY, W. R. Electrophysiological studies of hippocampal connections and excitability. *Electroencephalography and Clinical Neurophysiology*, 1956, **8**, 245–262.

GREEN, J. D., AND ARDUINI, A. A. Hippocampal electrical activity in arousal. *Journal of Neurophysiology*, 1954, **17**, 533–557.

HALLOWITZ, R. A., AND MacLEAN, P. D. Effects of vagal volleys on unit activity of medial thalamic nuclei in squirrel monkeys (Saimiri sciureus). *Federation Proceedings*, 1974, **33**, 342.

HARMAN, P. J., AND BERRY, C. M. Neuroanatomical distribution of action potentials evoked by photic stimuli in cat fore- and midbrain. *Journal of Comparative Neurology*, 1956, **105**, 395–416.

HAYMAKER, W., ANDERSON, E., AND NAUTA, W. J. H. *The hypothalamus*. Springfield, Ill.: Charles C Thomas, 1969.

HERRICK, C. J. A sketch of the origin of the cerebral hemispheres. *Journal of Comparative Neurology*, 1921, **32**, 429–454.

HESS, W. R., AND BRÜGGER, M. Das subkortikale Zentrum der affektiven Abwehrreaktion. *Helvetica Physiologica et Pharmacologica Acta*, 1943, **1**, 33–52.

HOCKMAN, C. H., TALESNIK, J., AND LIVINGSTON, K. E. Central nervous system modulation of baroceptor reflexes. *American Journal of Physiology*, 1969, **217**, 1681–1689.

HÖKFELT, T., LJUNGDAHL, A., FUXE, K., AND JOHANSSON, O. Dopamine nerve terminals in the rat limbic cortex: Aspects of the dopamine hypothesis of schizophrenia. *Science*, 1974, **184**, 177–179.

HOROWITZ, M. J., ADAMS, J. E., AND RUTKIN, B. B. Visual imagery and brain stimulation. *Archives of General Psychiatry*, 1968, **19**, 469.

HUGHES, J. R. Studies on the supracallosal mesial cortex of unanesthetized conscious mammals. I. Cat. B. Electrical activity. *Electroencephalography and Clinical Neurophysiology*, 1959, **11**, 459–470.

HUMPHREY, T. Correlations between the development of the hippocampal formation and the differentiation of the olfactory bulb. *Alabama Journal of Medical Science*, 1966, **3**, 235–269.

INGVAR, D. H., AND HUNTER, J. Influence of visual cortex on light impulses in the brain stem of the unanesthetized cat. *Acta Physiologica Scandinavica*, 1955, **33**, 194–218.

ISAACSON, R. L., AND WICKELGREN, W. O. Hippocampal ablation and passive avoidance. *Science*, 1962, **138**, 1104–1106.

JOHNSTON, J. B. The development of the dorsal ventricular ridge in turtles. *Journal of Comparative Neurology*, 1916, **26**, 481–505.

JOHNSTON, J. B. Further contributions to the study of the evolution of the forebrain. *Journal of Comparative Neurology*, 1923, **35**, 337–481.

JONES, E. G., AND POWELL, T. P. S. An anatomical study of converging sensory pathways within the cerebral cortex of the monkey. *Brain*, 1970, **93**, 793–820.

JOUVET, M., MICHEL, F., AND MOUNIER, D. Analyse électroencéphalographique comparée du sommeil physiologique chez le chat et chez l'homme. *Revue Neurologique*, 1960, **103**, 189–205.

JUORIO, A. V., AND VOGT, M. Monoamines and their metabolites in the avian brain. *Journal of Physiology (London)*, 1967, **189**, 489–518.

KAWAKAMI, M., SETO, K., TERASAWA, E., YOSHIDA, K., MIYAMOTO, T., SEKIGUCHI, M., AND HATTORI, Y. Influence of electrical stimulation and lesion in limbic structure upon biosynthesis of adrenocorticoid in the rabbit. *Neuroendocrinology*, 1968, **3**, 337–348.

KLÜVER, H., AND BUCY, P. C. Preliminary analysis of functions of the temporal lobes in monkeys. *Archives of Neurology and Psychiatry, (Chicago)*, 1939, **42**, 979–1000.

LEWIS, P. R., AND SHUTE, C. C. D. The cholinergic limbic system: Projections to hippocampal formation, medial cortex, nuclei of the ascending cholinergic reticular system, and the subfornical organ and supra-optic crest. *Brain*, 1967, **90**, 521–540.

LISK, R. D. Diencephalic placement of estradiol and sexual receptivity in the female rat. *American Journal of Physiology*, 1962, **203**, 493–496.

MACLEAN, P. D. Psychosomatic disease and the "visceral brain": Recent developments bearing on the Papez theory of emotion. *Psychosomatic Medicine*, 1949, **11**, 338–353.

MACLEAN, P. D. Some psychiatric implications of physiological studies on frontotemporal portion of limbic system (visceral brain). *Electroencephalography and Clinical Neurophysiology*, 1952, **4**, 407–418.

MACLEAN, P. D. Chemical and electrical stimulation of hippocampus in unrestrained animals. I. Methods and electroencephalographic findings. *AMA Archives of Neurology and Psychiatry*, 1957*a*, **78**, 113–127.

MACLEAN, P. D. Chemical and electrical stimulation of hippocampus in unrestrained animals. II. Behavioral findings. *AMA Archives of Neurology and Psychiatry*, 1957*b*, **78**, 128–142.

MACLEAN, P. D. The limbic system with respect to self-preservation and the preservation of the species. *Journal of Nervous and Mental Disease*, 1958, **127**, 1–11.

MacLean, P. D. The limbic and visual cortex in phylogeny: Further insights from anatomic and microelectrode studies. In R. Hassler and H. Stephan (Eds.), *Evolution of the forebrain.* Stuttgart:
Georg Thieme, 1966, pp. 443–453.

MacLean, P. D. A chronically fixed stereotaxic device for intracerebral exploration with macro- and
micro-electrodes. *Electroencephalography and Clinical Neurophysiology,* 1967, **22,** 180–182.

MacLean, P. D. Ammon's Horn: A continuing dilemma. Foreword in S. Ramón y Cajal, *The structure
of Ammon's horn.* Translated from the Spanish by L. Kraft. Springfield, Ill.: Charles C Thomas,
1968, xix + 78 pp.

MacLean, P. D. The triune brain, emotion, and scientific bias. In F. O. Schmitt (Ed.), *The
neurosciences: Second study program.* New York: Rockefeller University Press, 1970, pp. 336–349.

MacLean, P. D. Cerebral evolution and emotional processes: New findings on the striatal complex. *Annals of the New York Academy of Sciences,* 1972*a,* **193,** 137–149.

MacLean, P. D. Implications of microelectrode findings on exteroceptive inputs to the limbic cortex. In
C. H. Hockman (Ed.), *Limbic system mechanisms and autonomic function.* Springfield, Ill.: Charles
C Thomas, 1972*b,* pp. 115–136.

MacLean, P. D. Effects of pallidal lesions on species-typical display behavior of squirrel monkey. *Federation Proceedings,* 1973*a,* **32,** 384.

MacLean, P. D. A triune concept of the brain and behaviour, Lecture I. Man's reptilian and limbic
inheritance; Lecture II. Man's limbic brain and the psychoses; Lecture III. New trends in man's evolution. In T. Boag and D. Campbell (Eds.), *The Hincks memorial lectures.* Toronto: University of
Toronto Press, 1973*b,* pp. 6–66.

MacLean, P. D. The brain's generation gap: Some human implications. *Zygon-Journal of Religion and
Science,* 1973*c,* **8,** 113–127.

MacLean, P. D. Influence of limbic cortex on hypothalamus: New anatomic and microelectrode findings.
In K. Lederis and K. E. Cooper (Eds.), *Recent studies in hypothalamic function.* Basel: Karger,
1974*a,* pp. 216–231.

MacLean, P. D. Bases neurologiques du comportement d'imitation chez le singe-écureuil. In E. Morin
and M. Piattelli-Palmarini (Eds.), *L'Unite de l'Homme: Invariants biologiques et universaux cul-
turels.* Paris: Editions du Seuil, 1974*b,* pp. 186–212.

MacLean, P. D. The triune brain. *Medical World News/Psychiatry,* October, 1974*c,* **2,** 55–60.

MacLean, P. D., and Arellano, Z., AP Basal lead studies in epileptic automatisms. *Electroencephalography and Clinical Neurophysiology,* 1950, **2,** 1–16.

MacLean, P. D., and Creswell, G. Anatomical connections of visual system with limbic cortex of
monkey. *Journal of Comparative Neurology,* 1970, **138,** 265–278.

MacLean, P. D., and Ploog, D. W. Cerebral representation of penile erection. *Journal of
Neurophysiology,* 1962, **25,** 29–55.

MacLean, P. D., and Pribram, K. H. Neuronographic analysis of medial and basal cerebral cortex I.
Cat. *Journal of Neurophysiology,* 1953, **16,** 312–323.

MacLean, P. D., Horwitz, N. H., and Robinson, F. Olfactory-like responses in pyriform area to non-
olfactory stimulation. *Yale Journal of Biology and Medicine,* 1952, **25,** 159–172.

MacLean, P. D., Yokota, T., and Kinnard, M. A. Photically sustained on-responses of units in posterior hippocampal gyrus of awake monkey. *Journal of Neurophysiology,* 1968, **31,** 870–883.

Malamud, N. The epileptogenic focus in temporal lobe epilepsy from a pathological standpoint. *Archives
of Neurology,* 1966, **14,** 190–195.

Marg, E. The accessory optic system. *Annals of the New York Academy of Sciences,* 1964, **117,** 35–52.

Margerison, J. H., and Corsellis, J. A. N. Epilepsy and the temporal lobes: A clinical, electroencephalographic and neuropathological study of the brain in epilepsy, with particular reference to the
temporal lobes. *Brain,* 1966, **89,** 499.

Mason, J. W. The central nervous system regulation of ACTH secretion. In *Reticular formation of the
brain.* Boston: Little, Brown, 1958, pp. 645–662.

Meyer, M., and Allison, A. C. Experimental investigation of the connexions of the olfactory tracts in
the monkey. *Journal of Neurology, Neurosurgery, and Psychiatry,* 1949, **12,** 274–286.

Michael, R. P. Oestrogens in the central nervous system. *British Medical Bulletin,* 1965, **21,** 87–90.

Mishkin, M. Visual discrimination performance following partial ablations of the temporal lobe. II.

Ventral surface vs. hippocampus. *Journal of Comparative Physiology and Psychology*, 1954, **47**, 187–193.

MISHKIN, M. Cortical visual areas and their interactions. In A. G. Karczmar and J. C. Eccles (Eds.), *Brain and human behavior*. Heidelberg: Springer-Verlag, 1972, pp. 187–208.

MISHKIN, M., AND PRIBRAM, K. H. Visual discrimination performance following partial ablations of the temporal lobe. I. Ventral vs. lateral. *Journal of Comparative Physiology and Psychology*, 1954, **47**, 14–20.

MOREST, D. K. Connexions of dorsal tegmental nucleus in rat and rabbit. *Journal of Anatomy*, 1961, **95**, 1–18.

MOREST, D. K. Experimental study of the projections of the nucleus of the tractus solitarius and the area postrema in the cat. *Journal of Comparative Neurology*, 1967, **130**, 277–299.

MURRAY, M. Degeneration of some intralaminar thalamic nuclei after cortical removals in the cat. *Journal of Comparative Neurology*, 1966, **127**, 341–368.

MYERS, R. E. Projections of the superior colliculus in monkey. *Anatomical Record*, 1963, **145**, 264.

NAUTA, W. J. H. Neural associations of the amygdaloid complex in the monkey. *Brain*, 1962, **85**, 505–520.

O'LEARY, J. L., AND BISHOP, G. H. Margins of the optically excitable cortex in the rabbit. *Archives of Neurology and Psychiatry (Chicago)*, 1938, **40**, 482–499.

PAPEZ, J. W. A proposed mechanism of emotion. *Archives of Neurology and Psychiatry (Chicago)*, 1937, **38**, 725–743.

PARENT, A., AND OLIVIER, A. Comparative histochemical study of the corpus striatum. *Journal of Hirnforschung*, 1970, **12**, 75–81.

PENFIELD, W., AND ERICKSON, T. C. *Epilepsy and cerebral localization*. Springfield, Ill.: Charles C Thomas, 1941, 623 pp.

PENFIELD, W., AND JASPER, H. *Epilepsy and the functional anatomy of the human brain*. Boston: Little, Brown, 1954, 896 pp.

PENFIELD, W., AND PEROT, P. The brain's record of auditory and visual experience: A final summary and discussion. *Brain*, 1963, **86**, 596–696.

PFAFF, D. W. Autoradiographic localization of radioactivity in rat brain after injection of tritiated sex hormones. *Science*, 1968a, **161**, 1355–1356.

PFAFF, D. W. Uptake of ^{3}H-estradiol by the female rat brain: An autoradiographic study. *Endocrinology*, 1968b, **82**, 1149–1155.

POLETTI, C. E., KINNARD, M. A., AND MacLEAN, P. D. Hippocampal influence on unit activity of hypothalamus, preoptic region, and basal forebrain in awake, sitting squirrel monkeys. *Journal of Neurophysiology*, 1973, **36**, 308–324.

POLYAK, S. *The vertebrate visual system*. Chicago: University of Chicago Press, 1957, 1390 pp.

PORTER, R. W. The central nervous system and stress-induced eosinopenia. *Recent Progress in Hormone Research*, 1954, **10**, 1–27.

PRIBRAM, K. H. The limbic systems, efferent control of neural inhibition and behavior. *Progress in Brain Research*, 1967, **27**, 318–336.

PRIBRAM, K. H., AND MacLEAN, P. D. Neuronographic analysis of medial and basal cerebral cortex. II. Monkey. *Journal of Neurophysiology*, 1953, **16**, 324–340.

PUTNAM, T. J. Studies on the central visual system. II. A comparative study of the form of the geniculostriate visual system of mammals. *Archives of Neurology and Psychiatry*, 1926, **16**, 285–300.

RAMÓN Y CAJAL, S. *Studies on the cerebral cortex (limbic structures)*. Translated from the Spanish by L. M. Kraft. London: Lloyd-Luke Ltd., Chicago: Year Book Publishers, 1955, xii + 179 pp.

RAMÓN Y CAJAL, S. *The structure of Ammon's horn*. Translated from the Spanish by L. M. Kraft. Springfield, Ill.: Charles C Thomas, 1968, xxii + 78 pp.

ROBINSON, F., AND LENNOX, M. A. Sensory mechanisms in hippocampus, cingulate gyrus and cerebellum of the cat. *Federation Proceedings*, 1951, **10**, 110–111.

ROSE, J. E., AND MALIS, L. I. Geniculo-striate connections in the rabbit. II. Cytoarchitectonic structure of the striate region and of the dorsal lateral geniculate body; organization of the geniculo-striate projections. *Journal of Comparative Neurology*, 1965, **125**, 121–139.

SANIDES, F., AND VITZTHUM, H. G. Zur Architektonik der menschlichen Sehrinde und den Prinzipien ihrer Entwicklung. *Deutsche Zeitschrift für Nervenheilkunde*, 1965, **187**, 680–707.

SANO, K., AND MALAMUD, N. Clinical significance of sclerosis of the cornu ammonis: Ictal "psychic phenomena." *Archives of Neurology and Psychiatry (Chicago)*, 1953, **70**, 40–53.

SCHÄFER, E. A. *Text-book of physiology*. Vol. 2. Edinburgh and London: Young J. Pentland, 1900, 1365 pp.

SHERRINGTON, C. S. *The integrative action of the nervous system*. 2nd ed. New Haven: Yale University Press, 1947, 433 pp.

SIEGEL, A., AND FLYNN, J. P. Differential effects of electrical stimulation and lesions of the hippocampus and adjacent regions upon attack behavior in cats. *Brain Research*, 1968, **7**, 252–267.

SIMPSON, D. A. The efferent fibres of the hippocampus in the monkey. *Journal of Neurology, Neurosurgery, and Psychiatry*, 1952, **15**, 79–92.

SNYDER, M., AND DIAMOND, I. T. The organization and function of the visual cortex in the tree shrew. *Brain, Behavior and Evolution*, 1968, **1**, 244–288.

SPATZ, H. Vergangenheit und Zukunft des Menschenhirns: Sonderdruck. *Jahrbuch der Akademie des Wissenschaften und der Literatur in Wiesbaden*, 1964, pp. 228–242.

STEPHAN, H. Die kortikalen Anteile des limbischen Systems. Sonderdruck aus "Der Nervenarzt," 35. Jahrgang, 9. Heft, September 1964, S. 396–401, Berlin: Springer-Verlag.

STUMPF, W. E. Estradiol-concentrating neurons: Topography in the hypothalamus by dry-mount autoradiography. *Science*, 1968, **162**, 1001–1003.

SUDAKOV, K., MacLEAN, P. D., REEVES, A. G., AND MARINO, R. Unit study of exteroceptive inputs to claustrocortex in awake, sitting, squirrel monkey. *Brain Research*, 1971, **28**, 19–34.

THOMPSON, J. M., WOOLSEY, C. N., AND TALBOT, S. A. Visual areas I and II of cerebral cortex of rabbit. *Journal of Neurophysiology*, 1950, **13**, 277–288.

TURNER, W. The convolutions of the brain: A study in comparative anatomy. *Journal of Anatomy and Physiology*, 1890, **25**, 105–153.

UNGERSTEDT, U. Stereotaxic mapping of the monoamine pathways in the rat brain. *Acta Physiologica Scandinavica*, 1971, Supplement 367.

VALENSTEIN, E. S., AND NAUTA, W. J. H. A comparison of the distribution of the fornix system in the rat, guinea pig, cat and monkey. *Journal of Comparative Neurology*, 1959, **113**, 337–363.

VANEGAS, H., AND FLYNN, J. P. Inhibition of cortically-elicited movement by electrical stimulation of the hippocampus. *Brain Research*, 1968, **11**, 489–506.

VAN HOESEN, G. W., PANDYA, D. N., AND BUTTERS, N. Cortical afferents to the entorhinal cortex of the rhesus monkey. *Science*, 1972, **175**, 1471–1473.

VITZTHUM, H. G., AND SANIDES, F. Entwicklungsprinzipien der menschlichen Sehrinde. In R. Hassler and H. Stephan (Eds.), *Evolution of the forebrain*. Stuttgart: Georg Thieme Verlag, 1966, pp. 435–442.

YOKOTA, T., AND MacLEAN, P. D. Fornix and fifth-nerve interaction on thalamic units in awake, sitting squirrel monkeys. *Journal of Neurophysiology*, 1968, **31**, 358–370.

YOKOTA, T., REEVES, A. G., AND MacLEAN, P. D. Differential effects of septal and olfactory volleys on intracellular responses of hippocampal neurons in awake, sitting monkeys. *Journal of Neurophysiology*, 1970, **33**, 96–107.

II
Neurochemistry and Endocrinology

8

Monoamine Neurons Innervating the Hippocampal Formation and Septum: Organization and Response to Injury

Robert Y. Moore

1. Introduction

The development of the concept of chemical transmission at the synapse and its application to the central nervous system has resulted in remarkable advances in our understanding of brain structure, function, and pathology. Among these, it has provided a basis for characterizing neuronal systems other than on morphological criteria. The utility of this characterization is best exemplified by our current knowledge of catecholamine and indolamine neuron systems in the mammalian brain. The presence of an indolamine (5-hydroxytryptamine, or serotonin) and catecholamines (norepinephrine, dopamine) in brain in a nonuniform pattern was first established 20 years ago (Twarog and Page, 1953; Amin *et al.*, 1954; Bogdanski *et al.*, 1957; Bertler and Rosengren, 1959) and led to the proposal that these compounds serve as chemical mediators of synaptic transmission (Amin *et al.*, 1954; Vogt, 1954; Brodie and Shore, 1957). Subsequent studies in which brain amine content was analyzed following placement of localized destructive lesions indicated an association of specific neuron systems, particularly the medial forebrain bundle, with

Robert Y. Moore • Department of Neurosciences, University of California, San Diego, La Jolla, California. The preparation of this chapter and some of the work discussed in it was supported in part by grants NS-12080 and HD-04583 from the National Institutes of Health, USPHS.

these putative neurotransmitters (Heller *et al.*, 1962; Heller and Moore, 1968). It was not until the development of a specific and sensitive histochemical procedure for the intraneuronal localization of indolamines and catecholamines, the Falck–Hillarp method (Falck *et al.*, 1962; Carlsson *et al.*, 1962; Falck, 1962; Corrodi and Jonsson, 1967; Björklund *et al.*, 1972), however, that significant advances were made in analyzing the specific neuronal systems which produce serotonin and catecholamines in brain. Information has accumulated rapidly in this area and our understanding of the organization and function of these neuron systems is now well advanced (for review, see Anden *et al.*, 1966; Fuxe *et al.*, 1970; Ungerstedt, 1971; Lindvall and Björklund, 1974*b*). It is now generally accepted that the serotonin and catecholamine neuron systems innervating the telencephalon arise from cell bodies located entirely within the brain stem (Dahlström and Fuxe, 1965). Consequently, the projections of these systems upon the telencephalon, including the hippocampal formation and septum, represent direct afferent input from brain stem nuclei.

The purpose of this chapter is to summarize our current understanding of the organization of serotonin and catecholamine innervation of the hippocampus and septum and its response to injury. The latter is particularly intriguing since a number of recent studies (for review, see Moore *et al.*, 1974) suggest that central monoamine neurons are remarkably capable of plasticity in response to injury. The innervation of the septum is considered herein for two reasons. First, the hippocampal formation and septum arise from adjacent regions of the medial telencephalic vesicle and are closely related, both structurally and functionally. Second, some of the studies of plasticity in monoamine neurons have been carried out on the septum and form a model for investigations which are yet to be done on the hippocampal formation.

2. Organization of Hippocampal and Septal Monoamine Innervation

2.1. The Hippocampal Formation

2.1.1. *Norepinephrine Innervation.* The norepinephrine content of the hippocampal formation is significantly higher (about 0.5 μg/g, Table 1) than that of neocortical areas (0.15–0.25 μg/g, Thierry and Glowinski, 1973). In contrast to the neocortex, where there is direct dopaminergic innervation (Thierry and Glowinski, 1973), the dopamine content of the hippocampal formation (0.13 $\pm$ 0.04 μg/g) is within the range expected for a precursor of norepinephrine in a norepinephrine-producing neuron (Costa *et al*, 1972). The norepinephrine innervation of the hippocampal formation, like that of other telencephalic areas, appears to arise entirely from the neurons of the pontine nucleus, locus coeruleus (Ungerstedt, 1971; Lindvall and Björklund, 1974*b*; Pickel *et al.*, 1974*b*; Jones and Moore, 1975). Lesions destroying the locus coeruleus, or its projection in the medial forebrain bundle, produce large decreases in norepinephrine content in the rat hippocampal formation (Table 1). The anatomy of the projection has been studied using three methods, the Falck–Hillarp method (Blackstad *et al.*, 1967), the glyoxylic acid fluorescent his-

TABLE 1

Hippocampal Norepinephrine Content: Effects of
Locus Coeruleus and Medial Forebrain Bundle Lesions[a]

Lesion location	Hippocampal norepinephrine content (ng/g ± SE)		Percent difference
	Lesion side	Control side	
Locus coeruleus	141 ± 70	536 ± 62	−74
Medial forebrain bundle	94 ± 58	503 ± 87	−81

[a] Unilateral locus coeruleus or medial forebrain bundle lesions were made stereotaxically using a radiofrequency current. Animals were killed 30 days later and the hippocampal formations were removed and assayed for catecholamine (norepinephrine and dopamine) content using an isotopic assay method (Coyle and Henry, 1973). Dopamine values were in the range expected for a precursor in a norepinephrine neuron (about 20% of norepinephrine content; cf. Costa et al., 1972). Dopamine neurons do innervate neocortex and a restricted portion of the entorhinal area (Thierry and Glowinski, 1973; Hokfelt et al., 1974; Lindvall et al., 1974).

tochemical method (Moore, Lindvall, and Björklund, unpublished observations), and an autoradiographic tracing method (Jones and Moore, 1975). The observations obtained using these methods are all quite similar and will be described together. Fibers leaving the locus coeruleus run rostrally in the midbrain tegmentum in the dorsal norepinephrine bundle (Ungerstedt, 1971), which joins the ascending component of the medial forebrain bundle in the caudal diencephalon. The fibers of this projection continue forward in the medial forebrain bundle to reach the hippocampal formation by two routes. The first is via the diagonal band and the fornix. The second is via the cingulum (Lindvall and Björklund, 1974b; Jones and Moore, 1975). Within the hippocampal formation, fibers from the cingulum distribute principally to the CA1 zone, whereas fibers from the fornix distribute predominantly to the remaining hippocampal zones and the area dentata. The distribution of norephrine axons within the hippocampal formation is shown diagramatically in Fig. 1. All of the axons have the appearance described by Lindvall and Björklund (1974b) for locus coeruleus axons. As seen in the glyoxylic acid material (Figs. 2 and 3), the innervation is formed by a plexus of very fine, smooth fibers, approximately 0.4–0.6 μm in diameter, with fairly regularly and closely spaced varicosities approximately 1–3 μm in diameter.

The topography of the innervation will be described using Blackstad's (1956) terminology for the components of the hippocampal formation. There is a fairly dense lamina of axons in the superficial layers of the subicular complex which is continuous with a lamina of axons in that portion of the stratum lacunosum-moleculare bordering stratum radiatum. Fibers radiate out of this into the stratum radiatum to form a fine terminal plexus in CA1. There is also a fairly dense innervation of stratum lacunosum-moleculare but there are very few fibers present in the stratum pyramidale or stratum oriens. As the lamina of fibers in stratum lacunosum-moleculare reaches the CA2 zone, it begins to disperse and form a moderately dense ter-

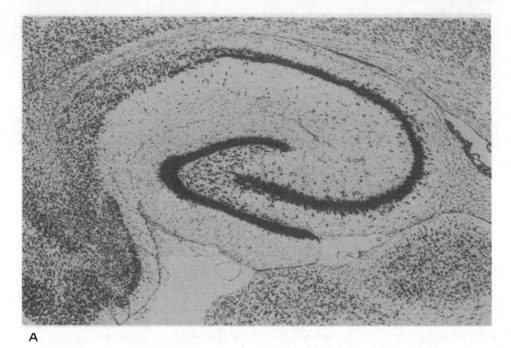

A

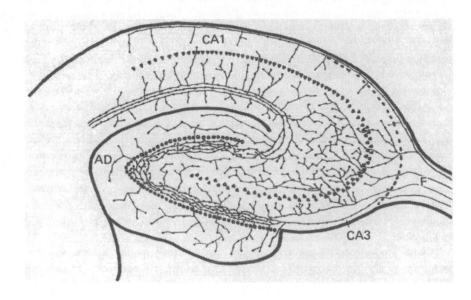

B

FIG. 1. Norepinephrine innervation of the hippocampal formation. A: Photograph of a horizontal section through the hippocampal formation (rat brain, cresyl violet stain, bar equals 100 μm). B: Drawing of the same section diagrammatically illustrating the distribution of locus coeruleus norepinephrine neuron axons in the hippocampal formation. Abbreviations for this and subsequent figures are as follows: AD, area dentata; CA1 and CA3, hippocampal zones; F, fimbria; T, thalamus; so, stratum oriens; sp, stratum pyramidale; sr, stratum radiatum; slm, stratum lacunosum-moleculare.

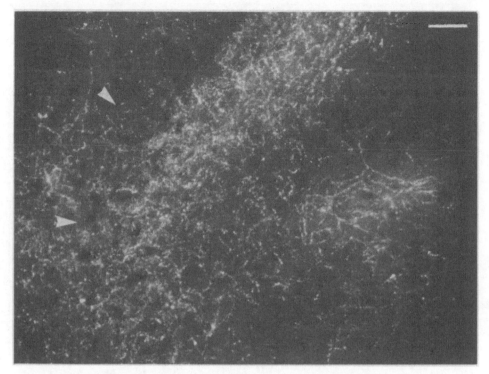

Fig. 2. CA3 zone norepinephrine innervation. Photograph of a glyoxylic acid preparation showing an extensive plexus of norepinephrine axons, particularly in the stratum radiatum. The arrows designate the stratum pyramidale. Bar equals 50 μm.

minal plexus scattered throughout the stratum radiatum of CA2 and CA3 (Fig. 2). Within CA3, fibers enter the stratum oriens from the fornix and distribute in a moderately dense plexus in the stratum oriens and some contribute to the plexus in the stratum radiatum by continuing through the stratum pyramidale. Other fornix fibers continue along the stratum oriens to enter the area dentata. Some form a very loose plexus within the molecular layer, but the most striking innervation is an extremely dense plexus of fibers just below the granule cell layer within the hilus (Fig. 3). This pattern of innervation is quite uniform throughout the dorsal and ventral parts of the hippocampal formation. The norepinephrine innervation appears to reach the hippocampal formation largely from fornix and cingulum, but it is not possible to determine from the data now available how much is contributed by fibers entering through the entorhinal area. It has been suggested by Storm-Mathisen and Guldberg (1974) that a major component of the norepinephrine innervation to the hippocampal formation traverses the entorhinal area to reach it. This can only be resolved by further study. It does appear, however, that approximately 25% of the hippocampal norepinephrine innervation, like that of other telencephalic structures, arises from the contralateral locus coeruleus (Jones and Moore, 1975). This observa-

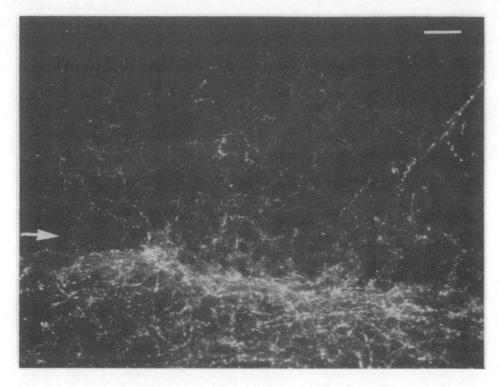

FIG. 3. Area dentata norepinephrine innervation. There is an extensive axonal plexus shown in this photograph of a glyoxylic acid preparation just beneath the granule cell layer (arrow). Bar equals 50 μm.

tion explains why unilateral locus coeruleus lesions never produce complete depletion of hippocampal norepinephrine (Table 1).

2.1.2. Serotonin Innervation. The serotonin content of the hippocampal formation is also significantly higher than that of neocortex (Moore and Heller, 1967; Lorens and Guldberg, 1974; Jacobs *et al.*, 1974) (Table 2). There is some discrepancy in the values reported, ranging from a low of 0.2–0.3 μg/g (Lorens and Guldberg, 1974; Jacobs *et al.*, 1974) to a high of 0.6–0.8 μg/g (Moore and Heller, 1967; Krieger, 1974; Moore and Halaris, 1975); whether this represents differences among strains of rats, differences in methodology, the effects of killing the rats at different points along a circadian rhythm of hippocampal serotonin content (Krieger, 1974), or some combination of these is not clear. The serotonin innervation of the hippocampal formation originates from neurons within the midbrain raphe nuclei (Dahlström and Fuxe, 1965; Anden *et al.*, 1966). Studies by Lorens and Guldberg (1974) and Jacobs *et al.* (1974) indicate that the innervation arises almost exclusively in the median raphe (principally nucleus centralis superior) with the dorsal raphe contributing little, if any. Axons arising from midbrain raphe neurons enter the ventral tegmental area and ascend in the medial forebrain bundle to reach the hippocampal formation via the fornix and the cingulum (Fuxe *et al.*, 1970; Björklund *et al.*, 1973; Moore and Halaris, 1975).

This distribution of the serotonin innervation within the hippocampal formation has not been analyzed extensively using histochemical methods. With the Falck–Hillarp method, terminals are evident in the hippocampal formation (Fuxe, 1965), but it has not been possible to study their exact distribution because of methodological problems. Similarly, the glyoxylic acid method which is so sensitive for catecholamines (Lindvall and Björklund, 1974*a*) has not been successfully applied as yet to the analysis of indolamine neuron systems. It has long been recognized that the fluorescent histochemical methods are not as successful for the analysis of serotonin neuron systems as they are for catecholamine neuron systems (Corrodi and Jonsson, 1967; Kuhar *et al.*, 1972; Björklund *et al.*, 1973). Data have been obtained, however, using the autoradiographic tracing technique (Cowan *et al.*, 1972) which demonstrate that the serotonin innervation of the hippocampal formation essentially overlaps the norepinephrine innervation in its terminal distribution (Conrad *et al.*, 1974; Moore and Halaris, 1975). This is shown in Fig. 4, which can be compared with Fig. 1.

The following description is taken from material in which tritiated amino acid was injected into the midbrain raphe and histological material was prepared by the autoradiographic technique. Fibers arising from the cingulum complex pass around the splenium of the corpus callosum and turn rostrally and ventrally to enter the hippocampal formation in two bundles. The largest of these runs through the molecular layer of the subiculum, where there appear to be some terminals, and enters the stratum lacunosum-moleculare of the hippocampal CA1 zone. The smaller bundle runs through deep layers of the subicular complex to enter the alveus and distribute as a rather sparse innervation to the stratum oriens of CA1. The fibers entering CA1 in the stratum lacunosum-moleculare form a dense band which is approximately 100 μm wide. This band occupies the portion of stratum lacunosum-moleculare which directly abuts on the stratum radiatum. The band appears to contain terminals as

TABLE 2

Hippocampal Serotonin Content: Effects of Raphe and
Medial Forebrain Bundle Lesions

Lesion location	Hippocampal serotonin content (ng/g $\pm$ SE)		Percent difference
	Lesion side	Control side	
Midbrain raphe[a]	180 $\pm$ 44	632 $\pm$ 40	−72
Medial forebrain bundle[b]	220 $\pm$ 50	710 $\pm$ 50	−68

[a] Data from Moore and Halaris (1975).
[b] Data from Moore and Heller (1967).

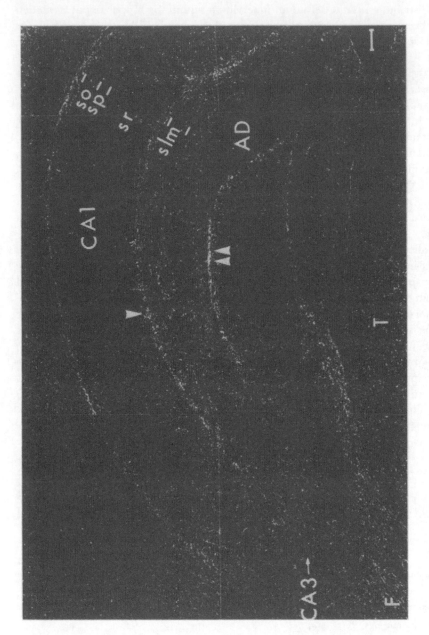

FIG. 4. Serotonin innervation of the hippocampal formation. Photographic montage of a frontal section of the hippocampal formation of a rat which had received an injection of tritiated proline into the midbrain raphe nuclei. The section was prepared by an autoradiographic technique. The small white dots are silver grains. The single arrow designates a dense band of silver grains in the CA1 stratum lacunosum-moleculare and the double arrow a dense band beneath the granule cell layer in the polymorph layer of the area dentata. Bar equals 100 μm.

well as fibers of passage. As it courses along the stratum lacunosum-moleculare of CA1, fibers leave to terminate in the stratum radiatum and in the stratum lacunosum-moleculare, particularly in the zone adjacent to radiatum. At the border of the CA2 zone, approximately at the point where the hippocampal fissure ends, the dense lamina of fibers in the stratum lacunosum-moleculare disperses to distribute as a moderately dense innervation of the stratum radiatum of CA2 and CA3. In these zones, the labeling of the stratum radiatum is fairly uniform and fibers continue around to enter the adjacent component of the molecular layer of the area dentata. Fornix fibers also contribute to the innervation of the stratum radiatum in CA2 and CA3, where there is a moderate innervation and run between cells of the pyramidal layer to enter the stratum radiatum. Fornix fibers also pass along the stratum oriens to innervate the area dentata. There is a sparse to moderate innervation in the area dentata molecular layer, but the most striking innervation of the entire hippocampal formation is found in the hilar zone. This is a very dense band of innervation, approximately 65 μm wide, along the ventral border of the dentate granule cells which extends the entire length of the area dentata from dorsal to ventral.

This distribution of innervation, derived from autoradiographic studies, is open to the criticism that it may not represent the distribution of axons of serotonin neurons. The midbrain raphe contains a mixed population of which some neurons are serotonin producing and others are not. An injection of tritiated amino acid into the raphe nuclei would result in axonal transport of labeled protein in both serotonin and nonserotonin neurons. Consequently, in the autogradiographic studies some animals were used which had been pretreated with the selective neurotoxin, 5,6-dihydroxytryptamine (Björklund *et al.*, 1973). This pretreatment prevented axonal transport of labeled protein to the hippocampal formation and autoradiograms from brains pretreated with the neurotoxin did not exhibit any of the specific labeling described above (Moore and Halaris, 1975).

The norepinephrine and serotonin neurons innervating the hippocampal formation have a very distinct and overlapping distribution. In the hippocampus itself, there is a lamina of fibers which extends from the subiculum along the entire CA1 zone in the stratum lacunosum-moleculare at its border with the stratum radiatum. This lamina appears to include both fibers of passage and terminals. Fibers leave it to terminate in adjacent layers, particularly the stratum radiatum, where they presumably innervate the hippocampal pyramidal cell dendrites. The innervation appears to be most dense in the stratum lacunosum-moleculare of CA1 in the zone along the apical dendrite where the Schaffer collaterals distribute. In CA2 and CA3, the distribution of serotonin and norepinephrine terminals along the surface of apical dendrites of pyramidal cells would appear to be more even. Within the area dentata, the pattern of innervation differs significantly from this. In the molecular layer, there is sparse to moderate innervation throughout, suggesting a rather uniform terminal distribution to granule cell dendrites. In the hilar area, however, there is a densely innervated zone just beneath the granule cell layer, which, presumably, represents terminals on interneurons. It should be emphasized that this distribution of terminals differs from that of any other known input to the hippocampal formation.

2.2. Septum

2.2.1. Norepinephrine Innervation.

The norepinephrine content of the septum is shown in Table 3. This is also higher than that for neocortical areas (see above) but is lower than that for the hippocampal formation (Table 1). This refers to the content of the medial and lateral septal nuclei. Like the norepinephrine innervation to other telencephalic areas, this arises from the locus coeruleus. Lesions transecting ascending norepinephrine axons within the medial forebrain bundle produce significant decreases in septal norepinephrine content (Table 3). The distribution of the norepinephrine innervation within the septum was first described by Fuxe (1965). It was extensively analyzed by Moore et al. (1971) using the Falck–Hillarp method. In addition, there have been recent studies using the glyoxylic acid histochemical method (Moore, unpublished observations). In glyoxylic acid material, there is a sparse to moderate innervation of fluorescent axons in the medial septal nucleus which is oriented predominantly in a dorsal–ventral direction. These axons have the appearance of norepinephrine-containing axons as described by Lindvall and Björklund (1974b). The major innervation to the primary septal nuclei is evident in the lateral septal nucleus (Fig. 5). Here there is a very dense innervation to the region of the lateral septal nucleus bordering the medial septal nucleus. This forms a rather broad band of terminals which are extremely dense and extend from the rostral portion of the nucleus to the point where the fornix is a dominant feature of the medial portion of the septum. Lateral to this band of terminals, there are scattered terminals which appear to predominantly provide axodendritic contact. In the most lateral part of the lateral septal nucleus, there are numerous fine terminals located around cell bodies. This is in a very dense plexus around individual cells. Similar patterns of pericellular innervation are formed adjacent to the densely innervated zone in the lateral septal nucleus and also in the medial septal nucleus (Figs. 6 and 7). Caudally there are scattered terminals through the posterior part of the lateral septal nucleus.

TABLE 3

Septal Norepinephrine and Serotonin Content:
Effect of Medial Forebrain Bundle Lesions

Group	Monoamine content (μg/g $\pm$ SE)			
	Norepinephrine[a]	Percent difference from sham	Serotonin[b]	Percent difference from sham
Sham operated	0.39 ± 0.03	—	1.50 ± 0.04	—
Medial forebrain bundle lesions	0.10 ± 0.01	-74	0.77 ± 0.12	-49

[a] Data from Moore et al. (1971).
[b] Data from Moore and Heller (1967).

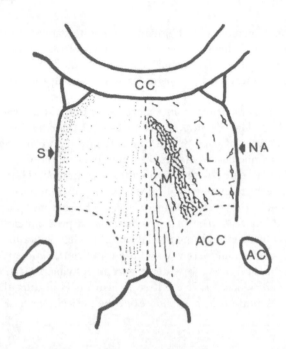

Fig. 5. Diagram of the norepinephrine (NA) and serotonin (S) afferents to the medial and lateral septal nuclei in the rat. The serotonin innervation, on the left, comes from an autoradiographic study (Halaris *et al.*, 1975) and the norepinephrine data from a Falck–Hillarp study (Moore *et al.*, 1971) and from unpublished glyoxylic acid material. The open ovals with extended lines on the norepinephrine side are intended to represent the distribution of the densely innervated cells as shown in Fig. 7.

2.2.2. Serotonin Innervation. The serotonin content of the septal area (Moore and Heller, 1967; Moore, 1974) (Table 3) is significantly higher than that of the hippocampal formation (Table 2) or other telencephalic areas (Moore and Heller, 1967; Lorens and Guldberg, 1974; Jacobs *et al.*, 1974). A number of studies have indicated that lesions in the midbrain raphe or medial forebrain bundle produce marked decreases in telencephalic, including septum, serotonin content (Moore and Heller, 1967; Heller and Moore, 1968; Kuhar *et al.*, 1972; Lorens and Guldberg, 1974; Jacobs *et al.*, 1974) (Table 3). Nevertheless, it has not been possible to map the distribution of serotonin neurons to the septal area using the Falck–Hillarp method (Fuxe, 1965), even when modifications are introduced to enhance serotonin fluorescence (Kuhar *et al.*, 1972; Moore, unpublished observations). Some few terminals are evident in the lateral nucleus in enhanced material, but these fade rapidly and the material is difficult to work with.

The development of an autoradiographic tracing technique (Cowan *et al.*, 1972) has allowed an analysis of midbrain raphe projections on the septum (Conrad *et al.*, 1974; Halaris *et al.*, 1975). Projections arising from the raphe nuclei, predominantly raphe dorsalis and central superior, ascend in the ventral tegmental area to enter the diencephalon in the medial forebrain bundle. They run forward in the medial forebrain bundle through the rostral diencephalon and turn medially and dorsally into

the septum via the diagonal band. The pattern of innervation to the medial and la-
terial septal nuclei is as follows. In more rostral parts of the septum, there is a dense
plexus of fibers throughout the medial septal nucleus. The lateral septal nucleus
contains a scattered innervation except in its most lateral portion where there is a
dense innervation (Fig. 5). As the fibers of the fornix appear medially, the innervation in the lateral septal nucleus becomes more dense but is still heaviest at the lateral
border of the nucleus (Halaris *et al.*, 1975). Many of the fibers in the medial septal
nucleus are clearly fibers of passage and continue into the fornix. Some of the fibers
in the medial septal nucleus would appear not to be serotonin producing as they
persist in animals pretreated with 5,6-dihydroxytryptamine (Halaris *et al.*, 1975).
This treatment appears to eliminate all of the raphe innervation in the lateral septal
nucleus. Consequently, the serotonin innervation of the septum appears to differ from
the norepinephrine innervation in that it is distributed predominantly to lateral cells
of the lateral septal nucleus, with the medial septal nucleus containing predominantly
fibers of passage. In addition, there is very little input to the interstitial nucleus of the
stria terminalis, which receives a very dense norepinephrine projection from the locus
coeruleus (Jones and Moore, 1975). There is also no indication that the serotonin
input to the lateral septal nucleus distributes in the pericellular basket arrangements,

FIG. 6. Norepinephrine innervation of the septal area. Photograph of a glyoxylic acid preparation through
the lateral septal nucleus showing many fibers scattered through the neuropil, presumably making
axodendritic contacts, but many others closely associated with cell bodies and proximal dendrites of septal
neurons. Bar equals 50 μm.

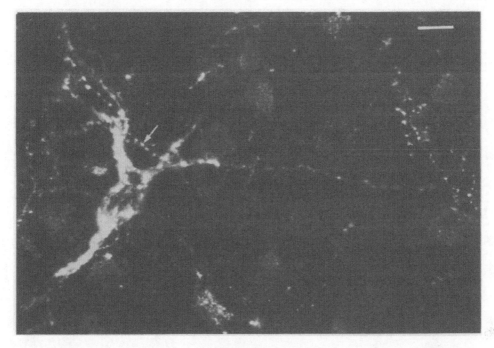

FIG. 7. Norepinephrine innervation of a lateral septal nucleus neuron, glyoxylic acid method. The cell body is encased in fluorescent fibers. A typical, thin axon with regularly spaced varicosities crosses the cell (arrow). Bar equals 30 μm.

which are, in part, characteristic of the norepinephrine input. It is likely, however, that both the serotonin and norepinephrine inputs to the septum are derived from collaterals of axons continuing to the hippocampal formation and other telencephalic areas.

3. Response to Injury of Hippocampal and Septal Monoamine Innervation

3.1. Regeneration of the Transected Axon

There are two forms of regenerative axonal growth which may occur in response to injury (Fig. 8). The first of these is regenerative sprouting: growth from a transected axon. The second, collateral sprouting, involves growth from an intact axon in response either to injury of a collateral of the axon or to loss of other neuronal elements from an area innervated by the axon.

It has long been recognized that severed central nervous system axons are capable of exhibiting regenerative sprouting (Ramón y Cajal, 1928). Unfortunately, such sprouting appears to have very little functional significance, and this remains a major problem in neurobiology. Among the questions that have arisen are whether

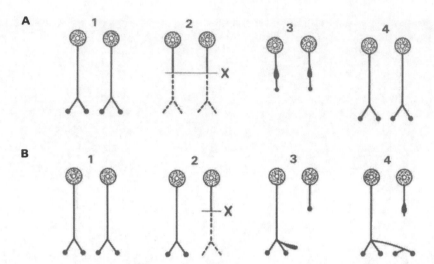

FIG. 8. Paradigms for regeneration in the nervous system. A: Sprouting from transected axons. The axons of two neurons diagrammatically innervating a structure are severed. They regrow from the proximal axon to replicate the original innervation. B: Collateral reinnervation. One axon innervating a structure is severed. Its distal portion degenerates and the remaining axon undergoes collateral sprouting to reinnervate the structure. From Moore *et al.* (1974).

groups of central neurons differ in their capacity to demonstrate regenerative sprouting and whether regenerating axons maintain the capacity to establish functional synaptic contacts. Similarly, little is known of the conditions that promote sprouting and provide directed growth necessary to promote functional synaptic contacts, nor is there a great deal of information on the conditions that determine the usual situation, failure of functional restoration following injury to central axons. Recent studies indicate that central adrenergic neurons exhibit both regenerative sprouting and collateral sprouting. Since the regenerative sprouting studies form a basis for all regeneration experiments, they will be considered first.

The regenerative sprouting following transection of central adrenergic neurons was first noted by Katzman *et al.* (1971). In their studies, lesions were placed in the caudal medial forebrain bundle, transecting axons of the ascending norepinephrine and dopamine neuron systems. After transection of the axon of a central adrenergic neuron, there is accumulation of amine in the several stump proximal to the lesion which is evident within hours and persists for several days (Katzman *et al.*, 1971). Between 4 and 7 days after transection of the axon, small, delicate, new axons arise from the transected fibers and form an abundant plexus of axons and terminals in the zone surrounding the proximal stumps of the transected axons. These regenerating axons exhibit a remarkable capacity for growth. They will grow into the walls of blood vessels in the vicinity of the lesions and occasionally grow along the path of adjacent cranial nerves (Katzman *et al.*, 1971; Björklund and Stenevi, 1971). These observations, indicating the capacity of severed central adrenergic axons to exhibit regenerative sprouting, do not indicate their capacity to form new or functional

synaptic contacts. To test this capacity, transplants of peripheral, sympathetically innervated tissue were placed within ascending or descending pathways of brain stem
adrenergic neuron systems. In each case, the transplant is denervated by removal
from its normal site and implantation of the transplant into central nervous system
transects the adrenergic axons. The most successful transplantation experiments have
been carried out with the placement of iris into the medial forebrain bundle
(Björklund and Stenevi, 1971). As noted above, there are early accumulations of
amine in transected axons adjacent ot the transplant and subsequent sprouting of
delicate, varicose fibers. The transplant is free of adrenergic innervation for the first
few days, but by 2 wk many adrenergic fibers can be traced from the area of the transected adrenergic axons into the transplanted iris. Within the iris, these fibers form a
plexus which is identical to the normal, sympathetic innervation of the iris, and this
innervation persists for many months. These observations establish that the transected adrenergic axon is capable of innervating a denervated structure if it is within
the reach of growing axonal sprouts and that the pattern of innervation by the growing axons is determined by the tissue innervated. This is further emphasized by the
fact that a transplanted iris is innervated predominantly by norepinephrine axons,
whereas the dopamine and serotonin axons of the medial forebrain bundle, which
also show regenerative sprouting proximal to the transplant, do not significantly
enter it. In addition, other tissues such as diaphragm or uterus, which normally are
not innervated significantly by peripheral adrenergic neurons, are not innervated by
central adrenergic neurons when transplanted into the medial forebrain bundle
(Björklund and Stenevi, 1971). Thus all of the evidence currently available indicates
that central adrenergic axons are capable of regenerative growth and that this growth
can result in a highly organized innervation pattern. There is recent evidence that
such growth forms functional contacts (Björklund, personal communication). No
studies have been carried out to examine the regenerative growth of catecholamine or
serotonin neurons innervating the septum or hippocampal formation, but it has been
noted that severed norepinephrine axons in the septum do not appear to form
regenerative sprouts (Moore et al., 1971).

3.2. Collateral Reinnervation

Nearly all studies of central nervous system regeneration have examined the
regenerative capacity of transected axons. It is evident, from the observations reviewed above, that central adrenergic neurons are capable of vigorous regenerative
sprouting, but until recently there were few studies of collateral sprouting in these or
other central neuron systems. The most extensive studies of this phenomenon in the
central nervous system have been carried out by Raisman (1969b) and Raisman and
Field (1973) on the innervation of the septum in the rat. The basis for these studies
was formed by the observation that septal neurons were innervated by two primary
sources, the hippocampal formation via the fornix and the brain stem via the medial
forebrain bundle (Raisman, 1966). In ultrastructural studies, each of these had a
characteristic pattern of termination on septal neurons (Raisman, 1969a). Following
removal of one source of innervation, e.g., the hippocampal formation, there is initial

degeneration of terminals from that source but little alteration in the long-term appearance of the ultrastructure of the area. The major change is an increase in the number of axons exhibiting multiple synaptic contacts (Raisman, 1969b), which indicate that the intact innervation has provided collateral sprouting to fill denervated synaptic sites. This was confirmed in a very detailed experiment analyzing the time course over which changes occurred in the septum following removal of hippocampal afferents (Raisman and Field, 1973).

The septal area receives a dense adrenergic innervation, as outlined above, from the medial forebrain bundle, and a study was carried out to determine whether this innervation participated in the collateral reinnervation demonstrated by Raisman (1969b) and Raisman and Field (1973). In this experiment, rats were subjected to unilateral section of the fornix. Since the fornix projects almost exclusively ipsilaterally on the septal nuclei (Raisman, 1966), the contralateral side serves as a control (Fig. 9). The distribution of catecholamine fibers innervating the septum was studied using the Falck–Hillarp technique in such animals at a number of postoperative survival periods ranging from 3 to 100 days after removal of hippocampal afferents (Moore et al., 1971). In the first week after partial denervation of the septum, there is no alteration of catecholamine innervation except that a few scattered axons have accumulations of catecholamine as seen in transected axons. These are no longer evident at 15 days after denervation, but at this point there is a consistent increase in the number of apparent catecholamine fibers in the denervated septum (Fig. 10). This difference becomes maximal by 30 days and is consistently present through 100 days. No other area in the basal telencephalon or hypothalamus exhibits a similar change.

Several interpretations of this observation are possible. The first is that the

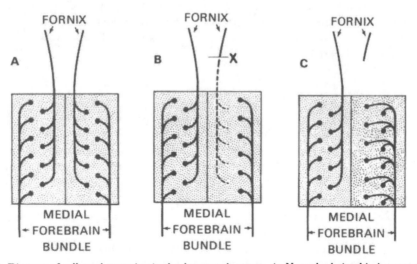

Fig. 9. Diagram of collateral sprouting in the denervated septum. A: Normal relationship between fornix and medial forebrain bundle input to the septum. B: Unilateral fornix section partially denervates one-half of the septum. C: Denervated area is reinnervated by medial forebrain bundle afferents. From Moore et al. (1974).

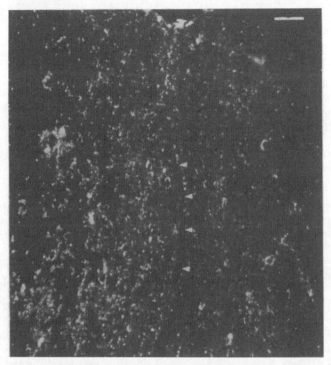

FIG. 10. Collateral sprouting in the denervated septum, photograph of Falck–Hillarp preparation made 100 days after a unilateral hippocampal ablation. The photograph shows the dorsal part of the lateral septal nucleus, denervated side on the left (arrows designate the midline) showing a much increased norepinephrine innervation. Bar equals 100 μm.

increase in apparent catecholamine innervation reflects collateral reinnervation of septal neurons by catecholamine fibers innervating the septum (Fig. 9). There is an increase in norepinephrine content in the denervated septum (Moore *et al.*, 1971), and the major change is evident in the heavy plexus of fibers placed in the medial portion of the lateral septal nucleus adjacent to the medial septal nucleus (Fig. 5). This interpretation is in accord with that of Raisman (1969*b*) and Raisman and Field (1973), but other alternatives are available. The most important of these are evident from consideration of the organization of innervation to both the septum and the hippocampal formation (Fig. 9). In all likelihood, the innervation to the septal nuclei is formed from collateral fibers that continue through the fornix to the hippocampal formation (Fig. 11). Section of these fibers at their entrance into the hippocampal formation could result in two phenomena taking place. One would be accumulation of amine in septal axons and terminals which were not previously demonstrable by the Falck–Hillarp method. That is, the amine content of the axons would be increased by a continuing normal rate of transport of amine distributed to a restricted terminal area. It is well known that accumulation of amine occurs in an axon proximal to section, including in the collaterals (Ungerstedt, 1971), but this phenomenon would not appear to explain the changes observed because the time course is prolonged. For the accumulation effect to be explanatory, the phenomenon

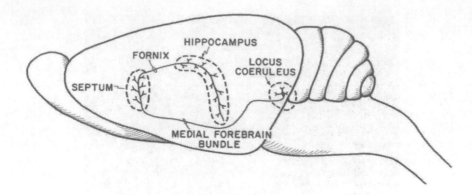

FIG. 11. Norepinephrine innervation of the septum and hippocampal formation. Axons arising from the locus coeruleus traverse the dorsal catecholamine bundle to reach the septum and hippocampus. As shown in this diagram, the septal innervation can be viewed as arising from collaterals of axons innervating the hippocampal formation.

should be short-lived. A second possibility is that the increase in apparent cate-cholamine innervation of the septum represents compensatory collateral growth of axons which have been transected. That is, loss of the hippocampal terminal field results in expansion of the terminal field in the septum. This phenomenon has been observed in the locus coeruleus norepinephrine innervation of the cerebellum (Pickel *et al.,* 1973, 1974*a*). Again, this appears to have a relatively limited time course which would not extend for the period of observation of changes observed by Moore *et al.* (1971). Transport of labeled protein from locus coeruleus to septum is increased following denervation, providing a further indication that the axonal plexus of locus coeruleus neurons is increased in the denervated septum (Moore, 1974). Some evidence is also available that serotonin neurons participate in the phenomenon (Moore, 1974) since serotonin content is increased in the denervated septum. As with the transected axon, we have no evidence at the present time that monoamine neurons innervating the hippocampus will form collateral reinnervation in response to denervation of the structure.

4. Conclusions

The septum and hippocampal formation are innervated by the norepinephrine neurons of the locus coeruleus and by serotonin neurons of the midbrain raphe nuclei. This innervation comprises a part of a general telencephalic innervation by these brain stem nuclei (Anden *et al.,* 1966; Ungerstedt, 1971; Jones and Moore, 1975; Halaris *et al.,* 1975), but the septum and hippocampal formation are among the most heavily innervated of all telencephalic structures. For both the norepinephrine and serotonin systems, the innervation to septum appears to arise as collateral fibers from axons continuing on to terminate in the hippocampal formation. In the septum, the norepinephrine and serotonin afferents have overlapping but distinct patterns of distribution. The norepinephrine input is largely concentrated over a zone of lateral septal nucleus adjacent to the medial septal nucleus. Within this zone, the input is extremely dense, matching the most heavily innervated areas of the hypothalamus. The septum also shows an unusual form of norepinephrine axon termination not seen elsewhere in the rat brain, a dense, pericellular "basket" plexus of fibers and terminals. The significance of this is as yet unknown. In contrast to the norepinephrine innervation, the serotonin afferents to the septum are most numerous in the lateral portion of the lateral septal nucleus and in the medial septal nucleus. The input to medial septal nucleus contains many fibers of passage and is clearly mixed with other brain stem input (Nauta, 1958; Morest, 1961) which is probably neither adrenergic nor serotonergic.

The only direct afferents to the hippocampal formation from the brain stem that have been demonstrated are those from the serotonin neurons of the median raphe and the norepinephrine neurons of the locus coeruleus. Unlike the situation in the septum, these inputs to hippocampal formation appear to have an identical pattern of terminal distribution and one which differs from all other known sources of input (Ramón y Cajal, 1911; Blackstad, 1956, 1967; Storm-Mathisen and Blackstad, 1964; Raisman et al., 1965). In the hippocampus, they distribute principally to the region occupied by the apical dendrites of pyramidal cells. The heaviest innervation in the CA1 zone appears to coincide with the distribution of Schaffer collaterals, but there is significant input to all layers except the stratum pyramidale. The input to the CA3 zone is distributed heavily throughout the stratum radiatum and stratum oriens. Within the area dentata, there is a dense input into a very restricted region of the hilar zone just beneath the granule cell layer. The terminals of monoamine axons in this zone would make contact primarily with dendrites of interneurons (Ramón y Cajal, 1911; Blackstad, 1967). These observations suggest that in both areas, hippocampal formation and septum, the serotonin and norepinephrine inputs have a functional significance differing from other afferent systems. There is little information available on specific functions at any level of analysis, but recent studies at the cellular level indicate that the norepinephrine input to hippocampal formation can alter hippocampal neuron firing patterns (Segal and Bloom, 1974a,b).

During the past 5 years, there has been increasing interest in morphological plasticity in the central nervous system. Much of the investigative work in this area has been directed toward two problems, developmental plasticity following either environmental or mechanical manipulation of the neonatal brain and regenerative responses to injury in the adult brain. In all likelihood, these are merely aspects of the same biological phenomenon since regeneration in the adult nervous system probably reflects a situation in which developmental processes are set in motion by injury. A detailed account of developmental plasticity is beyond the scope of this chapter. As noted above, regenerative responses in the adult nervous system take two forms, regenerative sprouting and collateral sprouting. Regenerative sprouting has been examined most extensively recently in adrenergic neurons (for review, see Moore et al., 1974). These neurons exhibit a remarkable capacity for regenerative sprouting, which clearly illustrates their ability to grow and form new terminal fields in the adult animal. Similarly, adrenergic neurons exhibit a significant capacity for collateral sprouting in a variety of situations (for review, see Moore et al., 1974), including in the septal area (Moore et al., 1971). A question which has arisen from these studies is whether the adrenergic neuron has a special capacity for regenerative responses that is not shared by other central neuron systems. Recent data indicate that this is not the case. Studies by Lynch and Cotman and their colleagues (Lynch *et al*, 1972, 1973*a,b*, 1974; Cotman *et al.*, 1973; Nadler *et al.*, 1973; Steward *et al*, 1974; Lynch and Cotman, this volume) and by Zimmer (1973*a,b*; 1974*a,b*) indicate that several inputs to the hippocampal formation may exhibit plasticity in either the developmental or adult situation.

Thus much of the recent work on morphological plasticity in the central nervous system has been carried out in two quite different experimental situations, either on the innervation of the hippocampal formation and septum or on a special class of central neurons, the adrenergic neuron. Consequently, it is not known whether these are exemplary of changes occurring in many neuron systems in many areas of the brain or whether they represent restricted responses of a few neuron systems in limited areas. Certainly, there are neuron systems which do not show significant regenerative or collateral sprouting (Ramón y Cajal, 1928; Kerr, 1972; Guillery, 1972) in the adult nervous system. The adrenergic neuron may well represent a class of neurons which remain relatively unspecified in terms of synaptic relationships within terminal fields (Moore *et al.*, 1974) and retain plasticity on this basis. Whether there are other neuron systems which may have similar properties, for example, the serotonin neurons, is unknown at present. Similarly, the organization of synaptic architecture in the hippocampal formation may retain a significant plasticity into adult life which is not shared by neuron systems in other areas. The intriguing aspect of this, and one which remains to be resolved, is whether morphological plasticity occurs in response to alterations in functional input as well as in response to injury. If this is the case, it could form a basis for some of the important behavioral functions in which the septum and hippocampal formation are believed to participate.

NOTE ADDED IN PROOF

Since this chapter was written, it has been shown by Lindvall (*Brain Research*, 1975, **87**, 89–95) that many of the catecholamine afferents to the septal area are from the dopamine neurons of the ventral tegmental area. Dopamine terminals are particularly numerous in the lateral septal nucleus and constitute much of what is identified in this chapter as norepinephrine terminals.

5. *References*

AMIN, A. H., CRAWFORD, T. B. B., AND GADDUM, J. H. The distribution of substance P and 5-hydroxytryptamine in the central nervous system of the dog. *Journal of Physiology* (*London*), 1954, **126**, 596–618.

ANDEN, N. E., DAHLSTROM, A., FUXE, K., LARSSON, K., OLSON, L., AND UNGERSTEDT, U. Ascending monoamine neurons to the telencephalon and diencephalon. *Acta Physiologica Scandinavica*, 1966, **67**, 313–326.

BERTLER, A., AND ROSENGREN, E. Occurrence and distribution of dopamine in brain and other tissues. *Experientia*, 1959, **15**, 10–11.

BJÖRKLUND, A., AND STENEVI, U. Growth of central catecholamine neurons into smooth muscle grafts in rat mesencephalon. *Brain Research,* 1971, **31,** 1–20.

BJÖRKLUND, A., FALCK, B., AND OWMAN, C. Fluorescence microscopic and microspectrofluorometric techniques for the cellular localization and characterization of biogenic amines. In S. A. Berson (Ed.), *Methods of investigative and diagnostic endocrinology.* Vol. I: *The thyroid and biogenic amines* (J. E. Rall and I. J. Kopin, Eds.). Amsterdam: Holland Publishing Co., 1972.

BJÖRKLUND, A., NOBIN, A., AND STENEVI, U. The use of neurotoxic dihydroxytryptamines as tools for morphologic studies and localized lesioning of central indoleamine neurons. *Zeitschrift für Zellforschung,* 1973, 145, 479–501.

BLACKSTAD, T. Commissural connections of the hippocampal region in the rat, with special reference to their mode of termination. *Journal of Comparative Neurology,* 1956, **105,** 417–538.

BLACKSTAD, T. Cortical gray matter. In H. Hyden (Ed.), *The neuron.* New York: Elsevier, 1967.

BLACKSTAD, T., FUXE, K., AND HÖKFELT, T. Noradrenaline nerve terminals in the hippocampal region of the rat and guinea pig. *Zeitschrift für Zellforschung,* 1967, **78,** 463–473.

BOGDANSKI, D. F., WEISSBACH, H., AND UDENFRIEUD, S. The distribution of serotonin, 5-hydroxytryptophan decarboxylase and monoamine oxidase in brain. *Journal of Neurochemistry,* 1957, **1,** 272–278.

BRODIE, B. B., AND SHORE, P. A. A concept for a role of serotonin and norepinephrine as chemical mediators in the brain. *Annals of the New York Academy of Sciences,* 1957, **66,** 631–641.

CARLSSON, A., FALCK, B., AND HILLARP, N. A. Cellular localization of brain monoamines. *Acta Physiologica Scandinavica,* 1962, Supplement 196, **56,** 1–28.

CONRAD, L. C. A., LEONARD, C. M., AND PFAFF, D. W. Connections of the median and dorsal raphe nuclei in the rat: An autoradiographic and degeneration study. *Journal of Comparative Neurology,* 1974, **156,** 179–206.

CORRODI, H., AND JONSSON, G. The formaldehyde fluorescence method for the histochemical demonstration of biogenic monoamines. *Journal of Histochemistry,* 1967, **15,** 65–78.

COSTA, E., GREEN, A. R., KOSLOW, S. H., LeFEVRE, H. F., REVUELTA, A. V., AND WANG, C. Dopamine and norepinephrine in noradrenergic axons: A study *in vivo* of their precursor product relationship by mass fragmentography and radiochemistry. *Pharmacological Reviews,* 1972, **24,** 167–190.

COTMAN, C. W., MATTHEWS, D. A., TAYLOR, D., AND LYNCH, G. Synaptic rearrangement in the dentata hyrus: Histochemical evidence of adjustments after lesions in immature and adult rats. *Proceedings of the National Academy of Sciences U.S.A.,* 1973, **70,** 3473–3477.

COWAN, W. M., GOTTLIEB, D. I., HENDRICKSON, A. E., PRICE, J. L., AND WOOLSEY, T. A. The autoradiographic demonstration of axonal connections in the central nervous system. *Brain Research,* 1972, **37,** 21–51.

COYLE, J. T., AND HENRY, D. Catecholamines in fetal and newborn rat brain. *Journal of Neurochemistry,* 1973, **21,** 61–68.

DAHLSTRÖM, A., AND FUXE, K. Evidence for the existence of monoamine-containing neurons in the central nervous system. I. Demonstration of monoamines in the cell bodies of brain stem neurons. *Acta Physiologica Scandinavica Supplement,* 1965, **232,** 1–55.

FALCK, B. Observations on the possibilities of the cellular localization of monoamines by a fluorescence method. *Acta Physiologica Scandinavica,* 1962, Supplement 197, **56,** 1–25.

FALCK, B., HILLARP, N. A., THIEME, G., AND TORP, A. Fluorescence of catechol amines and related compounds condensed with formaldehyde. *Journal of Histochemistry and Cytochemistry,* 1962, **10,** 348–354.

FUXE, K. Evidence for the existence of monoamine neurons in the central nervous system. IV. Distribution of monoamine nerve terminals in the central nervous system. *Acta Physiologica Scandinavica,* 1965, Supplement 247, **64,** 37–85.

FUXE, K., HÖKFELT, T., AND UNGERSTEDT, U. Morphological and functional aspects of central monoamine neurons. *International Review of Neurobiology,* 1970, **13,** 93–126.

GUILLERY, R. W. Experiments to determine whether retinogeniculate axons can form translaminar collateral sprouts in the dorsal lateral geniculate nucleus of the cat. *Journal of Comparative Neurology,* 1972, **146,** 407–420.

HALARIS, A. E., JONES, B. E., AND MOORE, R. Y. Projections of serotonin neurons of the midbrain raphe in the rat. 1975, in preparation.

HELLER, A., AND MOORE, R. Y. Control of brain serotonin and norepinephrine by specific neural systems. *Advances in Pharmacology*, 1968, **6A**, 191–206.

HELLER, A., HARVEY, J. A., AND MOORE, R. Y. A demonstration of a fall in brain serotonin following central nervous system lesions in the rat. *Biochemical Pharmacology*, 1962, **11**, 859–866.

HÖKFELT, T., LJUNGDAHL, A., FUXE, K., AND JOHANSSON, O. Dopamine nerve terminals in rat limbic cortex: Aspects of the dopamine hypothesis of schizophrenia. *Science*, 1974, **184**, 177–179.

JACOBS, B. L., WISE, W. D., AND TAYLOR, K. M. Differential behavioral and neurochemical effects following lesions of the dorsal or medial raphe nuclei in rats. *Brain Research*, 1974, **79**, 353–361.

JONES, B. E., AND MOORE, R. Y. Ascending projections of the locus coeruleus in the rat: Autoradiographic analysis. 1975, in preparation.

KATZMAN, R., BJÖRKLUND, A., OWMAN, C., STENEVI, U., AND WEST, K. Evidence for regenerative sprouting of central catecholamine neurons in the rat mesencephalon following electrolytic lesions. *Brain Research*, 1971, **25**, 579–596.

KERR, F. W. The potential of cervical primary afferents to sprout in the spinal nucleus of V following long term trigeminal denervation. *Brain Research*, 1972, **43**, 547–560.

KRIEGER, D. T. Food and water restriction shifts corticosterone, temperature, activity and brain amine periodicity. *Endocrinology*, 1974, **95**, 1195–1201.

KUHAR, M. J., ADHAJANIAN, G. K., AND ROTH, R. H. Tryptophan hydroxylase activity and synaptosomal uptake of serotonin in discrete brain regions after midbrain raphe lesions: Correlations with serotonin levels and histochemical fluorescence. *Brain Research*, 1972, **44**, 165–176.

LINDVALL, O., AND BJÖRKLUND, A. The organization of ascending catecholamine neuron systems in the rat brain as revealed by the glyoxylic acid fluorescence technique. *Acta Physiologica Scandinavica Supplement*, 1974a, **412**, 1–48.

LINDVALL, O., AND BJÖRKLUND, A. The glyoxylic acid fluorescence histochemical method: A detailed account of the methodology for the visualization of central catecholamine neurons. *Histochemistry*, 1974b, **39**, 97–127.

LINDVALL, O., BJÖRKLUND, A., MOORE, R. Y., AND STENEVI, U. Mesencephalic dopamine neurons projecting to neocortex. *Brain Research*, 1974, **81**, 325–331.

LORENS, S. A., AND GULDBERG, H. C. Regional 5-hydroxytryptamine following selective midbrain raphe lesions in the rat. *Brain Research*, 1974, **78**, 45–56.

LYNCH, G., MATTHEWS, D. A., MOSKI, S., PARKS, T., AND COTMAN, C. Induced acetylcholinesterase-rich layer in rat dentate gyrus following entorhinal lesions. *Brain Research*, 1972, **42**, 311–319.

LYNCH, G. S., MOSKO, S., PARKS, T., AND COTMAN, C. W. Relocation and hyperdevelopment of the dentate commissural system after entorhinal lesions in immature rats. *Brain Research*, 1973a, **49**, 57–61.

LYNCH, G. S., STANFIELD, B., AND COTMAN, C. W. Developmental differences in post-lesion axonal growth in the hippocampus. *Brain Research*, 1973b, **59**, 155–168.

LYNCH, G. S., STANFIELD, B., PARKS, T., AND COTMAN, C. W. Evidence for selective post-lesion axon growth in the dentate gyrus of the rat. *Brain Research*, 1974, **69**, 1–11.

MOORE, R. Y. Growth of adrenergic neurons in the adult mammalian nervous system. In K. Fuxe, L. Olson, and Y. Zotterman (Eds.), *Dynamics of degeneration and growth in neurons*. Oxford: Pergamon Press, 1974.

MOORE, R. Y. AND HALARIS, A. E. Hippocampal innervation by serotonin neurons of the midbrain raphe in the rat. *Journal of Comparative Neurology*, 1975, in press.

MOORE, R. Y., AND HELLER, A. Monoamine levels and neuronal degeneration in the rat brain following lateral hypothalamic lesions. *Journal of Pharmacology and Experimental Therapeutics*, 1967, **156**, 12–22.

MOORE, R. Y., BJÖRKLUND, A., AND STENEVI, U. Plastic changes in the adrenergic innervation of the rat septal area in response to denervation. *Brain Research*, 1971, **33**, 13–35.

MOORE, R. Y., BJÖRKLUND, A., AND STENEVI, U. Growth and plasticity of adrenergic neurons. In F. O. Schmitt, and F. G. Worden (Eds.), *The neurosciences: Third study program*. Cambridge, Mass.: MIT Press, 1974.

MOREST, D. K. Connexions of the dorsal tegmental nucleus in rat and rabbit. *Journal of Anatomy*, 1961, **95**, 229–246.

NADLER, J. V., COTMAN, C. W., AND LYNCH, G. S. Altered distribution of choline acetyltransferase and actylcholinesterase activities in the developing rat dentate gyrus following entorhinal lesion. *Brain Research*, 1973, **63**, 215–230.

NAUTA, W. J. H. Hippocampal projections and related neural pathways to the mid-brain in the cat. *Brain*, 1958, **81**, 319–340.

PICKEL, V. M., KREBS, H., AND BLOOM, F. E. Proliferation of norepinephrine-containing axons in rat cerebellar cortex after peduncle lesions. *Brain Research*, 1973, **59**, 169–178.

PICKEL, V. M., SEGAL, M., AND BLOOM, F. E. Axonal proliferation following lesions of cerebellar peduncles: A combined fluorescence microscopic and autoradiographic study. *Journal of Comparative Neurology*, 1974a, **155**, 43–60.

PICKEL, V. M., SEGAL, M., AND BLOOM, F. E. A radioautographic study of the efferent pathways of the nucleus locus coeruleus. *Journal of Comparative Neurology*, 1974b, **155**, 15–42.

RAISMAN, G. The connections of the septum. *Brain*, 1966, **89**, 317–348.

RAISMAN, G. A comparison of the mode of termination of the hippocampal and hypothalamic afferents to the septal nuclei as revealed by electron microscopy of degeneration. *Experimental Brain Research*, 1969a, **7**, 317–343.

RAISMAN, G. Neuronal plasticity in the septal nuclei of the adult rat. *Brain Research*, 1969b, **14**, 25–48.

RAISMAN, G., AND FIELD, P. M. A quantitative investigation of the development of collateral reinnervation after partial deafferentation of the septal nuclei. *Brain Research*, 1973, **50**, 241–264.

RAISMAN, G., COWAN, W. M., AND POWELL, T. P. S. The extrinsic afferent, commissural and association fibers of the hippocampus. *Brain*, 1965, **88**, 963–996.

RAMÓN Y CAJAL, S. *Histologie du systeme nerveux de l'homme et des vertebres*. Vol. II. Paris: Maloine, 1911, pp. 733–761.

RAMÓN Y CAJAL, S. *Degeneration and regeneration of the nervous system*. London: Oxford University Press, 1928.

SEGAL, M., AND BLOOM, F. E. The action of norepinephrine in the rat hippocampus. I. Iontophoretic applications. *Brain Research*, 1974a, **72**, 79–97.

SEGAL, M., AND BLOOM, F. E. The action of norepinephrine in the rat hippocampus. II. The activation of the input pathway. *Brain Research*, 1974b, **72**, 99–114.

STEWARD, O., COTMAN, C. W., AND LYNCH, G. S. Growth of a new fiber projection in the brain of adult rats: Re-innervation of the dentate gyrus by the contralateral entorhinal cortex following ipsilateral entorhinal lesions. *Experimental Brain Research*, 1974, **20**, 45–66.

STORM-MATHISEN, J., AND BLACKSTAD, T. W. Cholinesterase in the hippocampal region: Distribution and relation to architectonics and afferent systems. *Acta Anatomica*, 1964, **56**, 216–253.

STORM-MATHISEN, J., AND GULDBERG, H. C. 5-Hydroxytryptamine and noradrenaline in the hippocampal region: Effect of transaction of afferent pathways on endogenous levels, high affinity uptake and some transmitter-related enzymes. *Journal of Neurochemistry*, 1974, **22**, 793–803.

THIERRY, A. M., AND GLOWINSKI, J. Existence of dopaminergic nerve terminals in the rat cortex. In E. Usdin and S. H. Snyder (Eds.), *Frontiers in catecholamine research*. New York: Pergamon Press, 1973.

TWAROG, B. M., AND PAGE, I. H. Serotonin content of some mammalian tissues and urine. *American Journal of Physiology*, 1953, **175**, 157–161.

UNGERSTEDT, U. Stereotaxic mapping of the monoamine pathways in the rat brain. *Acta Physiologica Scandinavica Supplement*, 1971, **367**, 1–48.

VOGT, M. The concentration of sympathin in different parts of the central nervous system under normal conditions and after the administration of drugs. *Journal of Physiology (London)*, 1954, **123**, 451–481.

ZIMMER, J. Extended commissural and ipsilateral projections in postnatally deentorhinated hippocampus and fascia dentata demonstrated in rats by silver impregnation. *Brain Research*, 1973a, **64**, 293–311.

ZIMMER, J. Changes in the Tinn sulfide silver staining pattern of the rat hippocampus and fascia dentata following early postnatal deafferentation. *Brain Research*, 1973b, **64**, 313–326.

ZIMMER, J. Proximity as a factor in the regulation of aberrant axonal growth in postnatally deafferented fascia dentata. *Brain Research*, 1974a, **72**, 137–142.

ZIMMER, J. Long term synaptic reorganization in rat fascia dentata deafferented at adolescent and adult stages: Observations with the Tinn method. *Brain Research*, 1974b, **76**, 336–342.

9

Neurotransmitters and the Hippocampus

DONALD W. STRAUGHAN

1. Introduction

There are substantial reasons for believing that the majority of the transmissions across synapses in the mammalian central nervous system (CNS) are chemically and not electrically mediated. Thus a variety of substances are present in the CNS of mammals which are known to be transmitters either in the mammalian peripheral nervous system or in the nervous system of invertebrates. These substances include norepinephrine (NE), dopamine, 5-hydroxytryptamine (5-HT), acetylcholine (ACh), and the amino acids γ-aminobutyric acid (GABA) and glutamate, and are located to a substantial extent in nerve endings and often within a vesicular fraction. Additionally, under the electron microscope, "gap," "tight," or "close" junctions are relatively uncommon, and the morphological features of central synapses closely resemble those in the periphery, where chemical transmission is known to occur. Chemical rather than electrical synaptic transmission is required in the presence of small nerve endings and restricted synaptic area. Also, chemical transmission explains the characteristic occurrence of "synaptic delays" in the mammalian CNS. However, electrotonic transmission appears to occur occasionally in 10% of the cells in the mesencephalic nucleus of the Vth cranial nerve (Baker and Llinas, 1971) and in giant neurons of the lateral vestibular nucleus of the rat (Korn et al., 1973).

The hippocampus with its simple layered structure and defined connections should be a model area for the identification of neurotransmitters and the receptors involved, and for a general study of synaptic mechanisms. The available evidence on

DONALD W. STRAUGHAN • Wellcome Professor and Chairman, Department of Pharmacology, The School of Pharmacy, London, England.

chemical transmitters in the hippocampus will be reviewed in this chapter. It is intended to complement the excellent review on the localization of transmitters in the hippocampus by Storm-Mathisen and Fonnum (1972). In addition, the biochemical aspects of hippocampal transmitters have been dealt with elsewhere in this volume. It is hoped that detailed knowledge of transmitters in this area will contribute to our knowledge of brain mechanisms and provide a basis for understanding drug action with attendant implications for pathophysiology and therapeutics.

2. Techniques

A number of criteria have been suggested by which the central nervous system transmitter might be identified, and these have been reviewed critically by Werman (1966, 1969). Two important, and possibly the most essential, pieces of evidence in the confident identification of a neurotransmitter are demonstration of its *release* and demonstration of physiological and pharmacological *identity of action* between the putative transmitter applied onto neurons and that released by nervous stimulation.

2.1. Release Studies

The putative transmitter or its metabolites should be released in an identifiable form by selective stimulation of an appropriate nervous pathway. Release studies thus involve techniques for the collection of the putative transmitter, its identification and quantification, and lastly demonstration of the sites of release (if feasible) and the specificity of release. In this last connection, the release should be calcium dependent like all known neurosecretions, and not accompanied by the release of appropriate marker substances.

Wherever conventional release studies are made for the identification of transmitters in the CNS, they involve certain common assumptions. The first is that it is possible for the substance or its metabolites to spill over in sufficient quantities to allow certain identification and measurement. In the case of ACh and monoamines, normal transmitter economy can be influenced by inhibiting the degradative enzyme (acetylcholinesterase) or the reuptake mechanisms, respectively. However, for the putative transmitter amino acids it is not practical at the present time to block the removal mechanisms. This makes for great difficulties in demonstrating the overflow of endogenous GABA following its release in response to nerve stimulation as the reuptake mechanisms are particularly effective. Second, release studies invariably assume the identity of the transmitter at the outset of the experiment, and often only one transmitter condidate is measured. This makes for problems in determining the site of release as it is difficult to confine stimulation to chemically and anatomically homogeneous nerve endings, so release of one putative transmitter could occur indirectly through the involvement of a second neuron and the action of another transmitter. In hippocampal studies, there are particular difficulties in confining the stimulus (for instance, to the medial septal nucleus). Thus it may be hard to avoid retrograde excitation of pyramidal cell axons and stimulus spread to the lateral septal nucleus. In the latter case, widespread effects throughout the neuraxis are difficult to avoid and in particular respiratory arrest is particularly likely to occur in spontaneously breathing animals (Smith, unpublished observations: see also Kozlovskaya and Valdman, 1970). Third, when artifact is excluded, collection of a substance from brain tissue in increased amounts following stimulation of a nervous pathway does not immediately prove its function. The substance may have transmitter, "modulator," or metabolic functions in neurons. Additionally, the substance being measured could come from glia and not from nerve endings. Finally, with present techniques—the need to apply stimulation for long periods and diffusional delays—it is not possible to show a direct temporal correlation between an individual impulse and the release of transmitter, as is possible with electrophysiological techniques.

The collection of putative transmitter substances from superficial areas such as the cerebral cortex *in vivo* can be achieved with the cortical cup (Elliot *et al.*, 1950; Mitchell, 1963; Eccleston *et al.*, 1969; Iversen *et al.*, 1971). For deeper-lying structures *in vivo,* collection can be achieved with the push-pull cannula, or, if the structure under study is adjacent to the ventricles, by ventricular perfusion techniques. The cortical cup method has been adapted in my laboratory for use in the hippocampus by Smith (1972, 1974) (see Fig. 1). Smith's experiments utilized rabbits, anesthetized with urethane. The dorsal surface of the hippocampus in the CA1 region is exposed by removal of the overlying cortex, and posterior part of the septum. A small oval perspex cup is positioned gently on the dorsal hippocampus and made leakproof by applying silicone grease to the outer rim. The internal diameters of the cup in the longitudinal and transverse axes are 7.3 by 3.8 mm, so it covers an

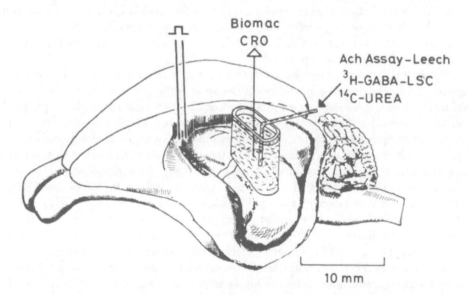

FIG. 1. Schematic diagram of hippocampal cortical cup experiments in urethane-anesthetized rabbits as used by Smith (1972). The perspex cup measures 7.3 by 3.8 mm in its internal diameters, covers 0.25 cm², and contains 0.5 ml Ringer–Locke solution. A bipolar stimulating electrode is shown on the septum.

area of approximately 0.25 cm². In experiments where ACh release is being studied, the Ringer–Locke solution in the cup contains eserine sulfate, 10^{-4} g/ml, throughout the experiment, and is allowed to stand for 30 min before a collection is made. To control against nonspecific release of ACh, a marker is required and the hippocampus is labeled by adding [^{14}C]urea (5 μCi specific activity, 62 mCi/mmol) to the cup fluid for a 45-min equilibration period. During subsequent collection periods, ACh and [^{14}C]urea diffusing from the hippocampus into the cup fluid can be collected at rest (spontaneous release), following electrical stimulation of afferent pathways and during the topical and systemic administration of drugs. Endogenous ACh diffusing from the hippocampus into the cup fluid can be readily measured by biological assay using the eserinized dorsal muscle of the leech, which is usually sensitive to about ½–1 ng ACh/ml.

Unfortunately, other putative transmitters cannot be measured as readily as ACh, so exogenous labeled GABA or monoamine is often used for *in vivo* and *in vitro* experiments. The tissue is preincubated with label for a fixed period and the subsequent efflux of label into fresh Ringer–Locke solution is followed by the rapid and sensitive technique of liquid scintillation spectrometry. These techniques require that the labeled putative transmitter be taken up predominantly and exclusively into nerve endings, where it mixes with endogenous pools of transmitter and behaves as such. Unfortunately, it is now clear that at some sites the label is accumulated in nonneuronal elements. Indeed, Brown and his colleagues have shown that the superior cervical ganglion accumulates [^{3}H]GABA *in vitro* into glial cells (Young *et al.*, 1973), from which it can be released by K^+ depolarization or electrical stimulation of the ganglia (Bowery and Brown, 1972). The extent to which this nonneuronal accumulation of labeled putative transmitter occurs in the CNS is variable and is the subject of much study at the present time. In future release studies, it might be advisable in place of GABA to use labeled 2,4-diaminobutyric acid (DABA), which selectively accumulates in the neuronal pool.

Whatever the site of release, with the *in vivo* cup technique electrical stimulation causes a selective increase in the release of ACh from cerebral cortex (Collier and Murray-Brown, 1968) or labeled NE from the olfactory bulb (Brenells, 1973, 1974) without a corresponding increase in the efflux of marker. There is evidence that other *in vivo* techniques involving the push-pull cannula or ventricular perfusion do not always permit this selective release of transmitter to be shown. Because of this, it seems unlikely that they can be usefully applied to the study of the hippocampus. Thus in classical experiments using the push-pull cannula technique, Chase and Kopin (1968) showed that olfactory stimulation caused a release of labeled 5-HT or NE from the olfactory bulb, accompanied by an increased release of markers such as urea and inulin. Similarly, in push-pull cannula experiments in the amygdaloid nucleus, Winson and Gerlach (1971) showed [^{3}H]NE and [^{14}C]urea release was increased by stress and drugs. Perfusion of the lateral ventricles as a means of studying the release of endogenous transmitters from the hippocampus (Beleslin *et al.*, 1964) does not localize the site of release to the hippocampus as opposed to the other structures lining the ventricles. Further, it is not practicable to test for calcium-dependent release in such ventricular perfusion studies, and although inert markers are rarely used, behavioral and drug stimuli have been described as causing the release of both marker and labeled NE into the ventricles (Sparber and Tilson, 1972).

There are two *in vitro* methods available for studying the release of putative transmitters from central nervous tissue. The first widely used method involves making simple slices of brain tissue with a mechanical chopper, and has been reviewed by Katz and Chase (1970). This technique has been used to study ACh release from slices of rat hippocampus by Large and Milton (1971). It has the general disadvantage that the integrity of nervous pathways is not preserved, so selective stimulation of a particular input is not possible and gross electrical field stimulation or potassium-induced depolarization is needed. An improvement of the simple slice method pioneered in McIlwain's laboratory (Yamamoto and McIlwain, 1966; Richards and Sercombe, 1968) maintains the integrity of intrinsic pathways. It has permitted elegant neurochemical and neurophysiological studies in the olfactory or prepyriform cortex where it has begun to be used for release studies by Bradford and Richards. Since this method has been used for neurophysiological studies of hippocampal pyramidal cell excitation by mossy fiber stimulation (Yamamoto, 1972), there seems no obvious reason why it should not also be applied to the study of transmitter release from the intact hippocampal layer *in vitro*.

2.2.   Identity of Action

Extracellular application of the putative transmitter to a neuron should mimic the transmitter released by nerve activity *physiologically,* changing cell activity in the same direction through similar conductance changes in the postsynaptic membrane, and *pharmacologically,* being similarly affected by specific blocking agents as well as by drugs affecting transmitter uptake and/or destruction. The essential technique involved is microiontophoresis of drugs onto central neurons from extracellular glass micropipettes, combined with simultaneous recording of cell activity through either extracellular or intracellular pipettes. This technique has been widely used for central nervous system studies and has been applied in a limited fashion to the hippocampus. Excellent descriptions of the method and associated problems have been given by Curtis (1964), Salmoiraghi and Weight (1967), Krnjević (1972), and Kelly *et al.* (1975).

The microiontophoretic technique is outlined in Fig. 2. Where possible, drugs are chosen which ionize well in aqueous solution so that they can be ejected or retained at will in a fairly precise manner by the application of currents of appropriate magnitude and direction through the micropipette barrel with respect to an indifferent electrode (usually a silver plate placed in the neck muscles of the animal). The method can also be used for substances which are poorly ionized in solution by dissolving them in a dilute of solution of NaCl. The passage of tip-positive currents will then expel the drug of interest through a combination of bulk flow and electro-osmosis.

In terms of avoiding the interpretive hazards which follow the intracerebral, intraventricular, or topical injection of putative transmitters, the microiontophoretic technique has many clear advantages in localizing drug action to a very small area of

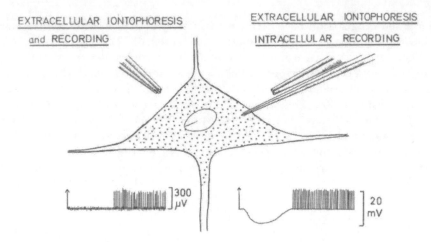

Fig. 2. Schematic diagram of multibarreled extracellular micropipette used for iontophoresis and recording, and of multibarreled extracellular glass micropipette used for iontophoresis glued to an intracellular recording electrode. Typical extra- and intracellular records following the application of an inhibitory stimulus (arrow) are shown. Extracellular recording yields information on neuronal spikes and field potentials and finds particular use in the pharmacological identification of excitatory and inhibitory neurotransmitters and the classification of receptors. Intracellular recording yields information on resting and synaptic potentials, equilibria potential, membrane resistance, and the ion species involved in conductance changes. It finds particular use in the physiological and pharmacological identification of inhibitory transmitters. Reproduced with kind permission of *Archivos de Farmacologia y Toxicologia*, from Straughan (1975).

brain tissue and avoiding tissue saturation. However, microiontophoresis is not without problems, particularly as the drugs may affect neurons remote to the one under study but connected to it through nervous pathways. This certainly does occur in rabbit olfactory bulb and could occur in other areas of the nervous system, although definitive evidence is lacking. Thus, in the olfactory bulb, Salmoiraghi *et al.* (1964) showed that iontophoretic application of NE and stimulation of the lateral olfactory tract depressed the firing of mitral cells and that both of these effects could be blocked by the iontophoretic application of adrenoceptor antagonists. They presumed that there were norepinephrine-releasing terminals ending directly on the mitral cells and that norepinephrine was likely to be an inhibitory transmitter at this site. However, subsequent histochemical studies showed that there were no norepinephrine-containing nerve terminals on the mitral cells but they were present on the adjacent granule cells (Dahlström *et al.*, 1965). The locally applied norepinephrine is now assumed to have diffused away from the immediate site of iontophoretic ejection to excite these neighboring granule cells, with the consequent release of an inhibitory transmitter, probably GABA, from their endings on the mitral cells. As indicated later, there are still outstanding problems as to whether the use of particular anesthetics, or local pH changes, modifies the responses of neurons to putative transmitters. The extent to which any of these problems is relevant to studies in the hippocampus remains to be determined. Two special difficulties of iontophoretic studies in the hippocampus are (1) the difficulty in finding cells and maintaining recording, which may reflect the powerful inhibitory mechanisms at this site, and (2) the extreme sensitivity to iontophoretic excitants such as glutamate and ACh as manifested by ready development of seizure discharges in the hippocampus.

3. Transmitter Substances

3.1. Acetylcholine

The presence of ACh in the hippocampus, with its synthetic and degradative enzymes, has been known for some years. In itself, this provides no certain evidence for ACh being a transmitter as it also occurs in the embryonic heart before innervation and in the placenta, where there are no cholinergic nerves. However, histochemical studies in rat (Shute and Lewis, 1961, 1963) and cat (Krnjević and Silver, 1965) provide strong evidence for a system of acetylcholinesterase-containing nerve fibers extending into the hippocampus from the medial septal nucleus via the fimbria. Evidence for such a pathway also exists in neuroanatomical studies (Raisman, 1966), and neurophysiological studies (Andersen et al., 1961). In the cat, as shown in Fig. 3, acetylcholinesterase-containing nerve fibers have a striking laminar distribution which spares the regions of pyramidal and granule cell bodies. The bulk of this acetylcholinesterase staining in the hippocampus, as in other central areas, is believed to occur within cholinergic nerves. Thus the distribution of the synthetic enzyme for ACh, choline acetyltransferase, measured biochemically closely matches that of acetylcholinesterase measured biochemically or histochemically in microdissection studies (Storm-Mathisen and Fonnum, 1972). Following lesions in the fimbria or medial septal nucleus, the levels of both enzymes fall substantially, suggesting that practically all the cholinergic elements in the hippocampus come from the septal nucleus.

3.1.1. *Release Studies.* At present, the only evidence from release studies to confirm the existence of these cholinergic nerves in the hippocampus comes from the experiments of Christine Smith, (1972, 1974, and unpublished observations)—and is shown in Fig. 4. In the absence of electrical stimulation, there was a spontaneous release of ACh at a rate of 0.16 ng/cm^2/min from the dorsal hippocampus of anesthetized rabbits into modified cortical cups containing eserinized Ringer–Locke solution. This was smaller than the spontaneous release figure of 0.5 ng ACh/cm^2/min given by Randić and Padjen (1967) for comparable experiments in cat cortex not utilizing topical atropine to enhance release. It seems likely that this difference is a function of the small collection area of the hippocampal cup, for when an identical oval cup was used on the cerebral cortex Smith found the spontaneous ACh release to be very similar to that from the hippocampus. Stimulation of the surface of the septum caused an average 2½-fold increase in the efflux of ACh from the hippocampus. This increase was not an artifact, as it was not accompanied by an increase in the efflux of the marker substance [^{14}C]urea. Evidence of pathway specificity in these release ex-

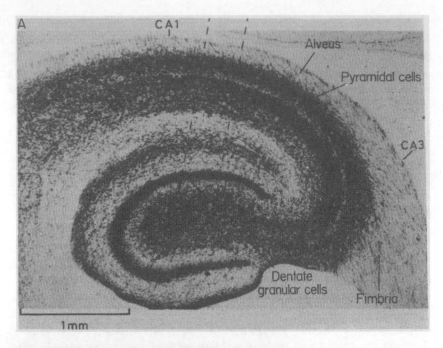

A

CA1

Alveus

Pyramidal cells

CA3

Dentate
granular cells

Fimbria

1mm

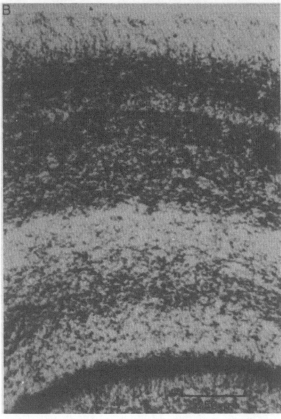

B

— Alveus

— Str. oriens

— Str. pyramidale

— Str. radiatum

— Str. lacunosum-moleculare

— Lam. supragranulare (mole
— Str. granulare

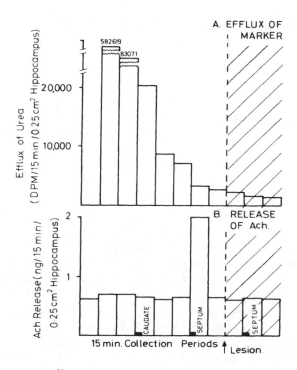

FIG. 4. Simultaneous efflux of [^{14}C]urea (A: upper histogram) and ACh (B: lower histogram) from CA1 region of hippocampus of urethane-anesthetized rabbit into perspex cup containing Ringer–Locke solution with eserine sulfate, 10^{-4} g/ml. Stimulation of the caudate nucleus through a bipolar silver ball electrode with 1-ms pulses of 3 mA intensity for 5 min at 100 Hz provoked no increase, whereas stimulation of the surface of the septum at the same parameters increased ACh release from 0.6 to 2 ng/0.25 cm^2/15 min but had no effect on [^{14}C]urea efflux. After acute section of the fimbria (↑ and shaded area), septal stimulation failed to increase ACh release. Reproduced with kind permission from Smith (1974).

periments was given by showing that stimulation of the caudate nucleus (with the same parameters used for the septum) failed to evoke an increased release of ACh. Further, the stimulated efflux of ACh was abolished by acute lesion of the septohippocampal pathway. An important piece of evidence still lacking is the demonstration that the stimulated efflux of ACh from the hippocampus is calcium dependent, as in the cerebral cortex (Randić and Padjen, 1967). Preliminary experiments in which all the calcium in the Ringer–Locke solution was removed did not abolish the stimulated release

FIG. 3. Sagittal section of cat hippocampus stained for acetylcholinesterase by the method of Krnjević and Silver (1965). A: Low-power view, scale 1 mm. The area between the dotted lines is shown in greater detail in B. B: High-power view, scale 250 nm, from the alveus to the dentate granular layer. Note relatively dense staining in the stratum oriens and stratum radiatum corresponding to basal and apical dendrites adjacent to the pyramidal cell layer and the dense staining in the lamina supragranulare corresponding to granule cell apical dendrites. The layers of pyramidal and granular cell bodies are unstained (section stained and loaned to authors by Drs. K. Krnjević and Silver). Reproduced with kind permission from *Journal of Physiology,* Biscoe and Straughan (1966).

of ACh. However, it should be remembered that these hippocampal cups have a very small surface area, which might militate against achieving a sufficient reduction of calcium ions in the extracellular fluid of the brain underlying the cup to reduce or abolish transmitter release from nerve endings. In addition, the small amounts of ACh released at rest and after stimulation make it difficult to see small changes. Wider application of this cup technique would be helped by more sensitive methods for measuring ACh, and by boosting the ACh release by adding atropine sulfate to the Ringer–Locke solution throughout the experiment. In addition, use of unanesthetized animals should bring about an increased release of ACh and other putative transmitters, to judge from results obtained in the cerebral cortex (Mitchell, 1963).

Few studies have been made on the effects of drugs on the release of putative transmitters. However, Smith (1972) has shown that the resting efflux of ACh from the hippocampus is increased at least 3½ times by atropine sulfate in the cup fluid (10^{-5} g/ml). Also, leptazol (30 mg/kg intravenously) markedly increased (2½ times) and morphine sulfate (10^{-4} g/ml topically) abolished ACh efflux. It is not known to what extent, if at all, any of these drugs act preferentially in the hippocampus, and the mechanisms need to be elucidated. In these respects, it will be apparent that ACh release from the hippocampus has many similarities to ACh release from the more complex-structured cerebral cortex (Mitchell, 1963). Intriguing differences are, however, suggested by the results of Large and Milton (1971), who studied the effects of various narcotic analgesic drugs as well as their antagonists on the potassium-stimulated release of ACh from slices of rat cerebral cortex and hippocampus *in vitro*. In both areas, the potassium-stimulated ACh release was calcium dependent but 1 mM morphine increased the potassium-evoked release of ACh from the hippocampus and significantly decreased the release of ACh from cerebral cortical slices. On the other hand, nalorphine inhibited the release of ACh from hippocampus slices but increased the release of ACh from cortical slices. The explanation for these regional differences and the *in vivo/in vitro* differences in the effects of morphine on release is unknown. In addition, the concentrations of morphine and nalorphine used may not be meaningful in terms of understanding the actions of these drugs in the whole brain. A particular problem in interpreting Large and Milton's results is the comparatively high levels of K^+ (15 mM) which they used for their "unstimulated" baseline.

3.1.2. Identity of Action. Acetylcholine and its antagonists have been applied microiontophoretically into the environment of units in the hippocampal cortex of the anesthetized cat by Stefanis (1964), Herz and Nacimiento (1965), and Biscoe and Straughan (1966). A substantial number of units were excited by ACh, in contrast to the results described by Herz and Nacimiento, and these cholinoceptive cells were concentrated into two main groups, one superficial and one deep, corresponding broadly to the hippocampal cortex proper and the dentate gyrus. A more precise depth distribution was not achieved because of the difficulties of recording depth exactly and averaging findings from different microelectrode penetrations, possibly at slightly different angles. However, as shown in Fig. 5, the depth distribution of cholinoceptive cells differed significantly from the depth distribution of units sensitive

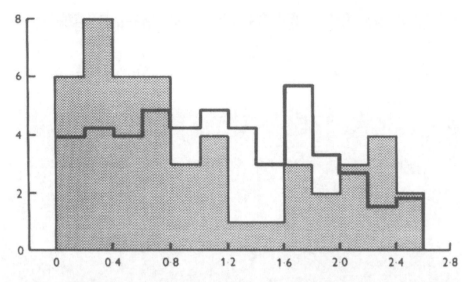

Fig. 5. Depth distribution histogram of all the 161 cells excited by glutamate (unshaded area, thick line) and of the 49 cells among them which were excited by ACh (shaded area, thin line). Abscissa, depth of cells from alvear surface in millimeters; ordinate, number of cells per 0.2 mm. To make the histograms comparable dimensionally, the proportion of cells excited by glutamate at each depth was replotted on a sample size of 49. There was a significant difference between the two histograms: $x^2 = 12.831$, $\phi = 5$, $0.05 > P > 0.025$. Reproduced with kind permission from *Journal of Physiology*, Biscoe and Straughan, (1966).

to glutamate. From Salmoiraghi and Stefanis' (1967) account, many of the cells excited by ACh lay within the pyramidal layer of CA1, CA2, and CA3 and could be excited antidromically through electrical stimulation of the fimbria. Characteristically, the time course of the ACh excitation of hippocampal neurons is slow in onset and recovery and contrasts with the rapid time course of glutamate excitation in the same cells, or the ACh excitation of Renshaw cells in the spinal cord (see Fig. 6).

The time course of ACh excitation in the hippocampal cortex is not dissimilar from that seen in the cerebral and pyriform cortices (Krnjević and Phillis, 1963a; Legge et al., 1966). In the cerebral cortex, the latency of the onset of the "slow" ACh excitation can be reduced considerably by the use of larger expelling currents (Johnson et al., 1969a; Kelly et al., 1975). The delays are probably not due to simple diffusion of ACh to distant receptor sites, but reflect relative insensitivity to ACh and perhaps the need to activate wide areas of dendritic membrane. The higher ejecting current provides higher ACh concentrations in the biophase at an earlier time in the ejection. However, the mechanisms underlying the slow muscarinic responses in brain and associated membrane changes set up by ACh are probably very different from those involved in fast nicotinic ACh responses, for example, in the Renshaw cell. Even in the periphery, muscarinic responses to ACh are characteristically slow. Absolute sensitivity to ACh cannot be reasonably judged from iontophoretic experiments; however, on the basis of the iontophoretic currents required, ACh is 4 or 5

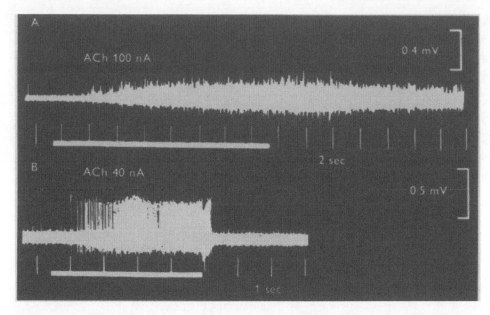

FIG. 6. Photographic record of action potentials recorded from (A) a hippocampal cell in response to iontophoretic ACh, 100 nA, compared with (B) the response of a Renshaw cell in the spinal cord to ACh, 40 nA. Note difference in time scales and onset and recovery. Reproduced with kind permission from *Journal of Physiology,* Biscoe and Straughan, (1966).

times less potent than glutamate on hippocampal compared to cerebral cortical neurons. This again provides a contrast with Renshaw cells, where similar ejecting currents of ACh and glutamate appear to be approximately equieffective.

The use of ACh antagonists should be helpful in terms of characterizing the ACh receptors of hippocampal cells and the pharmacological identification of synaptically released transmitters (see Fig. 7). Although atropine was a more effective antagonist than dihydro-β-erythroidine (DHβE) at the currents used, the hippocampal cholinoceptive receptors are not exclusively muscarinic since ACh and methacholine excitation was blocked occasionally by dimethyl-d-tubocurarine (Biscoe and Straughan, 1966). Apparently more consistent success was achieved in blocking ACh responses with DHβE in hippocampal pyramidal cells by Stefanis (1964). In all these experiments, the antagonists were applied by microiontophoresis, and more quantitative experiments with atropine and DHβE are required in order to select the best antagonist for future experiments. It is necessary to establish the extent to which each antagonist selects between graded doses of the principal agonist, in this case ACh, and an unrelated agonist, e.g., glutamate. This overcomes the main objection to single agonist dose studies—that the chosen dose may be supramaximal and mask antagonist induced effects.

In summary, from antagonist studies the ACh receptors on hippocampal cells appear to be mainly but not exclusively muscarinic. They are perhaps more analogous with those seen on cholinoceptive cells in cingulate gyrus (Krnjević, 1965), thalamus (Andersen and Curtis, 1964; Curtis and Davis, 1963), and cerebellum (McCance and Phillis, 1964) than with those in feline neocortex (Krnjević and Phillis, 1963*b*). The predominantly muscarinic nature of the hippocampal cholinoreceptors and the probable mainly atropine-sensitive nature of cholinergic transmission provide a good basis for psychopharmacological studies. Thus anticholinergic drugs produce substantial behavioral changes—e.g., impairment of passive avoidance and spontaneous alternation—not dissimilar from those produced by septal lesions.

The mechanism by which ACh excites hippocampal neurons following receptor activation is not known since the relevant experiments combining intracellular recording with extracellular microiontophoresis of ACh have not been made. However, it is likely that the mechanism is similar to that found in cerebral cortical neurons. Here Krnjević *et al.* (1971) showed that the muscarinic excitatory action of ACh was associated with a depolarization, with a mean reversal level of −86 mV and an

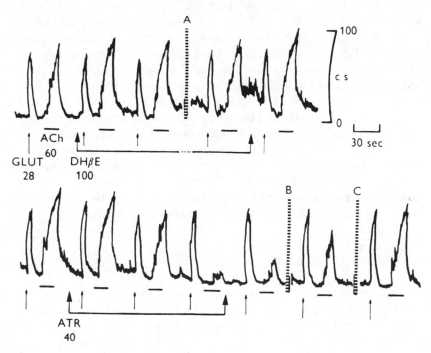

Fɪɢ. 7. Ratemeter record showing effects of dihydro-β-erythroidine (DHβE), 100 nA, and atropine (ATR), 40 nA, on a hippocampal unit excited by alternate applications of glutamate, 28 nA, for 5 s (arrows) and ACh, 60 nA, for 10 s (short lines). In the upper trace, ACh excitation was not affected by DHβE applied continuously for 5 min (note 2-min gap in record at A). However, atropine almost completely blocked ACh but not glutamate excitation within 3 min. Recovery of ACh excitation began within 1 min, reached 60% recovery within 3 min (after gap B), and had completely recovered within 7.5 min (after gap C). Reproduced with kind permission from *Journal of Physiology,* Biscoe and Straughan (1966).

increase in membrane resistance. They concluded that ACh probably acts through a reduction of the resting K^+ conductance of cortical neurons but has an additional inhibitory effect on the delayed K^+ current of the action potential. Since the equilibrium potential for ACh in cortical neurons was not sensitive to intracellular injections of Cl^- they concluded that ACh depolarization could not be attributed to a reduction in Cl^- conductance. These actions of ACh on cortical neurons, although clearly different from the actions of ACh at the neuromuscular junction, are compatible with the mechanisms involved in slow muscarinic excitation of sympathetic ganglia. Here the slow effect of ACh is associated with the rise in membrane resistance (Kobayashi and Libet, 1968, 1970) following a reduction in K^+ permeability (Weight and Votava, 1970). It may be difficult to achieve such clear-cut changes in hippocampal neurons with ACh, for, if the dendritic distribution of acetylcholinesterase-containing preterminal nerve fibers indicates the main areas for ACh receptors, then many of these may be sufficiently remote from the microelectrode impaling the soma to allow membrane changes induced by the transmitter to be "seen" in their entirety.

There is suggestive evidence from a variety of microelectrode and microiontophoretic studies that ACh (or a substance very closely related to it) is released as an excitatory transmitter from septal afferents ending on hippocampal pyramidal cells. Thus Brücke *et al.* (1963) noted that eserine sensitized units in hippocampus toward the excitatory effects of septal stimulation, while Stefanis (1964) described eserine as potentiating the excitatory responses of hippocampal pyramidal cells to both ACh application and antidromic stimulation via the septum and entorhinal cortex. In addition, DHβE blocked the response of pyramidal cells to applied ACh and to entorhinal stimulation, although it was less effective against the firing induced by septal and fimbrial stimulation (Salmoiraghi and Stefanis, 1967). More details of these latter experiments are required, for the entorhinal input, unlike the septal, is unlikely to be cholinergic and should not be particularly sensitive to cholinoceptor antagonists. More detailed quantitative studies are needed to exclude nonspecific depressant or even presynaptic actions in the antagonist.

3.2. Inhibitory Amino Acids

Current studies suggest that neutral amino acids may play important inhibitory transmitter roles in the mammalian central nervous system. There is much support for GABA as being the major supraspinal inhibitory transmitter, glycine being the major spinal inhibitory transmitter, but little or no evidence for taurine or β-alanine as inhibitory transmitters. One major difficulty in the study of amino acids as transmitters is the fact that while they occur in substantial quantities in nervous tissue, the largest amounts appear to be associated with a nontransmitter or metabolic pool. In the absence of suitable direct histochemical methods for demonstrating the presence of GABA, Bloom and Iversen (1971) used electron microscope autoradiography to show the localization of [^{3}H]GABA in nerve terminals of rat cerebral cortex following preincubation of brain tissue with labeled amino acid. Following its uptake, the subcellular distribution of [^{3}H]GABA is known to be very similar to that of endogenous GABA and its synthetic enzyme, glutamate acid decarboxylase (GAD) (Neal and Iversen, 1969). Bloom and Iversen found that there was a distinct population of nerve terminals, amounting to about 30% of all the terminals present, which could be labeled by [^{3}H]GABA and which were presumed to be GABA-nergic nerve terminals. Although powerful uptake processes for GABA exist in glial cells, only small amounts of labeled GABA were associated with them in the autoradiograms. This technique should be capable of application to the hippocampus since this possesses uptake processes for [^{3}H]GABA similar to those found in the cerebral cortex (Bond, 1973).

As an alternative to measuring the distribution of GABA, the activity of GAD can be measured, thus avoiding some of the problems associated with GABA measurements. From a variety of studies in crustacean and mammalian CNS, it is known that GAD activities correspond closely with GABA levels and that they have a fairly similar regional and subcellular distribution (see review by Iversen, 1972).

In the hippocampus, Storm-Mathisen and Fonnum (1972) have concluded that nearly all, or possibly all, the GAD activity (and hence GABA) is present within intrinsic or local neurons, as in many other areas of the CNS. Thus GAD activity was unaffected by deafferentation through lesions of the perforant path from the entorhinal cortex and/or intrinsic mossy fibers from dentate granule cells. Further, GAD activity was reduced only about 16% by fimbrial lesions, which reduced choline acetyltransferase activity in all zones by about 85–90%. This small loss of GAD activity was attributed to possible trophic changes following loss of the major septal and commissural afferents running in the fimbria. The pyramidal and granular cell layers showed a uniformly high activity of GAD compatible with the axosomatic inhibitory terminals concentrated in these areas. However, the localization of possible inhibitory nerve terminals from the regional distribution of GAD is not completely clear-cut in view of the high GAD activity found in the stratum moleculare, where present electrophysiological knowledge has not yet described a substantial system of inhibitory nerves, although this area does contain a system of local neurons of unknown function. These biochemical findings complement essentially the neurophysiological studies showing that interneurons in the stratum oriens mediate the inhibition of pyramidal cells (Andersen et al., 1964).

3.2.1. Release Studies. One major problem in studying the release of GABA from CNS tissue is that the amounts of endogenous amino acid which can be collected either at rest or during stimulation are very small. This is probably the result of the highly efficient uptake systems previously referred to and it is not practicable at the present time to inhibit them. A number of drugs can inhibit GABA uptake, e.g., chlorpromazine and p-chloromercuriphenylsulfate, but these are not specific. DABA is also a fairly potent inhibitor of the uptake of GABA (IC$_{50}$ 50 μM) but not glycine or monoamines and appears to distinguish between neuronal and glial uptake sites. However, the actions of DABA need to be clarified somewhat further before it can be used routinely as a tool for boosting GABA efflux from brain tissue. The small amounts of endogenous GABA leaking from brain can be measured with ultrasensitive amino acid analyzers or dansylation techniques, although measurement of exogenous

[^{3}H]GABA or [^{3}H]DABA efflux from brain (following preincubation) is simpler. Certainly in the cerebral cortex, electrical stimulation of either the surface or the lateral geniculate nucleus, which has mainly inhibitory effects, causes a release of [^{3}H]GABA paralleled by endogenous GABA from the surface of the cerebral cortex into cortical cups (Mitchell and Srinivasen, 1969; Iversen et al., 1971). However, in rat cerebral cortex, Roberts (1973) claims that direct cortical stimulation in vivo, while not affecting urea efflux, caused a small and fairly similar increase in the release of all the exogenously labeled amino acids tested, including [^{3}H]GABA, [^{14}C]glutamate, [^{14}C]serine, and [^{14}C]glycine. The statistical significance of these results was not given. The neurally evoked release of GABA was calcium dependent. Unfortunately, attempts to reproduce such results in hippocampal cortex in vivo have been unsuccessful to date. Thus Smith (unpublished results) studied the release of [^{3}H]GABA and [^{14}C]urea into cups placed on the surface of the CA1 area of hippocampus in urethane-anesthetized rabbits. The rate of efflux of labeled GABA was not affected by stimulation of septal and commissural afferent pathways. Although bipolar local stimulation at about 100 Hz for 7 min did increase the release of GABA in some experiments, there was often a simultaneous release of the marker [^{14}C]urea. The reasons for this failure are not known, but in similar release studies in my department in the rabbit olfactory bulb (Muckart, 1971; Brenells, 1973, 1974) stimulation caused the selective release of labeled NE or labeled GABA, but not urea or inulin, whereas higher-intensity pulses caused the release in addition of labeled urea probably through a nonspecific increase in cell membrane permeability.

3.2.2. *Identity of Action.* In common with cells in all other areas of the nervous system, hippocampal cells are readily depressed by iontophoretic GABA (Stefanis, 1964; Biscoe and Straughan, 1966). Curtis et al. (1970) have shown that identified pyramidal cells are depressed by GABA, which was approximately twice as potent as the other putative transmitter candidates glycine and β-alanine on an iontophoretic current basis. They also demonstrated that the effects of GABA and β-alanine were blocked by iontophoretic applications of the alkaloid bicuculline at a time when the depressant responses to application of glycine were unaffected. This is shown in Fig. 8. However, when strychnine was applied, it reduced the depressant effects of GABA, glycine, and β-alanine with no obvious selectivity toward glycine. This is most uncharacteristic and requires further experiments, for in all other areas of the CNS strychnine is a powerful and selective glycine antagonist. Nevertheless, the results imply the existence of at least two receptors for inhibitory amino acids in hippocampal pyramidal cells as in other areas of brain. One receptor is activated by GABA and β-alanine and readily blocked by bicuculline and is analogous to that in cerebral cortex, cerebellar cortex, and ventrobasal thalamus (Curtis et al., 1971). The second receptor for glycine and related amino acids is less readily blocked by bicuculline. These differ from those encountered at the crustacean neuromuscular junction, where β-alanine and glycine, respectively, have about 0.03 and 0.0001 the effectiveness of GABA (Dudel, 1965) and where bicuculline but not picrotoxin is virtually ineffective as an antagonist (Takeuchi and Onodera, 1972; Earl and Large, 1972).

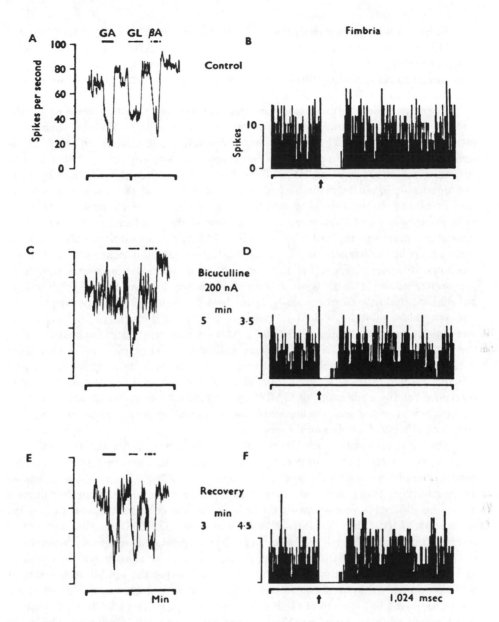

FIG. 8. Effects of iontophoretic bicuculline, 200 nA, on amino acid depression and neural inhibition of a hippocampal pyramidal cell. Ratemeter records A, C, and E show depression of firing by iontophoretic GABA (GA, solid line, 20 nA), glycine (GL, dotted line, 40 nA), and β-alanine (βA, broken line, 40 nA). Records B, D, and F are poststimulus histograms showing the inhibition of firing following electrical stimulation of the fimbria (3-V, 0.1-ms pulses, at 0.5 Hz, marked with an arrow). A and B are control records. C and D were taken 5 and 3.5 min, respectively, after ejection of bicuculline had begun. E and F were taken 3 and 4.5 min after bicuculline had been stopped. Reproduced with kind permission from *British Journal of Pharmacology,* Curtis *et al.* (1970).

In attempts to further characterize the inhibitory amino acid receptors, Segal *et al.* (1973) have described the effects of some GABA analogues applied iontophoretically on rat hippocampal neurons. In particular, they noted that the 1 : 2 *trans* and 1 : 3 *cis* spatial conformers of aminocyclohexane carboxylic acid were effective depressants and sensitive to block by bicuculline.

There is some pharmacological evidence for identity of action between GABA and the inhibitory transmitter released onto pyramidal cells by fimbrial stimulation (also shown in Fig. 8). Thus iontophoretic bicuculline attenuates both effects at the same time but does not stop the cell from responding to a control inhibitory agonist such as glycine (Curtis *et al.*, 1970). This shows that two candidate transmitters in the hippocampus can be distinguished by an antagonist and that the antagonist acts mainly at the receptor level, and not on the conductance channel. In that the endogenously released inhibitory transmitter was slightly less susceptible to block by bicuculline than was the artificially applied GABA, this example satisfies expectations regarding differences in the susceptibility of extrajunctional and junctional receptors. Further support for the view that the main hippocampal inhibitory transmitter is not glycine is seen from the relative insensitivity of inhibition to strychnine administered systemically. Thus basket cell inhibition of pyramidal cells and synaptic inhibition of dentate granular cells were not altered by intravenous strychnine (Andersen *et al.*, 1963, 1966). More extensive quantitative studies along the lines pioneered by Curtis' group are still needed. Thus bicuculline sometimes enhances inhibition as well as shortening it in sites such as cerebral cortex and olfactory bulb (Curtis and Felix, 1971; Felix and McLennan, 1971). In addition, data regarding relative shift ratios of GABA and glycine dose-response curves by GABA antagonists correlated with the degree of block of inhibition would be useful as in Kelly and Renaud's (1973) studies on the cuneate.

Physiological evidence for identity of action between GABA and the inhibitory transmitter released onto hippocampal pyramidal cells in terms of conductance change, equilibria potential, and ion species is lacking. Despite the claims of Werman (1966, 1969), such tests are not conclusive if different candidate transmitters activating different receptors work through similar conductance changes to give the same end effect. Thus in intracellular studies in Deiters' nucleus and spinal interneurons, iontophoretic GABA and glycine produced identical conductance changes and were only distinguished pharmacologically by antagonists (Bruggencate and Engberg, 1971; Curtis *et al.*, 1968). It is known that the inhibitory transmitter in hippocampal pyramidal cells causes an increased Cl^- conductance (Kandel *et al.*, 1961), as might be expected of GABA (or glycine), but studies like those of Krnjević and his colleagues are required (Krnjević and Schwartz 1968; Dreifuss *et al.*, 1969). In elegant studies in the cerebral cortex, they have shown that GABA and the inhibitory transmitter released by epicortical stimulation normally induce hyperpolarizing IPSPs in cortical neurons. This hyperpolarization follows an increase in the conductance of chloride ions (Fig. 9). In addition, voltage–current plots (Fig. 10) show that both iontophoretically applied GABA and the inhibitory transmitter cause a similar decrease in membrane resistance and have almost identical equilibrium potentials.

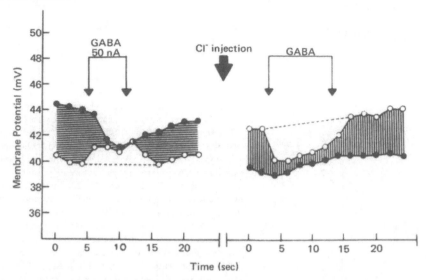

FIG. 9. Reversal of IPSPs (evoked by epicortical stimulation) and of the effect of iontophoretic GABA by the intracellular injection of chloride ions in a neocortical neuron. Open circles show the resting potentials; closed circles show the potentials at the peak of the IPSP. Shaded areas show the amplitude of the IPSPs and the broken lines show the probable sequence of changes in the absence of GABA. About 0.5 pmol of chloride was injected at the time indicated by the large arrow. The record on the left shows iontophoretic GABA and the IPSPs to be hyperpolarizing and the amplitude of the IPSPs to be markedly attenuated in the presence of GABA. The injection of chloride moved the resting potential in a hyperpolarizing direction and iontophoretic GABA and inhibitory stimulation now caused depolarizing potentials. Reproduced with kind permission from Krnjević and Schwartz (1968).

3.3. Other Putative Transmitters

3.3.1 Monoamines. The rat hippocampus has substantial levels of other transmitter candidates such as glutamate at 170 μg/g, NE at 0.43 μg/g, and 5-HT at 0.32 μg/g. Histochemical studies (see Fuxe, 1965; Blackstad *et al.*, 1967) suggest that the NE and 5-HT in the hippocampus is contained within nerve fibers originating in the brain stem. The NA fibers come from cells in the locus coeruleus of the pons, while the 5-HT fibers arise from cells in the raphe nucleus of the midbrain.

Both systems of fibers ascend in the medial forebrain bundle and in the main pass through a dorsal route (fimbria, superior fornix, and cingulum) to reach the hippocampus. The distribution of NE terminals in the hippocampus has been described (Fuxe, 1965). There is a high density of NE-containing terminals in the stratum radiatum of the dorsal hippocampus CA3 area and the ventral hippocampus. The NE terminals are moderately dense in the stratum lacunosum but not the stratum radiatum of the CA1 and CA2 areas of the dorsal hippocampus. In the dentate gyrus, there is a high density of NE terminals in the zone just below the granular layer. In

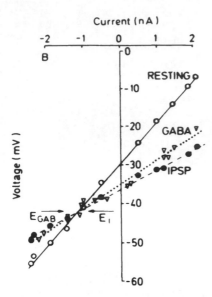

FIG. 10. Voltage–current lines obtained before, during, and after iontophoretic application of GABA, 140 nA, onto a cortical neuron in a Dial-anesthetized cat. Positive and negative 20-ms pulses at several steady current intensities (abscissa) were injected into the neuron through a bridge circuit and the voltage change produced by each pulse was measured (ordinate). Voltage–current lines in the resting membrane are shown by solid lines and open circles, at the peak of the IPSP by dashes and solid circles, and during the action of GABA by dots and inverted triangles. At rest (zero current), the resting potential was about −30 mV and both GABA and the IPSP hyperpolarized the membrane by about 5–35 mV. During the action of GABA and IPSP, the resistance of the cell membrane was reduced in similar fashion as indicated by the reduced slope of the voltage–current lines, which intersect the resting voltage–current line at the same equilibrium potentials, E_{GAB} and E_I. Reproduced with kind permission from *Experimental Brain Research*, Dreifuss *et al.* (1969).

this zone as in the stratum radiatum, most of the terminals form axodendritic junctions. It is not known if the NE input makes synaptic contact with the dendrites of pyramidal cells in the stratum radiatum or whether it is on some other cell.

Histochemical studies of the distribution of 5-HT-containing nerve terminals in the hippocampus have been difficult to make. However, through a variety of maneuvers, the intensity of the 5-HT fluophore can be intensified. In this way, it has been shown that the 5-HT terminals in the hippocampal formation have a different distribution from the NE terminals. The 5-HT terminals are found mainly in the stratum lacunosum-moleculare, areas 26 and 31 of the hippocampal gyrus, and the subiculum (unpublished observations cited in Fuxe and Jonsson, 1974).

From biochemical studies, it is clear that the bulk of the NE and 5-HT in the hippocampus is derived from these afferents and is not due to intrinsic systems. Thus after unilateral lesions of the dorsal afferents, 5-HT was reduced by about 85%, and NE reduced by about 70% in the hippocampus on the operated side (Storm-Mathisen and Guldberg, 1974). With lesions in the medial forebrain bundle, there is about a 70% loss of NE and 5-HT in the hippocampus (Heller *et al.*, 1966; Moore and Heller, 1967).

Monoamine-containing nerve fibers appeared to constitute only a minor part of the inputs reaching the hippocampus via the fimbria, and their function in the hippocampus is at present unknown. Thus it is not known if these monoamines can be released selectively by stimulation, nor whether the presence of a monoamine intracellularly in nerves inevitably means that it must act as a transmitter rather than have some as yet undefined supporting or metabolic role. It seems unlikely that the cortical cup technique described previously will permit sufficient amounts of endogenous monoamine to be recovered for satisfactory measurement. It will be worth trying exchange labeling of the NE and 5-HT stores in the hippocampus, by preincubation with labeled amine and marker and measuring the effects on the efflux of label of discrete stimulation of monoamine tracts. This technique has been satisfactory for showing a calcium-dependent and selective release of labeled NE but not marker following afferent stimulation from the olfactory bulbs into cup fluid (Muckart, 1971; Brenells, 1973, 1974).

NE and 5-HT have been applied to hippocampal neurons in the anesthetized cat (Stefanis, 1964; Herz and Nacimiento, 1965; Biscoe and Straughan, 1966; and Salmoiraghi and Stefanis, 1967). NE was a weak depressant of neuronal firing and comparable currents of 5-HT were usually more effective in depressing firing. In a small proportion of units, 5-HT caused an excitation, which in some cases progressed to a seizure discharge similar to that seen with L-glutamate or ACh (Biscoe and Straughan, 1966). In anesthetized rat, NE and 5-HT were again essentially depressant on presumed pyramidal cells and no excitations were seen (Segal and Bloom, 1974a). However, the NE depressant effect developed and recovered slowly, while the 5-HT depressions had a shorter latency of onset (2–15 s) and did not persist after the drug application had been stopped.

It must be emphasized that as far as the hippocampus is concerned it is not known if monoamine excitation or depression or both are genuine. Thus apparent neuronal depression by NE through excitation of distant inhibitory neurons linked synaptically to the cell under study might conceivably occur in the hippocampus as in the olfactory bulb—but we have no information on this point. Indeed, following large iontophoretic ejections in other areas of brain *in vivo*, monoamines could be detected histochemically at a distance of several hundred micrometers (Candy *et al.*, 1974). On the other hand, NE depression of spinal motoneurons and cerebellar Purkinje cells does appear to occur directly in the cell under study (Engberg and Marshall, 1971; Hoffer *et al.*, 1973). In the cerebral cortex, Phillis and his colleagues have suggested that monoamine depressions are the genuine effect, and that excitations are artifactual (Frederickson *et al.*, 1971). They suggested that these excitations are the consequence of drug-induced acid pH changes in the environment of the neuron. They mimicked monoamine excitations by ejection of H^+ from acid solutions and noted that monoamine excitation of cortical neurons was infrequent when NE or 5-HT was expelled from relatively alkaline solutions between pH 4 and 5. This problem is not completely resolved, for Roberts and his colleagues in Edinburgh have been unable to confirm these pH effects (Bevan *et al.*, 1973). If NE excitations in cortex are artifactual like 5-HT excitations, it is difficult to see how they are selectively blocked by appropriate antagonists (Johnson *et al.*, 1969b). In addition, in cerebral cortex, brain stem and spinal motoneurons, depressions are generally more resistant to block by antagonists than excitations, suggesting that they are produced by different mechanisms and that monoamines do not just have a single action on cerebral neurons. Also, monoamine excitations are still seen with some frequency in the brain stem even when the ions are expelled from the more alkaline solutions in the pipettes (Boakes *et al.*, 1971).

The pharmacology of the monoamine receptors in the hippocampus has not been studied in the cat. However, in rat hippocampus NE depressions could be blocked by the β-adrenoceptor antagonist Sotalol (MJ1999), although the specificity of the effect is unknown since the authors did not use a control depressant agonist in their studies (Segal and Bloom, 1974a). Only a small number of cells were tested with α-adrenoceptor antagonists—phentolamine exerted local anesthetic effects and hence could not be tested against NE, while dibenamine was without effect, but it is not known if the dose was adequate. By analogy with other central areas, selectivity between NE and 5-HT effects in the hippocampus might be poor, particularly when antagonists based on lysergic acid are used.

The mechanisms by which NE and 5-HT depress hippocampal neurones is not defined. However, in the cerebellum NE causes a hyperpolarization associated with a decreased conductance and increased membrane resistance, probably also involving cAMP as an intracellular mediator or "second messenger" (Hoffer et al., 1973). Similar mechanisms may operate in rat hippocampus since the extracellular application of cAMP was depressant and NE depressions were enhanced by the concurrent application of the phosphodiesterase inhibitor papaverine and reduced by the concurrent application of PGE_1. However, these were also single agonist studies and the effects of these agents on 5-HT or GABA depressions were not tested.

The physiological significance of monoamine-induced responses in the hippocampus is hard to establish; thus many tissues have receptors for substances not necessarily released into them as transmitters, and in the cerebral cortex the number of neurons responding to monoamines appears to be greatly in excess of the number receiving a significant monoamine nervous input. However, additional studies by Segal and Bloom (1974b) provide support for the view that NE is an inhibitory transmitter in rat hippocampus. Thus stimulation of the locus coeruleus (LC) produced a long-lasting inhibition of spontaneous activity in pyramidal cells. This inhibition was blocked reversibly during the iontophoretic application of Sotalol, potentiated by the uptake blocker desmethylimipramine, and absent in rats pretreated with 6-hydroxydopamine. Further, this NE-like inhibition from stimulation of the locus coeruleus was potentiated by papaverine and blocked by PGE_1. It seemed unlikely that GABA was involved in this inhibition since iontophoretic bicuculline had no effect on the inhibitory response to LC stimulation, although reducing the inhibitory response to fornix stimulation by half and completely abolishing the responses to iontophoretic GABA. It is possible that this NE inhibitory pathway is tonically active since the spontaneous firing rate of presumed pyramidal cells was significantly higher following 6-hydroxydopamine than in control animals (Segal and Bloom, 1974a).

3.3.2. *Excitatory Amino Acids.* From the elegant neurophysiological studies of Andersen and his colleagues (Andersen et al., 1966; Anderson and Lømo, 1966), it is clear that there are a variety of excitatory nerves in the hippocampus which are intrinsic and not dependent on long afferents in the fimbria. These include the collaterals of pyramidal axons exciting the inhibitory basket cells, the Schaffer collaterals from CA3 pyramidal axons running to CA1 neurons, and the commissural fibers derived from pyramidal cells in the opposite hippocampus. There are in addition other intrinsic excitatory nerve terminals including the perforant fibers from the entorhinal cortex to the dentate granule cells and the mossy fibers (dentate granule cell axons) to CA3 cells. The excitatory transmitters released from these intrinsic excitatory nerve endings within the hippocampus are not known at present, but it seems unlikely that they are either substantially cholinergic or monoaminergic. By Dale's law, all of the excitatory collaterals of pyramidal cell axons should release the same transmitter at their terminals in the lateral septal nucleus, the hypothalamus, and the subiculum. Attempts to identify the transmitters in these areas should therefore benefit studies in the hippocampus.

It is possible that the major unknown excitatory transmitter in the hippocampus and elsewhere within the brain is the acidic amino acid glutamate. However, confident identification of glutamate as a central transmitter is surrounded with special difficulties. It is even more ubiquitous in cerebral metabolism than GABA, so providing difficulties in separating metabolic and transmitter functions. Recent studies in the hippocampus by Crawford and Connor (1973) have shown that the concentration of glutamic acid and its precursor glutamine in the apical dendritic layer of CA3 (where the mossy fibers terminate) is about double that in the corresponding layer of CA1. However, the activity of the GAD was also more than twice as high in CA3 than in CA1 so the glutamate differences may only have reflected its role as the precursor of GABA.

As the precursor of GABA, it is difficult to achieve selective release of glutamate unless activation of inhibitory elements can be avoided. However, preliminary studies by Bradford and Richards have shown a selective release of glutamate and no other amino acid from the isolated olfactory cortex by stimuli which are known to be predominantly excitatory. Release studies have been made by Crawford and Connor in the cat hippocampus using somewhat smaller cortical cups (0.17 cm^2) than those used by Smith. Their results show that entorhinal but not local stimulation (10–40-V, 0.5-ms pulses at 3 Hz for 1 hr) doubled the endogenous release of glutamate from the resting level of approximately 0.17 nmol/min/cm^2 (Fig. 11). However, the specificity of this release is uncertain since the release of other amino acids and markers was not measured and no tests were made for calcium-dependent release.

Microiontophoretic identification of glutamate or other acidic amino acids as excitatory transmitters within the mammalian central nervous system has not as yet been achieved, although it is known that glutamate is an excitatory transmitter in crustaceans. These difficulties stem from the absence of really selective glutamate receptor antagonists. In this connection, however, recent studies in our laboratory by Davies and Watkins (1973a,b) and Clarke et al. (1974) show that in the cuneate nucleus and cerebral cortex HA966 (4-hydroxy-3-aminopyrolid-2-one) shows some selectivity between glutamate (and aspartate) induced excitations and those induced by ACh. Further, Davies and Watkins have also shown that HA966 is an effective depressant of synaptic excitation in the cuneate nucleus. More detailed and quantitative studies are still required, however, and it is by no means certain that this compound will prove generally useful in identifying amino acid excitatory transmitters, as its selectivity against ACh excitation may not be adequate. In addition, tests of physiological identity through a comparison of the membrane effects induced by excitatory transmitters and those by the iontophoretic application of acidic amino acids are frustrated by anatomical considerations. Intracellular recordings are usually derived

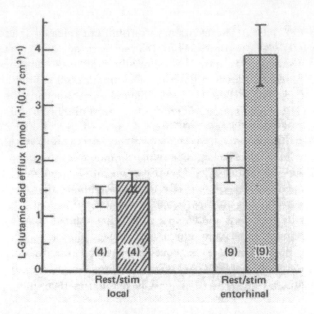

FIG. 11. Release of glutamic acid from cat hippocampus in the absence of stimulation (resting state) and in response to local and entorhinal electrical stimuli. Glutamate concentration (nmol) in superfusates (cup area 0.17 cm^2, volume 0.25 ml Ringer–Locke solution 13.6 μl/min) was determined by a microenzymatic method. Stimulus pulses (3 Hz, 0.5 ms duration, 10–40 V, for 1 h) were delivered to the entorhinal cortex or the hippocampal surface through bipolar electrodes (28-gauge stainless steel wire, 1.2 k ohm resistance, insulated except for the tip, less than 1 mm tip separation). The bars represent means ($\pm$ SEM); the number of animals is given in parentheses. Entorhinal stimulation significantly increased glutamate efflux when compared with release during rest ($P < 0.025$, paired t test) and during local stimulation ($P < 0.05$, two-sample t test). Reproduced with kind permission from *Nature,* Crawford and Connor (1973).

from the bodies of cells and thus changes in distant dendrites where the excitatory synapses predominantly terminate are not readily seen.

4.   Conclusions

Two important stages in the identification of a central transmitter are demonstration of (1) calcium-dependent and specific release in response to stimulation and (2) physiological and pharmacological identity of action between the effects of nerve stimulation and the putative transmitter applied to the neuron by iontophoresis. Although both approaches have particular disadvantages, the iontophoretic approach is considered to offer the more direct and conclusive evidence of function. With its simple layered structure and well-defined connections, the hippocampus should be a model area for the identification of neurotransmitters, the factors affecting their release, and their actions on the postsynaptic membrane. Such studies would provide a basis for determining whether and to what extent the actions of centrally acting drugs are exerted particularly in the hippocampus.

With some exceptions, our knowledge of hippocampal transmitters is generally sparse. From a variety of neurophysiological, histochemical, biochemical, and selective lesion studies, there is good evidence for a major system of acetylcholine-containing and a minor system of monoamine-containing nerve fibers running to the hippocampus. There is good evidence for a specific pathway for selective release of ACh but not [^{14}C]urea following septal stimulation. From iontophoretic studies, the ACh receptors appear to be predominantly muscarinic, although there is some evidence suggesting that nicotinic antagonists block the excitatory effects of ACh and fimbrial stimulation on hippocampal pyramidal cells, and more detailed studies are required.

At present, no studies have been made to show the release of monoamines from the hippocampus with afferent stimulation. Iontophoretic studies show that NE and 5-HT are generally depressant. Pharmacologically, the effect of NE on rat pyramidal cells resembles that of stimulation of the locus coeruleus. The 5-HT nerve input had not been stimulated nor the receptors characterized.

Two important putative amino acid transmitters, GABA and glutamate, are present in the hippocampus. Biochemical and lesions studies show the GABA system to be intrinsic to the hippocampus and make a strong case for GABA being associated with basket cell inhibition of the pyramidal cells. Local but not afferent stimulation can increase the efflux of labeled GABA from the hippocampus, but this release is often accompanied by that of labeled urea as a marker substance. Iontophoretic studies show GABA to be a more potent depressant than glycine or β-alanine. Bicuculline blocks the effects of GABA but not glycine, and at the same time reduces the inhibition of hippocampal cells induced by fimbrial stimulation. Although regional differences in the distribution of glutamate between CA1 and CA3 have been reported, similar differences were seen in the levels of the enzyme GAD, which synthesizes GABA from glutamate. Thus the significance of this observation in terms of glutamate as a transmitter is unknown. Stimulation of the entorhinal cortex produces a substantial increase in the efflux of glutamate from the hippocampus, but tests for selectivity in terms of other amino acids, including GABA and marker substances, as well as for calcium-dependent release have not been made. In iontophoretic studies, physiological tests for identity of action between glutamate and some of the unknown excitatory transmitters in the hippocampus have not yet been made. Pharmacological tests of identity are hampered by the absence of suitable tools, although some moderately selective excitatory amino acid antagonists are now becoming available which offer hope for future experiments.

Assuming that the hard evidence for a neurotransmitter function for glutamate, NE, and 5-HT becomes available, then in the hippocampus, as elsewhere in the CNS, there are at least five neurotransmitters. Whey are so many transmitters needed in the CNS? At a simplistic level, given location and wiring specificity, two transmitters, one excitatory and one inhibitory, might be thought to be sufficient. Multiple transmitters are not a purely teleological development since all the transmitter candidates in the mammalian brain with the exception of glycine seem to be transmitters in the nervous system of more primitive species.

Two arguments can be advanced for multiple transmitters: (1) they provide a safety factor by virtue of their heterogeneous origins or (2) they are necessary because they have functions in addition to the mere short-term transfer of information from presynaptic terminals to postsynaptic membrane. In this latter connection, the contrast between the fast actions of iontophoretic amino acids and the slower, more sustained effects seen with acetylcholine in supraspinal areas may provide a chemical rather than a purely circuit basis for sustained synaptic action in pathways of different significance. As in neocortex, the monoamine inputs to the hippocampus would not appear to provide a major system for excitation or inhibition. Applied iontophoretically, monoamines have relatively weak and long-lasting effects. In addition, if the mechanism for hippocampal inhibition by NE is a reduced conductance in the dendrites, then this is hardly the most efficient process. It may well be that these neuronal excitability changes seen with monoamines—i.e., the more typical transmitter functions—are not of great significance and are really secondary to a more important modulator role whereby monoamines bring about metabolic changes in the postsynaptic membrane possibly concerned with the imprinting of long-term changes.

5. References

ANDERSEN, P., AND CURTIS, D. R. The excitation of thalamic neurones by acetylcholine. *Acta Physiologica Scandinavica* 1964, **61**, 85–99.

ANDERSEN, P., AND LØMO, T. Mode of activation of hippocampal pyramidal cells by excitatory synapses on dendrites. *Brain Research*, 1966, **2**, 247–260.

ANDERSEN, P., BRULAND, H., AND KAADA, B. R. Activation of field CA1 of the hippocampus by septal stimulation. *Acta Physiologica Scandinavica,* 1961, **51**, 29–40.

ANDERSEN, P., ECCLES, J. C., LØYNING, Y., AND VOORHOEVE, P. E. Strychnine resistant inhibition in the brain. *Nature (London)*, 1963, **200**, 843–845.

ANDERSEN, P., ECCLES, J. C., AND LØYNING, Y. Pathway of postsynaptic inhibition in the hippocampus. *Journal of Neurophysiology,* 1964, **27**, 608–619.

ANDERSEN, P., BLACKSTAD, T. W., AND LØMO, T. Location and identification of excitatory synapses on hippocampal pyramidal cells. *Experimental Brain Research,* 1966, **1**, 236–248.

BAKER, R., AND LLINAS, R. Electronic coupling between neurones in the rat mesencephalic nucleus. *Journal of Physiology (London)*; 1971, **212**, 45–64.

BELESLIN, D., CARMICHAEL, E. A., AND FELDBERG, W. The origin of acetylcholine appearing in the effluent of perfused cerebral ventricles of the cat. *Journal of Physiology (London)*, 1964, **173**, 368–376.

BEVAN, P., BRADSHAW, C. M., ROBERTS, M. H. T., AND SZABADI, E. The excitation of neurons by noradrenaline. *Journal of Pharmacy and Pharmacology,* 1973, **25**, 309–314.

BISCOE, T. J., AND STRAUGHAN, D. W. Microelectrophoretic studies of neurones in the cat hippocampus. *Journal of Physiology (London)*, 1966, **183**, 341–359.

BLACKSTAD, T. W., FUXE, K., AND HÖKFELT, T. Noradrenaline nerve terminals in the hippocampal region of the rat and the guinea pig. *Zeitschrift für Zellforschung,* 1967, **78**, 463–473.

BLOOM, F. E., AND IVERSEN, L. L. Localizing ^{3}H-GABA in nerve terminals of rat cerebral cortex by electron microscopic autoradiography. *Nature (London)*, 1971, **229**, 628–630.

BOAKES, R. J., BRADLEY, P. B., BROOKES, N., CANDY, J. M., AND WOLSTENCROFT J. H. Actions of noradrenaline, other sympathomimetic amines and antagonists on neurones in the brain stem of the cat. *British Journal of Pharmacology,* 1971, **41**, 462–479.

BOND, P. A. The uptake of (γ^3H) aminobutyric acid by slices of various regions of rat brain and the effect of lithium. *Journal of Neurochemistry,* 1973, **20**, 511–517.

BOWERY, N. G., AND BROWN, D. A. γ-aminobutyric acid uptake by sympathetic ganglia. *Nature (London), New Biology,* 1972, **238**, 89–91.

BRADFORD, H. F., AND RICHARDS, C. D. Unpublished observations.

BRENELLS, A. B. An *in vivo* method for studying release of putative neurotransmitters from the rabbit olfactory bulbs. *British Journal of Pharmacology,* 1973, **47,** 667–668P.

BRENELLS, A. B. Spontaneous and neurally evoked release of labelled noradrenaline from rabbit olfactory bulbs *in vivo. Journal of Physiology (London),* 1974, **240,** 279–293.

BRÜCKE, F., GOGOLAK, G., AND STUMPF, C. Mikroelektrodenuntersuchung der Reizzantwort und der Zelltätigkeit im Hippocampus bei Septumreizung. *Pflügers Archiv für die Gesamte Physiologie,* 1963, **276,** 456–470.

BRUGGENCATE, G. TEN, AND ENGBERG, I. Iontophoretic studies in Deiters' nucleus of the inhibitory actions of GABA and related amino acids and the interactions of strychnine and picrotoxin. *Brain Research,* 1971, **25,** 431–448.

CANDY, J. M., BOAKES, R. J., KEY, B. J., AND NORTON, E. Correlation of the release of amines and antagonists with their effects. *Neuropharmacology,* 1974. **13,** 423–430.

CHASE, T. N., AND KOPIN, I. J. Stimulus-induced release of substances from olfactory bulb using the push-pull cannula. *Nature (London),* 1968, **217,** 466–467.

CLARKE, G., FORRESTER, P. A., AND STRAUGHAN, D. W. A quantitative analysis of the excitation of single cortical neurones by acetylcholine and L-glutamic acid applied microiontophoretically, *Neuropharmacology,* 1974, **13,** 1047–1055.

COLLIER, B., AND MURRAY-BROWN, N. Validity of a method measuring transmitter release from the central nervous system. *Nature (London),* 1968, **218,** 484–485.

CRAWFORD, I. L., AND CONNOR, J. D. Localization and release of glutamic acid in relation to the hippocampal mossy fibre pathway. *Nature (London),* 1973, **244,** 442–443.

CURTIS, D. R. Microelectrophoresis. In W. L. Nastuk (Ed.), *Physical techniques in biological research.* Vol. 5. New York: Academic Press, 1964, pp. 144–190.

CURTIS, D. R., AND DAVIS, R. The excitation of lateral geniculate neurones by quarternary ammonium derivatives. *Journal of Physiology (London),* 1963, **165,** 62–82.

CURTIS, D. R., AND FELIX, D. The effect of bicuculline upon synaptic inhibition in the cerebral and cerebellar cortices of the cat. *Brain Research,* 1971, **34,** 301–321.

CURTIS, D. R., HÖSLI, L., JOHNSTON, G. A. R., AND JOHNSTON, I. H. The hyperpolarization of spinal motoneurones by glycine and related aminoacids. *Experimental Brain Research,* 1968, **5,** 235–258.

CURTIS, D. R., FELIX, D., AND MCLENNAN, H. GABA and hippocampal inhibition. *British Journal of Pharmacology,* 1970, **40,** 881–883.

CURTIS, D. R., DUGGAN, A. W., FELIX, D., JOHNSTON, G. A. R., AND MCLENNAN, H. Antagonism between bicuculline and GABA in the cat brain. *Brain Research,* 1971, **33,** 57–73.

DAHLSTRÖM, A., FUXE, K., OLSEN, L., AND UNGERSTEDT, U. On the distribution and possible function of monoamine nerve terminals in the olfactory bulb of rabbit. *Life Sciences,* 1965, **4,** 2071–2074.

DAVIES, J., AND WATKINS, J. C. Antagonism of synaptic and aminoacid induced excitation in the cuneate nucleus of the cat by HA-966. *Journal of Neuropharmacology,* 1973a, **12,** 637–640.

DAVIES, J., AND WATKINS, J. C. Microelectrophoretic studies on the depressant action of HA-966 on chemically and synaptically excited neurones in the cat cerebral cortex and cuneate nucleus. *Brain Research,* 1973b, **59,** 311–322.

DREIFUSS, J. J., KELLY, J. S., AND KRNJEVIĆ, K. Cortical inhibition and γ-aminobutyric acid. *Experimental Brain Research,* 1969, **9,** 137–154.

DUDEL, J. Presynaptic and postsynaptic effects of inhibitory drugs in the crayfish neuromuscular junction. *Pflugers Archiv für die Gesamte Physiologie,* 1965, **283,** 104–118.

EARL, J., AND LARGE, W. A. The effects of bicuculline, picrotoxin and strychnine on neuromuscular inhibition in hermit crabs (*Eupagurus bernhardus*). *Journal of Physiology (London),* 1972, **224,** 45–46P.

ECCLESTON, D., RANDIĆ, M., ROBERTS, M. H. T., AND STRAUGHAN, D. W. Release of amines and amine metabolites from brain by nerve stimulation. In G. Hooper (Ed.), *Metabolism of amines in the brain.* London: Macmillan, 1969, pp. 29–33.

ELLIOT, K. A. C., SWANK, R. L., AND HENDERSON, N. Effects of anaesthetics and convulsants on acetylcholine content of brain. *American Journal of Physiology,* 1950, **162,** 469–474.

ENGBERG, I., AND MARSHALL, K. C. Mechanism of noradrenaline hyperpolarization in spinal cord motoneurones of the cat. *Acta Physiologica Scandinavica,* 1971, **83,** 142–144.

Felix, D., and McLennan, H. The effect of bicuculline on the inhibition of mitral cells of the olfactory bulb. *Brain Research,* 1971, **25,** 661–664.

Frederickson, C. A., Jordon, L. M., and Phillis, J. W. The action of noradrenaline on cortical neurons: Effects of pH. *Brain Research,* 1971, **35,** 556–560.

Fuxe, K. Evidence for the existence of monoamine neurons in the central nervous system. IV. Distribution of monoamine nerve terminals in the central nervous system. *Acta Physiologica Scandinavica,* 1965, Supplement 247, **64,** 37–84.

Fuxe, K., and Jonsson, H. Further mapping of central 5-HT neurons: Studies with dihydroxytryptamines. In E. Costa, G. L. Gessa, M. Sandler (Eds.), *Advances in biochemical psychopharmacology.* Vol. 10. Amsterdam: North Holland, 1974, pp. 1–12.

Heller, A., Seiden, L. S. and Moore, R. Y. Regional effects of lateral hypothalamic lesions on brain norepinephrine in the cat. *International Journal of Neuropharmacology,* 1966, **5,** 91–101.

Herz, A., and Nacimiento, A. Uber die Wirking von Pharmaka auf Neurone des Hippocampus nach mikroelektrophoretischer Verabfolgung. *Archiv für Experimentelle Pathologie und Pharmacologie,* 1965, **250,** 258–259.

Hoffer, B. J., Siggins, G. R., Oliver, A. P., and Bloom, F. E. Activation of the pathway from locus coeruleus to rat cerebellar Purkinje neurons: Pharmacological evidence of noradrenergic central inhibition. *Journal of Pharmacology and Experimental Therapeutics,* 1973, **184,** 553–569.

Iversen, L. L. The uptake, storage, release, and metabolism of GABA in inhibitory nerves. In S. H. Snyder (Ed.), *Perspectives in neuropharmacology: A Tribute to J. Axelrod.* London: Oxford University Press, 1972, pp. 75–111.

Iversen, L. L., Mitchell, J. F., and Srinivasan, V. The release of γ-aminobutyric acid during inhibition in the cat visual cortex. *Journal of Physiology (London),* 1971, **212,** 519–534.

Johnson, E. S., Roberts, M. H. T., and Straughan, D. W. The influence of rate of neuronal firing on the time-course of drug responses. *Journal of Physiology (London),* 1969a, **203,** 78P.

Johnson, E. S., Roberts, M. H. T., Sobieszek, A., and Straughan, D. W. Noradrenaline sensitive cells in cat cerebral cortex. *International Journal of Neuropharmacology,* 1969b, **8,** 549–566.

Kandel, E. R., Spencer, W. A., and Brinley, F. J. Electrophysiology of hippocampal neurons. 1. Sequential invasion and synaptic organization. *Journal of Neurophysiology,* 1961, **24,** 225–242.

Katz, R. I., and Chase, T. N. Neurohumoral mechanisms in the brain slice. In *Advances in pharmacology and chemotherapy.* Vol. 8. New York: Academic Press. 1970.

Kelly, J. S., and Renaud, L. P. On the pharmacology of ascending, descending and recurrent postsynaptic inhibition of the cuneothalamic relay cells in the cat. *British Journal of Pharmacology,* 1973, **48,** 396–408.

Kelly, J. S., Simmonds, M. A., and Straughan, D. W. Microelectrode techniques. In P. B. Bradley (Ed.), *Methods in brain research.* Vol. 1. New York: Wiley, 1975, pp. 333–377.

Kobayashi, H., and Libet, B. Generation of flow postsynaptic potentials without increases in ionic conductance. *Proceedings of the National Academy of Sciences, U.S.A.,* 1968, **60,** 1304–1311.

Kobayashi, H., and Libet, B. Actions of noradrenaline and acetylcholine on sympathetic ganglion cells. *Journal of Physiology (London),* 1970, **208,** 353–372.

Korn, H., Sotelo, C. O., and Crepel, F. Electrotonic coupling between neurons in the rat lateral vestibular nucleus. *Experimental Brain Research,* 1973, **16,** 255–275.

Kozlovskaya, M. M., and Valdman, A. V. Behavioral and EEG reactions evoked by stimulation of the medial and lateral septal zones of the brain in the rabbit. *Pavlov Journal of Higher Nervous Activity,* 1970, **20,** 1022–1030.

Krnjević, K. Cholinergic innervation of the cerebral cortex. In D. R. Curtis and A. K. McIntyre (Eds.), *Studies in physiology.* New York: Springer-Verlag, 1965, pp. 144–151.

Krnjević, K. Microiontophoresis. In R. Fried (Ed.). *Methods of neurochemistry.* New York: Marcel Dekker, 1972, pp. 129–172.

Krnjević, K., and Phillis, J. W. Acetylcholine-sensitive cells in the cerebral cortex. *Journal of Physiology (London),* 1963a, **166,** 296–327.

Krnjević, K., and Phillis, J. W. Pharmacological properties of acetylcholine-sensitive cells in the cerebral cortex. *Journal of Physiology (London),* 1963b, **166,** 328–352.

KRNJEVIĆ, K., AND SCHWARTZ, S. The inhibitory transmitter in the cerebral cortex. In *Structure and functions of inhibitory neuronal mechanisms*. Oxford: Pergamon Press, 1968, pp. 419–427.

KRNJEVIĆ, K., AND SILVER, A. A histochemical study of cholinergic fibres in the cerebral cortex. *Journal of Anatomy*, 1965, **99**, 711–759.

KRNJEVIĆ, K., PUMAIN, R., AND RENAUD, L. The mechanism of excitation by acetylcholine in the cerebral cortex. *Journal of Physiology (London)*, 1971, **215**, 247–268.

LARGE, W. A., AND MILTON, A. S. Effects of morphine, levorphanol, nalorphine and naloxone on the release of acetylcholine from slices of rat cerebral cortex and hippocampus. *British Journal of Pharmacology*, 1971, **41**, 398P.

LEGGE, K. F., RANDIĆ, M., AND STRAUGHAN, D. W. The pharmacology of neurones in the pyriform cortex. *British Journal of Pharmacology and Chemotherapy*, 1966, **26**, 87–107.

McCANCE, I., AND PHILLIS, J. W. The action of acetylcholine on cells in cat cerebellar cortex. *Experientia*, 1964, **20**, 1–5.

MITCHELL, J. F. The spontaneous and evoked release of acetylcholine from the cerebral cortex. *Journal of Physiology (London)*, 1963, **165**, 98–116.

MITCHELL, J. F., AND SRINIVASAN, V. Release of ^{3}H-γ-aminobutyric acid from the brain during synaptic inhibition. *Nature (London)*, 1969, **244**, 663–666.

MOORE, R. Y., AND HELLER, A. Monoamine levels and neuronal degeneration in brain following lateral hypothalamic lesions. *Journal of Pharmacology and Experimental Therapeutics*, 1967, **156**, 12–22.

MUCKART, A. B. Neurally evoked release of noradrenaline from the olfactory bulb. *British Journal of Pharmacology*, 1971, **42**, 641–642P.

NEAL, M. J., AND IVERSEN, L. L. Subcellular distribution of endogenous and [^{3}H]γ-aminobutyric acid in rat cerebral cortex. *Journal of Neurochemistry*, 1969, **16**, 1245–1252.

RAISMAN, G. The connexions of the septum. *Brain*, 1966, **89**, 317–348.

RANDIĆ, M., AND PADJEN, A. Effect of calcium ions on the release of acetylcholine from the cerebral cortex. *Nature (London)*, 1967, **215**, 990.

RICHARDS, C. D., AND SERCOMBE, R. Electrical activity observed in guinea-pig olfactory cortex maintained *in vitro*. *Journal of Physiology (London)*, 1968, **197**, 667–683.

ROBERTS, P. J. Glutamate, GABA and the direct cortical response in the rat. *Brain Research*, 1973, **49**, 451–455.

SALMOIRAGHI, G. C., AND STEFANIS, C. N. A critique of iontophoretic studies of central nervous system neurons. *International Review of Neurobiology*, 1967, **10**, 1–30.

SALMOIRAGHI, G. C., AND WEIGHT, F. Micromethods in neuropharmacology: An approach to the study of anaesthetics. *Anesthesiology*, 1967, **28**, 54–64.

SALMOIRAGHI, G. C., BLOOM, F. E., AND COSTA, E. Adrenergic mechanisms in rabbit olfactory bulb. *American Journal of Physiology*, 1964, **207**, 1417–1424.

SEGAL, M., AND BLOOM, F. E. The action of norepinephrine in the rat hippocampus. I. Iontophoretic studies. *Brain Research*, 1974a, **72**, 79–97.

SEGAL, M., AND BLOOM, F. E. The action of norepinephrine in the rat hippocampus. II. Activation of the input pathway. *Brain Research*, 1974b, **72**, 99–114.

SEGAL, M., SIMS, K., MAGGIORA, L., AND SMISSMAN, E. Analogues of gamma-aminobutyrate on rat hippocampal neurones. *Nature (London), New Biology*, 1973, **245**, 88–89.

SHUTE, C. C. D., AND LEWIS, P. R. The use of cholinesterase techniques combined with operative procedures to follow nervous pathways in the brain. *Bibliotheca Anatomica*, 1961, **2**, 34–49.

SHUTE, C. C. D., AND LEWIS, P. R. Cholinesterase-containing systems of the brain of the rat. *Nature (London)*, 1963, **199**, 1160–1164.

SMITH, C. M. The release of acetylcholine from rabbit hippocampus. *British Journal of Pharmacology*, 1972, **45**, 172P.

SMITH, C. M. Direct evidence for the existence of a cholinergic septo-hippocampal pattern. *Life Sciences*, 1974, **14**, 2159–2166.

SPARBER, S. B., AND TILSON, H. A. Schedule controlled and drug induced release of norepinephine-7-^{3}H into the lateral ventricle of rats. *Neuropharmacology*, 1972, **11**, 453–464.

STEFANIS, C. Hippocampal neurons: Their responsiveness to microelectrophoretically administered endogenous amines. *Pharmacologist*, 1964, **6**, 171.

Storm-Mathisen, J., and Fonnum F. Localization of transmitter candidates in the hippocampal region. In P. B. Bradley and R. W. Brimblecombe (Eds.), *Progress in brain research*. Vol. 36: *Biochemical and pharmacological mechanisms underlying behaviour*. Amsterdam: Elsevier, 1972.

Storm-Mathisen, J., and Guldberg, H. C. 5-Hydroxytryptamine and noradrenaline in the hippocampal region: Effect of transection of afferent pathways on endogenous levels, high affinity uptake and some transmitter-related enzymes. *Journal of Neurochemistry*, 1974, **22**, 793–803.

Straughan, D. W. Convulsant drugs and inhibitory mechanisms in the mammalian central nervous system. *Archivos de Farmacologia y Toxicologia*, 1973, **1**, 7–36.

Takeuchi, A., and Onodera, K. Effect of bicuculline on the GABA receptor of the crayfish neuromuscular junction. *Nature (London), New Biology*, 1972, **236**, 55–56.

Weight, F. F., and Votava, J. Slow synaptic excitation in sympathetic ganglion cells: Evidence for synaptic inactivation of potassium conductance. *Science (New York)*, 1970, **170**, 755–758.

Werman, R. Criteria for identification of a central nervous system transmitter. *Comparative Biochemistry and Physiology*, 1966, **18**, 745–766.

Werman, R. An electrophysiological approach to drug-receptor mechanisms. *Comparative Biochemistry and Physiology*, 1969, **30**, 997–1017.

Winson, J., and Gerlach, J. L. Stress-induced release of substances from the rat amygdala detected by the push-pull cannula. *Nature (London), New Biology*, 1971, **230**, 251–253.

Yamamoto, C. Activation of hippocampal neurons by mossy fibre stimulation in thin brain sections *in vitro*. *Experimental Brain Research*, 1972, **14**, 423–435.

Yamamoto, C., and McIlwain, H. Electrical activities in thin sections from the mammalian brain maintained in chemically defined media *in vitro*. *Journal of Neurochemistry*, 1966, **13**, 1333–1343.

Young, J. A. C., Brown, D. A., Kelly, J. S., and Schon, F. Autoradiographic localization of sites of [^{3}H]γ-aminobutyric acid accumulation in peripheral autonomic ganglia. *Brain Research*, 1973, **63**, 479–486.

10

Cholinergic Neurons: Septal–Hippocampal Relationships

1. Introduction

There is abundant evidence that acetylcholine functions as a central nervous system (CNS) neurotransmitter (Hebb, 1970). It is present throughout the CNS, although its concentration varies from region to region. There are reviews of the extensive biochemical studies of this compound (Hebb, 1972; Phillis, 1970a) and there have been numerous electrophysiological studies (Phillis, 1970b).

The cholinergic septal–hippocampal system is unique in that it is perhaps the only CNS cholinergic tract whose anatomy is well studied. Thus a system is present for a variety of experiments to determine many aspects of cholinergic function, including release, metabolism, and effects of drugs on these factors. This is a central theme of the following sections, which discuss histochemical methods, some anatomy, electrophysiological studies, and various biochemical experiments.

2. Localization of Cholinergic Neurons in the Mammalian Central Nervous System

The precise localization of neurotransmitter-specific neurons in the CNS is reliably known only in the case of catecholamine-containing and indoleamine-contain-

MICHAEL J. KUHAR • Departments of Pharmacology and Experimental Therapeutics and Psychiatry and the Behavioral Sciences, Johns Hopkins University School of Medicine, Baltimore, Maryland.

ing neurons. This knowledge is due to the reliable and reproducible histochemical fluorescence method of Falck and Hillarp (Falck et al., 1962). Unfortunately, there is no comparable histochemical method for acetylcholine. However, some papers have appeared demonstrating a method for the histochemical localization of choline acetyltransferase (ChAc; acetyl-CoA: choline O-acetyltransferase; E.C. 2.3.1.6), and there are well-established methods for the localizaion of acetylcholinesterase (AChE; E.C. 3.1.1.7). It is assumed that neurons containing acetylcholine and/or ChAc are cholinergic, i.e., function by releasing acetylcholine at their nerve terminals. However, there is some doubt, as is discussed below, whether this can be inferred for neurons that contain AChE. The following sections discuss in more detail the enzyme histochemical methods related to cholinergic neurons.

2.1. The Histochemical Method for Acetylcholinesterase

The publication of a histochemical technique for the localization of AChE (Koelle and Friendenwald, 1949) has stimulated an enormous volume of research in this area. The method is based on the precipitation of copper thiocholine (acetylthiocholine or butyrylthiocholine can be used as substrates) at sites of enzyme activity; copper thiocholine is then coverted to copper sulfide, which is seen in the microscope. The original technique has undergone many variations, including adaptions of the method to electron microscopy. Thus the localization of AChE-containing neuronal pathways in the brain has been examined in detail (Lewis and Shute, 1967; Shute and Lewis, 1967). Relative to the main topic of this chapter, and to be discussed in more detail below, Lewis and Shute (1967) reported an AChE-containing pathway from cells in the medial septal nucleus and the nucleus of the diagonal band traveling to the hippocampus via mainly the fimbria, dorsal fornix, and supracallosal striae.

Initially, it was thought that the localization of AChE to a given neuron would be indicative that the neuron was cholinergic. However, it appears that this postulate is not true. For example, a portion of the "ventral tegmental pathway" (Shute and Lewis, 1967), which appears to arise largely from AChE-staining cells of the pars compacta of the substantia nigra, is dopaminergic rather than cholinergic. Also, the cells of the dorsal and median raphe nuclei appear to stain for AChE but are known to be serotonergic (Dahlström and Fuxe, 1965; Ungerstedt, 1971). It may be that these cells are cholinoceptive.

Other biochemical studies would indicate that AChE is present in many neuronal structures that do not contain ChAc. Goldberg and McCaman (1967), for example, have shown in various regions of the cerebellum that AChE does not parallel the distribution of ChAc. Also, in experiments with cultures of neuroblastoma, Ammano et al. (1972) have shown that AChE content was rich in many clones that contained almost no ChAc. Further, in the periphery, AChE has been demonstrated in norepinephrine-containing nerve fibers in sympathetically innervated organs (Jacobowitz and Koelle, 1965). At present, it appears that AChE is present in cholinergic neurons, but not only in cholinergic neurons.

2.2.   The Histochemical Method for Choline Acetyltransferase

Papers have begun to appear dealing with the histochemical localization of ChAc (Burt, 1969, 1970; Kasa *et al.*, 1970*a,b*; Kasa and Morris, 1972). The basis of the method is that coenzyme A, as it is released from acetyl-CoA during the acetylation of choline, is precipitated *in situ* by lead. The insoluble lead mercaptide is then converted to lead sulfide, which can be visualized under the microscope. The main problem with the method in the earlier publications was that it did not appear to be substrate specific; that is, the staining did not depend on the presence the of choline in the incubation mixture (Burt, 1969; Kasa *et al.*, 1970*a*). However, it was subsequently reported that the presence of 1 mM diisopropylfluorophosphate (DFP) reduced the rate of the nonspecific hydrolysis to a very low level (Kasa and Morris, 1972). While there have not been numerous publications utilizing this method, and while the method may not yet be free of problems, one hopes for more progress in this area soon.

An immunohistochemical method for ChAc is of course possible, and is presently a major effort. A homogeneous preparation of ChAc has been reported (Chao and Wolfgram, 1973) and subsequent immunofluorescent localization of presumed cholinergic neurons in the brain has been reported (Eng *et al.*, 1974; McGeer *et al.*, 1975). At present, one must rely on a combined histochemical and biochemical approach to identify cholinergic tracts in the brain. This "combination" approach has worked well in establishing septal–hippocampal cholinergic relationships and is discussed in detail in the next section.

3.   Evidence for a Septal–Hippocampal Pathway

3.1.   Anatomy of the Hippocampus

A detailed discussion of the cytoarchitecture of the hippocampus is, of course, beyond the scope of this chapter. However, some of the more relevant anatomical features of the hippocampus that well be referred to in the following discussion will be discribed here. Figure 1 is a sketch of a horizontal section through the hippocampus and adjacent structures in the rat, diagrammatically showing fibers and cell bodies. The "hippocampus" as referred to throughout this chapter is actually the total hippocampal formation as shown in Fig. 1. This is made up of the subiculum, the hippocampus proper (divided into regio superior and regio inferior), and the area dentata (or dentate gyrus). The fimbria is a fiber tract containing afferents and efferents for the hippocampus. The alveus represents a sheet of fibers on the deep or ventricular surface of the hippocampus and contains the fibers that are continuous with those of the fimbria. The chief cell type of the hippocampus is the pyramidal cell, which is localized to the stratum pyramidale. The stratum oriens contains the basal dendrites of the pyramidal cells, while the stratum radiatum contains the main shafts of the apical dendrites of the pyramidal cells. The dentate gyrus may be regarded as very similar in structure to the hippocampus proper. It consists basically of a single layer of granule cells. The stratum moleculare contains the peripheral

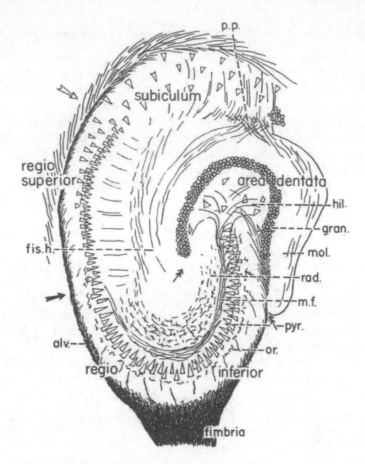

FIG. 1. Sketch of horizontal silver-impregnated section through hippocampus and adjacent structures in the rat, showing diagrammatically fibers and cell bodies. Open arrow, at limit between subiculum and hippocampus; solid arrow, at limit between the subfields regio superior and regio inferior of the hippocampus; double arrows, at limit between hippocampus and area dentata. Abbreviations: alv., alveus; fis. h., the obliterated hippocampal fissure; gran., granular layer of fascia dentata; hil., hilus fasciae dentatae (constituting, together with gran. and mol., the area dentata); m.f., layer of mossy fibers; mol., molecular layer of fascia dentata (forming, together with gran., the fascia dentata); or., stratum oriens (layer of basal dendrites of pyramidal cells; pyr., stratum pyramidale (layer of cell bodies); rad., stratum radiatum (layer of dendritic shafts). Reprinted with permission from Blackstadt and Kjaerheim (1961).

dendrites of the granule cells (Fig. 1). A more detailed discussion of the anatomy can be found in other chapters in this book as well as in the listed references.

3.2. Histochemical and Biochemical Studies

The precise localization of the neuronal elements that contain AChE in the septal–hippocampal system has been extensively examined. Lewis and Shute (1967) found that the AChE-containing afferents to the hippocampus appeared to arise from the medial septal nucleus and the nucleus of the diagonal band and traveled to the hippocampal formation via the medial supracallosal stria of Lancisi, the dorsal fornix, the alveus, and the fimbria (Fig. 2). The AChE-containing fibers appeared to be localized mainly to the outer side of the fimbria. The localization of AChE within the hippocampus itself has been shown in detailed fashion by Storm-Mathisen (1970) (Fig. 3). AChE activity was found to be localized to very discrete bands within the hippocampal formation (Fig. 3).

Lesion studies have been utilized to determine the polarity of the septal–hippocampal radiation. When lesions were placed in the fimbria, ChAc activity and AChE staining increased on the septal side of the transection, while both enzyme activities fell on the hippocampal side of the transection. Further, the increased intensity of AChE staining could be traced back to cell bodies located in the medial septal nucleus (Lewis et al., 1967; Lewis and Shute, 1967).

Shute and Lewis (1966) have performed an electron microscopic examination of AChE-containing neurons in the hippocampal formation of the rat. They observed AChE on axonal membranes in the alveus, on the pre- and postsynaptic membranes at certain synapses, and also in the granular endoplasmic reticulum and in the nuclear envelope of some cells in the stratum oriens, and to a small extent in pyramidal

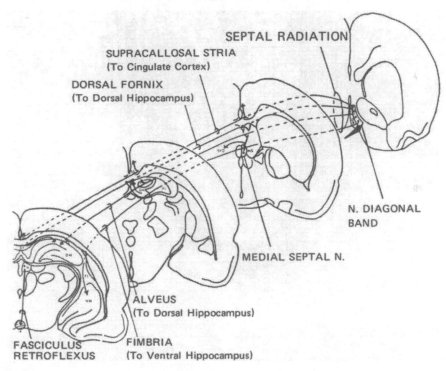

Fig. 2. Expanded diagram (natural spacing of transverse sections increased 5 times) showing the septal radiation of cholinergic fibers arising from the medial septal nucleus (MS) and the nucleus of the diagonal band (DB), and supplying the subfornical organ (SFO), the cingulate cortex via the supracallosal stria (SS), the dorsal hippocampus (DH) via the dorsal fornix (DF) and alveus (AL), and the ventral hippocampus (VH) via the fimbria (FI). The fasciculus retroflexus, with cholinesterase-containing fibers running from the habenular nuclei (H) to the interpeduncular nucleus (IP), is also included. Reprinted with permission from Lewis and Schute (1967).

cells. In general, these workers felt that the dense layers of AChE observed with the light microscope corresponded to regions mainly of dense cholinergic fibers, while the terminals that stained for AChE appeared to be scattered diffusely throughout the tissue. Further, the AChE-containing boutons appeared to be mainly axodendritic.

Another electron microscopic study of sucrose homogenates of hippocampus is in general agreement with the latter findings (Kuhar and Rommelspacher, 1974). Only a minority of the membranes in the homogenates were stained, and only a small fraction of these membranes were synaptosomal membranes. Most of the stained structures appeared to be broken unmyelinated axons and postsynaptic processes. The number of stained membranes was markedly reduced by prior placement of electrolytic lesions in the septum.

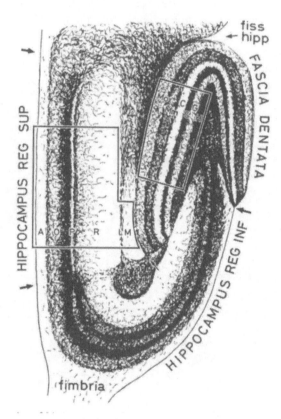

FIG. 3. Schematic drawing of hippocampus and area dentata of rat brain as seen in a horizontal section stained for AChE. The areas from which samples were dissected for quantitative analysis are circumscribed by solid lines. The zones dissected from hippocampus regio superior were the following: A, alveus; O, stratum oriens (divided in Oi and Oo); P, stratum pyramidale; R, stratum moleculare (only rostral part dissected). The zones dissected from the area dentata correspond to zones of different intensities of AChE staining: Mo, outer part of the stratum moleculare; Mm, middle part of the stratum moleculare; Co, portion of inner part of the stratum moleculare coinciding with the commissural afferent terminals; S, supragranular zone (innermost portion of molecular layer); G, granular zone; H, hilus fasciae dentatae. Reprinted with permission from Storm–Mathisen (1970).

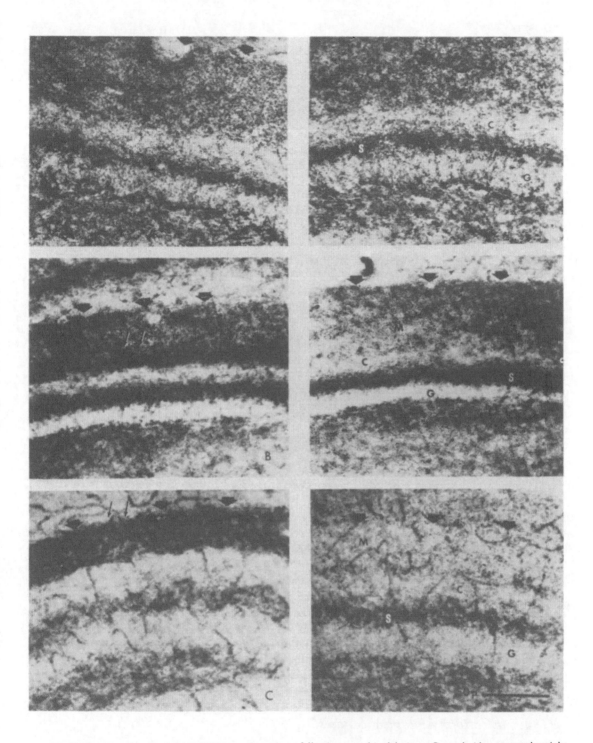

IG. 4. Dentate gyrus stained for AChE at various times following entorhinal lesions. Control sides are on the right.
mall arrows denote the new band. *A*: 15 days after entorhinal lesion. *B*: 30 days postoperative. *C*: 40 days
ostoperative. In *C*, promethazine was eliminated. Symbols: M, molecular layer; C, zone of commissural afferents in
ner one-third of M; S, normal, supragranular band of AChE staining; G, granule cell bodies; large arrows, obliterated
ippocampal fissure. Reprinted with permission from Lynch *et al.* (1972).

This brings us to the question of whether the observed AChE-containing neuronal elements are cholinergic. In studies involving the use of lesions, it appears that any lesion which interrupts a significant portion of the AChE-containing afferents to the hippocampal formation also affects to a significant extent the ChAc activity and the acetylcholine levels within the hippocampal formation. Lewis *et al.* (1967) demonstrated that if the fimbria were transected the AChE as well as ChAc activity fell dramatically on the operative side, and both fell in parallel in time. When lesions were placed in the septum, a very large decrease of acetylcholine levels, ChAc activity, and AChE activity was observed in the hippocampus (Kuhar *et al.*, 1973; McGeer *et al.*, 1969; Srebro *et al.*, 1973; Pepeu *et al.*, 1971; Storm-Mathisen and Fonnum, 1969).

Fonnum (1970) has provided very pertinent data with regard to the question of the cholinergic specificity of the AChE stain within the hippocampal region of the rat. Horizontal sections of the hippocampus were divided into several discrete layers corresponding to the localization of AChE within the hippocampus (see Fig. 3). The pyramidal cell layer, a layer actually containing more than just pyramidal cells, was found to contain the highest relative ChAc and AChE activity (Table 1). On subdivision of this layer, it was found that the highest ChAc and AChE activity was not in the precise zone containing only the pyramidal cell bodies, but rather in the narrow, infrapyramidal zone of the stratum oriens, which contains the basal dendrites of the pyramidal cells. It is relevant that Shute and Lewis (1966) in an electron microscopic study observed AChE-staining nerve terminals on the dendritic spines of the pyramidal cells in this layer. When the area dentata was divided in a similar way, it was found that the highest ChAc activity was in the supragranular zone and the hilus fasciae dentatae (Table 1). An important observation of Fonnum's was that the ratio of AChE activity to ChAc activity in the various regions was remarkably constant (Table 1), suggesting that both enzymes were localized to the same neuronal structures.

It is noteworthy that the AChE-containing fibers traveling to the hippocampus branch off to form connections in the cerebral cortex (Lewis and Shute, 1967; see also Fig. 2). In accordance with the notion that these are cholinergic fibers, it has been found that septal lesions which interrupt the hippocampal projections also result in reductions of ChAc activity in the cerebral cortex (Kuhar *et al.*, 1973).

Thus it appears that AChE in the septal–hippocampal pathway may be a valuable marker for cholinergic structures. This conclusion is based on the finding that ChAc activity is distributed very similarly to acetylcholinesterase activity in all layers of both the hippocampus proper and the dentate gyrus. The concomitant large fall in AChE and ChAc activity after fimbrial transection and after septal lesion is also consistent with the notion that the two enzymes are localized in the same fibers.

4. Electrophysiological Studies

There has been some electrophysiological examination of the effects of iontophoretically applied acetylcholine to units in the hippocampal cortex of the

TABLE 1

Comparison of ChAc and AChE Activities in Layers of Hippocampus Regio Superior and Area Dentata[a]

	Hippocampus regio superior				Area dentata		
Zone	AChE	ChAc	Ratio AChE/ChAc	Zone	AChE	ChAc	Ratio AChE/ChAc
A	550	6.5	85	Mo	2280	28.5	80
Oi	1960	19.9	98	Mm	2610	30.1	87
Oo	2780	31.2	89	Co	1910	23.6	81
P	3730	43.8	85	S	4530	61.7	74
Ri	2540	28.7	89	G	3380	48.9	69
Rm	1570	15.8	99	H	4140	53.7	77
Ro	1030	15.6	66	All layers area dentata	3070	40.6	77
L	1480	20.4	74	All layers hippocampus regio superior	1890	26.3	76
M	1160	15.1	78				
All layers regio superior	1750	21.7	81				

[a] The areas dissected and the nomenclature are shown in Fig. 3. The results are mean values from four to 12 measurements from one to seven different animals. ChAc values are expressed as μmoles ACh synthesized/h/g dry wt and are taken from this chapter. AChE values are calculated from Storm-Mathisen (1970) and are expressed as μmoles acetylthiocholine hydrolyzed/h/g dry wt (hippocampus regio superior) or as μmoles ACh hydrolyzed/h/g dry wt (area dentata). Data from Fonnum (1970).

anesthetized cat. A consistent finding appears to be that acetylcholine is excitatory in its action. Biscoe and Straughan (1966) and Steiner (1968) generally agree that approximately 60% of neurons examined respond to acetylcholine in an excitatory fashion. A difference in the two reports is that Steiner (1968) indicates that the neurons were activated with a relatively short latency (less than 1 s), while Biscoe and Straughan (1966) report a latency of several seconds. Guerrero-Figueroa et al. (1965) report a latency of 16 s or more. Biscoe and Straughan (1966) found that most cholinoceptive units were in the superficial layer of the hippocampus, an area corresponding to the location of the pyramidal cells and their dendritic processes. Thus acetylcholine was found to be most active in regions that appear to contain cholinergic neuronal elements by histochemical methods.

Biscoe and Straughan (1966) also addressed themselves to the question of the pharmacological nature of cholinergic receptors in the hippocampus. They found that atropine selectively blocked the excitation of cholinoceptive cells by acetylcholine but did not affect the excitation induced by glutamate. Since dihydro-β-erythroidine was without effect on these cells, it appears that their cholinergic receptors are predominantly muscarinic. However, a very small number of hippocampal cells were observed in which dimethyl-($+$)-tubocurarine blocked acetylcholine excitation, indicating that some of the receptors may not be exclusively muscarinic. The problem of the strict division of acetylcholine responses into muscarinic and nicotinic is not uncomplicated and has been discussed in the literature (see Biscoe and Straughan, 1966, for references).

Relevant to these findings, autoradiographic localization of tentative cholinergic muscarinic receptors in the hippocampus has been performed at the light microscopic level (Kuhar and Yamamura, 1975; Yamamura et al., 1974). Very high autoradiographic grain densities were observed in the stratum oriens, stratum radiatum, and stratum moleculare. These are the regions containing the dendrites of the pyramidal and granule cells and suggest that many cholinergic synapses are axodendritic. This is in agreement with the findings that AChE-staining boutons are mainly axodendritic (Shute and Lewis, 1966). The high density of grains in these regions is undoubtedly connected with the finding that most cells are sensitive to iontophoretic application of acetylcholine.

5. Sprouting of Cholinergic Neurons After Cortical Injury

It has been demonstrated that cholinergic neuronal terminals in the dentate gyrus may sprout and fill vacated synaptic space (Lynch et al., 1973; see also Lynch and Cotman, this volume). Within the dentate gyrus, two heavily staining bands of AChE are normally seen (Fig. 3). One is just below the granule cell layer that is in the hilus of the dentate gyrus, and another is immediately above the cell body zone, occupying an inner portion of the molecular layer indicated as the supragranular zone. Above the supragranular zone lies the portion of the stratum moleculare containing the terminals of the commissural afferents. Above this layer lies the zone of the molecular layer, which receives the terminals from the entorhinal afferents (designated approximately as Mm and Mo in Fig. 3).

Lynch *et al.* (1973) theorized that after destruction of the entorhinal cortex and after subsequent degeneration of entorhinal afferents in the molecular layer, AChE-containing neurons might spread to fill the vacated synaptic space. What they observed after entorhinal lesions was that an intense band of AChE appeared in the outer layers of the molecular layer. This region normally gives only a very slight stain for AChE. The increased staining for AChE was restricted to those portions of the molecular layer which normally contain the terminals of entorhinal afferents. This postlesion induction of AChE staining depended on the integrity of the septal–hippocampal fibers, as a septal lesion caused the disappearance of all AChE in the hippocampus, and no increase of AChE could be observed in animals with both entorhinal and septal lesions. While at this stage of investigation it was not possible to state whether new neuronal terminals had generated and whether these terminals were functional, the data are in accordance with these prospects.

The above findings demonstrate an interesting capacity of these cholinergic septal–hippocampal neurons. It is not thought that this capacity is unique, as other neuronal elements within the hippocampus may sprout to fill vacated synaptic space (Lynch *et al.*, 1973). Also, evidence that monoamine neurons sprout is presented in another chapter in this volume by Moore. These findings are an example of what is becoming increasingly clear, namely that CNS neurons have the capacity to undergo structural changes in response to, in this case, a rather drastic perturbation.

6. Biochemical Experiments

The biochemical measurement of acetylcholine has not been simple, and usually, in the past, at least, it has been necessary to rely on a bioassay. Prompted by the difficulties of these procedures, a number of investigators have developed analytical methods. These include gas chromatographic methods (Hanin and Jenden, 1969; Schmidt *et al.*, 1970), an enzymatic-fluorometric assay (Browning, 1972), and enzymatic-isotopic methods (Feigenson and Saelens, 1969; Goldberg and McCaman, 1973; Reid *et al.*, 1971; Shea and Aprison, 1973). The utilization of these various specific, sensitive, new methods will undoubtedly increase our knowledge of cholinergic systems in the brain.

The septal–hippocampal cholinergic pathway provides a convenient CNS tract for a variety of pharmacological and biochemical studies. While the full potential of this pathway has not yet been utilized and explored, some of the experiments performed with this system will be briefly discussed.

In a series of experiments, it was possible to test various features of the disposition and uptake of choline, the precursor of acetylcholine. Until recently, it was unclear whether cholinergic nerve terminals possessed a selective uptake mechanism for choline. The existence of the septal–hippocampal pathway provided a convenient way to explore this question. If cholinergic neuronal terminals possessed a selective uptake mechanism for choline, then choline uptake into synaptosomes should be drastically reduced several days after placement of lesions in the septum. In fact, there was a very large decrease in choline uptake into hippocampal synaptosomes which corresponded to the decrease in acetylcholine levels and cholineacetylase activity in the hippocampus after placement of septal lesions. This decrease was observed only in the hippocampus and cortex, regions that receive cholinergic projections, and not in other regions of the brain such as the striatum and cerebellum (Kuhar *et al.*, 1973). This choline uptake mechanism appears to have a high affinity for choline (Yamamura and Snyder, 1972). In similarly prepared animals, there was also a significant reduction in choline levels within the hippocampus after placement of lesions, suggesting that a significant portion of the free choline in the brain may be localized to cholinergic neurons (Sethy *et al.*, 1973).

There has been a study of acetylcholine release from the hippocampus utilizing a modification of the "cortical cup" technique (Smith, 1972). This technique involves placing a medium-filled open-ended cylinder against the surface of brain to form a superfusion chamber which collects biochemicals diffusing out of the brain. After rabbits were anesthetized with urethane, the cortex overlying the septum and hippocampus was removed by suction and a cortical cup was placed on the hippocampus. There was a resting release of acetylcholine and the release was greatly increased by stimulation of the septum. Various drugs were utilized and it was found that topically applied morphine decreased release of acetylcholine, while topically applied atropine increased release.

A similar study has measured acetylcholine levels in the hippocampus after prolonged septal stimulation (Rommelspacher and Kuhar, 1974). An interesting finding was that acetylcholine levels do not change despite high rate of stimulation. This suggests that the synthetic apparatus in these neurons is capable of supplying acetylcholine for release even under very active conditions. Only when hemicholinium-3, a drug which blocks the supply of choline to the neuron (Simon *et al.*, 1975), was injected into the lateral ventricles was a rapid depletion of acetylcholine observed.

Besides being able to activate cholinergic neurons by stimulation, it is possible also to accomplish the opposite, i.e., stop impulse flow acute experiments by placing lesions in the septum. Since it is possible to both increase and decrease impulse flow, one can perform a wide variety of experiments. One type of experiment would be to examine the relationship of impulse flow and the action of various cholinergic drugs. For example, administration of hemicholinium-3 causes a gradual, dose-dependent, and reversible depletion of brain acetylcholine. By utilizing the techniques of stimulation and lesions, it can be shown that the action of hemicholinium-3 is impulse flow dependent (Rommelspacher *et al.*, 1974; Rommelspacher and Kuhar, 1974).

Thus the presence of a manipulable cholinergic tract in the brain provides investigators with a powerful tool. The utilization of this and other tracts will greatly expand our understanding of cholinergic mechanisms in the future.

7. References

Amano, T., Richelson, E., and Nirenberg, M. Neurotransmitter synthesis by neuroblastoma clones. *Proceedings of the National Academy of Sciences,* 1972, **69,** 258–263.

Biscoe, T. J., and Straughan, D. W. Micro-electrophoretic studies of neurones in the cat hippocampus. *Journal of Physiology (London),* 1966, **183,** 341–359.

Blackstadt, T. W., and Kjaerheim, A. Special axo-dendritic synapses in the hippocampal cortex: Electron and light microscopic studies on the layer of mossy fibers. *Journal of Comparative Neurology,* 1961, **117,** 133–146.

Browning, E. T. Fluorometric assay for choline and acetylcholine. *Analytical Biochemistry,* 1972, **46,** 624–638.

Burt, A. M. The histochemical demonstration of choline acetyltransferase activity in the spinal cord of the rat. *Anatomical Record,* 1969, **163,** 162.

Burt, A. M. A histochemical procedure for the localization of choline acetyltransferase activity. *Journal of Histochemistry and Cytochemistry,* 1970, **18,** 408–415.

Chao, L. P., and Wolfgram, F. Purification and some properties of choline acetyltransferase (E.C. 2.3.1.6) from bovine brain. *Journal of Neurochemistry,* 1973, **20,** 1075–1082.

Dahlström, A., and Fuxe, K. Evidence for the existence of monoamine-containing neurons in the central nervous system. I. Demonstration of monoamines in the cell bodies of brain stem neurons. *Acta Physiologica Scandinavica,* 1962, Supplement 1965, **232,** 1–55.

Eng, L. F., Uyeda, C. T., Chao, L. P., and Wolfgram, Antibody to bovine choline acetyltransferase and immunofluorescent localization of the enzyme in neurons. *Nature (London),* 1974, **250,** 243–245.

Falck, B., Hillarp, N., Thieme, G., and Torp, A. Fluorescence of catecholamines and related compounds condensed with formaldehyde. *Journal of Histochemistry and Cytochemistry,* 1962, **10,** 348–354.

Feigenson, M. E., and Saelens, J. K. An enzyme assay for acetylcholine. *Biochemical Pharmacology,* 1969, **18,** 1479–1486.

Fonnum, F. Topographical and subcellular localization of choline acetyltransferase in the rat hippocampal region. *Journal of Neurochemistry,* 1970, **17,** 1029–1037.

Goldberg, A. M., and McCaman, R. E. A quantitative microchemical study of choline acetyltransferase and acetylcholinesterase in the cerebellum of several species. *Life Sciences,* 1967, **6,** 1493–1500.

Goldberg, A. M., and McCaman, R. E. The determination of picomole amounts of acetylcholine in mammalian brain. *Journal of Neurochemistry,* 1973, **20,** 1–8.

Guerrero-Figueroa, R., Gonzalez, G., Barros, A., Guerrero-Figueroa, E., DeBalbian Verster, F., and Heath, R. G. Cholinergic synapes in the hippocampus and effects of topical application of GABA aminooxyacetic acid and acetylcholine on hippocampal epileptogenic tissues of the cat. *Acta Neurologica Latinoamericana,* 1965, **11,** 185–204.

Hanin, I., and Jenden, D. Estimation of choline esters in brain by a new gas chromatographic procedure. *Biochemical Pharmacology,* 1969, **18,** 837–845.

Hebb, C. CNS at the cellular level: Identity of transmitter agents. In V. E. Hall, A. C. Giese, and R. R. Sonnenschein (Eds.), *Annual Review of Physiology,* 1970, **32,** 165–192.

Hebb, C. Biosynthesis of acetylcholine in nervous tissue. *Pharmacological Reviews,* 1972, **52,** 918–947.

Jacobowitz, D., and Koelle, G. B. Histochemical correlations of acetylcholinesterase and catecholamines in postganglionic autonomic nerves of the cat, rabbit and guinea-pig. *Journal of Pharmacology and Experimental Therapeutics,* 1965, **148,** 225–237.

Kasa, P., and Morris, D. Inhibition of choline acetyltransferase and its histochemical localization. *Journal of Neurochemistry,* 1972, **19,** 1299–1304.

Kasa, P., Mann, S. P., and Hebb, C. O. Localization of choline acetyltransferase: Histochemistry at the light microscope level. *Nature (London),* 1970a, **226,** 812–814.

Kasa, P., Mann, S. P., and Hebb, C. O. Localization of choline acetyltransferase: Ultrastructural localization in spinal neurons. *Nature (London),* 1970b, **226,** 814–816.

Koelle, G. G., and Friedenwald, J. S. A histochemical method for localizing cholinesterase activity. *Proceedings of the Society for Experimental Biology and Medicine (New York),* 1949, **70,** 617–622.

KUHAR, M. J., AND ROMMELSPACHER, H. Acetylcholinesterase-staining synaptosome from rat hippocampus: Relative frequency and tentative estimation of internal concentration of free or "labile bound" acetylcholine. *Brain Research,* 1974, **77,** 85–96.

KUHAR, M. J., AND YAMAMURA, H. Cholinergic muscarinic receptors in rat brain: Light autoradiographic localization by the specific binding of a potent antagonist. *Nature (London),* 1975, **253,** 560–561.

KUHAR, M. J., SETHY, V. H., ROTH, R. H., AND AGHAJANIAN, G. K. Choline: Selective accumulation by central cholinergic neurons. *Journal of Neurochemistry,* 1973, **20,** 581–593.

LEWIS, P. R., AND SHUTE, C. C. D. The cholinergic limbic system: Brain, *Journal of Neurology,* 1967, **90,** 521–540.

LEWIS, P. R., SHUTE, C. C. D., AND SILVER, A. Confirmation from choline acetylase of a massive cholinergic innervation to the rat hippocampus. *Journal of Physiology (London),* 1967, **191,** 215–224.

LYNCH, G., MATTHEWS, D. A., MASKO, S., PARKS, T., AND COTMAN, C. Induced acetylcholinesterase-rich layer in rat dentate gyrus following entorhinal lesions. *Brain Research,* 1972, **42,** 311–318.

LYNCH, G. S., MASKO, S., PARKS, T., AND COTMAN, C. W. Relocation and hyperdevelopment of the dentate gyrus commissural system after entorhinal lesions in immature rats. *Brain Research,* 1973, **50,** 174–178.

McGEER, E. G., WADA, J. A., TERAO, A., AND JUNG, E. Amine synthesis in various brain regions with caudate or septal lesions. *Experimental Neurology,* 1969, **24,** 277–284.

McGEER, P. L., McGEER, E. G., SINGH, V. K., AND CHASE, W. H. Choline acetyltransferase localization in the central nervous system by immunohistochemistry. *Brain Research,* 1974, **81,** 373–379.

PEPEU, G., MULAS, A., RUFFI, A., AND SOTGIU, P. Brain acetylcholine levels in rats with septal lesions. *Life Sciences,* 1971, **10,** 181–184.

PHILLIS, J. W. The metablolism of acetylcholine. In *The pharmacology of synapses.* London: Pergamon Press, 1970*a,* pp. 8–37.

PHILLIS, J. W. Pharmacological studies on neurones in the brain and spinal cord. Part I. Cholinergic mechanisms. In *The pharmacology of synapses.* London: Pergamon Press, 1970*b,* pp. 149–185.

REID, W. R., HAUBRICH, O. R., AND KRISHNA, G. Enzymic radioassay for acetylcholine and choline in brain. *Analytical Biochemistry,* 1971, **42,** 390–395.

ROMMELSPACHER, H., AND KUHAR, M. J. Effects of electrical stimulation on acetylcholine levels in central cholinergic nerve terminals. *Brain Research,* 1974, **81,** 243–251.

ROMMELSPACHER, H., GOLDBERG, A. M., AND KUHAR, M. J. Action of hemicholinium on cholinergic nerve terminals after alteration of neuronal impulse flow. *Neuropharmacology,* 1974, **13,** 1015–1023.

SCHMIDT, D. E., SZILAGYI, P. I. A., ALKON, D. L., AND GREEN, J. P. A method for measuring nanogram quantities of acetylcholine by pyrolysis–gas chromatography. *Journal of Pharmacology and Experimental Therapeutics,* 1970, **174,** 337–345.

SETHY, V., ROTH, R. H., KUHAR, M. J., AND VAN WOERT, M. H. Choline and acetylcholine: Regional distribution and effect of degeneration of cholinergic nerve terminals in the rat hippocampus. *Neuropharmacology,* 1973, **12,** 819–824.

SHEA, P. A., AND APRISON, M. H. Levels of acetylcholine and choline in rat brain as determined by a sensitive two-enzyme system. *Federation Proceedings,* 1973, **32,** 430.

SHUTE, C. C. D., AND LEWIS, P. R. Electron microscopy of cholinergic terminals and acetylcholinesterase-containing neurons in hippocampal formation of the rat. *Zeitschrift für Zellforschung and Mikroskopische Anatomie,* 1966, **69,** 334–343.

SHUTE, C. C. D., AND LEWIS, P. R. The ascending cholinergic reticular system. *Brain; Journal of Neurology,* 1967, **90,** 497–519.

SIMON, J. R., MITTAG, T. AND KUHAR, M. J. The inhibition of synaptosomal uptake of choline by various choline analogs. *Biochemical Pharmacology,* 1975, **24,** 1139–1142.

SMITH, C. M. The release of acetylcholine from the rabbit hippocampus. *British Journal of Pharmacology,* 1972, **45,** 172.

SREBRO, B., ODERFELD-NOWAK, B., KLODOS, I., DABROWSKA, J., AND NARKIEWICZ, O. Changes in acetylcholinesterase activity in hippocampus produced by septal lesions in the rat. *Life Sciences,* 1973, **12,** 261–270.

STEINER, F. A. Influence of microelectrophoretically applied acetylcholine on responsiveness of hippocampal and lateral geniculate neurons. *Pfluegers Archiv; European Journal of Physiology,* 1968, **303,** 173–180.

STORM-MATHISEN, J. Quantitative histochemistry of acetylcholinesterase in rat hippocampal region correlated to histochemical staining. *Journal of Neurochemistry,* 1970, **17,** 739–750.

STORM-MATHISEN, J., AND FONNUM, F. Neurotransmitter synthesis in excitory and inhibitory synapses of rat hippocampus. In R. Paoletti, R. Fumagalli, and C. Galli (Eds.), *Second International Meeting of the International Society for Neurochemistry.* Milan: Tamburini, 1969, p. 382.

UNGERSTEDT, U. Stereotoxic mapping of the monoamine pathways in the rat brain. *Acta Physiologica Scandinavica Supplement,* 1971, **267,** 1–48.

YAMAMURA, H., AND SNYDER, S. H. Choline: High-affinity uptake by rat brain synaptosomes. *Science,* 1972, **178,** 626–628.

YAMAMURA, H. I., KUHAR, M. J., AND SNYDER, S. H. *In Vivo* identification of muscarinic cholinergic receptor binding in rat brain. *Brain Research,* 1974, **80,** 170–176.

11

Putative Glucocorticoid Receptors in Hippocampus and Other Regions of the Rat Brain

BRUCE S. MCEWEN, JOHN L. GERLACH, AND DAVID J. MICCO

1. Introduction

It has been known for some time from neuroendocrine and behavioral studies that the brain is both a master controller of endocrine function and a target for these endocrine secretions. There exist complex feedback interactions between endocrine secretions and the brain which not only control the secretion of hypothalamic and pituitary hormones but also influence neural activity underlying behavior. It is only quite recently that we have come to appreciate the role of the entire limbic brain, and not just the hypothalamus, in these endocrine–brain interactions.

Our own involvement in this relevation arose from studies of the fate of injected radioactive adrenal steroids, particularly corticosterone, when they entered the brain from the blood. These studies were begun, under the impetus of recent advances in molecular biology of steroid hormone action, to look for intracellular hormone receptors in brain tissue. We expected to find such putative receptors in the hypothalamus, where effects of adrenal steroids on ACTH secretion have been demonstrated (Davidson *et al.,* 1968; Grimm and Kendall, 1968). Much to our surprise, the brain region which binds the most corticosterone is not the hypothalamus but the hippocampus. It is the purpose of this chapter to review our studies in the context of recent

BRUCE S. MCEWEN, JOHN L. GERLACH, AND DAVID J. MICCO • The Rockerfeller University, New York, N.Y.

advances in the molecular aspects of steroid hormone action and in relation to the known neuroendocrine and behavioral effects of glucocorticoids and gonadal steroids.

2.   Cellular Mechanisms of Hormone Action

Hormones are chemical messengers which trigger chemical changes in receptive cells. Such cells contain receptor sites which recognize the chemical structure of the hormone and initiate the cellular response. Steroid hormones enter the target cells and bind to specific proteins located, probably, in the cytoplasm. The hormone next moves into the cell nucleus, possibly still in combination with the putative receptor protein, and there initiates changes in genomic function, which in turn modify the physiological characteristics of the target cell through altered RNA and protein formation (see Fig. 1A). These changes include alterations in RNA synthesis and specific messenger formation (McGuire and O'Malley, 1968; Comstock *et al.*, 1972; Palacios *et al.*, 1973) and changes in amino acid incorporation and specific protein formation (O'Malley *et al.*, 1967; Gorski *et al.*, 1971), including the induction of enzymes characteristic of that particular cell type (Baxter and Tomkins, 1970; de Vellis *et al.*, 1971). That the intracellular hormone binding sites do appear to be actual receptors for these actions is based on several types of evidence: (1) the binding sites are found only in tissues where the hormones have an effect (e.g., uterus for estradiol); (2) the steroids which bind best to the binding sites are those which are most effective either as agonists or antagonists (Swaneck *et al.*, 1969; Alberti and Sharp, 1970; Korenman, 1970; Rousseau *et al.*, 1972; Katzenellenbogen and Katzenellenbogen, 1973); (3) in

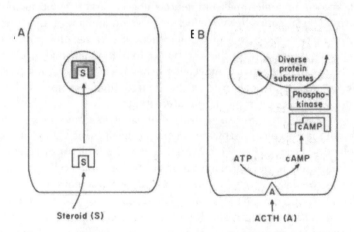

Fig. 1. Schematic diagram of hormone action at the cellular level. A: Entry of steroid hormone (S) into target cell, binding to an intracellular "receptor," and transfer to the cell nucleus. B: Attachment of amino acid or polypeptide hormone, illustrated by ACTH (A), to cell surface receptor, triggering intracellular formation of adenosine 3′,5′-monophosphate (cAMP). From *Frontiers in Neuroendocrinology 1973*, edited by W. F. Ganong and L. Martini. Copyright 1973 by Oxford University Press, Inc. Reprinted by permission. (See McEwen and Pfaff, 1973.)

certain conditions (e.g., glucocorticoid-insensitive lymphoma cells) where a target tissue is insensitive to the usual hormone, the intracellular binding sites are grossly deficient (Bullock *et al.*, 1971; Shyamala, 1972; Rosenau *et al.*, 1972).

In contrast to steroid hormones, amino acid and polypeptide hormones, such as epinephrine and ACTH, appear to interact with "receptors" located at the surface of the target cells, and in so doing initiate the conversion of ATP to adenosine 3',5'-monophosphate (cAMP) within the target cell (Fig. 1B). cAMP, formed as a result of hormone–receptor interaction, is responsible for a variety of changes within the target cells, including facilitation of glycogen breakdown, stimulation of protein biosynthesis, and alterations in genomic activity (Robison *et al.*, 1971). In certain cases (e.g., the induction of liver tyrosine aminotransferase) the effects of cAMP-mediated hormone action and adrenal steroids are similar, but these effects have been shown to be additive or synergistic and probably proceed by different mechanisms (Wicks *et al.*, 1969). In the glioma cell in tissue culture (de Vellis *et al.*, 1971), cAMP-mediated induction of lactic dehydrogenase by epinephrine and glucocorticoid induction of glycerolphosphate dehydrogenase proceed independently of one another, although both appear to be mediated at the genomic level. It thus appears that steroid hormone action does not necessarily involve cyclic AMP mediation.

3. Studies of Putative Estrogen Receptors in Brain

During the past decade, hormone binding sites have been demonstrated in a wide variety of target cells by means of *in vivo* and *in vitro* studies with radioactive hormones of high specific activity (for reviews, see Jensen and Jacobson, 1962; Gorski *et al.*, 1968; Liao and Fang, 1969; Swaneck *et al.*, 1969; Wilson and Glyona, 1970; Jensen and DeSombre, 1971; Litwack and Singer, 1972; O'Malley *et al.*, 1972). These studies, which began with classical endocrine target tissues such as uterus, prostate, kidney, and liver, are presently being extended to include the central nervous system. Interest in hormone receptors in brain has been stimulated by the recognition of direct effects of hormones on the brain and the study of these effects by hormone implantation in specific brain regions (Lisk, 1967; Johnston and Davidson, 1972). Many steroid hormones have been shown to enter the brain from the blood (for reviews, see McEwen *et al.*, 1972*b*; McEwen and Pfaff, 1973). Estradiol is probably the most extensively investigated of all the hormones. Scintillation counting of the regional distribution of [3H]estradiol in rat brain established that the hypothalamus and preoptic area, besides pituitary, concentrate the hormone more than any other brain region (Eisenfeld and Axelrod, 1965; Kato and Villee, 1967; McEwen and Pfaff, 1970). Cell fractionation studies demonstrated that these structures concentrate [^{3}H]estradiol at the level of the cell nucleus (Chader and Villee, 1970; Kato *et al.*, 1970*b*; Zigmond and McEwen, 1970; Vertes and King, 1971; Mowles *et al.*, 1971) and that the amygdala also shows cell nuclear concentration of [^{3}H]estradiol (Zigmond and McEwen, 1970). In addition, soluble "receptor" proteins for estradiol have been identified in these same brain regions (Kahwanango *et al.*, 1969; Eisenfeld, 1970; Kato *et al.*, 1970*a*; Notides, 1970). Their function is presumably to transfer hormone to the cell nucleus. Autoradiographic evidence supports the regional and subcellular distribution of estradiol binding sites and extends it by showing that neurons are the primary binding sites (Pfaff, 1968; Stumpf, 1968, 1970; Anderson and Greenwald, 1969; Pfaff and Keiner, 1972). Autoradiography has also provided a more detailed map of the cell groupings within the preoptic area, hypothalamus, midbrain, and amygdala (Pfaff and Keiner, 1972, 1973), which can be correlated with known neural pathways (Pfaff *et al.*, 1973; McEwen and Pfaff, 1973). The regional localization of estradiol binding agrees well with the sites whre implants of estradiol are effective in influencing gonadotropin secretion, lordosis behavior (Lisk, 1967), locomotor activity, and food intake (Wade and Zucker, 1970). This agreement has led to the concepts that the brain is regionally differentiated with respect to steroid hormone action and that the estradiol binding sites in cell nuclei function as receptors. In this connection, it is worthwhile noting that facilitation of lordosis behavior by estradiol in the rat requires 16–24 h to develop (Green *et al.*, 1970), suggesting an induction period is required for the biochemical and neurological effects of estradiol to manifest themselves. The induction of lordosis behavior (Quadagno *et al.*, 1971) and of increased locomotor activity (Stern and Jankowiak, 1972) is attenuated by intracranially applied actinomycin D, an inhibitor of RNA synthesis.

4. *Glucocorticoid Interaction with the Rat Brain: Preferential Uptake by Hippocampus*

Corticosterone, the principal glucocorticoid of the rat (Bush, 1962), is secreted in response to ACTH secreted by the pituitary. A growing body of evidence (see below) indicates that this hormone acts on the brain not only to influence ACTH secretion but also to affect behavior. When injected systemically, radioactive corticosterone, in contrast to estradiol, does not localize preferentially in the diencephalon of the rat brain, but rather in the hippocampus (McEwen *et al.*, 1969, 1972*b*; Knizely, 1972). This surprising finding is illustrated in Fig. 2. Tissue uptake studies revealed that both hippocampus and septum of adrenalectomized rats (see Fig. 2a) concentrate systemically injected [^{3}H]corticosterone from blood (Fig. 2b). Hippocampal radioactivity is distributed more or less uniformly along the length of hippocampus (Fig. 2b). However, the cellular organization of the hippocampus is similar along its length but changes laterally, in which direction are found the architectonic fields of Ammon's horn (Lorente de Nó, 1933, 1934) and the dentate gyrus (Fig. 2c). Indeed, as revealed by autoradiography, the primary differentiation of [^{3}H]corticosterone binding occurs within these fields.

Intense concentration of radioactivity due to [^{3}H]corticosterone is associated with pyramidal neurons in zones CA1 and CA2 of Ammon's horn. Slightly less intense radioactivity is seen over neuron cell bodies in CA3 and CA4 of Ammon's horn and over scattered granule neurons of the dentate gyrus. This distribution is shown in Fig. 3, which is an unstained autoradiogram in which the only contrast is provided by the concentration of reduced silver grains under neuron cell bodies. At a higher magnification, Fig. 4 depicts this intense concentration of [^{3}H]corticosterone in pyramidal

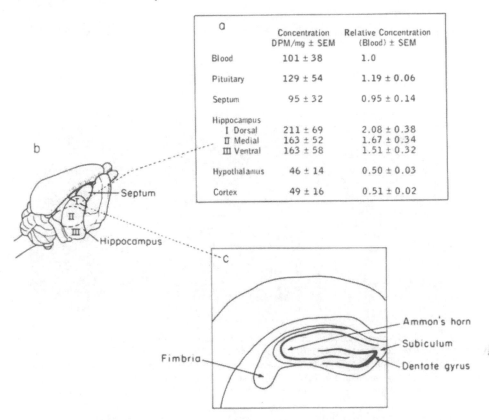

a	Concentration DPM/mg ± SEM	Relative Concentration (Blood) ± SEM
Blood	101 ± 38	1.0
Pituitary	129 ± 54	1.19 ± 0.06
Septum	95 ± 32	0.95 ± 0.14
Hippocampus		
I Dorsal	211 ± 69	2.08 ± 0.38
II Medial	163 ± 52	1.67 ± 0.34
III Ventral	163 ± 58	1.51 ± 0.32
Hypothalamus	46 ± 14	0.50 ± 0.03
Cortex	49 ± 16	0.51 ± 0.02

Fig. 2. Hippocampal structure in relation to [³H]corticosterone uptake. (a) [³H]Corticosterone uptake in dorsal, medial, and ventral hippocampus compared with other brain regions (see McEwen *et al.*, 1969). (b) Position of hippocampus in the rat brain showing dorsal (I), medial (II), and ventral (III) portions. (c) Cross-section of dorsal hippocampus showing cellular layers. An autoradiogram of [³H]corticosterone uptake in a similar cross-section is presented in Fig. 3. From *Structure and Function of Nervous Tissue*, Vol. 5, edited by G. H. Bourne. Copyright 1972 by Academic Press, Inc. Reprinted by permission. (See McEwen *et al.*, 1972*b*.)

neurons of the CA2 zone of an unstained autoradiogram. Whereas neuron cell bodies are distinctly labeled in hippocampus and other brain regions, there is also radioactivity associated with adjacent neuropil. At present, we are unable to determine the cellular localization of neuropil labeling, and it may indeed represent the labeling of glial cells.

The ability of the hippocampus to concentrate corticosterone has been observed in the hamster using cell fractionation (Kelley and McEwen, unpublished observations) and in the Pekin duck using autoradiography (Rhees *et al.*, 1972). The rhesus monkey is another species in which the hippocampus binds corticosterone. As in the rat hippocampus, this binding is detected by autoradiography (Fig. 5) to be greatest in the pyramidal neurons of Ammon's horn (Gerlach *et al.*, 1974). While other species have

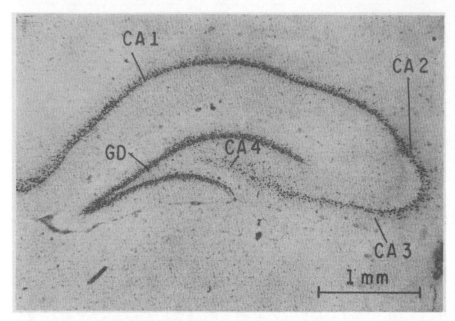

Fig. 3. Unstained autoradiogram of the dorsal hippocampus of an adrenalectomized male rat showing uptake of systemically injected [³H]corticosterone as black silver grains. All contrast is due entirely to silver grains, which delineate the hippocampus anatomically. A sagittal frozen section was exposed for 608 days. Shown are the longitudinal fields CA1 to CA4 of the pyramidal neuron layer in Ammon's horn, and the granule neuron layer in the dentate gyrus (GD). See Gerlach and McEwen (1972) for procedure.

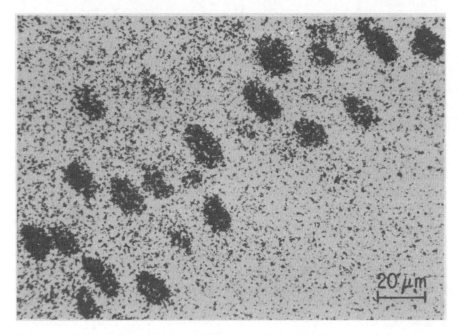

Fig. 4. Clusters of silver grains in this unstained autoradiogram depict a group of CA2 pyramidal neurons which concentrated [³H]corticosterone. See Fig. 3 for details.

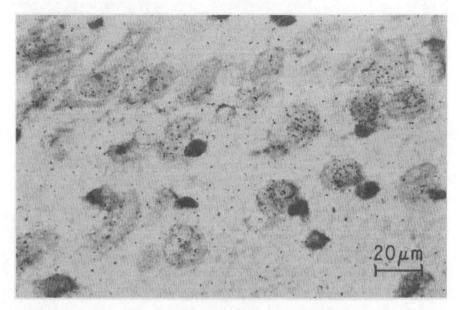

Fig. 5. This rhesus monkey hippocampus autoradiogram shows, for the first time, uptake of radioactivity by pyramidal neurons of Ammon's horn, 140 min after an intravenous perfusion of 0.6 mCi of 1,2,6,7-[^{3}H]corticosterone (84 Ci/mM) in an adrenalectomized adult female (body weight 3.8 kg). Ten minutes after the perfusion, the radioactivity in the blood fell from a peak and maintained a high and constant level while measured over the next 90 min. Following an exposure for 121 days, the greatest concentration of silver grains appears beneath pyramidal neurons, while a lesser concentration lies beneath granule neurons of the dentate gyrus (not shown) Pyronin Y stain. See Gerlach *et al.* (1974).

not been studied, the information so far suggests that the ability of the hippocampus to bind corticosterone may be general in vertebrates.

While radioactivity associated with neurons in rat hippocampus is greater than that observed in any other brain region thus far examined, other brain regions do contain neurons which concentrate radioactivity injected as [^{3}H]corticosterone. Labeling in some of these areas is shown in Fig. 6. These areas include the anterior hippocampus (Fig. 6a), the induseum griseum (Fig. 6b), the lateral septum (Fig. 6d; see also Fig. 2a), the neocortex (Fig. 6e), and the medial (Fig. 6g) and cortical (Fig. 6h) regions of the amygdala.

5. Subcellular Localizations of [^{3}H]Corticosterone Binding Sites in Hippocampus

Both cell fractionation and autoradiographic studies revealed the subcellular sites of [^{3}H]corticosterone accumulation in the hippocampus. One hour after the injection of [^{3}H]corticosterone into an adrenalectomized rat, 20% or more of the radioactivity in the hippocampus was associated with the cell nuclei, while approximately 60% was present in the soluble, cytosol fraction obtained by high-speed centrifugation of a homogenate (McEwen *et al.*, 1970a). Of this cytosol radioactivity, around 40–50%

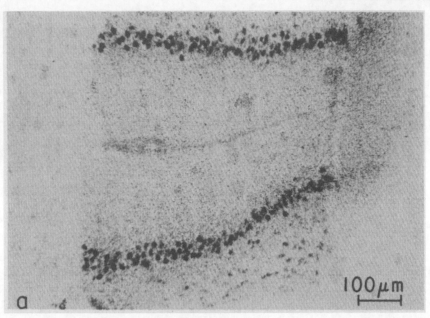

FIG. 6. Localization of systemically injected [³H]corticosterone bound to cells in selected anatomical regions in brains of adrenalectomized male rats. Autoradiograms (a,b,d,e,g,h) show cells which bind the hormone. Diagrams of coronal sections of rat brain (c,f,i), which are modified from König and Klippel (1963), show structures, most of which bind [³H]corticosterone, as illustrated by autoradiograms in this chapter. Methyl green–pyronin Y stain. (a) HIA, hippocampus pars anterior (exposed 608 days).

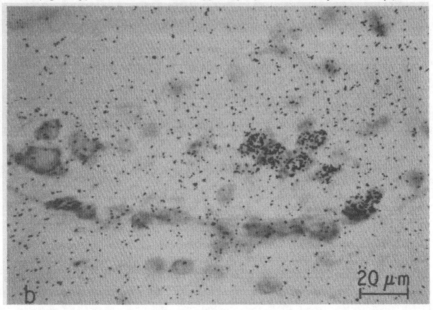

FIG. 6(b). IG, Induseum griseum (exposed 235 days).

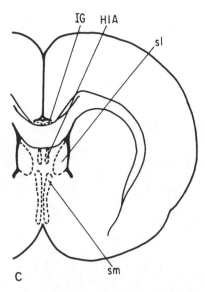

IG   HIA

sl

C   sm

Fɪɢ. 6(c). In addition to the structures shown in (a) and (b), (c) includes sl, nucleus septi lateralis; sm, nucleus septi medialis.

was associated with a macromolecular fraction obtained by passing the cytosol through a column of Sephadex G25 (McEwen *et al.*, 1972*a*). Approximately 70% of the cell nuclear radioactivity could be extracted by 0.4 M NaC1, and approximately 40% of this was bound to macromolecules after passage through a Sephadex G200 column (McEwen and Plapinger, 1970). Both cell nucleus and cytosol binding of [³H]corticosterone was subject to competition by injection of unlabeled corticosterone, indicating that these sites have limited capacity for the hormone (McEwen *et al.*, 1970*a*, 1971; Grosser *et al.*, 1971). Autoradiographic data (Gerlach and McEwen, 1972) correspond to the results of cell fractionation experiments by revealing a marked concentration of [³H]corticosterone over the region of the cell nucleus in hippocampal neurons (Fig. 7c,d). Autoradiography does not permit quantitative estimation or even identification of soluble "cytosol" sites of corticosterone binding, and it is unclear what this cellular origin is, although it is generally assumed that such cytosol sites are cytoplasmic in origin.

6.   Soluble Glucocorticoid Binding Proteins in Brain and Other Target Tissues

The properties of these cytosol binding sites for corticosterone indicate that they are typical of glucocorticoid binding sites in other tissues and distinctly different from the serum corticosterone binding factor, transcortin. For example, the brain protein is present after perfusion of the brain at sacrific with Dextran–saline solution (McEwen *et al.*, 1972*a*). Moreover, it is precipitated by protamine sulfate, a polycationic protein,

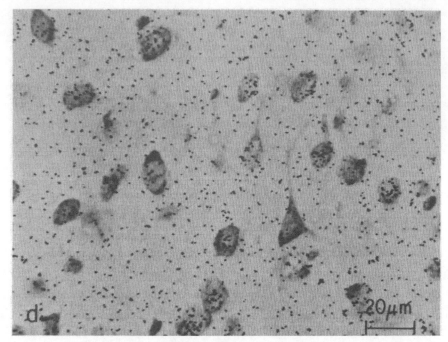

FIG. 6(d). sl, Nucleus septi lateralis (exposed 235 days).

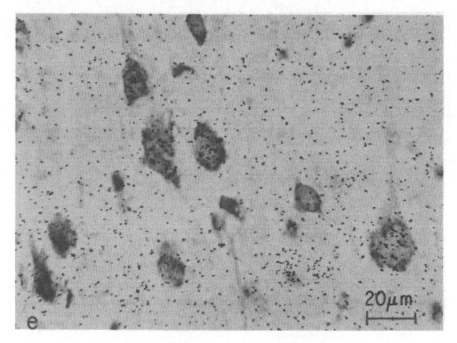

FIG. 6(e). NCX, neocortex (exposed 248 days).

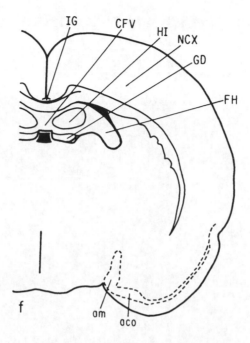

Fig. 6(f). In addition to the structures shown in (d) and (e), (f) includes aco, nucleus amygdaloideus corticalis; am, nucleus amygdaloideus medialis; CFV, commissura fornicus ventralis (commissura hippocampi ventralis); FH, fimbria hippocampi; GD, gyrus dentatus; HI, hippocampus; IG, induseum griseum.

under conditions in which very little of the serum binding protein is precipitated (Fig. 8), even when the serum and cytosol extracts are mixed together before addition of protamine sulfate (McEwen *et al.*, 1972*a*). The brain binding protein migrates differently from the serum binding proteins in glycerol density gradients and can be separated completely from the serum binding protein on polyacrylamide gels containing glycerol McEwen *et al.*, 1972*a*) (Fig. 9). Finally, brain cytosol protein binds dexamethasone (Chytil and Toft, 1972; McEwen and Wallach, 1973), while transcortin, as is well known, does not bind this steroid (Florini and Buyske, 1961; Peets *et al.*, 1969). Brain corticosteroid binding protein also binds another potent synthetic glucocorticoid, triamcinolone acetonide (Chytil and Toft, 1972).

Brain corticosterone binding proteins appear similar to those found in other target tissues, such as liver (Beatto and Feigelson, 1972), liver hepatoma cells in tissue culture (Rousseau *et al.*, 1972), thymus (Schaumburg, 1972*a*; Abraham and Sekevis, 1973; Kaiser *et al.*, 1973), and kidney (Funder *et al.*, 1973; Feldman *et al.*, 1973). Thymus contains a soluble corticosterone protein which is labile even at 0–4°C but can be stabilized by glycerol and temperatures below 0°C (Schaumburg, 1972*a*). This protein contains sulfhydryl groups essential to hormone binding and has high affinity for glucocorticoids, including corticosterone and dexamethasone, but is also able to bind mineralocorticoids and progesterone (Schaumburg, 1972*a*). The brain protein shares

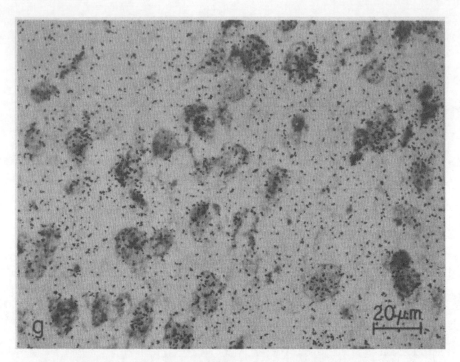

FIG. 6(g). am, Nucleus amygdaloideus medialis (exposed 248 days).

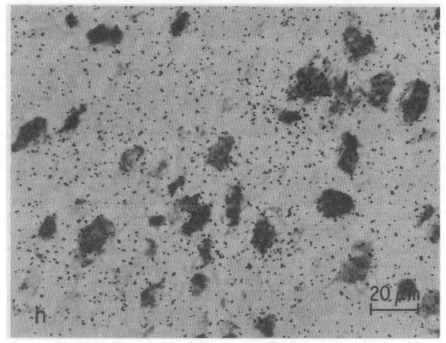

FIG. 6(h). aco, Nucleus amygdaloideus corticalis (exposed 248 days).

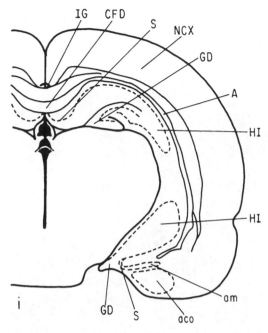

Fig. 6(i). In addition to the structures shown in (g) and (h), (i) includes A, alveus hippocampi; CFD, commissura fornicus dorsalis (commissura hippocampi dorsalis); GD, gyrus dentatus; HI, hippocampus; IG, induseum griseum; NCX, neocortex; S, subiculum.

with that from thymus those properties of lability, essential sulfhydryl groups, and broad specificity for C-21 steroids (McEwen *et al.*, 1972a; McEwen and Wallach, 1973; Grosser *et al.*, 1973). An interesting aspect of specificity is the binding to these cytosol proteins of the synthetic 17-hydroxysteroid triamcinolone acetonide. This steroid associates very slowly with the protein at 0–4°C, more slowly than other glucocorticoids, and apparently results in a more stable complex since surcrose density gradient profiles are more stable than those with corticosterone (Schaumburg, 1972a,b; Kaiser *et al.*, 1973). Corticosteroid binding sites having properties in common with the thymus, liver, and brain binding proteins have been reported in chick embryo neural retina (Chader and Reif-Lehrer, 1972), fetal lung (Ballard and Ballard, 1972), and placenta (Wong and Burton, 1973), as well as a number of tissue culture lines (Hackney and Pratt, 1971; Tucker *et al.*, 1971). It is conceivable that all of these tissue binding proteins may be related, if not even identical, to one another for a given species. Proof of such a situation, however, requires the isolation and chemical or immunochemcial study of the various tissue binding proteins.

7.   Study of Cellular Mechanism of [^{3}H]Corticosterone Entry into Hippocampal Cell Nuclei

We were able to demonstrate [^{3}H]corticosterone binding to hippocampal nuclei *in vitro* by incubating 300-μm slices of hippocampal tissue with labeled hormone in a

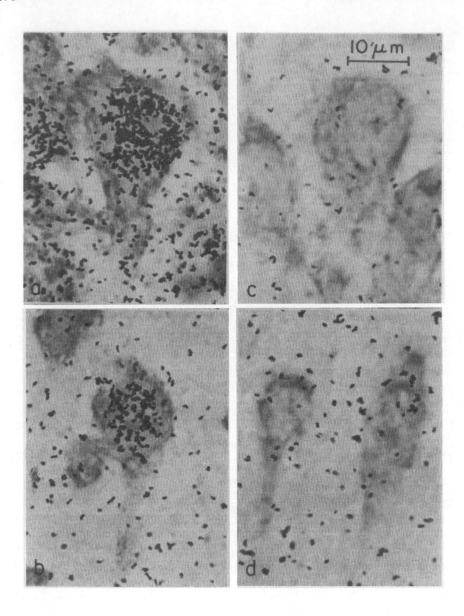

FIG. 7. Autoradiograms of [³H]corticosterone uptake by hippocampal neurons of adrenalectomized male rats. For details, see Gerlach and McEwen (1972). (a,b) Control uptake of [³H]corticosterone (exposed 248 days). (c,d) Uptake of [³H]corticosterone in presence of 3 mg unlabeled corticosterone injected to compete for binding sites (exposed 317 days). From *Frontiers in Neuroendocrinology 1973,* edited by W. F. Ganong and L. Martini. Copyright 1973 by Oxford University Press, Inc. Reprinted by permission. (See McEwen and Pfaff, 1973.)

Krebs–Ringer bicarbonate buffer (McEwen and Wallach, 1973). This technique circumvents the expense and experimental limitations of *in vivo* uptake studies and permits studies of the intracellular mechanism of hormone uptake. *In vitro* uptake into the cell nuclei is optimal at around 25°C, being lower at both 0° and 37°C. Nuclear uptake of hormone appeared to be independent of ATP energy supplied via either glycolysis or the respiratory chain. At 25°C, nuclear uptake reaches its maximum level when slices are incubated for 30 min with [^{3}H]corticosterone or [^{3}H]cortisol (Fig. 10). Nuclear uptake has a definite saturation point, while tissue uptake does not in the concentration range studied (Fig. 11). Nuclear uptake *in vitro* is highest in hippocampal tissue slices, lower in amygdala, still lower in cerebral cortex, and follows for these and other structures the same rank ordering obtained *in vivo* uptake experiments. This is summarized in Table 1. This result establishes that corticosterone uptake into the nucleus is an inherent property of the various brain regions and not the consequence of differences in blood flow, which might be suspected of playing a role in *in vivo* uptake studies.

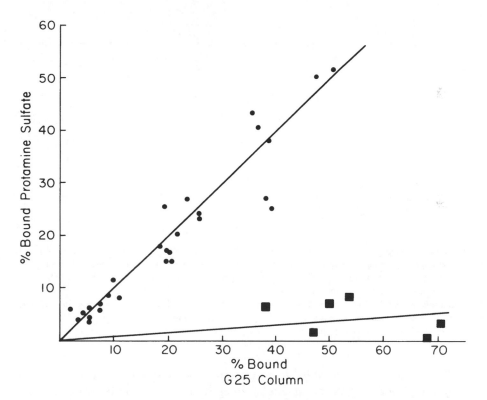

FIG. 8. Relationship between bound radioactivity in cytosol from perfused whole brain and bound radioactivity in serum. Abscissa, binding determined by gel filtration on Sephadex G25 columns. Ordinate, binding determined by precipitation with protamine sulfate. Circles, cytosol; squares, serum. For details, see McEwen *et al.* (1972*a*). Copyright by The Endocrine Society. Reprinted by permission.

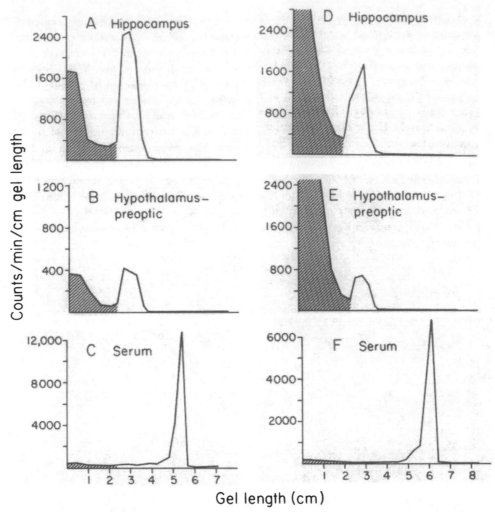

FIG. 9. Polyacrylamide gel separations in presence of 10% glycerol of binding proteins in brain cytosol and serum. Cytosol samples were prepared so as to have the same protein concentration for each brain region. A: Hippocampus, labeled *in vivo*. B: Hypothalamus–preoptic area, labeled *in vivo*. C: Serum, labeled *in vivo*. D: Hippocampus, labeled *in vitro* for 1 h with 10^{-9} M [^{3}H]corticosterone. E: Hypothalamus–preoptic area, labeled *in vitro* as in D. F: Serum, labeled *in vitro* as in D. Shaded zone indicates 4% polyacrylamide stacking gel; unshaded area 10% polyacrylamide gel. For details, see McEwen *et al.* (1972a). Copyright 1972 by The Endocrine Society. Reprinted by permission.

The specificity of uptake of [^{3}H]corticosterone was tested in hippocampal slices *in vitro* by comparing it with the uptake of other ^{3}H-labeled steroids (McEwen and Wallach, 1973). Structures of some of these steroids are shown in Fig. 12. Figure 13 (top) compares the uptake of [^{3}H]corticosterone, dexamethasone, cortisol, progesterone, and estradiol, each at a final concentration of 1×10^{-8} M, and expresses the result as femtomoles of steroid accumulated per milligram of protein. Uptake by hippocampal tissue (open bar) is inversely related to the polarity of the steroid, being highest for the least polar steroid, progesterone, and lowest for the most polar, cortisol. Cell nuclear uptake of these five steroids follows a different pattern, with corticosterone highest, followed by dexamethasone, cortisol, progesterone, and estradiol [Fig. 13 (top), black bar]. The relationship between cell nuclear uptake of the five steroids is made clearer when the nuclear uptake is expressed relative to uptake in the tissue [N/WH; Fig. 13 (bottom)]. The N/WH ratios for the three glucocorticoids, corticosterone, dexamethasone, and cortisol, are all much higher than the ratios for progesterone and estradiol.

While some radioactive progesterone does end up in the cell nucleus, this uptake is not saturable by unlabeled progesterone. However, the soluble corticosteroid binding

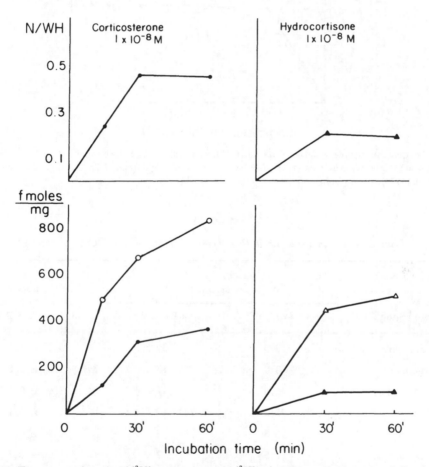

Fig. 10. Time course of uptake of [^{3}H]corticosterone and [^{3}H]hydrocortisone by hippocampal slices (open circles) and cell nuclei (solid circles) isolated from the slices. Uptake is expressed as femtomoles steroid/mg protein. Triangles, ratio of nuclear to whole homogenate concentration of radioactivity (N/WH). Incubation at 25°C at indicated steroid concentration and time. From McEwen and Wallach (1973). Reprinted by permission.

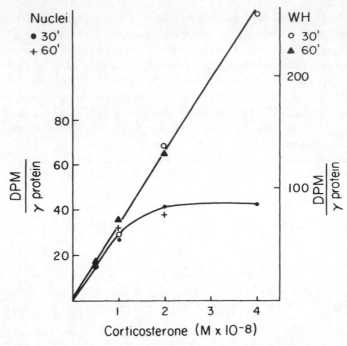

FIG. 11. Concentration dependence of uptake of [³H]corticosterone by hippocampal cell nuclei and slices from which nuclei were isolated (WH). Incubation for 30 min or 60 min at 25°C. From McEwen and Wallach (1973). Reprinted by permission.

TABLE 1

Comparison of *in Vitro* and *in Vivo* Regional Uptake of [³H]Corticosterone[a]

Brain region	*In vitro* nuclei[b] (dpm/μg protein)		*In vivo* nuclei[c] (dpm/μg protein) average ± SEM
	Expt. 1	Expt. 2	
Hippocampus	19.5	30.3	19.8 ± 4.4
Amygdala	8.8	17.3	5.6 ± 1.2
Cerebral cortex	6.3	11.9	2.4 ± 0.9
HPOA	5.2	10.5	1.8 ± 0.4
Midbrain	5.8	9.5	1.4 ± 0.4
Cerebellum	3.8	7.0	1.0 ± 0.3

[a] This table originally appeared in McEwen and Wallach (1973).
[b] 300-μm slices incubated at 25°C for 30 min in the presence of 2×10^{-8} M [³H]corticosterone.
[c] Data from McEwen *et al.* (1970b).

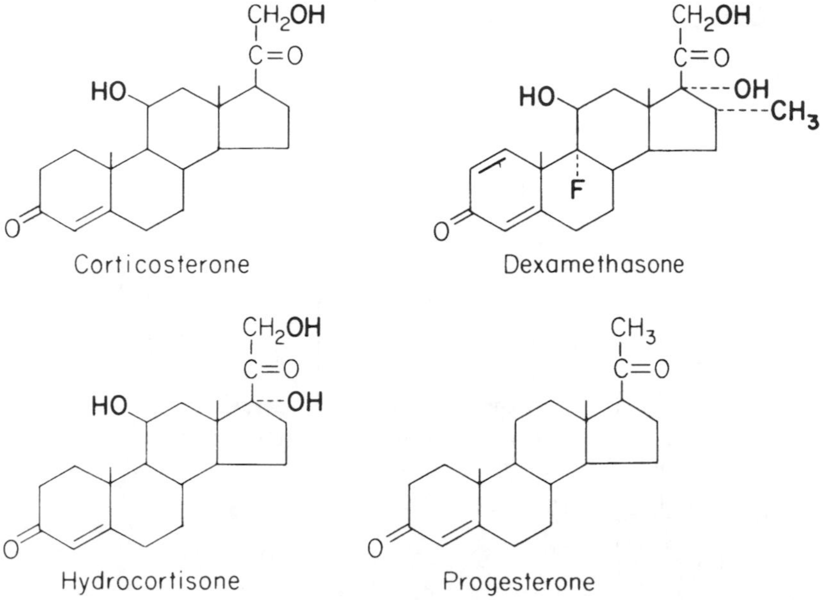

FIG. 12. Structures of some steroids used in study of binding specificity of hippocampal cell nuclei.

protein in hippocampus does have limited-capacity binding sites for progesterone (McEwen and Wallach, 1973). We used this characteristic of progesterone binding to test the role of cytosol binding of corticosterone in the transfer of [³H]corticosterone into the cell nuclei. We found that unlabeled progesterone, although not as effective as corticosterone or cortisol, is able to reduce binding of [³H]corticosterone into hippocampal cell nuclei (Fig. 14). It thus appears likely that cytosol binding of [³H]corticosterone is a step in the transfer of the hormone to cell nuclear binding sites and that when cytosol sites are occupied by progesterone, transfer to the cell nucleus does not occur. Progesterone has been shown to block induction by glucocorticoids of various enzymes in chick neural retina (Chader and Reif-Lehrer, 1972), hepatoma tissue culture cells (Rousseau et al., 1972), and thymus (Makman et al., 1967), and this may be a reflection of its ability to block entry of hormone into cell nuclei.

8. Factors Affecting Amount of Glucocorticoid Binding Protein in Hippocampus

Virtually all of the experiments demonstrating [³H]corticosterone binding sites in rat brain were carried out on animals adrenalectomized at least 3 days previously. It has been repeatedly demonstrated that injected [³H]corticosterone is not extensively bound in animals with intact adrenals, and this was interpreted as indicating competition for binding sites by endogenous corticosterone (McEwen et al., 1969, 1970a;

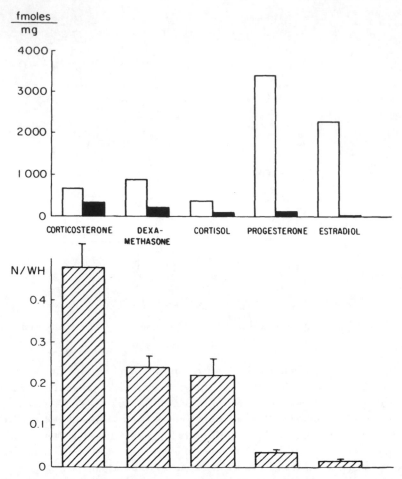

FIG. 13. Top: Uptake of various [³H]steroids by hippocampal cell nuclei (black bar) and slices from which nuclei were isolated (open bar) expressed as femtomoles of steroid/mg protein. Steroid concentration 1 × 10⁻⁸ M. Bottom: Ratio of nuclear to slice concentration (N/WH). Standard error of the mean is indicated by line. For [³H]corticosterone and [³H]progesterone, number of determinations was ten; for [³H]cortisol (hydrocortisone), [³H]dexamethasone, and [³H]estradiol, number of determinations was six. Incubation 25°C for 30 min. Some of data and procedural details may be found in McEwen and Wallach (1973).

Grosser *et al.*, 1971). Some hippocampal nuclear and cytosol binding could be detected in normal animals, with an average corticosterone blood level at death of around 13 μg%. This binding could be reduced to even lower levels by large doses of unlabeled hormone injected in other normal animals concurrently with isotopic steroid (McEwen *et al.*, 1974). Within 2 h after bilateral adrenalectomy, as plasma corticosterone levels fell, increased nuclear and cytosol binding could be demonstrated (McEwen *et al.*, 1974).

It is known that plasma transcortin binding activity increases after adrenalectomy, first as endogenous steroid disappears and subsequently as more transcortin is produced by the liver (Westphal *et al.,* 1963; Seal and Doe, 1965; Ács *et al.,* 1967). In order to distinguish between two such phases of increased corticosterone binding to brain and to see if the amount of available glucocorticoid binding sites in brain actually increases after adrenalectomy, binding studies were conducted with hippocampal tissue slices from animals adrenalectomized at various time intervals prior to being killed (McEwen *et al.,* 1974). As can be seen in Fig. 15, nuclear binding of [^{3}H]corticosterone increases rapidly in the first 2 h after adrenalectomy and is the same at 2 h after adrenalectomy as at 10–12 h after the operation. This binding undergoes a second increase between 11 and 18 h and reaches after several days a level 60% higher than at 2 or 11 h after adrenalectomy.

The reported changes in transcortin levels and the increases in brain binding of corticosterone occur over a simialr time course. Yet, as noted above, blood and brain binding proteins are different from one another, and the blood protein originates in another tissue, the liver. Thus the changes after adrenalectomy in amount of these binding proteins must reflect a common regulatory mechanism operating in liver and brain. While adrenalectomy is not encountered nonpathologically in nature, chronic

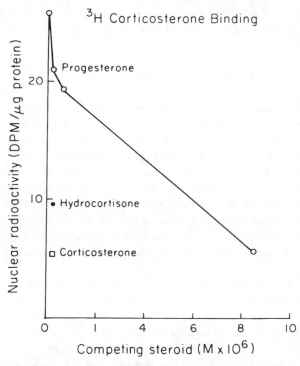

FIG. 14. Competition for nuclear uptake of [^{3}H]corticosterone in hippocampal slices *in vitro* by indicated concentrations of unlabeled progesterone, cortisol, and corticosterone. Data and procedural details may be found in McEwen and Wallach (1973).

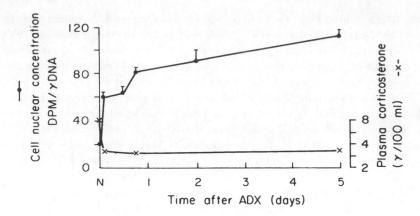

FIG. 15. Increase in cell nuclear [³H]corticosterone binding capacity in hippocampus as a function of time after bilateral adrenalectomy. Data from McEwen et al. (1974).

elevation of corticosteroids, as in chronic stress, or reduction in corticoid levels during long periods of absence of stress are both possible. In such "physiological" circumstances, there might be compensatory changes in levels of transcortin and in level of the brain and other tissue binding proteins. This possibility remains to be investigated.

9.   Adrenal Steroid and Brain Chemistry

It is possible in the case of adrenal steroids to point to evidence that certain enzymes in brain are regulated via hormonal effects on the genome. The clearest example is that of glycerolphosphate dehydrogenase, which is regulated by corticosteroids both in the brain (de Vellis and Inglish, 1968) and in tissue culture of glial cells (de Vellis et al., 1971). Induction of the enzyme by cortisol in tissue culture is blocked by actinomycin D (de Vellis et al., 1971). These findings raise the question of whether glial cells also contain a binding system for corticosteroids of the type found in neurons. As noted above, autoradiography shows the heavy limited-capacity labeling of neurons by [³H]corticosterone while binding in the neuropil is not apparent, but it is impossible to rule out from autoradiography the possibility that some glial cell binding does exist.

Another neural enzyme influenced by adrenal glucocorticoids is phenyl-ethanolamine-N-methyltransferase (PNMT), the enzyme responsible for converting norepinephrine to epinephrine (for review, see Pohorecky and Wurtman, 1971). This enzyme is concentrated in the adrenal medulla, although low levels have been found in olfactory areas and in hypothalamus (Pohorecky et al., 1969). Hypophysectomy decreases and dexamethasone treatment restores PNMT activity in the adrenal medulla, although naturally occurring glucocorticoids such as corticosterone are ineffective when administered exogenously in doses sufficient to replace glucocorticoid in adrenalectomized animals (Pohorecky and Wurtman, 1971). ACTH restores PNMT activity more effectively than natural glucocorticoids, even though the ACTH effect is mediated by glucocorticoid secretion (Pohorecky and Wurtman, 1971). This peculiar situation might be explained by preferred access of corticoids secreted by the adrenal cortex to the adjacent adrenal medulla (Pohorecky and Wurtman, 1971) or it might also be related to undetermined synergistic effects of other secretions stimulated by ACTH.

Another example of corticosteroid induction of an enzyme in neural tissue is the premature induction by cortisol of glutamine synthetase activity in the chick neural retina (Moscona and Piddington, 1966; Reif-Lehrer and Amos, 1968). This induction has been shown to involve both RNA and protein formation (Reif-Lehrer and Amos, 1968) and to proceed optimally at the low concentrations of cortisol (10^{-8} to 10^{-7} M) which are typical of hormone induction systems (Piddington, 1967). A corticosteroid binding protein resembling that in brain has been found in chick neural retina (Chader and Reif-Lehrer, 1972).

There are a number of other examples of corticosteroid regulation of cerebral metabolism. Parvez and Parvez (1973) reported that acute blockade of corticosteroid secretion with Metopirone increased the activity of monoamine oxidase in brain and pituitary. Adrenalectomy was not equally effective. However, cortisol inhibits the activity of the enzyme *in vitro*. The authors attribute the effects of Metopirone to a depletion of corticosteroids, which they postulate act directly on monamine oxidase as a controller of catalytic efficiency of the enzyme. Another regulatory action of corticosteroids occurs on serotonin synthesis in the midbrain (Azmitia and McEwen, 1969; Azmitia *et al.*, 1970; Millard *et al.*, 1972). Adrenalectomy decreases and glucocorticoids increase serotonin formation from [^{3}H]tryptophan, both *in vitro* in homogenates and *in vivo*. Although the effect was originally interpreted as an induction of the rate-limiting enzyme, tryptophan hydroxylase (Azmitia and McEwen, 1969), present evidence favors the view that the hormone effect is independent of *de novo* synthesis of the enzyme. Among the new evidence leading to this conclusion is the observation that corticosteroids increase serotonin formation with such rapidity (1–4 h after injection) that *de novo* enzyme formation cannot account for the changes (Millard *et al.*, 1972; Azmitia and McEwen, 1974). However, corticosteroids do not appear to have a direct activating effect on the catalytic efficiency of tryptophan hydroxylase (Azmitia and McEwen, 1974).

There are a number of reports concerning adrenal hormone effects on the cerebral levels of glutamic and aspartic acids and their metabolites, glutamine, GABA, and asparagine. In general, decreases of amino acid pools observed after adrenalectomy are not simply reversals of changes seen after glucocorticoid treatment (Yuwiler, 1971). Moreover, there are somewhat contradictory reports concerning the direction of changes occurring after adrenalectomy. Woodbury *et al.* (1957) reported that adrenalectomy decreases whole brain content of glutamic acid, GABA, and glutamine, while aspartic acid levels are unchanged. On the other hand, Rindi and Ventura (1961) found that in adrenalectomized rats maintained on saline there is no decrease in either glutamic acid or GABA. It is conceivable that some of these effects are masked by the summation of opposite changes in various parts of the brain. Pandolfo and Macaione (1964) reported that activities of GABA transaminase and glutamate decarboxylase decreased in the cerebral cortex after adrenalectomy. Sutherland and Rikimaru (1964) reported that aspartic acid levels fell after adrenalectomy in cerebral cortex and cerebellum but not in subcortical regions of the brain. Clearly, new studies must be carried out on individual brain regions to resolve the conflicting data.

10.   Neurochemistry of the Hippocampus

There is very little information concerning biochemical changes which occur in hippocampus as a result of corticosteroid binding. One indication of possible cell nuclear action of corticosteroids is the report of Mühlen and Ockenfels (1969) that cell nuclei of pyramidal neurons in guinea pig hippocampus undergo increases in size after systemic administration of cortisone. Such changes, while suggestive of genomic effects, are by themselves difficult to interpret. Another, much less direct suggestion of possible genomic effects is the observation that cytidine incorporation into RNA is changed in hippocampus during a learning task involving stress (Bowman and Strobel, 1969; Kottler *et al.*, 1972). These investigators did not establish, by studying adrenalectomized rats, whether the adrenal gland plays any role in these changes.

There is much about the hippocampus which attracts the attention of neurochemists. The highly ordered cellular organization and the stratification into layers have stimulated both histochemical and microneurochemical investigation of the distribution of enzyme activities (Lewis *et al.*, 1967; Storm-Mathisen, 1970; Fonnom, 1970; Storm-Mathisen and Fonnom, 1971) and amino acid incorporation (Hydén and Lange, 1968, 1970). Cholinergic innervation of this structure is intense (Lewis *et al.*, 1967), and lesion-degeneration studies have indicated that this innervation derives in large part from the septum (Lewis *et al.*, 1967; Lynch *et al.*, 1972; Srebro *et al.*, 1973). Cholinesterase and choline acetyltransferase distribution patterns agree well with the known sites of septal input within the layered organization of the hippocampus (Storm-Mathisen, 1970; Fonnum, 1970; see Kuhar, this volume). The GABA system appears, unlike the cholinergic system, to originate within the hippocampus. The distribution of glutamic acid decarboxylase (GAD) within the layers of the hippocampus agrees in some degree with the distribution of basket cell bodies and terminals (Storm-Mathisen and Fonnum, 1971). GAD activity remains constant in hippocampus after lesioning of the septum, while cholinergic enzymes decrease dramatically (Storm-Mathisen and Fonnum, 1971).

In addition to these two putative neurotransmitter systems in hippocampus, innervation is known to exist from both catecholaminergic and serotonergic nerves arising from cell bodies in the brain stem and midbrain, respectively (Blackstad *et al.*, 1967; Fuxe *et al.*, 1970). This catecholaminergic innervation is part of a pathway from the brainstem that may control rewarded or goal-directed behavior (Stein, 1968). Dysfunction of norepinephrine-containing neurons in this pathway, and particularly those with terminals in the hippocampus, is hypothesized to contribute to the development of schizophrenic behaviors, by producing deficits in goal-directed thinking and in the capacity to experience pleasure (Stein and Wise, 1971; Wise and Stein, 1973). The post-mortum hippocampi of schizophrenic patients showed significantly reduced dopamine-β-hydroxylase (DBH) activity (by 51%) when compared to hippocampi from victims of heart attacks or accidents (Wise and Stein, 1973). The activity of DBH was also significantly reduced in the diencephalon (by 42%) and the pons–medulla region (by 30%). Since the hippocampus and diencephalon, the regions showing the greatest reduction in DBH activity, contain the noradrenergic terminals, while the pons–medulla contains noradrenergic cell bodies, Wise and Stein (1973) suggest that these terminals may be particularly susceptible to neuropathology associated with schizophrenic behaviors.

While many of the input pathways to the hippocampus are chemically identified, the output pathways are not chemically identified as to presumptive neurotransmitter. This applies as well to the dentate granule cells, whose principal chemical characteristic is the ability of their mossy fiber terminals to concentrate zinc ions (Maske, 1955; Timm, 1958; Von Euler, 1962; McLardy, 1964; Otsuka and Kawamoto, 1967; Ibata and Otsuka, 1969; Crawford and Connor, 1972). With respect to these high concentrations of zinc, as well as the possible relationship of the hippocampus to schizophrenia mentioned above, it is interesting to note the suggestion (Pfeiffer and Iliev, 1972; Pfeiffer *et al.*, 1973) that an etiological factor in some types of schizophrenia may involve a deficiency of zinc with a relative increase in copper, as determined by levels in plasma.

11. Adrenal Glucocorticoids and Limbic Control of Pituitary Function

As was indicated in an earlier section, localization of sites of action of estradiol within the brain led to studies which demonstrated hormone uptake and binding by these same brain regions. Stimulated by these studies, investigations of corticosterone uptake and binding by brain regions have shown a different pattern from that for sex hormones, with particularly heavy concentration of radioactive hormone in hippocampus (see above). We shall now consider some of the functional studies which point to a role for these corticosterone binding sites in neural control of pituitary function.

With respect to the regulation of ACTH secretion at rest and during stress, various investigators have reported that implantation of corticosteroids in hippocampus tends to enhance both basal and stress-induced ACTH secretion (Davidson and Feldman, 1967; Knigge, 1966; Bohus *et al.*, 1968; Kawakami *et al.*, 1968*b*). Since these effects are opposite to the generally inhibitory effects of direct hippocampal stimulation on ACTH release (Mason, 1958; Mason *et al.*, 1961; Okinaka, 1961; Mangili *et al.*, 1966; Kawakami *et al.*, 1968*a*), it has been suggested that corticosteroids might act on the hippocampus to inhibit a system in this structure which is concerned with shutting off ACTH secretion activated by stress (McEwen *et al.*, 1972*b*). Corticosterone injections do, in fact, have inhibitory effects on firing rates of single hippocampal neurons (Pfaff *et al.*, 1971). Other brain structures, such as the septum, amygdala, midbrain, and hypothalamus, have been implicated in the control of ACTH release, and direct effects of corticosteroids on these regions have been shown by implantation (for review, see Mangili et al., 1966; McEwen et al., 1972b). Thus the hippocampus is not the only structure governing ACTH release which is sensitive to adrenal steroid "feedback." It is interesting that studies of binding of [^{3}H]corticosterone show that all of these brain regions besides hippocampus which are implicated in ACTH release have some binding sites for this hormone (McEwen et al., 1972b).

Another aspect of pituitary–adrenal function in which the hippocampus figures prominently is the diurnal variation of ACTH release. Lesions in the hippocampus or sectioning of the fornix (main output from hippocampus) tend to abolish the normal diurnal rhythm of ACTH secretion (Mason, 1958; Nakadate and DeGroot, 1963; Moberg et al., 1971), although the permanence of this effect has recently been questioned (Lengvari and Halász, 1973). Maintenance of constant levels of adrenal steroid by implantation of cortisone in hippocampus was found to abolish the diurnal ACTH rhythm (Slusher, 1966), thus indicating a possible role for adrenal steroids in the maintenance of this diurnal cycle. This observation is supported by the finding that the diurnal cycle of ACTH secretion, although not totally abolished by bilateral adrenalectomy, shifts to a peak earlier in the day (Cheifetz et al., 1968; Hiroshige and Sakakura, 1971).

The diurnal variation of ACTH release is time-locked to the daily activity cycle of the species, whether nocturnal or diurnal, and this suggests an important relationship between adrenal steroids and the animal's physiology in activity and sleep. Corticosteroid effects on cerebral carbohydrate metabolism and other direct effects on neural processes related to activity and sleep may constitue an important part of the adrenals' contribution to an animal's homeostatic regulation (Woodbury, 1958). It is particularly interesting in this connection that in human males the peak secretion of ACTH has been shown to occur during paradoxical sleep (Hellman et al., 1970). In the rat, bilateral adrenalectomy abolishes the circadian rhythm of paradoxical sleep, as measured by hippocampal θ activity, under an artificial light–dark cycle, and cortisol injection in late afternoon reestablishes a diurnal variation in this parameter (Johnson and Sawyer, 1971). In humans, administration of a synthetic glucocorticoid, prednisone, reduces the percentage of paradoxical sleep (Gillin et al., 1972).

It is interesting in this connection that Johnson et al. (1971) have noted in the preoptic area a diurnal rhythm of multiunit activity which is abolished by bilateral adrenalectomy. Consistent with corticosteroid as a regulatory factor in diurnal variations of neural activity, Pfaff et al. (1971) observed a decrease in single-unit activity of dorsal hippocampal neurons after systemic injection of corticosterone (Fig. 16). This decrease often had a slow onset and lasted for at least 3 h after the injection, which suggests that it may be mediated by hormone-induced metabolic changes within the cell which outlast the circulating level of the hormone.

A number of investigators have noted changes in the direction of effect of local brain stimulation on ACTH secretion, depending on the corticosteroid blood level at time of stimulation: at low corticosteroid levels, stimulation elicits increased corticosteroid secretion; at high corticosteroid levels, stimulation elicits no change or causes decreased secretion. It may be significant that the direction of the brain stimulation effect on ACTH secretion in the presence of corticoids is opposite to the direction of the

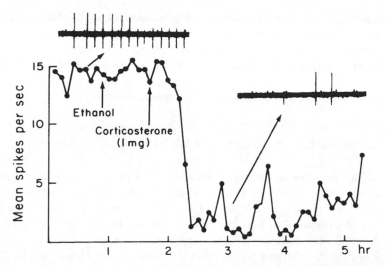

FIG. 16. Effect of corticosterone (injected intraperitoneally in 50 μl ethanol) on single-unit activity in the hippocampus, recorded by telemetry from a freely moving, hypophysectomized female rat. One-second samples from the record show unit activity before (left) and after (right) injection of corticosterone. From Pfaff *et al.* (1971). Copyright 1971 by the American Association for the Advancement of Science. Reprinted by permission.

brain stimulation effect in their absence. This change of stimulation effect with corticoid level has been observed when stimulation is applied in hippocampus (Kawakami *et al.*, 1968a), reticular activating system (Taylor, 1969), and amygdala (Matheson *et al.*, 1971). Since these brain regions are interconnected and interrelated, it is possible that these results reflect a single event common to all of these structures. In spite of the correlations with plasma corticosteroid levels, the causal relationship between hormone level and the stimulation effect is not yet established.

12. Adrenal Steroid Effects on Detection and Recognition of Sensory Stimuli

Besides affecting pituitary–adrenal reactivity, adrenal steroids have important effects on neural processes underlying behavior. One such effect, cited above, is that of glucocorticoids on the frequency of occurrence of paradoxical sleep. It is the purpose of this and the next two sections to review other hormone–brain interactions. People suffering from adrenocortical insufficiency have been shown to possess heightened sensitivity to a variety of sensory stimuli (Henkin, 1970a,b). This condition is unresponsive to mineralocorticoid replacement therapy but is totally reversed by glucocorticoids such as cortisol or prednisone (Henkin, 1970a,b). While *detection* thresholds decline dramatically in adrenocortical insufficiency, the ability to *recognize* the meaning of the stimulus (e.g., to recognize words) is worse than normal (Henkin, 1970a). This deficit is also reversed by glucocorticoids and not by mineralocorticoids.

Experimental analysis of these phenomena is made difficult by lack of a suitable animal model. Recently, however, the beginnings of such a model have become evident. Adrenalectomized rats have been shown to have reduced detection thresholds for odorants such as pyridine (Sakellaris, 1972). It will be most interesting to find out the degree to which limbic structures such as the hippocampus may be involved in glucocorticoid mediation of sensory detection and recognition.

13.   Pituitary–Adrenal System and Stress-Related Behavior

One of the most consistent physiological responses to noxious stimuli, referred to as "stressors" (Selye, 1936), is the activation of the pituitary–adrenal axis with the resultant increase in ACTH release and subsequent rise in plasma corticosteroids. "Psychological stressors," such as fear and anxiety, are among the most potent activators (Bush, 1962). It is now apparent that the increased levels of pituitary and adrenal hormones during stress have a robust effect on behavior (for review, see Di Giusto et al., 1971; Levine, 1968).

Mirsky et al. (1953) were the first to report the effects of the pituitary–adrenal system on learned behavior. They found that the administration of ACTH to monkeys could increase the maintenance of a conditioned avoidance response. It was subsequently shown that ACTH could delay extinction of a conditioned avoidance response (Murphy and Miller, 1955) and that this effect could be obtained without the adrenal glands present (De Wied, 1967; Miller and Ogawa, 1962).

Many studies have now demonstrated that ACTH can facilitate avoidance acquisition and retard the extinction of such behavior (for review, see de Wied, 1969). Further, de Wied and coworkers (de Wied et al., 1970; Greven and de Wied, 1967) have demonstrated that the administration of ACTH analogues materially devoid of corticotropic activities, such as the decapeptide ACTH 1–10 or the heptapeptide ACTH 4–10, can produce the same behavioral effects.

While the ACTH effects do not appear to be dependent on their action exerted on the adrenals, it is rather difficult to vary ACTH or adrenal steroids without at the same time varying the other. Given the limitations of these techniques (Di Giusto et al., 1971), the data appear to indicate that adrenal glucocorticoids may often have the opposite effect of ACTH on behavior. For example, while ACTH injections reliably retard avoidance extinction (Bohus and de Wied, 1966; de Wied and Pirie, 1968), it is well established that corticosteroids can facilitate (i.e., reduce resistance to) extinction (Bohus and Lissák, 1968; de Wied, 1967; Van Wimersma Greidanus, 1970). The effects of glucocorticoids on avoidance acquisition, however, are less clear. Thus injections of glucocorticoids have been found to both enhance (Levine and Brush, 1967), and retard (Endröczi, 1972, p. 40) shuttle-box avoidance acquisition.

In general, investigators have found more consistent effects of pituitary–adrenal manipulations during extinction of avoidance behavior rather than acquisition (Di Giusto et al., 1971). Weiss et al., (1969) have suggested that this may be due to the declining fear levels during extinction as opposed to the rather strong fear exhibited by most animals during acquisition of avoidance tasks. Indeed, their data indicated that when fear in a passive avoidance situation is weakened by changing environmental stimuli the influence of the pituitary–adrenal system is much greater than when fear is strong. Thus the pituitary–adrenal system probably plays a rather subtle role in fear responding—one which can easily be obscured by a ceiling effect. These authors have further reasoned that the slow-acting hormonal systems probably influence general levels of arousal, which are superimposed on the fast-acting neural mechanisms that provide immediate reaction to danger. This suggests that ACTH might help maintain an adequate state of arousal during early reactions to stress, while corticosteroids, lagging behind ACTH secretion, would restore the nervous system to a more normal state of excitability, a function attributed to corticosterone on the basis of other experiments by Woodbury (1958).

14. Pituitary–Adrenal Axis, Hippocampus, and Behavior

What brain regions mediate the hormone effects on behavior? In studies using implanted steroids, Bohus (1970) has reported that many structures of the limbic system are sensitive to corticosteroid implants in facilitating the extinction of conditioned avoidance responding. However, the recent discovery that glucocorticoids are selectively concentrated from the blood by neurons of the hippocampus in amounts greater than any other brain regions (see above) suggests that this structure may be especially involved with adrenal hormone effects on behavior. For example, shifting rats from a continuous reinforcement schedule to either an intermittent reinforcement schedule (Goldman *et al.*, 1973) or extinction (Coover *et al.*, 1971*a*) can produce increased plasma corticosteroid levels. Lesions of the hippocampus eliminate this stress response to nonreward (Coover *et al.*, 1971*b*), suggesting that the hippocampus is an essential part of the neural circuitry involved in this response. When frustrative nonreward occurs during appetitive acquisition, usually as a partial reinforcement schedule, resistance to extinction is increased—the well-known "partial reinforcement extinction effect" (Amsel, 1962). Similarly, a well-documented behavioral effect of hippocampal lesions is an increased resistance to extinction (Jarrard *et al.*, 1964; Kimble, 1968; Raphelson *et al.*, 1966). Taken together, these results suggest that the hippocampus may be involved in the adrenal response to nonreward and, further, that the higher plasma corticosteroid levels may feed back to depress the response-inhibiting characteristics of hippocampal function. Support for the latter notion has been reported by Pfaff *et al.* (1971), as stated earlier.

In more direct tests, Bohus (1971) has provided preliminary evidence that corticosteroid implants in the hippocampus produce poorer passive avoidance. Lesions of the hippocampus have the same effect (e.g., Isaacson and Wickelgren, 1962). In addition, Endröczi (1972) has found that corticosteroids tend to suppress general locomotor activity and that this suppression effect is dependent on an intact hippocampus. Since more detailed comparisons of this literature appear elsewhere in this volume, further examples need not be dealt with here. However, given the data outlined here and in the foregoing sections, there may be good reason to suspect that the behavioral syndromes following either hippocampal damage (Altman *et al.*, 1973; Douglas, 1967; Kimble,

1968) or pituitary–adrenal manipulations (Di Giusto *et al.*, 1971; Endröczi, 1972) may have a common intersect. Further studies are needed to demonstrate the effects of brain lesions and hormone manipulation within the same study, and, more importantly, to show whether the results are due to interference with a common mechanism or with two parallel systems. Two questions which need to be addressed are (1) what is the contribution of the pituitary–adrenal axis to the behaviors shown by lesion studies to depend on the hippocampus, and (2) can the known behavioral effects of manipulating adrenal glucocorticoid levels be attributed, in part, to their central action on the hippocampus? These questions are obviously convergent but must be approached individually, since each involves a separate literature and dissimilar tasks. In dealing with the first question, for example, it would seem reasonable to expect an interaction between the hippocampus and adrenal hormones on tasks involving a moderate amount of stress (Weiss *et al.*, 1969). To this end, it would seem useful to select from the behavioral literature pertaining to the hippocampus a group of tasks that span a continuum of pituitary–adrenal involvement. One end of this continuum might be anchored with relatively low-stress tasks such as alternation behavior and habituation of exploration, while the other end might include the more stressful avoidance tasks. Discrimination reversal learning, extinction, and other such tasks would fall somewhere in between. By use of such an approach, it may be possible to demonstrate that the proposed inhibitory role of the hippocampus (Altman *et al.*, 1973; Douglas, 1967; Kimble, 1968; Micco and Schwartz, 1971) depends to some extent on the action of the pituitary–adrenal hormones (Bohus, 1968; Endröczi, 1972).

Acknowledgments

Research described in this chapter was supported by research grants NS07080 and MH13189 from the United States Public Health Service by grant GB43558 from the National Science Foundation and by institutional grant RF70095 from the Rockefeller Foundation.

15. References

Abraham, A. D., and Sekeris, C. E. Corticosteroid binding macromolecules in the nucleus and cytosol of rat thymus cells. *Biochimica et Biophysica Acta,* 1973, **297**, 142–154.

Ács, A., Stark, E., and Csáki, L. The effect of long-term corticotrophin treatment on the corticosteroid-binding capacity of transcortin. *Journal of Endocrinology,* 1967, **39**, 565–569.

Alberti, K. G. M. M., and Sharp, G. W. G. Identification of four types of steroid by their interaction with mineralocorticoid receptors in the toad bladder. *Journal of Endocrinology,* 1970, **48**, 563–574.

Altman, J., Brunner, R. L., and Bayer, S. The hippocampus and behavioral maturation. *Behavioral Biology,* 1973, **8**, 551–596.

Amsel, A. Frustrative non-reward in partial reinforcement and discrimination learning: Some recent history and a theoretical extension. *Psychological Review,* 1962, **69**, 306–328.

Anderson, C. H., and Greenwald, S. S. Autoradiographic analysis of estradiol uptake in the brain and pituitary of the female rat. *Endocrinology,* 1969, **85**, 1160–1165.

AZMITIA, E. C., JR., AND McEWEN, B. S. Corticosterone regulation of tryptophan hydroxylase in midbrain of the rat. *Science,* 1969, **166,** 1274–1276.

AZMITIA, E. C., JR., AND McEWEN, B. S. Adrenocortical influence on rat brain tryptophan hydroxylase activity. *Brain Research,* 1974, **78,** 291–302.

AZMITIA, E. C., JR., ALGERI, S., AND COSTA, E. Turnover rate of *in vivo* conversion of tryptophan into serotonin in brain areas of adrenalectomized rats. *Science,* 1970, **169,** 201–203.

BALLARD, P. L., AND BALLARD, R. A. Glucocorticoid receptors and the role of glucocorticoids in fetal lung development. *Proceedings of the National Academy of Sciences,* 1972, **69,** 2668–2672.

BAXTER, J. D., AND TOMKINS, G. M. The relationship between glucocorticoid binding and tyrosine aminotransferase induction in hepatoma tissue culture cells. *Proceedings of the National Academy of Sciences U.S.A.,* 1970, **65,** 709–715.

BEATO, M., AND FEIGELSON, P. Glucocorticoid-binding proteins of rat liver cytosol. I. Separation and identification of the binding proteins. *Journal of Biological Chemistry,* 1972, **247,** 7890–7896.

BLACKSTAD, T. W., FUXE, K., AND HÖKFELT, T. Noradrenaline nerve terminals in the hippocampal region of the rat and the guinea pig. *Zeitschrift für Zellforschung,* 1967, **78,** 463–473.

BOHUS, B. Pituitary ACTH release and avoidance behavior of rats with cortisol implants in mesencephalic reticular formation and median eminence. *Neuroendocrinology,* 1968, **3,** 358–368.

BOHUS, B. Central nervous structures and the effect of ACTH and corticosteroids on avoidance behavior: A study with intracerebral implantation of corticosteroids in the rat. *Progress in Brain Research,* 1970, **32,** 171–183.

BOHUS, B. Adrenocortical hormones and central nervous function: The site and mode of their behavioral action in the rat. In V. H. T. James and L. Martini (Eds.), *Hormonal steroids.* Amsterdam: Excerpta Medica, 1971.

BOHUS, B., AND DE WIED, D. Inhibitory and facilitatory effects of two related peptides on extinction of avoidance behavior. *Science,* 1966, **153,** 318–320.

BOHUS, B., AND LISSÁK, K. Adrenocortical hormones and avoidance behavior of rats. *International Journal of Neuropharmacology,* 1968, **7,** 307–314.

BOHUS, B., NYAKAS, C., AND LISSÁK, K. Involvement of suprahypothalamic structures in the hormonal feedback action of corticosteroids. *Acta Physiologica Hungaricae,* 1968, **34,** 1–8.

BOWMAN, R. E., AND STROBEL, D. A. Brain RNA metabolism in the rat during learning. *Journal of Comparative and Physiological Psychology,* 1969, **67,** 448–456.

BULLOCK, L. P., BARDIN, C. W., AND OHNO, S. The androgen insensitive mouse: Absence of intranuclear androgen retention in the kidney. *Biochemical and Biophysical Research Communication,* 1971, **44,** 1537–1543.

BUSH, I. E. Chemical and biological factors in the activity of adrenocortical steroids. *Pharmacological Review,* 1962, **14,** 317–445.

CHADER, G. J., AND REIF-LEHRER, L. Hormonal effects on the neural retina: Corticoid uptake, specific binding and structural requirements for the induction of glutamine synthetase. *Biochimica et Biophysica Acta,* 1972, **264,** 186–196.

CHADER, G. J., AND VILLEE, C. A. Uptake of oestradiol by the rabbit hypothalamus. *Biochemical Journal,* 1970, **118,** 93–97.

CHEIFETZ, P., GAFFUD, N., AND DINGMAN, J. F. Effects of bilateral adrenalectomy and continuous light on the circadian rhythm of corticotropin in female rats. *Endocrinology,* 1968, **82,** 1117–1124.

CHYTIL, F., AND TOFT, D. Corticoid binding component in brain. *Journal of Neurochemistry,* 1972, **19,** 2877–2880.

COMSTOCK, J. P., ROSENFELD, G. C., O'MALLEY, B. W., AND MEANS, A. R. Estrogen-induced changes in translation, and specific messenger RNA levels during oviduct differentiation. *Proceedings of the National Academy of Sciences U.S.A.,* 1972, **69,** 2377–2380.

COOVER, G., GOLDMAN, L., AND LEVINE, S. Plasma corticosterone increases produced by extinction of operant behavior in rats. *Physiology and Behavior,* 1971*a*, **6,** 261–263.

COOVER, G., GOLDMAN, L., AND LEVINE, S. Plasma corticosterone levels during extinction of a lever-press response in hippocampectomized rats. *Physiology and Behavior,* 1971*b*, **7,** 727–732.

CRAWFORD, I. L., AND CONNOR, J. D. Zinc in maturing rat brain: Hippocampal concentration and localization. *Journal of Neurochemistry,* 1972, **19,** 1451–1458.

Davidson, J. M., and Feldman, S. Effects of extrahypothalamic dexamethasone implants on the pituitary–adrenal system. *Acta Endocrinologica,* 1967, **55,** 240–246.

Davidson, J. M., Jones, L. E., and Levine, S. Feedback regulation of adrenocorticotropin secretion in "basal" and "stress" conditions: Acute and chronic effects of intrahypothalamic corticoid implantation. *Endocrinology,* 1968, **82,** 655–663.

de Vellis, J., and Inglish, D. Hormonal control of glycerol phosphate dehydrogenase in the rat brain. *Journal of Neurochemistry,* 1968, **15,** 1961–1970.

de Vellis, J., Inglish, D., Cole, R., and Molson, J. Effects of hormones on the differentiation of cloned lines of neurons and glial cells. In D. Ford (Ed.), *Influence of hormones on the nervous system.* Basel: Karger, 1971.

de Wied, D. Opposite effects of ACTH and glucocorticoids on extinction of conditioned avoidance behavior. In L. Martini, F. Fraschini, and M. Motta (Eds.), *Hormonal steroids.* Amsterdam: Excerpta Medica, 1967.

de Wied, D. Effects of peptide hormones on behavior. In W. F. Ganong and L. Martini (Eds.), *Frontiers in neuroendocrinology.* New York: Oxford University Press, 1969.

de Wied, D., and Pirie, G. The inhibitory effect of ACTH-10 on extinction of a conditioned avoidance response: Its independence of thyroid function. *Physiology and Behavior,* 1968, **3,** 355–358.

de Wied, D., Witter, A., and Lande, S. Anterior pituitary peptides and avoidance acquisition of hypophysectomized rats. *Progress in Brain Research,* 1970, **32,** 213–220.

Di Giusto, E. L., Cairneross, K., and King, M. G. Hormonal influences on fear motivated responses. *Psychological Bulletin,* 1971, **75,** 432–444.

Douglas, R. J. The hippocampus and behavior. *Psychological Bulletin,* 1967, **67,** 416–442.

Eisenfeld, A. J. ^{3}H-Estradiol: *In vitro* binding to macromolecules from the rat hypothalamus, anterior pituitary and uterus. *Endocrinology,* 1970, **86,** 1313–1318.

Eisenfeld, A. J., and Axelrod, J. Selectivity of estrogen distribution in tissues. *Journal of Pharmacology and Experimental Therapeutics,* 1965, **150,** 469–475.

Endröczi, E. *Limbic system, learning, and pituitary–adrenal function.* Budapest: Akademiae Kiado, 1972.

Feldman, D., Funder, J. W., and Edelman, I. S. Evidence for a new class of corticosterone receptors in the rat kidney. *Endocrinology,* 1973, **92,** 1429–1441.

Florini, J. R., and Buyske, D. A. Plasma protein binding of triamcinolone-H^3 and hydrocortisone-4-C^{14}. *Journal of Biological Chemistry,* 1961, **236,** 247–251.

Fonnum, F. Topographical and subcellular localization of choline acetyltransferase in rat hippocampal region. *Journal of Neurochemistry,* 1970, **17,** 1029–1037.

Funder, J. W., Feldman, D., and Edelman, I. S. Glucocorticoid receptors in rat kidney: The binding of tritiated-dexamethasone. *Endocrinology,* 1973, **92,** 1005–1013.

Fuxe, K., Hökfelt, T., and Ungerstedt, U. Morphological and functional aspects of central monoamine neurons. *International Review of Neurobiology,* 1970, **13,** 93–126.

Gerlach, J. L., and McEwen, B. S. Rat brain binds adrenal steroid hormone: Radioautography of hippocampus with corticosterone. *Science,* 1972, **175,** 1133–1136.

Gerlach, J. L., McEwen, B. S., Pfaff, D. W., Ferin, M., and Carmel, P. W. Rhesus monkey brain binds radioactivity from ^{3}H estradiol and ^{3}H corticosterone, demonstrated by nuclear isolation and autoradiography. *Program of the Endocrine Society 56th Annual Meeting,* Atlanta, Ga., 1974, **Abstract 370.**

Gillin, J. C., Jacobs, L. S., Fram, D. H., and Snyder, F. Acute effect of a glucocorticoid on normal human sleep. *Nature (London),* 1972, **237,** 398–399.

Goldman, L., Coover, G., and Levine, S. Bidirectional effects of reinforcement shifts on pituitary adrenal activity. *Physiology and Behavior,* 1973, **10,** 209–214.

Gorski, J., Toft, D., Shyamala, G., Smith, D., and Notides, A. Hormone receptors: Studies on the interaction of estrogen with uterus. *Recent Progress in Hormone Research,* 1968, **24,** 45–80.

Gorski, J., De Angelo, A. B., and Barnea, A. Control of gene expression: An early response to estrogen. In V. H. T. James and L. Martini (Eds.), *Proceedings of the Third International Congress on Hormonal Steroids, Hamburg, 1970.* Basel: International Congress Series No. 219, Excerpta Medica, 1971.

GREEN, R., LUTTGE, W. G., AND WHALEN, R. E. Induction of receptivity in ovariectomized female rats by a single intravenous injection of estradiol-17β. *Physiology and Behavior*, 1970, **5**, 137–141.

GREVEN, H. M., AND DE WIED, D. The active sequence in the ACTH molecule responsible for inhibition of the extinction of conditioned avoidance behavior in rats. *European Journal of Pharmacology*, 1967, **2**, 14–16.

GRIMM, Y., AND KENDALL, J. W. A study of feedback suppression of ACTH secretion utilizing glucocorticoid implants in the hypothalamus: The comparative effects of cortisol, corticosterone, and their 21-acetates. *Neuroendocrinology*, 1968, **3**, 55–63.

GROSSER, B. I., STEVENS, W., BRUENGER, F. W., AND REED, D. J. Corticosterone binding by rat brain cytosol. *Journal of Neurochemistry*, 1971, **18**, 1725–1732.

GROSSER, B. I., STEVENS, W., AND REED, D. J. Properties of corticosterone-binding macromolecules from rat brain cytosol. *Brain Research*, 1973, **57**, 387–396.

HACKNEY, J. F., AND PRATT, W. B. Characterization and partial purification of the specific glucocorticoid-binding component from mouse fibroblasts. *Biochemistry*, 1971, **10**, 3002–3008.

HELLMAN, L., NAKADA, F., CURTIZ, J., WEITZMAN, E. D., KREAM, J., ROFFWARG, H., ELLMAN, S., FUKUSHIMA, D. K., AND GALLAGHER, T. F. Cortisol is secreted episodically by normal men. *Journal of Clinical Endocrinology and Metabolism*, 1970, **30**, 411–422.

HENKIN, R. I. The neuroendocrine control of perception. In *Perception and its disorders*. Vol. XLVIII. Research Publication of the Association for Research in Nervous and Mental Disease, 1970a.

HENKIN, R. I. The effects of corticosteroids and ACTH on sensory systems. *Progress in Brain Research*, 1970b, **32**, 270–293.

HIROSHIGE, T., AND SAKAKURA, M. Circadian rhythm of corticotrophin-releasing activity in the hypothalamus of normal and adrenalectomized rats. *Neuroendocrinology*, 1971, **7**, 25–36.

HYDÉN, H., AND LANGE, P. W. Protein synthesis in the hippocampal pyrimadal cells of rats during a behavioral test. *Science*, 1968, **159**, 1370–1373.

HYDÉN, H., AND LANGE, P. W. S-100 brain protein: Correlation with behavior. *Proceedings of the National Academy of Sciences U.S.A.*, 1970, **67**, 1959–1966.

IBATA, Y., AND OTSUKA, N. Electron microscopic demonstration of zinc in the hippocampal formation using Timm's sulfide–silver technique. *Journal of Histochemistry and Cytochemistry*, 1969, **17**, 171–175.

ISAACSON, R. L., AND WICKELGREN, W. O. Hippocampal ablation and passive avoidance. *Science*, 1962, **138**, 1104–1106.

JARRARD, L. C., ISAACSON, R. L., AND WICKELGREN, W. O. Effects of hippocampal ablation and inter-trial interval on runway acquisition and extinction. *Journal of Comparative and Physiological Psychology*, 1964, **57**, 442–444.

JENSEN, E. V., AND DESOMBRE, E. R. Effects of ovarian hormones at the subcellular level. In L. Martini and V. H. T. James (Eds.), *Current topics in experimental endocrinology*. Vol. 1. New York: Academic Press, 1971.

JENSEN, E. V., AND JACOBSON, H. I. Basic guides to the mechanism of estrogen action. *Recent Progress in Hormone Research*, 1962, **18**, 387–408.

JOHNSON, J. H., AND SAWYER, C. H. Adrenal steroids and the maintenance of a circadian distribution of paradoxical sleep in rats. *Endocrinology*, 1971, **89**, 507–512.

JOHNSON, J. H., TERKEL, J., WHITMOYER, D. I., AND SAWYER, C. H. A circadian rhythm in the multiple unit activity (MUA) of preoptic neurons in the female rat. *Anatomical Record*, 1971, **169**, 348.

JOHNSTON, P., AND DAVIDSON, J. M. Intracerebral androgens and sexual behavior in the male rat. *Hormones and Behavior*, 1972, **3**, 345–357.

KAHWANAGO, I., HEINRICHS, W. L., AND HERRMANN, W. L. Isolation of oestradiol "receptors" from bovine hypothalamus and anterior pituitary gland. *Nature (London)*, 1969, **223**, 313–314.

KAISER, N., MILHOLLAND, R. J., AND ROSEN, F. Glucocorticoid-binding macromolecules in rat and mouse thymocytes. *Journal of Biological Chemistry*, 1973, **248**, 478–483.

KATO, J., AND VILLEE, C. A. Preferential uptake of estradiol by the anterior hypothalamus of the rat. *Endocrinology*, 1967, **80**, 567–575.

KATO, J., ATSUMI, Y., AND INABA, M. A soluble receptor for estradiol in rat anterior hypophysis. *Journal of Biochemistry (Tokyo)*, 1970a, **68**, 759–761.

318   BRUCE S. McEWEN *et al.*

KATO, J., ATSUMI, Y., AND MURAMATSU, M. Nuclear estradiol receptor in rat anterior hypophysis. *Journal of Biochemistry (Tokyo)*, 1970b, **67**, 871–872.

KATZENELLENBOGEN, B. S., AND KATZENELLENBOGEN, J. A. Antiestrogens: Studies using an *in vitro* estrogen-responsive uterine system. *Biochemical and Biophysical Research Communication*, 1973, **50**, 1152–1159.

KAWAKAMI, M., SETO, L., TERASAWA, E., YOSHIDA, E., MIYAMOTO, T., SEKIGUCHI, M., AND HATTORI, Y. Influence of electrical stimulation and lesion in limbic structure upon biosynthesis of adrenocorticoid in the rabbit. *Neuroendocrinology*, 1968a, **3**, 337–348.

KAWAKAMI, M., SETO, K., AND YOSHIDA, K. Influence of corticosterone implantation in limbic structures upon biosynthesis of adrenocortical steroid. *Neuroendocrinology*, 1968b, **3**, 349–354.

KIMBLE, D. P. Hippocampus and internal inhibition. *Psychological Bulletin*, 1968, **70**, 285–295.

KNIGGE, K. M. Feedback mechanisms in neural control of adenohypophyseal function: Effect of steroids implanted in amygdala and hippocampus. *Abstracts, Second International Congress on Hormonal Steroids, Milan*, 1966, p. 208.

KNIZLEY, H., JR. The hippocampus and septal area as primary target sites for corticosterone. *Journal of Neurochemistry*, 1972, **19**, 2737–2745.

KÖNIG, J. F. R., AND KLIPPEL, R. A. *The rat brain*. Baltimore: Williams and Wilkins, 1963.

KORENMAN, S. G. Relation between estrogen inhibitory activity and binding to cytosol of rabbit and human uterus. *Endocrinology*, 1970, **87**, 1119–1123.

KOTTLER, P. D., BOWMAN, R. E., AND HAASCH, W. D. RNA metabolism in the rat brain during learning following intravenous and intraventricular injections of ^{3}H-cytidine. *Physiology and Behavior*, 1972, **8**, 291–297.

LENGVARI, I., AND HÁLASZ, B. Evidence for a diurnal fluctuation in plasma corticosterone levels after fornix transection in the rat. *Neuroendocrinology*, 1973, **11**, 191–196.

LEVINE, S. Hormones and conditioning. In W. J. Arnold (Ed.), *Nebraska symposium on motivation*. Lincoln, Neb.: University of Nebraska Press, 1968.

LEVINE, S., AND BRUSH, F. R. Adrenocortical activity and avoidance learning as a function of time after avoidance training. *Physiology and Behavior*, 1967, **2**, 385–388.

LEWIS, P. R., SHUTE, C. C. D., AND SILVER, A. Confirmation from choline acetylase analyses of a massive cholinergic innervation to the rat hippocampus. *Journal of Physiology (London)*, 1967, **191**, 215–224.

LIAO, S., AND FANG, S. Receptor proteins for androgens and the mode of action of androgens on gene transcription in ventral prostate. *Vitamins and Hormones*, 1969, **27**, 17–90.

LISK, R. D. Sexual behavior: Hormonal control. In L. Martini and W. F. Ganong (Eds.), *Neuroendocrinology*. Vol. 2. New York: Academic Press, 1967.

LITWACK, G., AND SINGER, S. Subcellular actions of glucocorticoids. In G. Litwack (Ed.), *Biochemical actions of hormones*. Vol. 2. New York: Academic Press, 1972.

LORENTE DE NÓ, R. Studies on the structure of the cerebral cortex. I. The area entorhinalis. *Journal of Psychology and Neurology*, 1933, **45**, 381.

LORENTE DE NÓ, R. Studies on the structure of the cerebral cortex. II. Continuation of the study of ammonic system. *Journal of Psychology and Neurology*, 1934, **46**, 113.

LYNCH, G., MATTHEWS, D. A., MOSKO, S., PARKS, T., AND COTMAN, C. Induced acetylcholinesterase-rich layer in rat dentate gyrus following entorhinal lesions. *Brain Research*, 1972, **42**, 311–318.

MAKMAN, M. H., NAKAGAWA, S., AND WHITE, A. Studies of the mode of action of adrenal steroids on lymphocytes. *Recent Progress in Hormone Research*, 1967, **23**, 195–219.

MANGILI, G., MOTTA, M., AND MARTINI, L. Control of adrenocorticotrophic hormone secretion. In L. Martini and W. F. Ganong (Eds.), *Neuroendocrinology*. Vol. 1. New York: Academic Press, 1966.

MASKE, H. Ueber den topochemischen Nachweis von Zink im Ammonshorn verschiedener Säugetiere. *Die Naturwissenschaften*, 1955, **42**, 424.

MASON, J. W. The central nervous system regulation of ACTH secretion. In H. H. Jasper *et al.* (Eds), *Reticular formation of the brain*. Boston: Little, Brown, 1958.

MASON, J. W., NAUTA, W. J. H., BRADY, J. V., ROBINSON, J. A., AND SACHAR, E. J. The role of limbic system structures in the regulation of ACTH release. *Acta Neurovegitativa*, 1961, **23**, 4–14.

MATHESON, G. K., BRANCH, B. J., AND TAYLOR, A. N. Effects of amygdaloid stimulation on pituitary–adrenal activity in conscious cats. *Brain Research*, 1971, **32**, 151–167.

McEwen, B. S., and Pfaff, D. W. Factors influencing sex hormone uptake by rat brain regions. I. Effects of neonatal treatment, hypophysectomy, and competing steroid on estradiol uptake. *Brain Research,* 1970, **21,** 1–16.

McEwen, B. S., and Pfaff, D. W. Chemical and physiological approaches to neuroendocrine mechanisms: Attempts at integration. In F. Ganong and L. Martini (Eds.), *Frontiers in neuroendocrinology.* New York: Oxford University Press, 1973.

McEwen, B. S., and Plapinger, L. Association of corticosterone-1, 2-H^3 with macromolecules extracted from brain cell nuclei. *Nature (London),* 1970, **226,** 263–264.

McEwen, B. S., and Wallach, G. Corticosterone binding to hippocampus: Nuclear and cytosol binding *in vitro. Brain Research,* 1973, **57,** 373–386.

McEwen, B. S., Weiss, J. M., and Schwartz, L. S. Uptake of corticosterone by rat brain and its concentration by certain limbic structures. *Brain Research,* 1969, **16,** 227–241.

McEwen, B. S., Weiss, J. M., and Schwartz, L. S. Retention of corticosterone by cell nuclei from brain regions of adrenalectomized rats. *Brain Research,* 1970a, **17,** 471–482.

McEwen, B. S., Zigmond, R. E., Azmitia, E. C., Jr., and Weiss, J. M. Steroid hormone interaction with specific brain regions. In R. E. Bowman and S. P. Datta (Eds.), *Biochemistry of brain and behavior.* New York: Plenum Press, 1970b.

McEwen, B. S., Magnus, C., and Wallach, G. Biochemical studies of corticosterone binding to cell nuclei and cytoplasmic macromolecules in specific regions of the brain. In C. H. Sawyer and R. A. Gorski (Eds.), *Steroid hormones and brain function.* Berkeley: University of California Press, 1971.

McEwen, B. S., Magnus, C., and Wallach, G. Soluble corticosterone-binding macromolecules extracted from rat brain. *Endocrinology,* 1972a, **90,** 217–226.

McEwen, B. S., Zigmond, R. E., and Gerlach, J. L. Sites of steroid binding and action in the brain. In G. H. Bourne (Ed.), *Structure and function of nervous tissue.* Vol. 5. New York: Academic Press, 1972b.

McEwen, B. S., Wallach, G., and Magnus, C. Corticosterone binding to hippocampus: Immediate and delayed influences of the absence of adrenal secretion. *Brain Research,* 1974, **70,** 321–334.

McGuire, W. L., and O'Malley, B. W. Ribonucleic acid polymerase activity of the chick oviduct during steroid-induced synthesis of a specific protein. *Biochimica et Biophysica Acta,* 1968, **157,** 187–194.

McLardy, T. Second hippocampal zinc-rich synaptic system. *Nature (London),* 1964, **201,** 92–93.

Micco, D. J., and Schwartz, M. Effects of hippocampal lesions upon the development of Pavlovian internal inhibition in rats. *Journal of Comparative and Physiological Psychology,* 1971, **76,** 371–377.

Millard, S. A., Costa, E., and Gal, E. M. On the control of brain serotonin turnover rate by end produce inhibition. *Brain Research,* 1972, **40,** 545–551.

Miller, R. E., and Ogawa, N. The effect of adrenocorticotrophic hormone (ACTH) on avoidance conditioning in the adrenalectomized rat. *Journal of Comparative and Physiological Psychology,* 1962, **55,** 211–213.

Mirsky, J., Miller, R., and Stein, R. Relation of adrenocortical activity and adaptive behavior. *Psychosomatic Medicine,* 1953, **15,** 574–584.

Moberg, G. P., Scapagnini, U., DeGroot, J., and Ganong, W. F. Effect of sectioning the fornix on diurnal fluctuation in plasma corticosterone levels in the rat. *Neuroendocrinology,* 1971, **7,** 11–15.

Moscona, A. A., and Piddington, R. Stimulation by hydrocortisone of premature changes in the development pattern of glutamine synthetase in embryonic retina. *Biochimica et Biophysica Acta,* 1966, **121,** 409–411.

Mowles, T. F., Ashkanazy, B., Mix, E., Jr., and Sheppard, H. Hypothalamic and hypophyseal estradiol-binding complexes. *Endocrinology,* 1971, **89,** 484–491.

Mühlen, K., and Ockenfels, H. Morphologische Veränderungen im Diencephalon und Telencephalon nach Störungen des Regelkreises Adenohypophyse-Nebennierenrinde. III. Ergebnisse beim Meerschweinchen nach Verabreichung von Cortison und Hydrocortison. *Zeitschrift für Zellforschung,* 1969, **93,** 126–141.

Murphy, J. V., and Miller, R. E. The effect of adrenocorticotrophic hormone (ACTH) on avoidance conditioning in the rat. *Journal of Comparative and Physiological Psychology,* 1955, **48,** 47–49.

Nakadate, G. M., and DeGroot, J. Fornix transection and adrenocortical function in rats. *Anatomical Record,* 1963, **145,** 338 (abst.).

NOTIDES, A. C. Binding affinity and specificity of the estrogen receptor of the rat uterus and anterior pituitary. *Endocrinology*, 1970, **87**, 987–992.

OKINAKA, S. Regulation der Hypophysen-Nebennierenfunktion durch das Limbic-Septum und der Mittelhirnanteil der Formatio-reticularis. *Acta Neurovegitativa*, 1961, **23**, 15–20.

O'MALLEY, B. W., MCGUIRE, W. L., AND KORENMAN, S. G. Estrogen stimulation of synthesis of specific proteins and RNA polymerase activity in the immature chick oviduct. *Biochimica et Biophysica Acta*, 1967, **145**, 204–207.

O'MALLEY, B. W., SPELSBURG, T. C., SCHRADER, W. T., CHYTIL, F., AND STEGGLES, A. W. Mechanisms of interaction of a hormone–receptor complex with the genome of a eukaryotic target cell. *Nature (London)*, 1972, **235**, 141–144.

OTSUKA, N., AND KAWAMOTO, M. Histochemische und autoradiographische Untersuchungen der Hippocampus-formation der Maus. *Histochemie*, 1967, **6**, 267–273.

PALACIOS, R., SULLIVAN, D., SUMMERS, N. M., KIELY, M. L., AND SCHIMKE, R. T. Purification of ovalbumin messenger ribonucleic acid by specific immunoadsorption of ovalbumin-synthesizing polysomes and Millipore partition of ribonucleic acid. *Journal of Biological Chemistry*, 1973, **248**, 510–548.

PANDOLFO, L., AND MACAIONE, S. Influence of adrenalectomy on activation of GABA transaminase and glutamate decarboxylase in cortex of rat. *Giornale di Biochimica*, 1964, **13**, 256–261.

PARVEZ, H., AND PARVEZ, S. The effects of metopirone and adrenalectomy on the regulation of the enzymes monoamine oxidase and catechol-*O*-methyl transferase in different brain regions. *Journal of Neurochemistry*, 1973, **20**, 1011–1020.

PEETS, E. A., STAUB, M., AND SYMCHOWICZ, S. Plasma binding of betamethasone-^{3}H, dexamethasone-^{3}H, and cortisol-^{14}C—A comparative study. *Biochemical Pharmacology*, 1969, **18**, 1655–1663.

PFAFF, D. W. Autoradiographic localization of radioactivity in rat brain after injection of tritiated sex hormones. *Science*, 1968, **161**, 1355–1356.

PFAFF, D. W., AND KEINER, M. Estradiol-concentrating cells in the rat amygdala as part of a limbic–hypothalamic hormone sensitive system. In B. Eleftheriou (Ed.), *The neurobiology of the amygdala*. New York: Plenum Press, 1972.

PFAFF, D. W., AND KEINER, M. Atlas of estradiol-concentrating cells in the central nervous system of the female rat. *Journal of Comparative Neurology*, 1973, **151**, 121–158.

PFAFF, D. W., SILVA, M. T. A., AND WEISS, J. M. Telemetered recording of hormone effects on hippocampal neurons. *Science*, 1971, **172**, 394–395.

PFAFF, D., LEWIS, C., DIAKOW, C., AND KEINER, M. Neurophysiological analysis of mating behavior responses as hormone-sensitive reflexes. In E. Stellar and J. Sprague (Eds.), *Progress in physiological psychology*. Vol. 5. New York: Academic Press, 1973, pp. 253–297.

PFEIFFER, C. C., AND ILIEV, V. A study of zinc deficiency and copper excess in the schizophrenias. *International Review of Neurobiology Supplement*, 1972, **1**, 141–165.

PFEIFFER, C. C., ILIEV, V., AND GOLDSTEIN, L. Blood histamine, basophil counts, and trace elements in the schizophrenias. In D. Hawkins and L. Pauling (Eds.), *Orthomolecular psychiatry: Treatment of schizophrenia*. San Francisco: W. H. Freeman, 1973.

PIDDINGTON, R. Hormonal effects on the development of glutamine synthetase in the embryonic chick retina. *Developmental Biology*, 1967, **16**, 168–188.

POHORECKY, L. A., AND WURTMAN, R. J. Adrenocortical control of epinephrine synthesis. *Pharmacological Review*, 1971, **23**, 1–35.

POHORECKY, L. A., ZIGMOND, M. J., KARTEN, H., AND WARTMAN, R. J. Enzymatic conversion of norepinephrine to epinephrine by the brain. *Journal of Pharmacology and Experimental Therapeutics*, 1969, **165**, 190–195.

QUADAGNO, D. M., SHRYNE, J., AND GORSKI, R. A. The inhibition of steroid-induced sexual behavior by intrahypothalamic actinomycin D. *Hormones and Behavior*, 1971, **2**, 1–10.

RAPHELSON, A. C., ISAACSON, R. L., AND DOUGLAS, R. J. The effect of limbic damage on the retention and performance of a runway response. *Neuropsychologia*, 1966, **4**, 253–264.

RIEF-LEHRER, L. AND AMOS, H. Hydrocortisone requirements for the induction of glutamine synthetase in chick embryo retina, *Biochemistry*, 1968, **J106**, 425–430.

RHEES, R. W., ABEL, J. H., JR., AND HAACK, D. W. Uptake of tritiated steroids in the brain of the duck (*Anas platyrhynchos*): An autoradiographic study. *General and Comparative Endocrinology*, 1972, **18**, 292–300.

RINDI, G., AND VENTURA, V. Influence of adrenalectomy, adrenal cortex hormones, and of cold on the γ-aminobutyric acid and glutamic acid content of rat brain. *Italian Journal of Biochemistry*, 1961, **10**, 135–146.

ROBISON, G. A., BUTCHER, R. W., AND SUTHERLAND, E. W. *Cyclic AMP*. New York: Academic Press, 1971.

ROSENAU, W., BAXTER, J. D., ROUSSEAU, G. G., AND TOMKINS, G. M. Mechanism of resistance to steroids: Glucocorticoid receptor defect in lymphoma cells. *Nature (London), New Biology*, 1972, **237**, 20–24.

ROUSSEAU, G. G., BAXTER, J. D., AND TOMKINS, G. M. Glucocorticoid receptors: Relations between steroid binding and biological effects. *Journal of Molecular Biology*, 1972, **67**, 99–115.

SAKELLARIS, P. C. Olfactory thresholds in normal and adrenalectomized rats. *Physiology and Behavior*, 1972, **9**, 495–500.

SCHAUMBURG, B. P. Investigations on the glucocorticoid-binding protein from rat thymocytes. II. Stability, kinetics and specificity of binding of steroids. *Biochimica et Biophysica Acta*, 1972a, **261**, 219–235.

SCHAUMBURG, B. P. Studies of the glucocorticoid-binding protein from thymocytes. III. pH dependence of the binding and density-gradient centrifugation of the protein. *Biochimica et Biophysica Acta*, 1972b, **263**, 414–423.

SEAL, U. S., AND DOE, R. P. Vertebrate distribution of corticosteroid-binding globulin and some endocrine effects on concentration. *Steroids*, 1965, **5**, 827–841.

SELYE, H. A syndrome produced by diverse noxious agents. *Nature (London)*, 1936, **138**, 32.

SHYAMALA, G. Estradiol receptors in mouse mammary tumors: Absence of the transfer of bound estradiol from the cytoplasm to the nucleus. *Biochemical and Biophysical Research Communication*, 1972, **46**, 1623–1630.

SLUSHER, M. A. Effects of cortisol implants in the brainstem and ventral hippocampus on diurnal corticosterone levels. *Experimental Brain Research*, 1966, **1**, 184–194.

SREBRO, B., ODERFELD-NOWAK, B., KLODOS, I., DEBROWSKA, J., AND NARKIEWICZ, O. Changes in acetylcholinesterase activity in hippocampus produced by septal lesions in the rat. *Life Sciences*, 1973, **12**, 261–270.

STEIN, L. Chemistry of reward and punishment. In D. H. Efron (Ed.), *Psychopharmacology: A review of progress: 1957–1967*. Washington, D.C.: U.S. Government Printing Office, 1968.

STEIN, L., AND WISE, C. D. Possible etiology of schizophrenia: Progressive damage to the noradrenergic reward system by 6-hydroxydopamine. *Science*, 1971, **171**, 1032–1036.

STERN, J. J., AND JANKOWIAK, R. Effects of actinomycin D implanted in the anterior hypothalamic–preoptic region of the diencephalon on spontaneous activity in ovariectomized rats. *Journal of Endocrinology*, 1972, **55**, 465–466.

STORM-MATHISEN, J. Quantitative histochemistry of acetylcholinesterase in rat hippocampal region correlated to histochemical staining. *Journal of Neurochemistry*, 1970, **17**, 739–750.

STORM-MATHISEN, J., AND FONNUM, F. Quantitative histochemistry of glutamate decarboxylase in the rat hippocampal region. *Journal of Neurochemistry*, 1971, **18**, 1105–1112.

STUMPF, W. E. Estradiol-concentrating neurons: Topography in the hypothalamus by dry mount autoradiography. *Science*, 1968, **162**, 1001–1003.

STUMPF, W. E. Estrogen-neurons and estrogen-neuron systems in the periventricular brain. *American Journal of Anatomy*, 1970, **129**, 207–217.

SUTHERLAND, V. C., AND RIKIMARU, M. The regional effects of adrenalectomy and ethanol on cerebral amino acids in the rat. *International Journal of Neuropharmacology*, 1964, **3**, 135–139.

SWANECK, G. E., HIGHLAND, E., AND EDELMAN, I. S. Stereospecific nuclear and cytosol aldosterone-binding proteins of various tissues. *Nephron*, 1969, **6**, 297–316.

TAYLOR, A. N. The role of the reticular activating system in the regulation of ACTH secretion. *Brain Research*, 1969, **13**, 234–246.

TIMM, F. Zur histochemie der Ammonshorngebietes. *Zeitschrift für Zellforschung*, 1958, **48**, 548–555.

TUCKER, H. A., LARSON, B. L., AND GORSKI, J. Cortisol binding in cultured bovine mammary cells. *Endocrinology*, 1971, **89**, 152–160.

VAN WIMERSMA GREIDANUS, T. J. B. Effects of steriods on extinction of an avoidance response in rats. *Progress in Brain Research,* 1970, **32,** 185–191.

VERTES, M., AND KING, R. J. B. The mechanism of oestradiol binding in rat hypothalamus: Effect of androgenization. *Journal of Endocrinology,* 1971, **51,** 271–282.

VON EULER, C. On the significance of the high zinc content in the hippocampal formation. In *Physiologie de l'hippocampe.* Paris: Centre National de la Recherche Scientifique, 1962.

WADE, G., AND ZUCKER, I. Modulation of food intake and locomotor activity in female rats by diencephalic hormone implants. *Journal of Comparative and Physiological Psychology,* 1970, **72,** 328–336.

WEISS, J. M., MCEWEN, B. S., SILVA, T. M. A., AND KALKUT, M. F. Pituitary–adrenal influences on fear responding. *Science,* 1969, **163,** 197–199.

WESTPHAL, U., WILLIAMS, W. C., JR., ASHLEY, B. D., AND DEVENUTO, F. Proteinbindung der Corticosteroide im Serum adrenalektomierter und hypophysektomierter Ratten. *Hoppe Seylers Zeitschrift für Physiologische Chemie,* 1963, **332,** 54–69.

WICKS, W. D., KENNEY, F. T., AND LEE, K.-L. Induction of hepatic enzyme synthesis *in vitro* by adenosine 3',5' monophosphate. *Journal of Biological Chemistry,* 1969, **244,** 6008–6013.

WILSON, J. D., AND GLYONA, R. E. The intranuclear metabolism of testosterone in the accessory organs of reproduction. *Recent Progress in Hormone Research,* 1970, **26,** 309–330.

WISE, C. D., AND STEIN, L. Dopamine-β-hydroxylase deficits in the brains of schizophrenic patients. *Science,* 1973, **181,** 344–347.

WONG, M. D., AND BURTON, A. F. Isolation and preliminary characterization of corticosterone–receptor complexes in mouse placental tissue. *Biochemical and Biophysical Research Communication,* 1973, **50,** 71–79.

WOODBURY, D. M. Relation between the adrenal cortex and the central nervous system. *Pharmacological Review,* 1958, **10,** 275–357.

WOODBURY, D. M., TIMIRAS, P. S., AND VERNADAKIS, A. Influence of adrenocortical steroids on brain function and metabolism. In H. Hoagland (Ed.), *Hormones, brain function, and behavior.* New York: Academic Press, 1957.

YUWILER, A. Stress. In A. Lajtha (Ed.), *Handbook of neurochemistry, Vol. 6,* New York: Plenum Press, 1971.

ZIGMOND, R. E., AND MCEWEN, B. S. Selective retention of oestradiol by cell nuclei in specific brain regions of the ovariectomized rat. *Journal of Neurochemistry,* 1970, **17,** 889–899.

12

The Hippocampus and the Pituitary–Adrenal System Hormones

Béla Bohus

1. Introduction

The pituitary–adrenal system is one of the most prominent systems of bodily adaptation. Noxious stimuli specified as "stressors" by Selye (1936) activate the pituitary–adrenal system. The concept of a "general adaptation syndrome" was introduced by Selye (1950) based on the "nonspecificity" of the pituitary–adrenal cortical responses to a variety of stimuli, mainly physical. Although Selye (1950) himself observed that "even mere emotional stress" such as immobilization activates the pituitary–adrenal axis, it was only later recognized that psychological stimuli are among the most potent of all stimuli affecting the pituitary–adrenal system (Mason, 1968). This recognition led Mason (1971) to suggest that the "primary mediator" underlying the pituitary–adrenal response to a variety of stimuli "may simply be the psychological apparatus involved in emotional or arousal reactions to threatening or unpleasant factors in the life situation as a whole." It is also recognized that not only anxiety, fear, and rage are effective to stimulate the pituitary–adrenal axis but also stimuli signaling hope or disappointment (Coover *et al.*, 1971*a*; Levine *et al.*, 1972). As suggested by Levine *et al.* (1972), changes in expectancies during well-established behavior such as during reinforcement shifts or withdrawal of reinforcement in rewarded behavioral situations result in the activation of the pituitary–adrenal system. If we consider that all these stimuli which activate the pituitary–adrenal axis elicit rather specific behavioral changes, the specificity of the "stress" concept as suggested

Béla Bohus • Rudolf Magnus Institute for Pharmacology, Medical Faculty, University of Utrecht, Utrecht, The Netherlands.

by Mason (1971) seems to be obvious. However, this notion considers only a one-way relationship between the central nervous system and the pituitary–adrenal system: the pituitary ACTH release is controlled by the central nervous activity.

In the course of clinical assessment of adrenocortical hormones and pituitary adrenocorticotropic hormone (ACTH) as therapeutic agents, it was frequently noticed that a number of psychological changes occurred, including mood alterations and disturbance or excitability changes in convulsive threshold. A number of experimental observations are available suggesting not only that the release of pituitary–adrenal system hormones accompanies behavioral changes but also that these hormones modulate the changes. Adrenocorticotropic hormone of pituitary origin enhances the retention of learned responses (de Wied, 1966; Bohus et al., 1968a; Weiss et al., 1970; Gray et al., 1971; Guth et al., 1971). The effect of ACTH is not mediated through the adrenal cortex, because this peptide is also effective in adrenalectomized rats (Miller and Ogawa, 1962; Bohus et al., 1968a), and fragments of this peptide as α- and β-MSH, ACTH 1–10, or ACTH 4–10, which are practically devoid of adrenocorticotropic activity, are also effective to modulate both avoidance and approach behavior (de Wied, 1966; Kastin et al., 1973; Garrud et al., 1974; Bohus et al., 1975). Corticosteroids, on the other hand, enhance extinction of conditioned responses, i.e., promote the elimination of nonreinforced behavior (de Wied, 1967; Bohus, 1970, 1971, 1973; de Wied et al., 1972; Garrud et al., 1974). Furthermore, extinction of either active or passive avoidance responses is affected by the removal of the adrenal cortex or the pituitary gland (de Wied, 1967; Bohus et al., 1968a; Lissák and Bohus, 1972; Weiss et al., 1970).

Certain "acquisition" effects of the pituitary–adrenal system hormones have also been demonstrated. ACTH may facilitate (Bohus and Endröczi, 1965; Beatty et al., 1970; Guth et al., 1971; Pagano and Lovely, 1972) and corticosteroids suppress (Bohus, 1971, 1973) the acquisition of avoidance and approach responses. Level of motivation, stage of conditioning, diurnal rhythmicity, etc., are factors of primary importance for an acquisition effect of these hormones. However, adrenalectomized rats are not inferior in acquisition behavior (Applezweig and Moeller, 1959; Bohus and Endröczi, 1965; de Wied et al., 1968) and hypophysectomized rats may also acquire conditioned responses in the absence of pituitary ACTH (Stone and King, 1954; Stone and Obias, 1955; de Wied, 1964; Lissák and Bohus, 1972).

Taken together, these studies suggest that an intact pituitary–adrenal system is not an essential requisite for learning a response, but the physiological retention or elimination of the learned response requires an intact pituitary–adrenal system function. These studies also suggest that the pituitary–adrenal system hormones affect those adaptive behavioral responses which themselves activate the output of these hormones. The pituitary–adrenal system hormones do not elicit a behavior, but specifically modify ongoing behavior. That is, the psychological response is of primary importance, which in turn is specifically affected by the hormones acting on the brain but in a rather selective manner.

In order to demonstrate such selective action of hormones in the brain, localization of the site of action is of primary importance. The aim of this chapter is to describe direct evidence indicating that the hippocampus is but one locus of action of the pituitary–adrenal system hormones in the mediation of hormone effects on the brain. It is necessary to stress this point since some attempts have been made to infer the role of the hippocampus in the mediation of hormone effects on behavior on the basis of certain similarities between the influence of hippocampal ablation and of the pituitary–adrenal system hormones (Antelman and Brown, 1972; Pagano and Lovely, 1972). The similarities do not necessarily imply the involvement of the hippocampus in the hormonal adjustment of behavior, but the dissimilarities, and there are a number of them, do not necessarily exclude a relationship between the hippocampus and endocrine functions. Accordingly, direct evidence suggesting a reciprocal connection between the septal–hippocampal and the pituitary–adrenal system function will be discussed and an attempt will be made to determine the role of the hippocampus in neurohumoral adaptation.

2. Septal–Hippocampal Function and Release of Pituitary–Adrenal System Hormones

Recognition that the limbic system plays an important role in the organization of emotional behavior (Papez, 1937; MacLean, 1949) and that emotional stimuli activate the pituitary–adrenal system function directed research toward the study of the influence of septal–hippocampal system function on hormone release. Porter (1954) was the first to report that the eosinopenic response, used as the index of pituitary–adrenal activity to noxious stimuli, was inhibited by hippocampal stimulation in the monkey. Measurements of the plasma 17-hydroxycorticosteroid level as a more direct index of adrenal activity after hippocampal stimulation in stressed monkeys led Mason (1957) to a similar conclusion. In a more extended study, we subsequently demonstrated that electrical stimulation of the dorsal hippocampus inhibits the activation of the pituitary–adrenal system by both neural (painful electrical shock) and humoral (epinephrine, histamine) stimuli in several species (Endröczi et al., 1959). An inhibitory influence of the hippocampus on the pituitary–adrenal system function was demonstrated by Rubin et al. (1966) in humans, using low-frequency electrical stimulation of CA2 and CA1 layers of the hippocampus.

Observations in animals bearing lesions in the hippocampus or in the afferent or efferent pathways of this structure also favored the hypothesis that the role of the hippocampus in controlling ACTH release is primarily inhibitory. I have demonstrated that lesions in the medial septal area and the anterior cingulate cortex result in an elevated output of adrenocortical hormones in the rat (Bohus, 1961). Sub- and supracallosal lesions in the cat induced similar changes (Endröczi and Lissák, 1960). Ablation of the hippocampus proper also resulted in increase of both basal and stress-induced activation of pituitary–adrenal system function in the rat and cat (Knigge, 1961; Knigge and Hays, 1964; Fendler et al., 1961). (However, see Chapter 14.)

Subsequent studies have made the picture rather complicated. Endröczi and Lissák (1960, 1962) showed that the effect of electrical stimulation of the hippocampus on the pituitary–adrenal activity depended on the stimulation frequency and the behavioral effects of stimulation. Low-freqeuncy stimulation suppressed the stress-

induced pituitary–adrenal activation, while higher frequency than 120 Hz did not influence the response. Stimulation with a frequency of 240 Hz, on the other hand,
resulted in a further activation of stress response. However, if the animals were habituated to the stimulation, no significant changes were observed in stress response.
Kawakami *et al.* (1968*a*) reported that a rather high-frequency stimulation of CA2 and
CA3 layers of the dorsal hippocampus for 1 h resulted in a facilitation of the basal,
nonstress pituitary–adrenal function of the rabbit. This was followed by a slight
decrease 4 h after the cessation of stimulation. Facilitation of the pituitary–adrenal
system by immobilization, on the other hand, was blocked by the same stimulation.
Casady *et al.* (1972) could not confirm this observation in the rat. Low-frequency
stimulation did not influence the early elevation of the plasma corticosteone level due to
ether stress but some inhibition did occur in a later period. Immobilization-induced activation of the pituitary–adrenal system was not affected in an early period, but later a
facilitation was observed. Dupont *et al.* (1972), however, reported that low-frequency
(10 Hz) stimulation of the gyrus dentatus in the rat inhibited pituitary–adrenal activation following nicking of the tail or cold stress. Inhibitory effect of hippocampal
stimulation has also been reported in conscious unrestrained pigeon by Bouillé and
Baylé (1973/1974).

     More recent observations on the influence of hippocampal ablation or septal le-
sions on the pituitary–adrenal system function are also sometimes in conflict with
earlier data. No changes in plasma corticosterone level were observed by Coover *et al.*
(1971*b*) in rats with surgical damage to the hippocampus in resting conditions or after
exposure to either stress or novelty. This lack of effect was also found by Lanier
et al. (1975). Endröczi (1972), on the other hand, reported that the plasma
corticosterone level was higher in hippocampectomized rats than in controls after exposure to a novel environment. A similar observation was made in rats bearing a
rostral septal lesion (Endröczi and Nyakas, 1971). An increase in plasma corticosterone level in rats bearing large bilateral lesions in the medial septal area was observed by Seggie and Brown (1973) after behavioral testing for "septal syndrome."
Uhlir *et al.* (1974) demonstrated that septal lesions lowered the stimulation threshold
for activation of the pituitary–adrenal axis. On the other hand, Usher *et al.* (1967) and
Usher and Lamble (1969) reported that the pituitary–adrenal response to mild neuro-
genic stressors such as intermittent air blast or shuttle-box avoidance was completely
blocked by lateral septal lesions. Later, they suggested that the quantity and severity of
stress may be of importance, because ACTH release due to food deprivation was not
blocked but even slightly enhanced by the same septal lesions (Usher *et al.*, 1974).
Available data on the plasma corticosterone level under resting conditions in septally
damaged rats are also rather conflicting. Both increased (Usher *et al.*, 1974) and not
changed baseline levels (Seggie *et al.*, 1974; Uhlir *et al.*, 1974) have been reported.

     A number of observations indicate that ablation of the hippocampus or disruption
of its afferent or efferent connections modifies the circadian rhythmicity of the
pituitary–adrenal system function. Thus fornix transsection abolished the daily
rhythm of plasma corticosteroid level: the normally low morning levels were increased
up to the level of the afternoon peak (Mason, 1957; Nakadate and De Groot, 1963;
Moberg *et al.*, 1971). Endröczi and Nyakas (1971) and Endröczi (1972) reported

similar changes in the rat after rostral septal lesions or surgical removal of the hip-pocampus. Wilson and Critchlow (1973) were unable to observe the disappearance of daily rhythmicity of plasma corticosterone level in rats bearing septal lesions involving the fornix as well. Taylor *et al.* (1973), however, reported that stimulation of the lateral septum depressed the afternoon elevation of plasma corticosterone level in the freely moving rat. Wilson and Critchlow (1973/1974, 1974) were unable to observe substantial changes in daily rhythmicity of plasma corticosterone level in rats with hip-pocampal or septal damage. Similar observations have been reported by Seggie *et al.* (1974) in septally lesioned rats.

Although the observations discussed above are rather conflicting, I would like to assume that the septal–hippocampal system plays some role, albeit a complex one, in the control of the pituitary–adrenal system activity and that this role is primarily of an inhibitory character. It seems that the presence or absence of this inhibitory action de-pends on the functional state of the hippocampus and on the hormonal status of the organism at a given moment. The functional state of the hippocampus may depend on the environmental stimuli and on what the animal is actually doing or should do (Ben-nett, 1971; Olds, 1972). Therefore, stringent control of the experimental situation and selection of appropriate stimulus conditions to affect the pituitary–adrenal activity are essential in further studies. Furthermore, lesion studies remain inconclusive unless the neuroanatomical characteristics of the hippocampal formation and its afferent and ef-ferent connections are more precisely considered. The importance of the hormonal status of the organism—that is, the circadian rhythmicity and the basal or stressed con-dition of the pituitary–adrenal system—is suggested but not yet fully characterized. The question of whether the hormones directly affect the functional activity of the septal–hippocampal system or whether the hippocampal influences on other brain structures depend on the "hormonal state" of these other areas is the subject of the next sections.

3. Pituitary–Adrenal System Hormones and Functional Activity of the Septal–Hippocampal complex: Endocrine Aspects

Introduction of pituitary–adrenal hormones as therapeutic agents led to some observations that these hormones may induce mood alterations or changes in brain excitability in humans (Cleghorn, 1957). Hormone actions on the brain were also indicated experimentally by Woodbury *et al.* (1957), who showed changes in elec-troconvulsive shock threshold in rats by administering adrenocortical hormones. However, vast evidence of a reciprocal connection between the central nervous struc-tures responsible for the control of pituitary ACTH release and adrenocortical hormone secretion was first provided by neuroendocrine research. This research was stimulated by the fact that the excess or absence of adrenocortical hormones modifies the release of pituitary ACTH and this feedback effect of corticosteroids may be modified by hypothalamic or extrahypothalamic lesions (Ganong, 1963). It was then demonstrated by local implantation of steroids in the brain that the feedback of corti-costeroids was exerted on discrete brain areas known to be involved in the activation of the pituitary–adrenal axis (Mangili *et al.*, 1966; Kendall, 1971). The observations that the septal–hippocampal system may exert inhibitory influences on the pituitary–adrenal function stimulated our interest in investigating whether the adrenal steroids also have a feedback effect on these structures (Bohus and Lissák, 1967; Bohus *et al.*, 1968*b*). The methodological approach was to increase the connection of corticosteroids locally by implanting minute amounts of crystalline corticosteroids stereotaxically in the septum and the hippocampus of the rat. The effect of intraseptal and intrahippocampal cortisol implants on the pituitary–adrenal system activity was then studied 48 h after the implantation in rats exposed to environmental change (transfer of the animals from the animal quarters to the experimental room) and in rats under nonstress conditions (decapitation took place in the animal room in order to avoid stress). After decapitation, the adrenals were removed and the *in vitro* corticosteroid production was determined and used as the index of ACTH release.

Bilateral implantation of cortisol in the ventral hippocampus enhanced the release of ACTH in response to environmental change (Table 1). Basal ACTH release was also elevated in rats bearing ventral hippocampal implants. Dorsal hippocampal implants, on the other hand, did not affect the pituitary–adrenal response to stress and the basal ACTH release also remained unaltered. Cortisol implants in the medial septal area suppressed stress-induced ACTH release, and a decrease in basal pituitary–adrenal function appeared compared to that of cholesterol-implanted controls. These control rats, however, showed hyperirritability. This kind of behavior develops shortly after septal destruction and is known as the "septal syndrome" (Brady and Nauta, 1953). Therefore, it is likely that cholesterol implants produced a lesion-like effect which was also reflected in higher pituitary ACTH release in both basal and stress conditions compared to that of other controls bearing cholesterol implants in the hippocampus or cerebral cortex. It is noteworthy that cortisol-implanted rats did not exhibit hyperirritability. It is unlikely that cortisol implants specifically blocked the development of the "septal syndrome." Rather, the different physicochemical properties of cholesterol and cortisol crystals may be responsible for the phenomenon. Seggie and Brown (1973) reported that administration of dexamethasone, a highly potent synthetic corticosteroid, did not affect the development of hyperirritability after septal lesion, but blocked the increased pituitary ACTH release of lesioned rats. Cortisol implants in the basal septal area (lateral septal nuclei, nucleus accumbens) suppressed stress-induced ACTH release, but the basal pituitary–adrenal system function remained unaltered (Table 1).

Elevated basal and stress-induced ACTH release in rats with cortisol implants in the ventral hippocampus reinforces the notion that hippocampal control of the pituitary–adrenal system is primarily of an inhibitory nature. A similar conclusion was drawn by Knigge (1966), using cortisone for implantation in the rat, and by Slusher (1966), who observed that morning values of plasma corticosterone level are increased by ventral hippocampal cortisol implants. Facilitation of basal pituitary–adrenal function by dorsal hippocampal implants of corticosterone in CA2 and CA3 layers of the cornu ammonis and part of the fascia dentata but not in the CA1 layer was observed by Kawakami *et al.* (1968*b*) 3 wk after implantation in the rabbit. In our experiments, the pituitary–adrenal function remained unchanged after dorsal hippocampal cortisol im-

TABLE 1

Effect of Intracerebral Cortisol Implantation on the Pituitary–Adrenal System Function Under Environmental Stress or Nonstress Conditions

Implants	Site	Pituitary–adrenal activity[a]	
		Environmental stress	Nonstress
Cortisol	Ventral hippocampus	26.4 ± 1.3^b (12)[c]	15.6 ± 1.1^d (17)
Cholesterol		20.8 ± 1.5 (10)	11.6 ± 1.3 (9)
Cortisol	Dorsal hippocampus	25.1 ± 1.1 (23)	10.0 ± 0.9 (7)
Cholesterol		26.2 ± 2.1 (18)	12.5 ± 1.4 (9)
Cortisol	Medial septum	21.6 ± 2.0^d (11)	9.7 ± 1.4^d (8)
Cholesterol		27.0 ± 1.2 (8)	14.3 ± 1.2 (8)
Cortisol	Basal septum	10.3 ± 0.7^b (11)	10.9 ± 1.6 (8)
Cholesterol		20.1 ± 0.9 (9)	12.0 ± 2.1 (7)
Cortisol	Frontal cortex	21.1 ± 1.0 (8)	10.7 ± 1.1 (8)
Cholesterol		20.7 ± 1.1 (8)	11.1 ± 0.9 (8)

[a] Adrenal corticosteroid production *in vitro* in µg/100 mg adrenal/h: mean ± standard error of the mean.
[b] $p < 0.01$ (*t* test).
[c] Number of observations.
[d] $p < 0.05$ (*t* test).

plantation. It is not yet clear whether simply the difference in time between implantation and assessment of pituitary–adrenal system function would explain the different observations.

Most of the experiments with lesions suggested that the medial septal area exerts an inhibitory effect on the pituitary–adrenal system function similar to that of the hippocampus. Unfortunately, present attempts with corticosteroid implantation cannot answer the question of whether or not this area is corticosteroid sensitive from endocrine aspects because of the different behavior of steroid- and control-implanted rats. The effect of basal-lateral septal implants, on the other hand, resembled the influence of electrolytic lesions in this area. The lateral septum is intimately connected with the hippocampus (see Powell and Hines, this volume). In terms of what is known about hippocampal influences on the pituitary–adrenal system, it seems unlikely that both lesions and implantations in this area interfere with the hippocampal input or output. Since the septal area is also intimately connected with the hypothalamus and tegmentum, it seems that alterations in basal septal function by lesions or corticosteroid implantation impair excitatory rather than inhibitory mechanisms of the central control of the pituitary–adrenal system.

It should be stressed that the hippocampus and the septum are not the only extrahypothalamic regions which appeared to be corticosteroid sensitive. Inhibition of stress-induced ACTH release was reported after the implantation of various corticosteroids into the mesencephalic reticular formation, medial thalamic nuclei, and amygdala (Endröczi, 1961; Bohus and Endröczi, 1964; Corbin et al., 1965; Davidson and Feldman, 1967; Bohus et al., 1968b; Dallman and Yates, 1968; Kendall, 1971). Therefore, the septum and the hippocampus represent but one part of a corticosteroid-sensitive system in the limbic–midbrain circuit. Strong evidence for such a corticosteroid-sensitive system has been provided by McEwen et al. (1969), who showed that the hippocampus and the septum preferentially take up labeled corticosteroids. Furthermore, autoradiographic experiments showed selective steroid uptake in all of those regions which were shown functionally as corticosteroid-sensitive areas (Stumpf, 1971).

Implantation experiments should be considered as model studies suggesting a potentiality of the given region to be affected by local corticosteroid concentration. The characteristics of the pituitary–adrenal system response are then determined by the interaction between excitatory and inhibitory mechanisms. This interaction may be of a temporal sequence; that is, the excitatory and inhibitory mechanisms may not compete but follow each other in order. Evidence in favor of this suggestion has been provided by experiments concerning influence of chlorpromazine, a centrally acting drug, on the function of the pituitary–adrenal system (Bohus and de Wied, 1967a). It was found that high doses of chlorpromazine given systemically induced a long-lasting release of the pituitary ACTH, but a partial inhibition of stress-induced activation of the pituitary–adrenal system occurred 12 h after the administration of the drug. This inhibitory action of the drug was prevented by lesions interrupting the basal hypothalamic projection of the hippocampal–fornix system, but the drug-induced facilitation remained unaffected. Furthermore, local injection of this drug in the mesencephalic reticular formation elevated basal pituitary–adrenal activity, but, similar to the systemic administration, blocked stress-induced ACTH release, presumably due to activating inhibitory (hippocampal) mechanisms as well. Accordingly, the steroid sensitivity or insensitivity of excitatory and inhibitory structures may follow a temporal sequence depending on the input modalities but also on some genuine characteristics of the steroid receptors. Such a highly integrated control mechanism could then assure the dynamic character of the pituitary–adrenal responsiveness for adaptive requirements. This suggestion seems to be highly speculative, but supported by a number of observations. Thus it has already been noted that electrical stimulation of the hippocampus elicits changes in the pituitary–adrenal system function depending on the functional activity of both the hippocampus and the pituitary–adrenal system. Furthermore, failure to affect the pituitary–adrenal system response to stressors such as ether or sound by intrahippocampal cortisol implants in the rat (Feldman, 1973) suggests the importance of the stimulus modalities. Ether stress activates the pituitary–adrenal system function even in the absence of afferents to the hypothalamus, and the response to sound seems to be organized in hypothalamic rather than in extrahypothalamic levels (Feldman and Dafny, 1970; Feldman, 1973). Temporal aspects of hippocampal inhibitory influences in adaptation to repeated stress were suggested by Kawakami *et al.* (1972). They observed that the increased multiple-unit activity in the hippocampus of immobilized rabbits was diminished after repeated immobilization and the pituitary–adrenal response was also diminished in a parallel fashion. Transsection of the fornix appeared to block the adaptation of the pituitary–adrenal response. As far as the characteristics of the uptake of corticosteroids in the hippocampus and septum are concerned, McEwen and Weiss (1970) have shown that the hippocampal receptor sites may easily be saturated even with physiological levels of corticosteroids while septal receptor sites cannot be saturated.

It is of special interest that ventral hippocampal implants which increased morning low levels of plasma corticosterone suppressed the afternoon crest values (Slusher, 1966). That is, the response of hippocampal cells to high local corticosteroid concentration depended on the circadian rhythm of metabolic activity of the neural tissue. Evidence in favor of such a mechanism was presented by Scapagnini *et al.* (1971), who showed that the 5-hydroxytryptamine content of the hippocampus, amygdala, and frontal cortex varied with the circadian rhythm of plasma corticosterone in the rat. Parachlorphenylalanine treatment caused a decrease in the 5-HT content of these structures together with an increase in morning levels of plasma corticosterone and a decrease in afternoon levels. In addition, Vernikos-Danellis *et al.* (1973) emphasized the role of the 5-HT system in the corticosteroid feedback mechanism regulating pituitary ACTH release. The enzyme tryptophan hydroxylase, which converts tryptophan to 5-HT in the brain, is also affected by the pituitary–adrenal function: adrenalectomy decreases, corticosterone administration increases the activity of this enzyme (Azmitia and McEwen, 1969). It is premature to draw a conclusion that a serotonergic system may modulate the hippocampal influences on the circadian rhythmicity and adaptive responsiveness of the pituitary–adrenal system function by altering the response of hippocampal cells to corticosteroid taken up selectively. However, the above-discussed studies and the series of experiments by Fuxe *et al.* (1973) are promising and may provide a system which is affected by external and internal stimuli and which may determine the endocrine and behavioral effects of corticosteroids on the brain.

4. Pituitary–Adrenal System Hormones and Functional Activity of the Septal–Hippocampal Complex: Behavioral Aspects

In the course of studying the site and mode of action of the pituitary–adrenal system hormones on adaptive behavior, two lines of observation stimulated our interest in investigating the involvement of the septal–hippocampal system in the hormonal modulation of behavior. First, behavioral observations were in favor of a suggestion that the mode of action of corticosteroids is to enhance internal inhibitory processes in the brain (Lissák and Endröczi, 1964; Bohus and Endröczi, 1965; Bohus and Lissák, 1968). A number of electrophysiological and behavioral observations favored the hypothesis that the hippocampus constitutes part of the neural systems necessary to generate brain processes which are equivalents of internal inhibition. Second, it has been demonstrated that ACTH also affects adaptive behavior independent of its action on the adrenal cortex (Miller and Ogawa, 1962; de Wied, 1966; Bohus et al., 1968a). Although behavioral effects of corticosteroids not mediated by the pituitary gland (de Wied, 1967; Bohus, 1968) were observed, it was also found that suppression of ACTH release by local implantation of corticosteroids in the median eminence of the hypothalamus results in a similar behavioral action as systemic corticosteroid administration (Bohus, 1968). Furthermore, as has been discussed before, hippocampal corticosteroid-sensitive mechanisms may alter pituitary ACTH release. Therefore, two questions were raised: first, whether the septal–hippocampal system as part of the internal inhibitory system is a locus of the behavioral action of corticosteroids and, second, whether the behavioral action of steroids depends on the changes of pituitary ACTH release.

To answer the first question, rats were trained to avoid an electrical shock by jumping onto a pole (Bohus, 1970). The acquisition period was continued until the rats had reached the criterion of learning, i.e., 24 or more avoidances during three consecutive sessions of ten trials each. Then crystalline cortisone acetate was implanted bilaterally in the hippocampus and in the medial septal nuclei. The rats were then reconditioned until they reached the learning criterion again. Extinction of the conditioned avoidance response was studied by presenting nonreinforced trials. As in intact rats treated with cortisone, intrahippocampal or intraseptal implantation of cortisone acetate enhanced extinction of the conditioned avoidance response. The latencies of the avoidance responses were, however, of the same order of magnitude in both corticosteroid-implanted and control rats (Fig. 1). The hippocampal implants were localized in CA1 and CA2 layers of the dorsal hippocampus. Ventral hippocampal implants were only slightly or not at all effective to enhance extinction. Corticosteroid implants in the ventricles almost completely inhibited avoidance performance during extinction and suppressed open-field activity as well. Effective hippocampal or septal implants did not influence open-field activity. Implants in the frontal or occipital cortex appeared to be ineffective on the avoidance and open-field behavior.

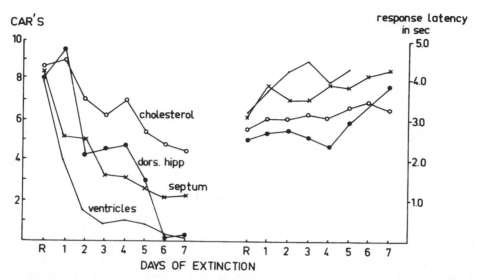

Fig. 1. Effect of dorsal hippocampal and medial septal implantation of cortisone acetate on the extinction of a conditioned avoidance response. Abbreviations: CAR, conditioned avoidance response; R, last session of reconditioning.

Other implantation experiments showed that forebrain structures such as the area hypothalamica anterior and the amygdala and some brain stem structures such as the rostral mesencephalic reticular formation and the posterior thalamic parafascicular area are also involved in the mediation of the behavioral effects of corticosteroids (Bohus, 1968, 1970; van Wimersma Greidanus and de Wied, 1969). On the basis of these observations, it has been suggested that the behavioral effect of corticosteroids is of a dual character, i.e., enhancement of forebrain inhibition and suppression of ascending reticular activation. This dual character of the effect seemed to be of a synergistic nature in facilitating extinction of a conditioned avoidance response (Bohus, 1970). There were, however, observations which seemed difficult to interpret in terms of a synergistic influence of the two systems. Contrary to expectations, corticosteroids suppressed passive avoidance behavior in both aversive–aversive and approach–aversive conflict situations (Bohus et al., 1970). Cortisone or dexamethasone administered before the single learning trial suppressed both immediate and long-term retention of passive avoidance responses. Accordingly, if corticosteroids affect forebrain inhibitory mechanisms, then observations of both active and passive avoidance performance suggest that the effect is to attenuate motivational influences by influencing the input to the forebrain structures. However, if the corticosteroid effect on forebrain areas were enhancement of internal inhibition, then better avoidance would be expected in rats bearing corticosteroid implants in the forebrain.

In order to investigate the role of the septal–hippocampal system in passive avoidance behavior affected by corticosteroids, water-deprived rats were trained to run for water reward in a three-compartment apparatus consisting of a start box, a runway compartment, and a goal box with the spout of a water container (Bohus, 1971). Acquisition training consisted of one trial on each of 4 consecutive days, during which the rat was free to drink in the goal box for 10 min. After these acquisition trials, crystalline cortisone acetate was implanted in the septal–hippocampal complex. Twenty-four hours after this, a further trial was given, but the spout and the grid floor of the box were electrified (0.14 mA, a.c.). The shock lasted throughout the 10-min trial.

In rats bearing cortisone implants in the hippocampus or medial septal area or those injected with cortisone acetate (2.0 mg/100 g body weight) 3 h prior to the shock trial, immediate passive avoidance was significantly suppressed compared to that of controls bearing cholesterol implants in the same areas (Table 2). Rats with cortisone implants in the frontal or orbital cortex exhibited similar passive avoidance behavior as the controls. After the first shock, which the animals received by touching the electrified spout, the latency of the first attempt to drink was significantly shorter in experimental animals. Since the shock continued during the 10-min test period, each further attempt was punished. Despite this punishment, the corticosteroid-implanted rats (hippocampal and septal groups) and the systemically treated animals made more attempts to drink, and between these attempts also a high number of approaches toward the spout were made. Although the passive avoidance behavior of the experimental rats was suppressed immediately after the shock, a repeated test 24 h later led to some surprising observations. Rats with cortisone implants in the dorsal hippocampus initiated drinking with shorter latencies than the controls. Rats with septal cortisone implants or with cholesterol implants, but treated with cortisone, showed longer latency to initiate drinking than controls (Table 3).

At least two basic problems were raised by these latter findings. First, the implantation of cholesterol in the septal region induced hyperirritability. However, such behavior was completely absent in cortisone-implanted rats and in cholesterol-implanted animals treated with cortisone acetate. Although a number of studies suggest that septal hyperreactivity is not correlated with other response measures (Fried, 1972a), the possibility cannot be excluded that the local injury caused by the implantation modifies the behavioral effects of corticosteroids both in the septum and in other areas. Second, cortisone-implanted or -treated rats made more attempts to drink during the first test and therefore received more shocks than the controls. Accordingly, this difference may be reflected in their performance in the repeated test 24 h later. Rats with cortisone implants in the hippocampus or animals bearing hippocampal cholesterol implants and treated with cortisone received more shocks than the controls but showed less retention of the response at the 24-h trial. However, the fact that septal injury leads to other behavioral changes besides hyperreactivity, such as increased exploratory activity and water intake, which are probably not due to the disruption of septal–hippocampal functions (Fried, 1972a) suggests that the shock effects cannot be equally treated in the septal and hippocampal groups. Therefore, two further experiments were conducted using a more stringent control of the implantation procedure and testing the rats either immediately after the first shock for 10 min or removing them from the conditioning apparatus immediately after the first shock and testing 24 h later. The other procedures were the same as in the previous study except that corti-

TABLE 2

Effect of Cortisone Acetate Implanted Intracerebrally or Administered Systemically on the Immediate Passive Avoidance Behavior of the Rat

Implant	Site	Treatment	N	Avoidance latency[a]	Repeated attempts to drink[b]	Approaches[b]
Cortisone acetate	Dorsal hippocampus	Saline	11	66.4[c]	6.1 ± 1.6[d]	35.5 ± 6.1[e]
Cholesterol		Saline	7	132.2	2.8 ± 1.0	21.6 ± 5.6
Cholesterol		Cortisone acetate	6	34.7[f]	7.3 ± 1.1[d]	39.5 ± 7.2[e]
Cortisone acetate	Medial septum	Saline	8	79.7[c]	4.1 ± 0.3[d]	46.6 ± 7.5[d]
Cholesterol		Saline	8	126.7	2.4 ± 1.0	26.8 ± 3.1
Cholesterol		Cortisone acetate	8	50.1[f]	4.5 ± 0.6[d]	39.4 ± 4.6[e]
Cortisone acetate	Cortex	Saline	8	141.4	3.1 ± 1.0	21.7 ± 3.5
Cholesterol		Saline	6	130.7	2.7 ± 1.2	23.1 ± 2.4
Cholesterol		Cortisone acetate	7	46.8[f]	7.9 ± 1.0[d]	41.5 ± 3.5[d]

[a] Median in seconds.
[b] Mean ± standard error of the mean.
[c] $p < 0.05$ (U test).
[d] $p < 0.01$ (t test).
[e] $p < 0.05$ (t test).
[f] $p < 0.01$ (U test).

TABLE 3

Effect of Cortisone Acetate Implanted Intracerebrally or Administered Systemically Prior to the Learning Trial on the Retention of Passive Avoidance Response 24 h After the Learning Trial

Implant	Site	Treatment	N	Avoidance latency[a]	Approach index[b]
Cortisone acetate	Dorsal hippocampus	Saline	11	110.0[c]	14.7 ± 2.1[d]
Cholesterol		Saline	7	214.7	6.1 ± 1.0
Cholesterol		Cortisone acetate	6	84.0[e]	19.4 ± 3.1[d]
Cortisone acetate	Medial septum	Saline	8	426.0[c]	11.7 ± 3.5
Cholesterol		Saline	8	259.0	7.6 ± 2.1
Cholesterol		Cortisone acetate	8	449.7[c]	15.9 ± 4.0
Cortisone acetate	Cortex	Saline	8	221.0	5.7 ± 3.0
Cholesterol		Saline	6	210.7	7.1 ± 2.1
Cholesterol		Cortisone acetate	7	96.7[e]	16.5 ± 3.1[f]

[a] Median in seconds.
[b] Number of approaches/avoidance latency × 100: mean ± standard error of the mean.
[c] $p < 0.05$ (U test).
[d] $p < 0.01$ (t test).
[e] $p < 0.01$ (U test).
[f] $p < 0.05$ (t test).

costerone, which is the main physiological product of the rat adrenal cortex, was used for intracerebral implantation or systemic treatment. Previous experiments have revealed that, at near physiological dose levels, corticosterone is more effective on passive avoidance behavior than other, "nonphysiological" corticosteroids for the rat such as cortisone and cortisol (Bohus, 1973). Besides cholesterol, deoxycorticosterone was used as a control substance. This steroid affects brain excitability as measured by convulsive threshold after electroshock, but the effect is opposite to that of cortisol (Woodbury et al., 1957). The pituitary–adrenal activity was also assessed in order to determine whether behavioral alterations are correlated with the changes in ACTH release.

The apparent difference between the previous observations using cortisone and the corticosterone experiments was that suppression of passive avoidance behavior by dorsal hippocampal implantation of corticosterone or by systemic administration of corticosterone (0.5 mg/100 g body weight) prior to the learning trial appeared to be significant only in the 24-h retention test. The immediate passive avoidance latencies were slightly shorter but not significantly different from those of controls. The corticosterone-implanted or -treated rats made more approaches toward the spout but no more attempts to drink than the controls. The extended control studies indicate that the operation procedures in the hippocampus did not influence the behavior and the responsiveness to systemic treatment (Tables 4 and 5).

The observations with septal implantations and treatments clearly indicated that each manipulation in the septal region affected the behavior of the rats and modified the response to local corticosterone implantation or systemic corticosterone treatment. Corticosterone-induced behavioral changes are therefore the result of interactions between local tissue damage (or stimulation?) and the hormone effect rather than a genuine corticosterone effect. Accordingly, the results with septal implantations cannot be conclusive.

The experiments with hippocampal implants reinforced the notion that the hippocampus should be considered as a corticosteroid-sensitive structure for behavioral purposes. Although the behavior of the rats in a conflict situation may be altered by changes in either avoidance or approach components, it seems unlikely that suppression of passive avoidance retention by corticosterone given prior to the learning trial is due to an increased approach rather than decreased avoidance tendency. The approach behavior of the rats was not affected by corticosterone implantation in the hippocampus. Open-field behavior and water intake pattern after 23.5 h water deprivation also remained unaltered. It is worthwhile to mention that corticosterone when administered during the acquisition of a space-discriminative conditioned drinking response impairs the acquisition when the drive intensity is low (12 h water deprivation) but not when it is high (Bohus, 1973). Therefore, an influence on approach behavior would tend to improve instead of suppress conflict-induced passive avoidance.

The issue of approach–avoidance tendencies in a conflict situation may be of importance to explain controversions between our observations and some more recent ones of Endröczi and Nyakas (1971, 1972) and Endröczi (1972). They observed that corticosteroids enhanced the immediate passive avoidance response of adrenalectomized rats in a thirst vs. fear conflict situation. However, the effective dose of steroids

TABLE 4

Effect of Corticosterone Implanted in the Dorsal Hippocampus and Medial Septum or Administered Systemically on the Immediate Passive Avoidance Behavior of Male Rats

Implant	Site	Treatment	N	Avoidance latency[a]	Repeated attempts to drink[b]	Approaches[b]
Corticosterone	Hippocampus	Saline	17	38.0	5.9 ± 1.1	19.9 ± 2.6[c]
Deoxycorticosterone		Saline	8	61.0	4.3 ± 1.2	16.4 ± 3.6
Cholesterol		Saline	10	53.5	6.4 ± 1.6	11.8 ± 0.9
Cholesterol		Corticosterone	8	58.2	5.2 ± 1.5	18.7 ± 3.2[d]
Sham		Saline	6	51.0	3.5 ± 1.0	14.5 ± 1.1
Sham		Corticosterone	8	41.7	5.7 ± 1.9	18.9 ± 1.6[d]
Corticosterone	Septum	Saline	8	29.0[e]	4.1 ± 1.0	17.8 ± 5.1
Deoxycorticosterone		Saline	8	61.5	5.4 ± 1.7	21.2 ± 2.9
Cholesterol		Saline	8	63.5	3.0 ± 0.9	19.0 ± 3.6
Cholesterol		Corticosterone	8	54.5	5.8 ± 1.7	16.7 ± 3.9
Sham		Saline	5	175.0	6.2 ± 2.1	19.4 ± 4.8
Sham		Corticosterone	8	141.6	6.7 ± 3.1	20.8 ± 4.5

[a] Median in seconds.
[b] Mean ± standard error of the mean.
[c] $p < 0.01$ (t test).
[d] $p < 0.05$ (t test).
[e] $p < 0.05$ (U test).

TABLE 5

Effect of Corticosterone Implanted in the Dorsal Hippocampus and Medial Septum or Administered Systemically Prior to the Learning Trial on the Retention of the Passive Avoidance and the Pituitary–Adrenal Responses 24 h After the Learning Trial

Implant	Site	Treatment	N	Avoidance latency[a]	Approach index[b]	Pituitary–adrenal activity[c]
Corticosterone	Hippocampus	Saline	8	57.2[d]	25.2 ± 3.1[e]	24.3 ± 1.2
Deoxycorticosterone		Saline	7	75.9	10.9 ± 3.7	22.9 ± 0.7
Cholesterol		Saline	8	90.9	7.7 ± 3.1	23.4 ± 1.7
Cholesterol		Corticosterone	8	49.7[d]	21.5 ± 3.0	21.7 ± 2.4
Sham		Saline	8	89.7	8.1 ± 3.1	26.2 ± 2.1
Sham		Corticosterone	7	40.7[d]	26.2 ± 1.1[e]	25.8 ± 2.1
Corticosterone	Septum	Saline	8	118.0[d]	20.4 ± 4.2	15.4 ± 3.1
Deoxycorticosterone		Saline	7	79.7[d]	14.7 ± 1.1	24.6 ± 3.8
Cholesterol		Saline	8	212.3	10.2 ± 3.3	23.8 ± 2.1
Cholesterol		Corticosterone	8	231.0	9.7 ± 1.7	23.7 ± 2.1
Sham		Saline	8	81.7	7.9 ± 3.6	24.1 ± 2.3
Sham		Corticosterone	7	46.7[d]	17.6 ± 4.1	23.5 ± 2.1

[a] Median in seconds.
[b] Number of approaches/avoidance latency × 100: mean ± standard error of the mean.
[c] Adrenal corticosteroid production in vitro in μg/100 mg adrenal/h: mean ± standard error of the mean.
[d] $p < 0.05$ (U test).
[e] $p < 0.01$ (t test).

also resulted in a long-lasting suppression of the exploratory activity of the water-deprived, adrenalectomized rats and delayed the latency to start to drink in a nonconflict
situation. Accordingly, facilitated passive avoidance by corticosteroids in their experiments seems to be rather due to the alteration in exploratory activity and thirst drive
than to a primary influence on aversive response. It is noteworthy that impaired
passive avoidance behavior and increased exploratory activity in rats with septal lesions or surgical hippocampectomy remained unaltered after corticosteroid administration (Endröczi and Nyakas, 1971; Endröczi, 1972).

The implantation studies in both active and passive avoidance situations revealed
the involvement of the hippocampus in the mediation of the corticosteroid effect on
adaptive behavior, but the role of the septal area remained questionable. The
observations on pituitary–adrenal system activity in the passive avoidance experiments
reinforced the notion that the behavioral effects of corticosteroids are due to a primary
action on the organization of behavior rather than an action on pituitary ACTH
release. The specificity of the hippocampus as but one site of the action of corticosteroids is shown by the fact that implantation of minute quantities of hormones into
this structure mimics the effects of systemic administration of larger quantities of
these hormones.

One of the strongest indications in favor of a physiological significance of
corticosteroid action on the hippocampus and the septal region is provided by the
observations of McEwen et al. (1969). They were the first to demonstrate that labeled
corticosterone is selectively accumulated in the hippocampus at higher concentration
than in the other brain regions when the steroid is given systemically to adrenalectomized rats. Corticosterone binds tightly to the cell nuclei in the hippocampus, and to
the cytosol fraction as well (McEwen et al., 1970; Stevens et al., 1971). The nuclear
binding is specific for corticosteroids since other steroids such as progesterone are
bound to the cytosol fraction but not to the nuclei (McEwen and Wallach, 1973).
Limited-capacity retention in the hippocampus on a more general entry of corticosteroids into the brain tissue is of special interest (McEwen et al., 1969; Knizley,
1972). Autoradiographic studies also revealed the uptake of labeled corticosteroids not
only in the pyramidal layer of the cornu ammonis and the stratum granulosum of the
gyrus dentatus but also in the septum and the precommissural and supracommissural
hippocampus (Stumpf, 1971; Gerlach and McEwen, 1972).

It is far from clear how the corticosteroids affect the septal–hippocampal system.
The observations with corticosteroid-implanted rats in both one-way active and passive
avoidance situations seem to be in favor of a deactivation hypothesis; that is, implanted
rats seem to behave similarly to hippocampus-damaged animals during one-way avoidance extinction (Olton and Isaacson, 1968) and during passive avoidance (Kimura,
1958; Isaacson and Wickelgren, 1962; Kimble, 1963). But opposite to the effects of
hippocampal ablation on the relearning of an already learned one-way avoidance
response (Olton and Isaacson, 1968), cortisone implantation in the septum and hippocampus does not affect the reacquisition of a conditioned avoidance behavior.
Furthermore, implantation of corticosterone in the hippocampus prior to the passive
avoidance learning did not prevent inhibition of the previously learned response during
the immediate passive avoidance retention test, but did affect retention 24 h after the

learning trial. Accordingly, corticosteroid-implanted rats cannot be fully considered as functionally hippocampectomized animals.

Another point worthy of discussion comes from the series of observations that either systemic administration or intraventricular or intrahippocampal infusion of corticosteroids markedly increases the excitability of the hippocampus to evoke propagated afterdischarges (Lissák and Endröczi, 1962; Feldman, 1966, 1971; Endröczi, 1969). Posttrial stimulation of the dorsal hippocampus resulting in propagated afterdischarges results in a retention deficit of passive avoidance or conditioned emotional responses (Barcik, 1970; Brunner et al., 1970; Nyakas and Endröczi, 1970; Vardaris and Schwartz, 1971). If locally implanted corticosteroids induce spontaneous afterdischarges, then the observed behavioral effects could be explained in terms of amnestic treatments. However, spontaneous afterdischarges were not reported by Endröczi (1969) after the injection of corticosteroids in the layer of apical dendrites of pyramidal cells in CA1 and CA2 layers of the dorsal hippocampus. Although no EEG records were taken in our implantation studies, the observed behavioral changes did not resemble "amnestic" effects. It is also questioned whether increased hippocampal convulsive susceptibility bears any physiological significance in the mechanism of action of corticosteroids. Corticosterone, which is the main physiological product of the rat adrenal cortex, does not substantially influence the convulsive threshold for electroshock in the rat (Woodbury et al., 1957). According to Endröczi (1969), the threshold to induce propagated hippocampal seizures by hippocampal stimulation was only slightly affected by this steroid, but cortisone or cortisol was highly effective to lower the threshold. Early clinical observations demonstrated slow-wave activity with paroxysmal bursts in patients' EEG records under corticosteroid therapy (Streifler and Feldman, 1953; Glaser, 1953). High doses of cortisol given intravenously resulted in similar changes of the EEG pattern of the rabbit (Feldman and Davidson, 1966) or even led to electrical and behavioral convulsions in the cat after intraventricular administration (Feldman, 1966). Cortisone implantation in the ventricles appeared to strongly suppress both conditioned and open-field behavior in our experiments. Accordingly, all these data call attention to the possibility that, at pharmacological levels, convulsive activity of steroids plays a role in the behavioral action of these hormones.

Increasing numbers of observations indicate nonconvulsive changes in the electrical activity of the septal–hippocampal system due to alterations in pituitary–adrenal system function. Kawakami et al. (1966) have shown that corticosteroids gradually increase the 4–8 cps component in the hippocampal rhythmic slow-wave activity of the freely behaving rabbit. According to Pfaff et al. (1971), corticosterone if given intraperitoneally decreases the unit activity as recorded from the pyramidal layer and dentate gyrus of the dorsal hippocampus of freely moving rats. Local injection of corticosterone into the hippocampus mimics the effect of systemic administration. More recently, Gray (personal communication, 1972) observed that corticosterone in a dose which accelerates the extinction of a food-reinforced runway response selectively lowers the threshold for evocation of θ rhythm in the hippocampus to the frequency of about 7 Hz. In an earlier experiment, Gray (1972) showed that septal driving of hippocampal activity with a frequency of 7.7 Hz decreased the resistance to extinction of a water-reinforced runway response. That is, the "corticosteroid-influenced" and the "extinction" frequencies are different. In the rat, the "corticosteroid-influenced" frequency generally accompanies the performance of fixed action patterns, such as eating, drinking, and grooming (Routtenberg, 1968). The frequency around 7.7 Hz is related to the response to novelty or frustrative nonreward (Gray and Ball, 1970), while frequencies of 8.5–10 Hz accompany the initiation of the performance of learned patterns of behavior (Elazar and Adey, 1967; Gray and Ball, 1970; Vanderwolf, 1969). Small-amplitude, irregular activity has been observed during a sudden break in behavior which may occur, for example, in a passive avoidance situation (Vanderwolf, 1971). Discussion of the "meaning" of hippocampal electrical activity cannot be the subject of this chapter. However, some preliminary conclusions can be drawn on the basis of the abovementioned data. It seems likely that corticosteroids affect the input to the hippocampus, probably by influencing both the septal "pacemaker" activity (Stumpf, 1965) and the hippocampal units generating slow-wave activity and therefore resulting in a lower-frequency response of the hippocampus. Activation of the reticular activating system is known to be accompanied by θ activity in the hippocampus (Green and Arduini, 1954) and the frequency of θ rhythm depends on the intensity of the stimulation of the midbrain reticular formation (Klemm, 1972). This activation of the septal–hippocampal system is clearly multisynaptic. Accordingly, the characteristics of the stimulus input into the septal–hippocampal system may be modified at several levels. Indeed, it has been observed that the septal–hippocampal system is not the one locus where corticosteroids act to modulate adaptive behavior. Changes in active and passive avoidance retention behavior were also observed after brain stem, midline thalamic, and forebrain implantation of corticosteroids (Bohus, 1968, 1970, 1973; van Wimersma Greidanus and de Wied, 1969). These changes are similar to those evoked by septal–hippocampal corticosteroid implants. Furthermore, a number of electrophysiological studies indicate that the response to sensory modalities in these areas is modified by systemic corticosteroid administration (Feldman and Dafny, 1970; Korányi and Endröczi, 1970). It is of course questionable whether such multiple corticosteroid-sensitive loci are functionally always active or selectively participate in the modulation of learned behavior. Implantation studies should be considered rather as experimental models which may provide information on the potentiality of certain loci to be corticosteroid sensitive from behavioral aspects. Meanwhile, the fact that the behavioral effect of corticosteroid is a function of drive intensity (Bohus, 1971, 1973) suggests that the corticosteroid sensitivity of these loci in the brain stem and the forebrain may be more prominent at a moderate level of activation.

5. The Septal–Hippocampal System: Is It the Substrate of the Behavioral Effect of ACTH and Related Peptides?

It has already been mentioned in the Introduction that adrenocorticotropic hormone of pituitary origin affects adaptive behavior of the rat. The influence of ACTH appears to be of extraadrenal nature and opposite to the effect of corticosteroids (de Wied, 1967; Bohus *et al.*, 1968a; Weiss *et al.*, 1970). Furthermore, these behavioral effects are not the genuine property of the whole ACTH molecule because peptides with practically no adrenocorticotropic activity which share the 4–10 amino acid sequence of the ACTH molecule such as α- or β-MSH, ACTH 1–10, and ACTH 4–10 affect behavior (de Wied, 1969; Kastin *et al.*, 1973) as greatly as the entire molecule. Gray *et al.* (1971) reported that administration of ACTH during the acquisition of a partially reinforced appetitive response appeared to block both the usual partial reinforcement acquisition and extinction effects, resembling the influence of amylobarbitone on this behavior. Gray and Ball (1970) showed that amylobarbitone raised the threshold of septal driving of θ frequencies to 7.5–8.5 Hz. This frequency range of hippocampal activity accompanied frustrative nonreward (partial reinforcement) in the freely behaving rats. Gray and Ball (1970) proposed that the behavioral effect of the drug was due to the disruption of the septal–hippocampal function. The electrical driving of θ rhythm through the medial septum with these frequencies elicits behavior opposite to that observed after amylobarbitone or ACTH (Gray, 1972). It has also been observed that ACTH 4–10 increases the threshold of septal θ driving in the higher-frequency ranges (Gray, personal communication). Changes in the electrical activity of the hippocampus following ACTH or ACTH-like peptide administration were reported by other investigators as well. Thus Kawakami *et al.* (1966) reported that ACTH administration gradually decreased the 4–8 Hz components of hippocampal slow-wave activity in freely behaving rabbits. Increases of unit activity in the hippocampus after ACTH administration were observed by Pfaff *et al.* (1971) in normal and hypophysectomized freely moving or urethane-anesthetized rats. However, the number of units investigated was quite low and only 50% showed an increased activity. More recently, a shift in the dominant frequency of hippocampal θ activity during prestimulus "facing" period of a conditioned food-reinforced operant response was observed by Urban *et al* (1974) in the dog after ACTH 4–10 administration. The peptide treatment shifted the dominant frequency into the direction of lower frequencies without affecting the conditioned behavior under mild food deprivation. Urban and de Wied (1975) further showed that administration of ACTH 4–10 produces a 0.5-Hz shift in dominant frequency and an increase in 7.5–9.0 Hz components of θ activities both in hippocampus and in posterior thalamus as induced by the electrical stimulation of the reticular formation in freely moving rats. Since similar changes could be observed by increasing the stimulus intensity, they suggested that ACTH 4–10 increases the excitability of the θ-generating system.

These observations may favor the suggestion that the hippocampus is also a locus of the behavioral action of ACTH or ACTH-like peptides. However, our attempts to localize the site of action of ACTH-like peptides revealed the parafascicular area of the posterior thalamus as the primary area where these peptides modify avoidance behavior. Destruction of this area by electrolytic lesions appeared to block the effect of θ-MSH or ACTH 4–10 on the extinction of either a shuttle-box or a pole-jumping active avoidance response (Bohus and de Wied, 1967*b*; van Wimersma Greidanus *et al.*, 1974). Intracerebral implantation of ACTH 4–10 in this area, on the other hand, mimicked the influence of systemically injected peptide on the extinction of a conditioned avoidance response: the extinction was significantly delayed (van Wimersma Greidanus and de Wied, 1971). Furthermore, it has been suggested that the opposite effects of ACTH and corticosteroids on conditioned avoidance behavior are probably mediated through the same thalamic area (Bohus, 1970). It was observed that implantation of cortisone in the centromedian–parafascicular area of the thalamus of the rat blocked the behavioral effect of systemically administered ACTH. On the other hand, ACTH delayed the extinction of the avoidance response when the cortisone implants were located in the forebrain including the medial septum. The observations of Bush *et al.* (1973) are along this line of data. They showed that impaired acquisition of a shuttle-box avoidance response in amygdala-lesioned rats was normalized by ACTH administration.

Taken together, these observations do not allow the conclusion that the septal–hippocampal complex is the locus of action of ACTH or ACTH-like peptides to modulate adaptive behavior. Rather, it seems more likely that changes in the hippocampal electrical activity are the reflection of alterations of the activity of the reticular activating system. Extensive research is going on in this area which may elucidate the role of the septal–hippocampal system in the central nervous effects of these peptides.

6.   *General Discussion and Conclusions*

The data presented in this chapter leave no doubt about a reciprocal connection between the septal–hippocampal and the pituitary–adrenal systems. The response of the pituitary–adrenal axis to changes of the external environment is modified by the hippocampus, but the actual role of the hippocampus in controlling pituitary ACTH release seems to be determined by some known and some unknown modalities of the external environment, and the *milieu intérieure* of the hippocampus itself. The other side of the coin is the endocrine and behavioral aspects of corticosteroid action on the septum and hippocampus through corticosteroid-sensitive elements which preferentially take up steroids. The endocrine aspect of this "feedback" of corticosteroids manifests itself by modulating the pituitary–adrenal response to environmental stimuli and probably by controlling the circadian rhythmicity of this endocrine system. The behavioral aspect of the corticosteroid effect on the hippocampus may represent the modulation of behavioral expressions to the adaptive needs.

Although several aspects of this reciprocal connection have been discussed in the appropriate sections, two main issues from the number of unsolved questions deserve further discussion. The observations on the endocrine and behavioral aspects of corticosteroid action on the hippocampus indicate a separation of "endocrine"- and "behavioral"-sensitive elements within the hippocampus. The "endocrine" elements seem to be located in the ventral hippocampus and the "behavioral" elements mainly in the dorsal areas.

Functional dissociation within the hippocampus as related to the anatomy of the system has been suggested by several authors. Deficits in active avoidance behavior may occur in the absence of deficits in passive avoidance behavior and *vice versa* in rats bearing restricted lesions in the ventral or dorsal hippocampus or with interrupted afferent or efferent connections (Nadel, 1968; Van Hoesen *et al.*, 1972). Rats bearing lesions in the ventral hippocampus or lateral septum are deficient in acquiring a shock-induced passive avoidance response but proficient in acquiring an illness-induced one (McGowan *et al.*, 1972). Other lesion (Fried, 1972*b*; Holdstock, 1972), chemical stimulation (Grant and Jarrard, 1968), and even electrophysiological stimulation (Adey, 1962; McGowan-Sass, 1973) indicate further the functional differentiation of the ventral and dorsal portion of the hippocampus. Neuroanatomical observations showing preferentially medial septal–dorsal hippocampal and lateral septal–ventral hippocampal reciprocal and differential entorhinal connections (Blackstad, 1956; Raisman *et al.*, 1965, 1966; Siegel and Tassoni, 1971*a,b*) seem to give a morphological basis for functional dissociation. Since the behavioral aspects of "behaviorally" sensitive corticosteroid receptors in the hippocampus were not fully investigated, further studies are necessary to give the final answer on whether the separation of "endocrine" and "behavioral" elements is indeed anatomical or functional within each region depending on a priority of behavioral events.

It is difficult to understand how the tiny corticosteroid implants within the large hippocampus proper exert their endocrine- and behavioral-modulating effects. The implanted corticosteroid is probably taken up by the surrounding cells, binds to a soluble protein of cytosol fraction, and finally reaches the specific neuronal cell nuclear binding sites. The bound hormone alters genomic activity, which leads to the production of proteins that affect the level or nature of neuronal function (McEwen *et al.*, 1972). That such proteins may be of great importance in the behavioral function of the hippocampus is indicated by the observations showing that local administration of protein synthesis inhibitors blocked learning functions (Flexner *et al.*, 1963; Agranoff *et al.*, 1967). Increased protein synthesis in the hippocampus and a formation of a brain-specific acid protein, S100, in the pyramidal nerve cells of the hippocampus as observed by Hydén and Lange (1970*a,b*, 1972) may be of interest in this context. Although direct evidence is missing, it seems possible that changes in protein formation due to a local effect of corticosteroids in a restricted number of cell nuclei are involved in the propagation of hormone effects to a larger cell population in the hippocampus proper.

The significance of septal–hippocampal function has for a long time been of interest in neurobiological research. A number of physiological, electrophysiological, and behavioral theories have been proposed during the past years. Some of them concern hippocampal function as a unitary process (emotion, memory, motivation, somatic activity, etc.); others propose a more complex organizing function (Papez, 1937; Green, 1964; Grastyán *et al.*, 1966; Douglas, 1967; Kimble, 1968; Isaacson, 1972; Olds, 1972; Jarrard, 1973; Altman *et al.*, 1973). Although a good deal of evidence has been presented here about the endocrine aspects of the septal–hippocampal function, I feel that an "endocrine" theory of hippocampal function is not appropriate at present. However, there is a common point where our ideas about the role of the pituitary–adrenal system hormones in behavioral adaptation and different hypotheses on hippocampal function meet. The adaptive role of the hormones seems to be to promote remembering if the contingency of the situation requires the retention or repeated retrieval, and then to enhance forgetting, unlearning, or extinction of the experience in order to made a place for acquisition of new adaptive responses when environmental

changes make them necessary. The large number of observations on hippocampal function suggest that the system must be intact in order to adjust behavior to new requirements under circumstances of environmental uncertainty. In other words, the hippocampal function by suppressing the older memories allows new actions to be taken,
new memories to be formed (Isaacson, 1974, 1975). Two important issues should,
however, be suggested. Our behavioral studies indicate that the hormones of the
pituitary–adrenal system do not evoke but rather modulate ongoing behavior. Accordingly, integrated central nervous activity controls the function of the pituitary–adrenal
system, the hormones of which in turn specifically modulate neural events that evoke
the hormonal response. The other issue is that "remembering" and "forgetting" may
be terms specifying behavioral expressions. The hormones seem to modulate the motivational properties of conditioned external and internal modalities. Thus increasing
the motivational value of conditioned stimuli results in doing the response again, while
decreasing this value results in not doing the previous pattern in order to make a place
for acquisition of new experiences. The hippocampus is but one substrate of this action
of the pituitary–adrenal system hormones: the little hormonal devil touches the large
hippocampus to serve the adaptation.

ACKNOWLEDGMENTS

A large part of the author's own research was done in the Institute of Physiology,
University Medical School, Pécs, Hungary. The enthusiastic participation of Drs.
Csaba Nyakas, János Grubits, and Gábor Kovács is gratefully acknowledged. I owe
special thanks to Drs. David de Wied and Ivan Urban for their helpful criticism while
I was preparing the manuscript.

7. References

ADEY, W. R. EEG studies of hippocampal system in the learning process. In J. Cadilhac (Ed.), *Physiologie
de l'hippocampe*. Coll. Int. CNRS No. 107. Paris: Editions CNRS, 1962, pp. 203–224.

AGRANOFF, B. W., DAVIS, R. E., CASOLA, L., AND LIM, R. Actinomycin D blocks formation of memory of
shock avoidance in goldfish. *Science*, 1967, **158**, 1600–1601.

ALTMAN, J., BRUNNER, R. L., AND BAYER, S. A. The hippocampus and behavioral maturation. *Behavioral
Biology*, 1973, **8**, 557–596.

ANTELMAN, S. M., AND BROWN, T. S. Hippocampal lesions and shuttlebox avoidance behavior: A fear
hypothesis. *Physiology and Behavior*, 1972, **9**, 15–20.

APPLEZWEIG, M. H., AND MOELLER, G. Anxiety, the pituitary–adrenocortical system and avoidance learning. *Acta Psychologica*, 1959, **15**, 602.

AZMITIA, E. C., AND McEWEN, B. S. Corticosterone regulation of tryptophan hydroxylase in midbrain of
the rat. *Science*, 1969, **166**, 1274–1276.

BARCIK, J. D. Hippocampal afterdischarges and conditioned emotional response. *Psychonomic Science*,
1970, **20**, 297–299.

BEATTY, P. A., BEATTY, W. W., BOWMAN, R. E., AND GILCHRIST, J. C. The effects of ACTH, adrenalectomy and dexamethasone on the acquisition of an avoidance response in rats. *Physiology and Behavior*, 1970, **5**, 939–944.

Bennett, T. L. Hippocampal theta activity and behavior: A review. *Communications in Behavioral Biology*, 1971, **6**, 37–48.

Blackstad, T. W. Commissural connections of the hippocampal region in the rat, with special reference to their mode of termination. *Journal of Comparative Neurology*, 1956, **105**, 417–538.

Bohus, B. The effect of central nervous lesions on pituitary–adrenocortical function in the rat. *Acta Physiologica Academiae Scientiarum Hungaricae*, 1961, **20**, 373–377.

Bohus, B. Pituitary ACTH release and avoidance behavior of rats with cortisol implants in mesencephalic reticular formation and median eminence. *Neuroendocrinology*, 1968, **3**, 355–365.

Bohus, B. Central nervous structures and the effect of ACTH and corticosteroids on avoidance behaviour: A study with intracerebral implantation of corticosteroids in the rat. In D. de Wied and J. A. W. M. Weijnen (Eds.), *Pituitary, adrenal and the brain*. Vol. 32 of *Progress in brain research*. Amsterdam: Elsevier, 1970, pp. 171–184.

Bohus, B. Adrenocortical hormones and central nervous function: The site and mode of their behavioural action in the rat. In V. H. T. James and L. Martini (Eds.), *Hormonal steroids: Proceedings of the Third International Congress on Hormonal Steroids*. Excerpta Medica International Congress Series No. 219. Amsterdam: Excerpta Medica, 1971, pp. 752–758.

Bohus, B. Pituitary–adrenal influences on avoidance and approach behavior of the rat. In E. Zimmermann, W. H. Gispen, B. H. Marks, and D. de Wied (Eds.), *Drug effects on neuroendocrine regulation*. Vol. 39 of *Progress in brain research*. Amsterdam: Elsevier, 1973, pp. 407–420.

Bohus, B., and de Wied, D. Facilitatory and inhibitory influences on the pituitary–adrenal system: A study with chlorpromazine. In A. Lunedei and M. Cagnoni (Eds.), Proceedings of the First International Symposium on Biorhythms in Clinical and Experimental Endocrinology. *Rassegna di Neurologia Vegetativa*, 1967a, **21**, 71–81.

Bohus, B., and de Wied, D. Failure of α-MSH to delay extinction of conditioned avoidance behavior in rats with lesions in the parafascicular nuclei of the thalamus. *Physiology and Behavior*, 1967b, **2**, 221–223.

Bohus, B., and Endröczi, E. Effect of intracerebral implantation of hydrocortisone on adrenocortical secretion and adrenal weight after unilateral adrenalectomy. *Acta Physiologica Academiae Scientiarum Hungaricae*, 1964, **25**, 11–19.

Bohus, B., and Endröczi, E. The influence of pituitary–adrenocortical function on the avoiding conditioned reflex activity in rats. *Acta Physiologica Academiae Scientiarum Hungaricae*, 1965, **26**, 183–189.

Bohus, B., and Lissák, K. The sites of feedback action of corticosteroids at extrahypothalamic level. *General and Comparative Endocrinology*, 1967, **9**, 434–435.

Bohus, B., and Lissák, K. Adrenocortical hormones and avoidance behaviour of rats. *International Journal of Neuropharmacology*, 1968, **7**, 301–306.

Bohus, B., Nyakas, C., and Endröczi, E. Effects of adrenocorticotropic hormone on avoidance behaviour of intact and adrenalectomized rats. *International Journal of Neuropharmacology*, 1968a, **7**, 307–314.

Bohus, B., Nyakas, C., and Lissák, K. Involvement of suprahypothalamic structures in the hormonal feedback action of corticosteroids. *Acta Physiologica Academiae Scientiarum Hungaricae*, 1968b, **34**, 1–8.

Bohus, B., Grubits, J., Kovacs, G., and Lissák, K. Effect of corticosteroids on passive avoidance behaviour of the rat. *Acta Physiologica Academiae Scientiarum Hungaricae*, 1970, **38**, 381–391.

Bohus, B., Hendrickx, H. H. L., van Kolfschoten, A. A., and Krediet, T. G. The effect of ACTH 4–10 on copulatory and sexually motivated approach behavior in the male rat. In M. Sandler and G. L. Gessa (Eds.), *Sexual behavior: Pharmacology and biochemistry*. New York: Raven Press, 1975, pp. 269–275.

Bouillé, C., and Bayle, J. D. Effects of limbic stimulations or lesions on basal and stress-induced hypothalamic–pituitary–adrenocortical activity in the pigeon. *Neuroendocrinology*, 1973/1974, **13**, 264–277.

Brady, J. W., and Nauta, W. J. H. Subcortical mechanisms in emotional behavior: Affective changes following septal forebrain lesions in the albino rats. *Journal of Comparative and Physiological Psychology*, 1953, **46**, 333–346.

Brunner, R. L., Rossi, R. R., Stutz, R. M., and Roth, T. G. Memory loss following posttrial electrical stimulation of the hippocampus. *Psychonomic Science*, 1970, **18**, 159–160.

BUSH, D. F., LOVELY, R. H., AND PAGANO, R. R. Injection of ACTH induces recovery from shuttle-box avoidance deficits in rats with amygdaloid lesions. *Journal of Comparative and Physiological Psychology*, 1973, **83**, 168–172.

CASADAY, R. L., BRANCH, B. J., AND TAYLOR, A. N. Effect of hippocampal stimulation upon stress responses in freely behaving rats. *Abstracts IVth International Congress of Endocrinology*. Excerpta Medica International Congress Series No. 256, 1972, p. 204.

CLEGHORN, R. A. Steroid hormones in relation to neuropsychiatric disorders. In H. Hoagland (Ed.), *Hormones, brain function, and behavior*. New York: Academic Press, 1957, pp. 3–25.

COOVER, G. D., GOLDMAN, L., AND LEVINE, S. Plasma corticosterone increases produced by extinction of operant behavior in rats. *Physiology and Behavior*, 1971a, **6**, 261–263.

COOVER, G. D., GOLDMAN, L., AND LEVINE, S. Plasma corticosterone levels during extinction of a lever-press response in hippocampectomized rats. *Physiology and Behavior*, 1971b, **7**, 727–732.

CORBIN, A., MANGILI, G., MOTTA, M., AND MARTINI, L. Effect of hypothalamic and mesencephalic steroid implantations on ACTH feedback mechanisms. *Endocrinology*, 1965, **76**, 811–818.

DALLMAN, M. F., AND YATES, F. E. Anatomical and functional mapping of central neural input and feedback pathways of the adrenocortical system. *Memoirs of the Society for Endocrinology*, 1968, **17**, 39–71.

DAVIDSON, J. M., AND FELDMAN, S. Effects of extrahypothalamic dexamethasone implants on the pituitary–adrenal system. *Acta Endocrinologica*, 1967, **55**, 240–246.

DE WIED, D. Influence of anterior pituitary on avoidance learning and escape behavior. *American Journal of Physiology*, 1964, **207**, 255–259.

DE WIED, D. Inhibitory effect of ACTH and related peptides on extinction of conditioned avoidance behavior. *Proceedings of the Society for Experimental Biology and Medicine*, 1966, **122**, 28–32.

DE WIED, D. Opposite effects of ACTH and glucocorticosteroids on extinction of conditioned avoidance behavior. *Proceedings of the Second International Congress on Hormonal Steroids*. Excerpta Medica International Congress Series No. 132, 1967, pp. 945–951.

DE WIED, D. Effects of peptide hormones on behavior. In W. F. Ganong and L. Martini (Eds.), *Frontiers in neuroendocrinology, 1969*. New York: Oxford University Press, 1969, pp. 97–140.

DE WIED, D., BOHUS, B., AND GREVEN, H. M. Influence of pituitary and adrenocortical hormones on conditioned avoidance behavior in rats. In R. P. Michael (Ed.), *Endocrinology and human behavior*. Oxford: Oxford University Press, 1968, pp. 188–199.

DE WIED, D., VAN DELFT, A. M. L., GISPEN, W. H., WEIJNEN, J. A. W. M., AND VAN WIMERSMA GREIDANUS, T. B. The role of pituitary–adrenal system hormones in active avoidance conditioning. In S. Levine (Ed.), *Hormones and behavior*. New York: Academic Press, 1972, pp. 135–171.

DOUGLAS, R. J. The hippocampus and behavior. *Psychological Bulletin*, 1967, **67**, 416–422.

DUPONT, A., BASTARACHE, E., ENDRÖCZI, E., AND FORTIER, C. Effect of hippocampal stimulation on the plasma thyrotropin (TSH) and corticosterone responses to acute cold exposure in the rat. *Canadian Journal of Physiology and Pharmacology*, 1972, **50**, 364–367.

ELAZAR, Z., AND ADEY, W. R. Spectral analysis of low frequency components in the electrical activity of the hippocampus during learning. *Electroencephalography and Clinical Neurophysiology*, 1967, **23**, 225–240.

ENDRÖCZI, E. Brain stem and hypothalamic substrate of motivated behaviour. In K. Lissák (Ed.), *Results in neurophysiology, neuroendocrinology, neuropharmacology and behaviour*. Vol. 2 of *Recent developments of neurobiology in hungary*. Budapest: Akadémiai Kiadó, 1969, pp. 27–46.

ENDRÖCZI, E. *Limbic system, learning and pituitary–adrenal function*. Budapest: Akadémiai Kiadó, 1972, p. 154.

ENDRÖCZI, E., AND LISSÁK, K. The role of the mesencephalon, diencephalon and archicortex in the activation and inhibition of the pituitary–adrenocortical system. *Acta Physiologica Academiae Scientiarum Hungaricae*, 1960, **17**, 39–55.

ENDRÖCZI, E., AND LISSÁK, K. Interrelations between paleocortical activity and pituitary–adrenocortical function. *Acta Physiologica Academiae Scientiarum Hungaricae*, 1962, **21**, 257–263.

ENDRÖCZI, E., AND NYAKAS, C. Effect of septal lesion on exploratory activity, passive avoidance learning and pituitaryadrenal function in the rat. *Acta Physiologica Academiae Scientiarum Hungaricae*, 1971, **39**, 351–360.

ENDRÖCZI, E., AND NYAKAS, C. Effect of corticosterone on passive avoidance learning in the rat. *Acta Physiologica Academiae Scientiarum Hungaricae,* 1972, **41,** 55–61.

ENDRÖCZI, E., LISSÁK, K., BOHUS, B., AND KOVACS, S. The inhibitory influence of archicortical structures on pituitary–adrenal function. *Acta Physiologica Academiae Scientiarum Hungaricae,* 1959, **16,** 17–22.

ENDRÖCZI, E., LISSÁK, K., AND TEKERES, M. Hormonal "feed-back" regulation of pituitary–adrenocortical activity. *Acta Physiologica Academiae Scientiarum Hungaricae,* 1961, **18,** 291–299.

FELDMAN, S. Convulsive phenomena produced by intraventricular administration of hydrocortisone in cats. *Epilepsia,* 1966, **7,** 271–282.

FELDMAN, S. Electrical activity of the brain following cerebral microinfusion of cortisol. *Epilepsia,* 1971, **12,** 249–262.

FELDMAN, S. The interaction of neural and endocrine factors regulating hypothalamic activity. In A. Brodish and E. S. Redgate (Eds.), *Brain–pituitary–adrenal interrelationships.* Basel: Karger, 1973, pp. 224–238.

FELDMAN, S., AND DAFNY, N. Effects of adrenocortical hormones on the electrical activity of the brain. In D. de Wied and J. A. W. M. Weijnen (Eds.), *Pituitary, adrenal and the brain.* Vol. 32 of *Progress in brain research.* Amsterdam: Elsevier, 1970, pp. 90–100.

FELDMAN, S., AND DAVIDSON, J. M. Effect of hydrocortisone on electrical activity, arousal thresholds and evoked potentials in the brains of chronically implanted rabbits. *Journal of the Neurological Sciences,* 1966, **3,** 462–472.

FENDLER, K., KARMOS, G., AND TELEGDY, G. The effect of hippocampal lesion on pituitary–adrenal function. *Acta Physiologica Academiae Scientiarum Hungaricae,* 1961, **20,** 293–297.

FLEXNER, J. B., FLEXNER, L. B., AND STELLAR, E. Memory in mice as affected by intracerebral puromycin. *Science,* 1963, **141,** 57–59.

FRIED, P. A. Septum and behavior: A review. *Psychological Bulletin,* 1972a, **78,** 292–310.

FRIED, P. A. The effect of differential hippocampal lesions and pre- and postoperative training on extinction. *Canadian Journal of Psychology,* 1972b, **26,** 61–70.

FUXE, K., HÖKFELT, T., JONSSON, G., LEVINE, S., LIDBRINK, P., AND LÖFSTRÖM, A. Brain and pituitary–adrenal interactions: Studies on central monoamine neurons. In A. Brodish and E. S. Redgate (Eds.), *Brain–pituitary–adrenal interrelationships.* Basel: Karger, 1973, pp. 239–269.

GANONG, W. F. The central nervous system and the synthesis and release of ACTH. In A. V. Nalbandov (Ed.), *Advances in neuroendocrinology.* Urbana: University of Illinois Press, 1963, pp. 92–149.

GARRUD, P., GRAY, J. A., AND DE WIED, D. Pituitary–adrenal hormones and extinction of rewarded behaviour in the rat. *Physiology and Behavior,* 1974, **12,** 109–119.

GERLACH, J. L., AND McEWEN, B. S. Rat brain binds adrenal steroid hormone: Radioautography of hippocampus with corticosterone. *Science,* 1972, **175,** 1133–1136.

GLASER, G. H. On the relationship between adrenal cortical activity and the convulsive state. *Epilepsia,* 1953, **2,** 7–14.

GRANT, L. D., AND JARRARD, L. E. Functional dissociation within the hippocampus. *Brain Research,* 1968, **10,** 392–401.

GRASTYÁN, E., KARMOS, G., VERECZKEY, L., AND KELLÉNYI, L. The hippocampal electrical correlates of the homeostatic regulation of motivation. *Electroencephalography and Clinical Neurophysiology,* 1966, **21,** 34–53.

GRAY, J. A. Effects of septal driving of the hippocampal theta rhythm on resistance to extinction. *Physiology and Behavior,* 1972, **8,** 481–490.

GRAY, J. A., AND BALL, G. G. Frequency-specific relation between hippocampal theta rhythm, behavior and amobarbital action. *Science,* 1970, **168,** 1246–1248.

GRAY, J. A., MAYES, A. R., AND WILSON, M. A barbiturate-like effect of adrenocorticotropic hormone on the partial reinforcement acquisition and extinction effects. *Neuropharmacology,* 1971, **10,** 223–230.

GREEN, J. D. The hippocampus. *Physiological Reviews,* 1964, **44,** 561–608.

GREEN, J. D., AND ARDUINI, A. Hippocampal electrical activity in arousal. *Journal of Neurophysiology,* 1954, **17,** 533–557.

GUTH, S., LEVINE, S., AND SEWARD, J. P. Appetitive acquisition and extinction effects with exogenous ACTH. *Physiology and Behavior,* 1971, **7,** 195–200.

HOLDSTOCK, T. L. Dissociation of function within the hippocampus. *Physiology and Behavior*, 1972, **8**, 659–667.

HYDÉN, H., AND LANGE, P. W. Protein synthesis in limbic structures during change in behavior. *Brain Research*, 1970a, **22**, 423–425.

HYDÉN, H., AND LANGE, P. W. S100 brain protein: Correlation with behavior. *Proceedings of the National Academy of Sciences U.S.A.*, 1970b, **67**, 1959–1966.

HYDÉN, H., AND LANGE, P. W. Protein changes in different brain areas as a function of intermittent training. *Proceedings of the National Academy of Sciences U.S.A.*, 1972, **69**, 1980–1984.

ISAACSON, R. L. Neural systems of the limbic brain and behavioural inhibition. In J. Halliday and R. Boakes (Eds.), *Inhibition and learning*. New York: Academic Press, 1972, pp. 497–528.

ISAACSON, R. L. Memory processes and the hippocampus. In D. A. Deutsch and A. J. Deutsch (Eds.), *Short-term memory*. New York: Academic Press, 1975, in press.

ISAACSON, R. L., AND WICKELGREN, W. O. Hippocampal ablation and passive avoidance. *Science*, 1962, **138**, 1104–1106.

JARRARD, L. E. The hippocampus and motivation. *Psychological Bulletin*, 1973, **79**, 1–12.

KASTIN, A. J., MILLER, L. M., NOCKTON, R., SANDMAN, C. A., SCHALLY, A. V., AND STRATTON, L. O. Behavioral aspects of melanocyte-stimulating hormone (MSH). In E. Zimmermann, W. H. Gispen, B. H. Marks, and D. de Wied (Eds.), *Drug effects on neuroendocrine regulation*. Vol. 39 of *Progress in brain research*. Amsterdam: Elsevier, 1973, pp. 461–470.

KAWAKAMI, M., KOSHINO, T., AND HATTORI, Y. Changes in the EEG of the hypothalamus and limbic system after administration of ACTH, SU-4885 and Ach in rabbits with special reference to neurohumoral feedback regulation of pituitary–adrenal system. *Japanese Journal of Physiology*, 1966, **16**, 551–569.

KAWAKAMI, M., SETO, K., TERASAWA, E., YOSHIDA, K., MIYAMOTO, T., SEKIGUCHI, M., AND HATTORI, Y. Influence of electrical stimulation and lesion in limbic structure upon biosynthesis of adrenocorticoid in the rabbit. *Neuroendocrinology*, 1968a, **3**, 337–348.

KAWAKAMI, M., SETO, K., AND YOSHIDA, K. Influence of corticosterone implantation in limbic structure upon biosynthesis of adrenocortical steroid. *Neuroendocrinology*, 1968b, **3**, 349–354.

KAWAKAMI, M., SETO, K., YANASE, M., AND MOHRI, M. A role of the hippocampus in the control of ACTH secretion. *Hormones*, 1972, **3**, 270–271.

KENDALL, J. W. Feedback control of adrenocorticotropic hormone secretion. In L. Martini and W. F. Ganong (Eds.), *Frontiers in neuroendocrinology 1971*. New York: Oxford University Press, 1971, pp. 177–207.

KIMBLE, D. P. The effects of bilateral hippocampal lesions in rats. *Journal of Comparative and Physiological Psychology*, 1963, **56**, 273–283.

KIMBLE, D. P. Hippocampus and internal inhibition. *Psychological Bulletin*, 1968, **70**, 285–295.

KIMURA, D. Effects of selective hippocampal damage on avoidance behavior in the rat. *Canadian Journal of Psychology*, 1958, **12**, 213–218.

KLEMM, W. R. Effects of electrical stimulation of brain stem reticular formation on hippocampal theta rhythm and muscle activity in unanesthetized, cervical- and midbrain-transected rats. *Brain Research*, 1972, **41**, 331–344.

KNIGGE, K. M. Adrenocortical response to immobilization in rats with lesion in hippocampus and amygdala. *Federation Proceedings*, 1961, **20**, 185.

KNIGGE, K. M. Feedback mechanisms in neural control of adenohypophyseal function: Effect of steroids implanted in amygdala and hippocampus. *Abstracts 2nd International Congress on Hormonal Steroids*. Excerpta Medica International Congress Series No. 111, 1966, 208.

KNIGGE, K. M., AND HAYS, M. Evidence of inhibitive role of hippocampus in neural regulation of ACTH release. *Proceedings of the Society for Experimental Biology and Medicine*, 1964, **114**, 67–69.

KNIZLEY, H., JR. The hippocampus and septal area as primary target sites for corticosterone. *Journal of Neurochemistry*, 1972, **19**, 2737–2745.

KORÁNYI, L., AND ENDRÖCZI, E. Influence of pituitary–adrenocortical hormones on thalamo-cortical and brain stem limbic circuits. In D. de Wied and J. A. W. M. Weijnen (Eds.), *Pituitary, adrenal and the brain*. Vol. 32 of *Progress in brain research*. Amsterdam: Elsevier, 1970, pp. 120–130.

LANIER, L. P., VAN HARTESVELDT, C., WEIS, B. J., AND ISAACSON, R. L. Effects of differential hippocampal damage upon rhythmic and stress-induced corticosterone secretion in the rat. *Neuroendocrinology,* 1975, in press.

LEVINE, S., GOLDMAN, L., AND COOVER, G. D. Expectancy and the pituitary–adrenal system. In *Physiology, emotion and psychosomatic illness: Ciba Foundation Symposium 8* (new series). Amsterdam: Elsevier, Excerpta Medica, North Holland: Associated Scientific Publishers, 1972, pp. 281–296.

LISSÁK, K., AND BOHUS, B. Pituitary hormones and avoidance behavior of the rat. *International Journal of Psychobiology,* 1972, **2,** 103–115.

LISSÁK, K., AND ENDRÖCZI, E. Some aspects of the effect of hippocampal stimulation on the endocrine system. In J. Cadilhac (Ed.), *Physiologie de l'hippocampe.* Coll. Int. CNRS, No. 107. Paris: Editions CNRS, 1962, pp. 463–473.

LISSÁK, K., AND ENDRÖCZI, E. Neuroendocrine interrelationships and behavioral processes. In E. Bajusz and G. Jasmin (Eds.) *Major problems in neuroendocrinology.* Basel: Karger, 1964, pp. 1–16.

MACLEAN, P. D. Psychosomatic disease and the "visceral brain": Recent developments bearing on the Papez theory of emotion. *Psychosomatic Medicine,* 1949, **11,** 338–353.

MANGILI, G., MOTTA, M., AND MARTINI, L. Control of adrenocorticotropic hormone secretion. In L. Martini and W. F. Ganong (Eds.), *Neuroendocrinology.* New York: Academic Press, 1966, pp. 297–370.

MASON, J. W. The central nervous regulation of ACTH secretion. In H. H. Jasper, L. D. Proctor, R. S. Knighton, W. C. Noshay, and R. T. Costello (Eds.), *Reticular formation of the brain.* Boston: Little, Brown, 1957, pp. 645–670.

MASON, J. W. A review of psychoendocrine research on the pituitary–adrenal cortical system. *Psychosomatic Medicine,* 1968, **30,** 576–607.

MASON, J. W. A re-evaluation of the concept of "non-specificity" in stress theory. *Journal of Psychiatric Research,* 1971, **8,** 323–333.

MCEWEN, B. S., AND WALLACH, G. Corticosterone binding to hippocampus: nuclear and cytosol binding *in vitro. Brain Research,* 1973, **57,** 373–386.

MCEWEN, B. S., AND WEISS, J. M. The uptake and action of corticosterone: Regional and subcellular studies on rat brain. In D. de Wied and J. A. W. M. Weijnen (Eds.), *Pituitary, adrenal and the brain.* Vol. 32 of *Progress in brain research.* Amsterdam: Elsevier, 1970, pp. 200–212.

MCEWEN, B. S., WEISS, J. M., AND SCHWARTZ, L. S. Uptake of corticosterone by rat brain and its concentration by certain limbic structures. *Brain Research,* 1969, **16,** 227–241.

MCEWEN, B. S., WEISS, J. M., AND SCHWARTZ, L. S. Retention of corticosterone by cell nuclei from brain regions of adrenalectomized rats. *Brain Research,* 1970, **17,** 471–482.

MCEWEN, B. S., ZIGMOND, R. E., AND GERLACH, J. L. Sites of steroid binding and action in the brain. In G. H. Bourne (Ed.), *Structure and function of the nervous system.* Vol. 5. New York: Academic Press, 1972, pp. 205–291.

MCGOWAN, B. K., HANKINS, W. G., AND GARCIA, J. Limbic lesions and control of the internal and external environment. *Behavioral Biology,* 1972, **7,** 841–852.

MCGOWAN-SASS, B. K. Differentiation of electrical rhythms and functional specificity of the hippocampus of the rat. *Physiology and Behavior,* 1973, **11,** 187–194.

MILLER, R. E., AND OGAWA, N. The effect of adrenocorticotrophic hormone (ACTH) on avoidance conditioning in the adrenalectomized rat. *Journal of Comparative and Physiological Psychology,* 1962, **55,** 211–213.

MOBERG, G. P., SCAPAGNINI, U., DE GROOT, J., AND GANONG, W. F. Effect of sectioning the fornix on diurnal fluctuation in plasma corticosterone levels in the rat. *Neuroendocrinology,* 1971, **7,** 11–15.

NADEL, L. Dorsal and ventral hippocampal lesions and behavior. *Physiology and Behavior,* 1968, **3,** 891–900.

NAKADATE, G., AND DE GROOT, J. Fornix transsection and adrenocortical function in rats. *Anatomical Record,* 1963, **145,** 338.

NEWMAN-TAYLOR, A., BRANCH, B. J., CASADY, R. L., AND TURNER, B. B. Septal inhibition of pituitary–adrenal activity in freely behaving rats. *Endocrinology,* 1973, **92,** Suppl. A-81.

NYAKAS, C., AND ENDRÖCZI, E. Effect of hippocampal stimulation on the establishment of conditioned fear response in rat. *Acta Physiologica Academiae Scientiarum Hungaricae,* 1970, **37,** 281–289.

OLDS, J. Learning and the hippocampus. *Revue Canadienne de Biologie,* 1972, **31,** Suppl. 215–238.

OLTON, D. S., AND ISAACSON, R. L. Hippocampal lesions and active avoidance. *Physiology and Behavior,* 1968, **3,** 719–724.

PAGANO, R. R., AND LOVELY, R. H. Diurnal cycle and ACTH facilitation of shuttlebox avoidance. *Physiology and Behavior,* 1972, **8,** 721–723.

PAPEZ, J. W. A proposed mechanism of emotion. *Archives of Neurology and Psychiatry,* 1937, **38,** 725–744.

PFAFF, D. W., SILVA, M. T. A., AND WEISS, J. M. Telemetred recording of hormone effects on hippocampal neurons. *Science,* 1971, **172,** 394–395.

PORTER, R. W. The central nervous system and stress-induced eosinopenia. *Recent Progress in Hormone Research,* 1954, **10,** 1–27.

RAISMAN, G. The connections of the septum. *Brain,* 1966, **89,** 317–348.

RAISMAN, G., COWAN, W. M., AND POWELL, T. P. S. The extrinsic afferent, commissural and association fibers of the hippocampus. *Brain,* 1965, **88,** 963–996.

RAISMAN, G., COWAN, W. M., AND POWELL, T. P. S. An experimental analysis of the efferent projection of the hippocampus. *Brain,* 1966, **89,** 83–108.

ROUTTENBERG, A. Hippocampal correlates of consumatory and observed behavior. *Physiology and Behavior,* 1968, **3,** 533–535.

RUBIN, R. T., MANDELL, A. J., AND CRANDALL, P. H. Corticosteroid responses to limbic stimulation in man: Localization of stimulus sites. *Science,* 1966, **153,** 767–768.

SCAPAGNINI, U., MOBERG, G. P., VAN LOON, G. R., DE GROOT, L., AND GANONG, W. F. Relation of brain 5-hydroxytryptamine content to the diurnal variation in plasma corticosterone in the rat. *Neuroendocrinology,* 1971, **7,** 90–96.

SEGGIE, J., AND BROWN, G. M. Effect of dexamethasone on affective behavior and adrenal reactivity following septal lesions in the rat. *Journal of Comparative and Physiological Psychology,* 1973, **83,** 60–65.

SEGGIE, J., SHAW, B., UHLIR, I., AND BROWN, G. M. Baseline, 24-hour plasma corticosterone rhythm in normal, sham-operated and septally-lesioned rats. *Neuroendocrinology,* 1974, **15,** 51–61.

SELYE, H. A syndrome produced by diverse nocuous agents. *Nature (London),* 1936, **138,** 32–33.

SELYE, H. *Stress: The physiology and pathology of exposure to stress.* Montreal: Acta Medica Publication, 1950.

SIEGEL, A., AND TASSONI, J. P. Differential efferent projections from the ventral and dorsal hippocampus of the cat. *Brain, Behavior and Evolution,* 1971a, **4,** 185–200.

SIEGEL, A., AND TASSONI, J. P. Differential efferent projections of the lateral and medial septal nuclei to the hippocampus in the cat. *Brain, Behavior and Evolution,* 1971b, **4,** 201–219.

SLUSHER, M. A. Effects of cortisol implants in the brainstem and ventral hippocampus on diurnal corticosteroid levels. *Experimental Brain Research,* 1966, **1,** 184–194.

STEVENS, W., GROSSER, B. I., AND REED, D. J. Corticosterone-binding molecules in rat brain cytosols: Regional distribution. *Brain Research,* 1971, **35,** 602–607.

STONE, C. P., AND KING, F. A. Effects of hypophysectomy on behavior in rats. I. Preliminary survey. *Journal of Comparative and Physiological Psychology,* 1954, **47,** 213–219.

STONE, C. P., AND OBIAS, M. D. Effects of hypophysectomy on behavior in rats. II. Maze and discrimination learning. *Journal of Comparative and Physiological Psychology,* 1955, **48,** 404–411.

STREIFLER, M., AND FELDMAN, S. On the effect of cortisone on the electroencephalogram, *Confinia Neurologica,* 1953, **13,** 16–27.

STUMPF, C. Drug action on the electrical activity of the hippocampus. *International Review of Neurobiology,* 1965, **8,** 77–138.

STUMPF, W. E. Autoradiographic techniques and the localization of estrogen, androgen and glucocorticoid in the pituitary and brain. *American Zoologist,* 1971, **11,** 725–739.

TAYLOR, A. N., BRANCH, B. J., CASADY, R. L., AND TURNER, B. B. Septal inhibition of pituitary-adrenal activity in freely behaving rats. *Endocrinology,* 1973, **92,** suppl. A-81.

UHLIR, I., SEGGIE, J., AND BROWN, G. M. The effect of septal lesions on the threshold of adrenal stress response. *Neuroendocrinology,* 1974, **14,** 351–355.

URBAN, I., AND DE WIED, D. Changes in excitability in the theta activity generating substrate by ACTH 4–10 in the rat. *Brain Research,* 1975, submitted.

URBAN, I., LOPES DA SILVA, F. H., STORM VAN LEEUWEN, W., AND DE WIED, D. A frequency shift in the hippocampal theta activity: An electrical correlate of central action of ACTH analogues in the dog? *Brain Research,* 1974, **69,** 361–365.

USHER, D. R., AND LAMBLE, R. W. ACTH synthesis and release in septal-lesioned rats exposed to air shuttle-avoidance. *Physiology and Behavior,* 1969, **4,** 923–927.

USHER, D. R., KASPER, P., AND BIRMINGHAM, M. K. Comparison of pituitary–adrenal function in rats lesioned in different areas of the limbic system and hypothalamus. *Neuroendocrinology,* 1967, **2,** 157–174.

USHER, D. R., LIEBLICH, I., AND SIEGEL, R. A. Pituitary–adrenal function after small and large lesions in the lateral septal area in food-deprived rats. *Neuroendocrinology,* 1974, **16,** 156–164.

VANDERWOLF, C. H. Hippocampal electrical activity and voluntary movement in the rat. *Electroencephalography and Clinical Neurophysiology,* 1969, **26,** 407–418.

VANDERWOLF, C. H. Limbic–diencephalic mechanisms of voluntary movement. *Psychological Reviews,* 1971, **78,** 83–113.

VAN HOESEN, G. W., WILSON, L. M., MacDOUGALL, J. M., AND MITCHELL, J. C. Selective hippocampal complex deafferentation and deefferentation and avoidance behavior in rats. *Physiology and Behavior,* 1972, **8,** 873–879.

VAN WIMERSMA GREIDANUS, T. B., AND DE WIED, D. Effects of intracerebral implantation of corticosteroids on extinction of an avoidance response in rats. *Physiology and Behavior,* 1969, **4,** 365–370.

VAN WIMERSMA GREIDANUS, T. B., AND DE WIED, D. Effects of systemic and intracerebral administration of two opposite acting ACTH-related peptides on extinction of conditioned avoidance behavior. *Neuroendocrinology,* 1971, **7,** 291–301.

VAN WIMERSMA GREIDANUS, T. B., BOHUS, B., AND DE WIED, D. Differential localization of the behavioral effects of lysine vasopressin and of ACTH 4–10: A study in rats bearing lesions in the parafascicular nuclei. *Neuroendocrinology,* 1974, **14,** 280–288.

VARDARIS, R. M., AND SCHWARTZ, K. E. Retrograde amnesia for passive avoidance produced by stimulation of dorsal hippocampus. *Physiology and Behavior,* 1971, **6,** 131–135.

VERNIKOS-DANELLIS, J., BERGER, P., AND BARCHAS, J. D. Brain serotonin and pituitary–adrenal function. In E. Zimmermann, W. H. Gispen, B. H. Marks, and D. de Wied (Eds.), *Drug effects on neuroendocrine regulation.* Vol. 39 of *Progress in brain research.* Amsterdam: Elsevier, 1973, pp. 301–310.

WEISS, J. M., McEWEN, B. S., SILVA, M. T., AND KALKUT, M. Pituitary–adrenal alterations and fear responding. *American Journal of Physiology,* 1970, **218,** 864–868.

WILSON, M. AND CRITCHLOW, V. Effects of septal lesions on rhythmic pituitary–adrenal function. *Federation Proceedings,* 1973, **32,** 296.

WILSON, M., AND CRITCHLOW, V. Effect of fornix transection or hippocampectomy on rhythmic pituitary–adrenal function in the rat. *Neuroendocrinology,* 1973/1974, **13,** 29–40.

WILSON, M., AND CRITCHLOW, V. Effect of septal ablation on rhythmic pituitary–adrenal function in the rat. *Neuroendocrinology,* 1974, **14,** 333–344.

WOODBURY, D. M., TIMIRAS, P. S., AND VERNADAKIS, A. Influence of adrenocortical steroids on brain function and metabolism. In H. Hoagland (Ed.), *Hormones, brain function, and behavior.* New York: Academic Press, 1957, pp. 27–54.

13

The Hippocampus and Hormonal Cyclicity

BRENDA K. McGOWAN-SASS AND PAOLA S. TIMIRAS

1. Introduction

In recent years, considerable importance has been attributed to the role of the hormonal environment in modulating the activity of the central nervous system (CNS) and, conversely, the role of the CNS in regulating endocrine and associated physiological functions. It has become evident that any final analysis of behavior attempting to deal with CNS alterations induced by lesion, stimulation, drugs, etc., must take into account the resultant effects on endogenous levels of hormones. In addition, not only do normal cyclical alterations in the hormonal environment have widespread effects on the CNS, but these effects vary over time. The endocrine–CNS interaction could be an important factor in the complex behavioral changes observed as a result of interruption of either of these systems. This chapter deals specifically with the relationship of the limbic system, especially the hippocampus and amygdala, to alterations in the hormonal environment, with primary emphasis on cyclical changes in gonadal and adrenal steroids.

2. Gonadal Hormones and Cyclicity

The cyclicity of female sexual activity in mammalian species other than humans was an obvious and early observation. Only periodically does the female cease to avoid and repulse male sexual advances and actively attempt to mate. That this ab-

BRENDA K. McGOWAN-SASS AND PAOLA S. TIMIRAS • Department of Physiology-Anatomy, University of California, Berkeley, California. This investigation was supported by Grant No. HD 101 from the National Institute of Health.

rupt change in her sexual behavior is brought on by a rise in the circulating blood levels of estrogen is substantiated by observations that treatment with estrogen antagonists (McClintock and Schwartz, 1968; Shirley *et al.*, 1968) or with antibodies to estrogen (Ferin *et al.*, 1969) eliminates cyclical mating behavior.

These regular endocrine changes have been extensively studied in the rat. When maintained under normal conditions of alternating light and darkness, female rats show regular cycles of either 4 or 5 days' duration, with spontaneous ovulation occurring during the last hours of the estrous stage of the cycle (Fig. 1). The major endocrine events can be summarized as follows: follicle stimulating hormone (FSH) from the pituitary stimulates the growth and maturation of one or several follicles in the ovary; FSH plus luteinizing hormone (LH) promotes the maturation of follicles; a burst of LH secretion (the "ovulatory surge") causes ovulation by inducing a rupture of the follicular wall with egg release. Estrogen is secreted by the developing follicles, while, following ovulation, the follicular cells undergo luteinization, and the resulting corpora lutea secrete mainly progesterone and some estrogen. Maintenance of the corpora lutea in the rat is accomplished by prolactin; however, prolactin does

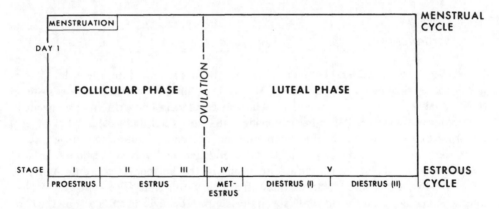

Fig. 1. Comparison of the menstrual cycle in the human and the estrus cycle in the rat. The average length of the menstrual cycle is from 22 to 40 days, day 1 designated as the first day of menstruation. Vaginal as well as cervical mucous changes have been described and correlated with the menstrual cycle in the human, but these changes are not as well delineated as those in the rat. The exact timing of ovulation in the human has not been firmly established, although tests developed by Farris (1948) place ovulation at 1 or 2 days before midcycle. The 4- to 5-day estrous cycle, described by Long and Evans (1922), is divided into five distinct stages which are distinguished by classification of cell types obtained from the vaginal mucosa. In stage I, proestrus (duration 12–15 h), the characteristic vaginal smear reveals epithelial cells only. Estrous stage II is characterized by the appearance of cornified cells and, behaviorally, by the appearance of heat. In estrous stage III, the vaginal smear exhibits even more abundant cornified cells which take on a "cheesy" appearance. Spontaneous ovulation occurs during the last hours of the cornified cell stages II and III, the duration of stages II and III together being approximately 30 h. Stage IV, metestrus, duration 6–8 h, is characterized by the appearance of leukocytes among the cornified cells. In stage V, diestrus, the cornified cells disappear altogether, leukocytes and some epithelial cells being present at this time. This stage lasts from 57 to 60 h and is sometimes broken up into two periods, diestrus I and diestrus II. In the nonpregnant rat, while the corpora lutea secrete estrogen and some progesterone, they do not secrete large amounts of progesterone; hence the corpora lutea may be referred to as "nonfunctional" (Gorbman and Bern, 1962).

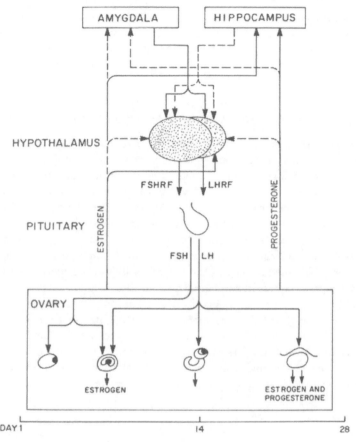

Fig. 2. Proposed and known interrelationships of limbic–hypothalamic–pituitary gonadotropin release and inhibition. – – –, Inhibition; ——, facilitation.

not have a luteotropic effect in humans, monkeys, cattle, or rabbits (Ganong, 1971). It remains to be determined what hormone, or combination of hormones, performs this function in these species (see Fig. 2). Some animals, notably the rabbit, ferret, and cat, remain in the estrous state until pregnancy or pseudopregnancy occurs. In these species, stimulation of the genitalia consequent to copulation, together with other sensory stimuli, provokes luteinizing hormone (LH) release from the pituitary, thereby causing rupture of the ovarian follicles.

The suspicion that sex steroids act not only on target organs but on the CNS as well—indeed, the CNS can be considered a target organ for steroids—is buttressed by early observations that female rats are about twice as active as males (Hitchcock, 1925/1926); that this activity, which correlates with the estrous cycle, is lost after ovariectomy (Wang, 1923); that female rats exposed to constant light will remain in "persistent estrus" without ovulating (Lawton and Schwartz, 1967); and that CNS depressants, such as pentobarbital, block or delay ovulation for 24 h (Everett, 1961). Beach hypothesized in 1948 that sex steroids alter thresholds within the nervous system, a finding later supported by electrophysiological studies (Woolley and Timiras, 1962a,b). A current hypothesis of endocrine action well substantiated in several areas of investigation (see below) is that the hippocampus and amygdala function reciprocally in regulating hypothalamic production of critical releasing factors for pituitary gonadotropic hormones.

2.1. Electrical Activity: Spontaneous, Evoked, and Multiple Unit

Interactions between sex steroids and the limbic system during the estrous cycle in normal female rats were reported by Terasawa and Timiras (1968a). These authors demonstrated that the dorsal hippocampus and medial and lateral amygdala show cyclical changes in excitability which correlate with stages of the estrous cycle. Threshold for electrically induced seizures of the dorsal hippocampus decreased from the morning of proestrus through the morning of estrus, with the lowest value on the night of proestrus. The threshold then increased to the original diestrous level, and remained low until the next proestrous morning, and so on, showing a definite cyclical pattern. The medial part of the amygdala followed a pattern similar to that of the dorsal hippocampus. In the lateral part of the amygdala, however, changes in the threshold occurred in an opposite direction from those observed in the dorsal hippocampus; i.e., the threshold was higher during the period from the morning of proestrus to the morning of estrus and peaked at the night of proestrus (Fig. 3). Ovariectomy abolished this cyclicity; however, injection of 10 μg estradiol benzoate to ovariectomized rats temporarily restored cyclical changes in threshold, although they were decreased in amplitude and duration over time. Administration of 2 mg

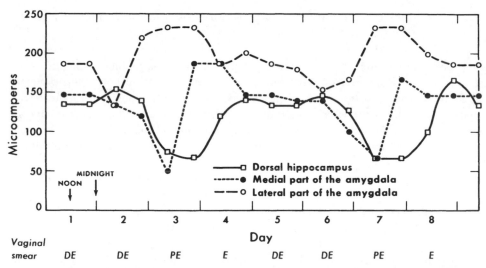

Fig. 3. Comparison of localized seizure threshold curves of three portions of the limbic system during two estrous cycles. Abbreviations: DE, diestrus; PE, proestrus; E, estrus. From Terasawa and Timiras (1968a).

progesterone effected an increase in dorsal hippocampal seizure threshold 7–12 h after administration. Again, the effect on the lateral amygdala, where threshold decreased 5–18 h after hormone administration, was opposite to that obtained in the dorsal hippocampus.

Other studies have correlated changes in electrical activity of the limbic system with stages in the estrous cycle (Kawakami *et al.*, 1966c). On-line frequency analysis of the EEG in the dorsal and ventral hippocampus of the cat showed that during the stage of maximum estrus, as compared to the pre-estrous stage, the 8–13 Hz regular sinusoidal waves (usually increased during the arousal or alert state) were decreased by 32–47%, the 4–8 Hz regular waves were increased by 25–35%, and the 20–30 Hz components were decreased. After copulation-induced ovulation, these EEG patterns were characterized by a marked domination of the rhythmic sinusoidal waves, and their analyzed EEG showed an increase of approximately 15% in the 4–8 and the 8–13 Hz band values, and almost no alterations in the amplitude of the other frequency-integrated band values. The authors have used these data to affirm hippocampal electrical activity drops during estrus and rises at the postcoital stage. In the amygdala, with the progress of estrus the integrated band values of the 2–8 and the 13–30 Hz components showed an increase as compared with those recorded during the period of pre-estrus. One hour following copulation-induced ovulation, the amplitude of both the 2–8 and the 20–30 Hz integrated band values was reduced. From these observations, the authors report that the level of activity in the amygdaloid complex increases at the estrous stage and decreases following copulation. Thus, in terms of electrical activity, the changes from estrogen-dominated stages to progesterone-dominated stages in relation to the amygdala and hippocampus assume a "seesaw" relationship.

Other electrical changes that correlate with the estrous cycle are cortical EEG arousal thresholds (Terasawa and Timiras, 1969) and EEG after-reaction thresholds. The after-reaction is represented by high-amplitude 8 Hz hippocampal slow waves (Kawakami and Sawyer, 1959). In general, arousal thresholds and EEG after-reaction thresholds parallel each other; when they diverge, the arousal threshold appears to be more closely associated with sexual behavior and the after-reaction threshold more intimately related to pituitary activation.

Similarly, multiple-unit activity within the limbic system has been correlated with the estrous cycle. In the normally cycling rat, the basal level of multiple-unit activity in the dorsal hippocampus and amygdala was found to reach its highest level at the proestrous and estrous stages of the cycle (Kawakami *et al.*, 1970). During and for some time following vaginal stimulation in the rat, neuron activity in the dorsal hippocampus, medial amygdala, lateral septum, and arcuate nucleus was enhanced, whereas the response in the ventromedial hypothalamus was inhibited (Kawakami and Kubo, 1971). These changes were correlated with the appearance of the EEG after-reaction, thus adding evidence for the hypothesis that the after-reaction is related to pituitary activation.

The foregoing studies focus on changes in electrical activity during the estrous cycle, whereas other studies investigate the regulatory role of these changes on hypothalamic–pituitary secretory activity. For example, in cats and rabbits, the amplitude of both negative and positive responses of the arcuate nucleus to hippocampal stimulation decreased during estrus and increased during the period when progesterone dominated over estrogen. These cyclical changes in evoked responses to hippocampal stimulation showed an inverse relationship to those of the amygdala (Kawakami and Terasawa, 1965; Kawakami et al., 1966c). Similarly, in the rat, Terasawa and Timiras (1969) demonstrated that in the arcuate nucleus evoked potentials to stimulation of the midbrain reticular formation increased during proestrus and estrus as compared to diestrus. In other experiments, stimulation of the dorsal hippocampus accelerated unit activity in the arcuate nucleus and inhibited unit activity in the ventromedial hypothalamus. In contrast, stimulation of the medial part of the amygdala accelerated ventromedial hypothalamic unit activity and inhibited arcuate nucleus unit activity (Kawakami and Kubo, 1971). Thus these authors conclude that the dorsal hippocampus and the medial amygdala influence the hypothalamus reciprocally; that is, the hippocampus accelerates and the amygdala depresses activity in the arcuate nucleus circuit; on the other hand, each exerts an opposite effect on ventromedial hypothalamic activity.

2.2. Effects of Lesion and Stimulation

It is well known that massive lesions of the temporal lobe including the amygdala and hippocampus induce hypersexual behavior in the adult monkey (Kluver and Bucy, 1939), cat (Schreiner and Kling, 1953), and rabbit (Beyer et al., 1964). Beyer et al. (1964) have shown that hypersexual behavior exhibited by the female is dependent on estrogen, inasmuch as it disappears following withdrawal of estrogens. In another study with female rabbits, Mena and Beyer (1968) reported milk secretion in ovariectomized, estrogen-primed rabbits following lesion of the temporal lobe including the amygdala and ventral hippocampus. In adult female rats, ablation of the hippocampus or transection of the fornix induced increases in serum LH and FSH (Terasawa and Kawakami, 1973). In immature animals, lesions of hippocampus induced delayed vaginal opening and gonadal atrophy (Riss et al., 1963) as well as irregularities of the estrous cycle (Conrey and De Groot, 1964), whereas lesions of the amygdala and stria terminalis resulted in precocious puberty (Elwers and Critchlow, 1960, 1961).

That the hypothalamus is essential to the cyclical release of LH, responsible for ovulation in several species, has been confirmed by studies indicating that stimulation of discrete areas of the hypothalamus is capable of inducing ovulation that had been experimentally blocked either by nembutal injection or by exposure to constant illumination (Critchlow, 1957, 1958; Kawakami and Terasawa, 1970; Kawakami et al., 1970; Terasawa et al., 1969; Velasco and Taleisnik, 1969a). Similar confirmation comes from the observation of Helasz and Pupp (1965) that complete hypothalamic deafferentation (the hypothalamic island preparation) prevents ovulation entirely. That the limbic system plays a role in this aspect of gonadal function is indicated by observations that ovulation can be induced by stimulation of the medial part of the amygdala in rats (Bunn and Everett, 1957; Kawakami et al., 1970; Velasco and Taleisnik, 1969a), in cats (Shealy and Peele, 1957), and in rabbits (Koikegami et al., 1954; Kawakami et al., 1966b). On the other hand, conflicting reports of the effects of hippocampal stimulation on this system continue to obscure its role. Kawakami *et al.* (1966*c*), working with the rabbit, have reported that ovulation occurs following stimulation of either the dorsal or the ventral hippocampus and that bilateral lesions of fornix, septum, and periventricular arcuate nucleus block this response. Whereas stimulation of the amygdala appears to enhance the biosynthesis of ovarian estrogen as well as some progesterone (Kawakami *et al.*, 1967), stimulation of the hippocampus, more than that of the amygdala and the periventricular arcuate nucleus, enhances progesterone formation. Because lesion of the arcuate nucleus blocked this effect, these researchers concluded that the effects of hippocampal stimulation on progesterone formation were mediated through the arcuate nucleus. Further, inasmuch as the increase of ovarian progesterone formation is in proportion to that following an exogenous dose of LH, these authors also concluded that the hippocampus regulates the discharge of LH from the anterior hypophysis via the hypothalamus.

Studies in the rat have yielded somewhat different results from those in the rabbit. Velasco and Taleisnik (1969*a*) and Kawakami *et al.* (1970) report that hippocampal stimulation does not induce ovulation in rats in continuous estrus. According to further studies undertaken by Velasco and Taleisnik (1969*b*), stimulation of the hippocampus inhibited ovulation in spontaneously ovulating rats, stimulation of the ventral hippocampus inhibiting ovulation in 80% of the rats and stimulation of the dorsal hippocampus inhibiting it in 40%. Furthermore, stimulation of ventral hippocampus 5 min prior to stimulation of the medial amygdala or preoptic area inhibited the ovulatory response usually observed. That this hippocampal stimulation effected a true inhibition was supported by further studies using picrotoxin, a drug known to block inhibitory synapses. When it was injected before hippocampal stimulation, all effects of the stimulation were blocked. The inhibitory effect of hippocampal stimulation was also blocked by section of the medial corticohypothalamic tract, a tract which degenerates following lesioning of the ventral hippocampus and which distributes to the arcuate nucleus in the hypothalamus. Additionally, these authors measured plasma levels of LH directly and found that in proestrous rats, 3 h after stimulation of area CA3 of the ventral hippocampus or of the subiculum, plasma LH was significantly lower. The elevated plasma LH characteristic of ovariectomized rats also was depressed after ventral hippocampal stimulation, providing further evidence that the ventral hippocampus exerts an inhibitory influence on LH release. These data were confirmed by Gallo *et al.* (1971), who showed further that ventral hippocampal stimulation which results in decreases of plasma LH concomitantly increases neuronal activity in the arcuate nucleus.

While it is difficult to interpret these conflicting findings, the species differences should not be overlooked. It should also be noted that Kawakami *et al.* (1966*c*) were not able to eliminate the possibility of hippocampal seizure discharges traveling to the adjacent amygdala. Until further studies are completed, it still appears likely that the triggering of ovulation by the release of LH from the pituitary is under the influence of the limbic system, and that, definitively in the rat and probably in man and other mammals, the amygdala and the hippocampus play an important stimulatory and inhibitory role, respectively.

3. Adrenocortical Hormones and Sensory Information Processing

Adrenal steroids are important in "setting" the levels of brain excitability and in the processing of sensory information. Adrenal hypofunction is associated with the onset of convulsions, and electroconvulsive seizures can be induced more readily in this physiological state (Hoagland, 1957); in addition, the absolute thresholds for several sensory systems are greatly decreased in hypocorticoidism, and these changes are reflected by alterations in cortical evoked potentials (Henkin, 1970; Ojeman and Henkin, 1967). Steroids that predominantly affect electrolyte metabolism (deoxycorticosterone, 11-deoxycortisol) raise or normalize the threshold for electrically induced seizures but have no effect on sensory thresholds. Those steroids mainly concerned with carbohydrate metabolism (cortisone, cortisol) will bring the sensory thresholds back to normal. but will further antagonize the seizure effects and actually increase brain excitability (Timiras *et al.*, 1956; Woodbury, 1954). Although under most circumstances brain excitability is correlated with electrolyte metabolism, the excitatory action of cortisone or cortisol is not mediated by changes in electrolyte metabolism since it is not greatly altered by chronic administration of these hormones (Woodbury, 1954; Donovan, 1968). In seeking the mechanisms of action for these sensory and central excitatory effects, it is of interest to note that the hippocampus has the lowest seizure threshold of any structure in the brain (Goodfellow and Niemer, 1961; Green, 1964), and it has been postulated that electrically induced seizures probably originate in the hippocampus, gradually involving other portions of the limbic system and finally spreading to the neocortical mantle (Saul and Feld, 1961). In addition, we know the hippocampus–fimbrial system is involved both in the maintenance of endogenous levels of corticosteroids and in the mediation of sensory information. Stimulation of the hippocampus, on one hand, has been shown to alter blood levels of glucocortticoids (see Chapters 12 and 14; also, Endröczi and Lissák, 1963; Kawakami *et al.*, 1968; Mason, 1958). On the other hand, stimulation of the fimbria has been shown to increase auditory and visual cortical evoked potentials, an effect which can be reversed by lesions (Golden and Lubar, 1971); also, stimulation of the dorsal hippocampus has been shown to alter firing rates of median eminence units responding to photic, auditory, and sciatic stimulation (Mandelbrod and Feldman, 1972). Thus we designed and carried out the following studies to assess the role of the hippocampus in the modulation of sensory signals through the regulation of glucocorticoid levels.

Evoked responses recorded concurrently from the dorsal and ventral hippocampus and from the thalamus were compared in rats under varying endogenous and exogenous levels of glucocorticoids. Somatosensory and visual stimuli were presented to both intact and adrenalectomized animals and changes in evoked potentials were observed for up to 4 h following intravenous injection of corticosterone. Adult Long-Evans male rats were anesthetized with α-chloralose (80 mg/kg for normal rats, 40 mg/kg for adrenalectomized rats) and tracheostomized, and the femoral vein was cannulated. Bipolar electrodes were stereotaxically placed and signals were amplified, tape-recorded, and averaged on a Mnemotron Computer of Average Transients. Somatosensory stimulation was delivered through two stainless steel needles inserted into the skin on the dorsal portion of the contralateral forepaw.

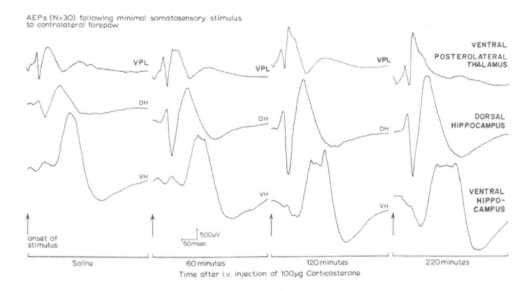

AEPs (N=30) following minimal somatosensory stimulus to contralateral forepaw

VENTRAL POSTEROLATERAL THALAMUS

DORSAL HIPPOCAMPUS

VENTRAL HIPPO-CAMPUS

onset of stimulus

500μV
50msec

Saline   60 minutes   120 minutes   220 minutes
Time after i.v. injection of 100μg Corticosterone

FIG. 4. Average evoked potentials to a somatosensory stimulus recorded from the ventral posterolateral thalamus and dorsal and ventral hippocampus in an intact rat following saline injection, and at 60, 120, and 220 min following intravenous injection of 100 μg corticosterone.

The stimulus used was just suprathreshold for eliciting evoked responses, and was delivered at 10-s intervals (0.6 Hz) to eliminate the possibility of habituation to the stimulus (Groves *et al.,* 1969). Shown on the left side of Fig. 4 is the average of 30 evoked responses of the intact subject to the somatosensory stimulus after an injection of physiological saline alone. The response from the ventroposterolateral thalamus has a latency of approximately 12 ms, the same as that recorded in the cat (Mc-Gowan-Sass and Eidelberg, 1972). The response from the dorsal hippocampus is also short latency, about 15–18 ms, and shows components similar to those recorded from the thalamus. The ventral hippocampus consistently shows a very large late component, about 100 ms, and shows no prominent short-latency oligosynaptic components. Following injection of corticosterone, changes occur in the amplitude of the evoked potentials, but no significant changes are evident in the form or the latency of the evoked responses. The amplitude changes are most marked in the dorsal hippocampus and the thalamus, with little or no amplitude changes in the ventral hippocampus.

When responses are compared between intact and adrenalectomized rats (Fig. 5), responses from the adrenalectomized rat appear somewhat less complex, and the response of the dorsal hippocampus lacks the typical short-latency components. Following corticosterone treatment, the amplitude of responses from the thalamus and the dorsal hippocampus increases in the intact rat, whereas in the adrenalectomized rat there is an increase in amplitude of response from the thalamus, and changes in complexity of the responses recorded from the hippocampus.

The responses of the dorsal and ventral hippocampus and the lateral geniculate to a visual stimulus delivered to the ipsilateral eye are shown in Fig. 6. The light

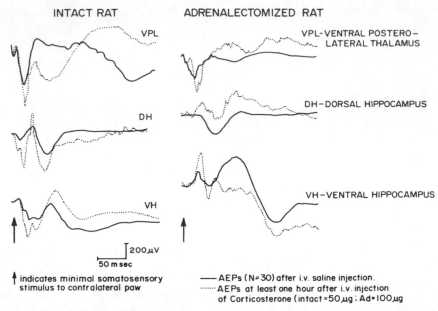

INTACT RAT   ADRENALECTOMIZED RAT

VPL   VPL-VENTRAL POSTERO-
                            LATERAL THALAMUS

DH   DH-DORSAL HIPPOCAMPUS

VH   VH-VENTRAL HIPPOCAMPUS

⌐200μV
50 m sec

↑ indicates minimal somatosensory   ——— AEPs (N=30) after i.v. saline injection.
stimulus to contralateral paw   ········· AEPs at least one hour after i.v. injection
                                          of Corticosterone (intact =50μg ; Ad=100μg)

FIG. 5. Comparison of somatosensory evoked potentials recorded from the ventral posterolateral thalamus and from the dorsal and ventral hippocampus in an intact and an adrenalectomized rat following intravenous injection of corticosterone.

source was an incandescent bulb which illuminated a fiberoptics tube of light projected directly into the eye, which had been dilated with atropine sulfate. The evoked potential was recorded for 1 second following a 500-ms light flash, so that both an "on" and an "off" response could sometimes be identified. Before treatment, the response from the ventral hippocampus was greater in amplitude and more complex than that from the dorsal hippocampus, and showed latencies similar to the lateral geniculate. The main components of the response from the dorsal hippocampus were delayed as compared to those from the ventral hippocampus.

As shown on the right-hand side of Fig. 6, following corticosterone treatment, all responses from the thalamus and from the hippocampus increased in amplitude, with those from the ventral hippocampus increasing considerably more than those from the dorsal hippocampus or the lateral geniculate. It is of particular interest that the "off" response in the ventral hippocampus, and to some extent in the dorsal hippocampus, increased markedly, inasmuch as these increases may indicate an augmentation of inhibitory synapses.

As shown in Fig. 6, 1 h following intravenous injection of 200 μg corticosterone, no significant changes in evoked potentials were observed. A subsequent additional dosage of 200 μg corticosterone administered 80 min after the inital injection, however, did induce an increase comparable to that observed in other intact subjects at a lower dosage. Thus it appears that the degree to which corticosterone injection elevates evoked potentials depends on the animals' endogenous levels of adrenal steroids: this is supported by the above-reported data as well as the observation that the intact animal, as compared to the adrenalectomized animal, requires a larger dose of the hormone to produce the same response (*cf*. Figs. 6 and 7).

Figure 7 shows the pattern of evoked responses to visual stimuli in an adrenalectomized rat. As shown, the responses in the saline-injected control are of small amplitude and not well defined, hippocampal potentials being especially small (in fact, detectable only when averaged). Following administration of corticosterone, marked increases in evoked potentials occurred in all leads. Once more, the increase in the amplitude of the evoked potential was more marked in the ventral hippocampus, and there is evidence that the "off" response was augmented considerably.

In Fig. 8, a summary comparison of the mean amplitude changes for 2 h following glucocorticoid administration for all subjects is presented. Both dorsal and ventral hippocampus increased more than the lateral geniculate following visual stimulation, and the ventral hippocampus showed a greater increase than the dorsal hippocampus. For the somatosensory stimulus, the dorsal hippocampus increased more than the ventral hippocampus and both parts of the hippocampus showed final levels higher than the thalamus, although the ventral hippocampus showed less increase during the first 30 min after corticoid administration than the corresponding increase

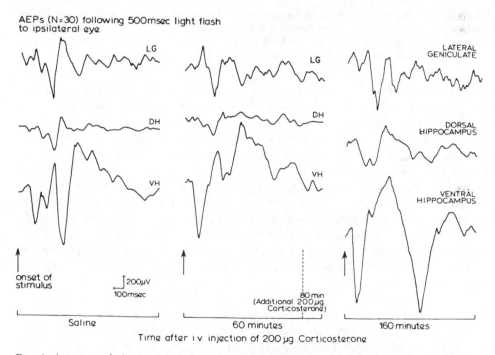

FIG. 6. Average evoked potentials in lateral geniculate (LG), dorsal hippocampus (DH), and ventral hippocampus (VH) following a 500-ms light flash to the ipsilateral eye in an intact rat. Note that 60 min after administration of 200 μg corticosterone no significant increase in amplitude of evoked responses was observed, although an additonal dose of 200 μg was sufficient to bring about a considerable increase in amplitude of evoked potentials in all leads.

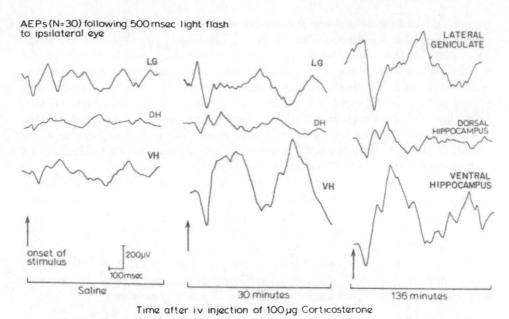

FIG. 7. Average evoked potentials in lateral geniculate (LG), dorsal hippocampus (DH), and ventral hippocampus (VH) following a 500-ms light flash to the ipsilateral eye in an adrenalectomized rat. Shown are responses after saline injection, and 30 and 136 min after intravenous injection of corticosterone. Note that the "off" response in LG and VH is considerably augmented following corticosterone injection.

in the thalamus. Regression analysis of the data for the visual stimulus indicated that the regression lines of lateral geniculate, ventral hippocampus, and dorsal hippocampus were significantly different from zero ($p < 0.01$) for all three brain regions; t tests on the differences between the three regions showed that the lateral geniculate was significantly different from the dorsal hippocampus ($p < 0.01$) and from the ventral hippocampus ($p < 0.001$), although the ventral hippocampus and dorsal hippocampus were not significantly different from each other on this measure. For the somatosensory stimulus, ventral posterolateral thalamus, dorsal hippocampus, and ventral hippocampus were significantly different from zero ($p < 0.02$, $p < 0.01$, $p < 0.01$, respectively). The ventral posterolateral thalamus was significantly different from both dorsal hippocampus ($p < 0.01$) and ventral hippocampus ($p < 0.01$), and ventral hippocampus was significantly different from dorsal hippocampus ($p < 0.005$).

The foregoing data indicate that the way in which the hippocampus processes afferent sensory data is closely related to the localized hormonal environment. In addition, sensory information is processed differentially within the hippocampus itself; that is, the dorsal portion of the hippocampus responds maximally to one type of cue, whereas the ventral hippocampus responds maximally to other types of stimuli. These data substantiate previous observations of differences in the electrical responsivity of the dorsal and ventral hippocampus (Adey and Walter, 1963; Elul, 1964; McGowan-Sass, 1973). This specificity of the hippocampus appears also in the interaction between its responses to glucocorticoids and the processing of sensory information; for example, different portions of the hippocampal arch show maximum amplitude changes in response to corticoids only when paired with specific signals.

Other researchers working in different species have observed changes in amplitude of evoked activity in nonspecific areas of the brain following corticoid administration. Koranyi and Endröczi (1970), working with chicks, have reported increases in the cortical evoked potential from stimulation of optic chiasm following hydrocortisone administration, and decreases in the same response following ACTH administration. Feldman (1962) reported an increase in evoked potentials in the reticular formation from stimulation of the sciatic nerve following cortisol administration in cats. Steroid-dependent changes have also been reported in the EEG of the dorsal hippocampus and basolateral amygdala (Kawakami *et al.*, 1966c) and in the responsiveness of single units in the hypothalamus (Feldman and Dafney, 1966), the fornix (Koranyi *et al.*, 1971), and the hippocampus (Taylor *et al.*, 1970).

4. Adrenocortical Hormones and Cyclicity

The above short review points clearly to the responsiveness of the hippocampal system to the exogenous administration of adrenal steroids. It must also be borne in

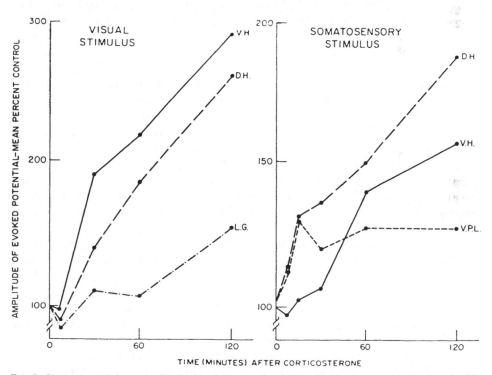

FIG. 8. Comparison of mean amplitude changes in dorsal and ventral hippocampus and thalamus for 2 h following corticosterone injection. All subjects (both adrenalectomized and intact rats) had electrodes in all three brain regions. (Visual stimulus, $N = 4$; somatosensory stimulus, $N = 5$.)

mind that the adrenal cortical secretion is cyclical in nature and that this cyclicity can be influenced by and reciprocally can influence many physiological systems. For example, besides the well-documented variations in sleep–wakefulness, there are diurnal variations in activity (Aschoff, 1966), in learning ability (Pagano and Lovely, 1972), in eating and drinking (personal observations), and in taste and related chemical senses (Henkin, 1970). These physiological changes are brought about by very small changes in the circulating adrenocortical hormone level (usually a factor of 2 or 3), whereas the resultant peripheral and behavioral changes appear much greater, as, for instance, in the case of taste sensitivity which increases daily by a factor of 100 (Henkin, 1970), thus indicating the sensitivity of the CNS to minute changes in steroid levels. These diurnal changes in steroid levels also are reflected in alterations in multiple-unit activity in several limbic areas, the basal level of hippocampal activity increasing in the afternoon in the rat (Kawakami et al., 1970).

Conversely, several authors have reported alterations in the diurnal variation of adrenal steroids with interruption of the hippocampal system. Mason (1958), working with the monkey, reported that either hippocampal ablation or fornix section altered the diurnal pattern, resulting in night values being equal to or greater than day values. In the rat, similar disruption of the diurnal rhythm of corticosterone has been reported with section of the fornix (Nakadate and De Groot, 1963), with implants of crystalline cortisol in the ventral hippocampus (Slusher, 1966), and with lesions of anterior hypothalamus bilaterally involving the periventricular zone and arcuate nuclei (Slusher, 1964). Moberg et al. (1971) confirmed that this disruption of the hippocampal system in the rat alters the pattern of steroid release but not the total amount of corticosterone secreted over a 24-h period.

Recent evidence in other areas of research has indicated that both the hippocampus and amygdala exhibit diurnal variations in 5-hydroxytryptamine content (Scapagnini et al., 1971), the magnitude of fluctuation in hippocampus being second only to the marked changes seen in the pineal gland (Quay, 1968). In the cat, Krieger and Rizzo (1969) report that drugs which affect CNS serotonin levels (Monase, PCPA) or antagonize serotonin action (cinanserin, cyproheptadine) can prevent the normal circadian rise in plasma corticoids. Other authors have shown that the maturation of circadian rhythm of brain serotonin is closely related to spontaneous activity and the maturation of corticosteroid secretion (Okada, 1971; Asano, 1971).

Separate experiments investigating the effects of the light–dark cycle on diurnal variation have indicated that it is an important factor in corticosteroid secretion. Investigations on the effects of maintaining rats under constant light conditions have indicated that the diurnal rhythm of corticosterone and corticotropin is abolished (Cheifetz et al., 1968), as well as that of pineal serotonin (Snyder et al., 1967), and that the total amount of both corticotropic hormones and adrenal hormones secreted over the 24-h period is decreased in these animals.

With these data in mind, the results reported above are of particular interest in indicating the overwhelming effectiveness of visual stimulation on the hippocampus following corticoid administration. It is possible to postulate, as others have (Moberg et al., 1971; Retienne and Schulz, 1970), that efferent impulses discharged over a purely inhibitory system, such as the hippocampus, could be responsible for diurnal variations in the adrenal and related systems. This inhibitory discharge would reduce the level of tropic hormone secretion, as, for instance, during periods of activity. During sleep or inactivity, once the inhibitory discharge stopped, ACTH would increase progressively until circulating corticosterone rose to a level that produced significant feedback inhibition of the pituitary. Interruption of such a system by lesion or neural isolation would be expected to produce a steady corticosterone level somewhere between the highest and the lowest levels normally seen, as, indeed, lesions of the hippocampal–fornix system do. On the other hand, constant stimulation of the system through normal afferent pathways (such as the visual system) would be expected not only to abolish the circadian rhythm but to decrease further the circulating level of steroids, as has been observed in experiments with constant illumination.

5. New Horizons

As we clarify the specific roles of the hippocampus and amygdala with respect to gonadal and adrenal systems, it must be borne in mind that in the dynamic functioning of the total organism their interactions are complex and interdependent. For instance, adrenal activation with consequent stimulation of the hippocampus and lowering of seizure threshold could alter the environment for subsequent release of gonadotropic and other pituitary hormones, a postulation that could be significant in light of recent findings that pituitary hormones other than ACTH are released on a diurnal basis. For example, prolactin secretion is characterized by a steadily rising multiple-peaked release during early sleep, and its levels remain elevated during sleep but fall upon awakening, in the human (Sassin *et al.,* 1972), in the male rat (Ronnekleiv *et al.,* 1973), and in the pseudopregnant female rat (Freeman and Neill, 1972); in the male rat, this pattern is abolished by pinealectomy (Ronnekleiv *et al.,* 1973). Human growth hormone also has been shown in adults to rise markedly following the onset of nocturnal sleep (Pawel *et al.,* 1972; Takebe *et al.,* 1969). Similarly, release of LH in normal men reportedly occurs as repetitive discharges during the night, with highest levels in the early morning (Nankin and Troen, 1972; Leyendecker and Saxena, 1970). Although this finding has not been confirmed by other researchers (Krieger *et al.,* 1972; Gallagher *et al.,* 1973), Gallagher *et al.* (1973) do report an increase of plasma LH concentration just prior to the end of the sleep period in women at mid-menstrual cycle.

In addition to their actions on the adult organism, hormones are very important in the developing organism and evidence indicates that the CNS is especially responsive to the hormonal environment in both a facilitatory and an inhibitory fashion. Data indicating that the sensitivity of the limbic system to gonadal hormones alters with development have been reported by Terasawa and Timiras (1968*b*), who followed changes in localized seizure thresholds in the hippocampus and amygdala of normally developing female rats. The amygdaloid and hippocampal seizure thresholds were found to gradually decrease from day 5, reaching their lowest values at days 18 and 20, respectively. At day 27, the amygdaloid threshold dropped sharply and remained low until day 40. Vaginal opening and first ovulation followed. Pregnant mare serum gonadotropin (PMSG), which hastens the onset of puberty, lowered the amygdaloid and hippocampal threshold when administered at 30 days of age. Other electrophysiological studies in this area have shown that brain maturation, as assessed by maximal electroshock seizure responses, was significantly hastened in rats when estradiol was administered on days 6–10 or 8–12 (Heim and Timiras, 1963; Heim, 1966), a finding consistent with the earlier report by Woolley and Timiras (1962a) that ovariectomy prior to puberty delayed brain maturation. Curry and Timiras (1972) report that neonatal administration of estradiol hastens the development of the transcallosal response, and Hudson et al. (1970) demonstrated that estradiol accelerates the decrease known to occur in putative inhibitory neurotransmitters with development. Adrenal steroids also affect CNS development, cortisol hastening brain maturation when administered between 8 and 16 days of age (Timiras et al., 1968; Vernadakis and Woodbury, 1963). Vernadakis and Woodbury (1971) note that cortisol has a biphasic effect on the nervous system: it delays maturation and decreases brain excitability when given acutely or chronically between the first and seventh days after birth, but enhances maturation of the nervous system and increases excitability when given acutely or chronically after the eighth postnatal day.

The above short review shows the direction that research on hormonal systems and the CNS (especially the limbic system) will be taking. As we draw together several areas of research elucidating the working of separate systems, we can begin to focus on the functioning of the total biological entity, recognizing that CNS–hormonal responses vary considerably depending on the maturity of the organism, the hormone in question and its interaction with other hormones, and the functional status of their target sites, all of which are influenced by diurnal and other cyclical rhythms.

Acknowledgment

The authors wish to thank Mrs. Laurel Cook for editorial assistance.

6. References

Adey, W. R., and Walter, D. O. Application of phase detection and averaging techniques in computer analysis of EEG records in the cat. *Experimental Neurology,* 1963, **7,** 186–209.

Asano, Y. The maturation of the circadian rhythm of and serotonin in the rat. *Life Sciences,* 1971, **10,** 1, 883–894.

Aschoff, J. Circadian activity rhythms in chaffinches (*Fringilla coelebs*) under constant conditions. *Japanese Journal of Physiology,* 1966, **16,** 363–370.

Beach, F. A. *Hormones and behavior.* New York: Hoeber, 1948.

Beyer, C., Yaschine, T., and Mena, F. Alterations in sexual behavior induced by temporal lobe lesions in female rabbits. *Boletin del Instituto de Estudios y Biologicos,* 1964, **22(3),** 379–386.

Bunn, J. P., and Everett, J. W. Ovulation in persistent-estrus rats after electrical stimulation of the brain. *Proceedings of the Society for Experimental Biology,* 1957, **96,** 369–371.

CHEIFETZ, P., GAFFUD, N., AND DINGMAN, J. F. Effects of bilateral adrenalectomy and continuous light on the circadian rhythm of corticotrophin in female rats. *Endocrinology*, 1968, **82**, 1117–1124.

CONREY, K., AND DE GOOT, J. Limbic system involvement in the regulation of luteinizing hormone release. *Federation Proceedings*, 1964, **23**, 109.

CRITCHLOW, B. V. Ovulation induced by hypothalamic stimulation in the rat. *Anatomical Record*, 1957, **127**, 283, abst. 233.

CRITCHLOW, V. Ovulation induced by hypothalamic stimulation in anesthetized rat. *American Journal of Physiology*, 1958, **195**, 171–174.

CURRY, J. J., AND TIMIRAS, P. S. Development of evoked potentials in specific brain systems after neonatal administration of estradiol. *Experimental Neurology*, 1972, **34(1)**, 129–139.

DONOVAN, B. T. Hormones and brain function. In *Mammalian neuroendocrinology*. New York: McGraw-Hill, 1968, pp. 166–176.

ELUL, R. Regional differences in the hippocampus of the cat. *Electroencephalography and Clinical Neurophysiology*, 1964, **16**, 470–502.

ELWERS, M., AND CRITCHLOW, V. Precocious ovarian stimulation following hypothalamic and amygdaloid lesions in rats. *American Journal of Physiology*, 1960, **198**, 381–385.

ELWERS, M., AND CRITCHLOW, V. Precocious ovarian stimulation following interruption of stria terminalis. *American Journal of Physiology*, 1961, **201**, 281–384.

ENDRÖCZI, E., AND LISSÁK, K. Effects of hypothalamic and brain stem structure stimulation on pituitary adrenocortical function. *Acta Physiologica Academiae Scientiarrum Hungaricae*, 1963, **24**, 67–78.

EVERETT, J. W. The mammalian female reproductive cycle and its controlling mechanisms. In W. C. Young (Ed.), *Sex and internal secretions*. Baltimore: Williams and Wilkins, 1961, pp. 497–555.

FARRIS, E. J. The prediction of the day of human ovulation by the rat test as confirmed by fifty conceptions. *American Journal of Obstetrics and Gynecology*, 1948, **56**, 347–352.

FELDMAN, S. Electrophysiological alterations in adrenalectomy. *Archives of Neurology*, 1962, **7**, 460–470.

FELDMAN, S., AND DAFNY, N. Effect of hydrocortisone on single cell activity in the anterior hypothalamus. *Israel Journal of Medical Science*, 1966, **2(5)**, 621–23.

FERIN, M., TEMPONE, A., ZIMMERING, P. E., AND VAN DE WIELE, R. L. Effect of antibodies to 17β-estradiol and progesterone on the estrous cycle of the rat. *Endocrinology*, 1969, **85**, 1070.

FREEMAN, M. E., AND NEILL, J. D. The pattern of prolactin secretion during pseudopregnancy in the rat: A daily nocturnal surge. *Endocrinology*, 1972, **90(5)**, 1292–1294.

GALLAGHER, T. F., YOSHIDA, K., ROFFWARG, H. D., FUKUSHIMA, D. K., WEITZMAN, E., AND HELLMAN, L. ACTH and cortisol secretory patterns in man. *Journal of Clinical Endocrinology and Metabolism*, 1973, **36**, 1061–1068.

GALLO, R. V., JOHNSON, J. H., GOLDMAN, B. D., WHITMOYER, D. I., AND SAWYER, C. H. Effects of electrochemical stimulation of the ventral hippocampus on hypothalamic electrical activity and pituitary gonadotropin secretion in female rats. *Endocrinology*, 1971, **89**, 704–713.

GANONG, W. F. *Review of medical physiology*, 5th ed. Los Altos, Calif.: Lange Medical Publications, 1971, p. 176.

GOLDEN, G. H., AND LUBAR, J. F. Effect of septal and fimbrial stimulation on auditory and visual cortical evoked potentials in the rat. *Experimental Neurology*, 1971, **30(3)**, 389–402.

GOODFELLOW, E. F., AND NIEMER, W. T. The spread of after-discharge from stimulation of rhinencephalon in cat. *Electrophysiology and Clinical Neurophysiology*, 1961, **13**, 710–721.

GORBMAN, A., AND BERN, H. A. *A textbook of comparative endocrinology*. New York: Wiley, 1962, p. 254.

GREEN, J. D. The hippocampus. *Physiological Review*, 1964, **44(4)**, 561–608.

GROVES, P. M., LEE, D., AND THOMPSON, R. F. Effects of stimulus frequency and intensity on habituation and sensitization in acute spinal cat. *Physiology and Behavior*, 1969, **4**, 383–388.

HEIM, L. M. Effect of estradiol on brain maturation: Dose and response relationships. *Endocrinology*, 1966, **78**, 1130.

HEIM, L. M., AND TIMIRAS, P. S. Gonad-brain relationship: Precocious brain maturation. *Endocrinology*, 1963, **72**, 598–606.

HELASZ, B., AND PUPP, L. Hormone secretion of the anterior pituitary gland after physical interruption of all nervous pathways to the hypophysiotrophic area. *Endocrinology*, 1965, **77**, 553–562.

HENKIN, R. I. The effects of corticosteroids and ACTH on sensory systems. In D. de Wied and J. A. W. M. Weijnen (Eds.), *Pituitary adrenal and the brain*. Vol. 32 of *Progress in brain research*. Amsterdam: Elsevier, 1970, pp. 270–294.

HITCHCOCK, F. A. The comparative activity of male and female albino rats. *American Journal of Physiology*, 1925/1926, **75**, 205–210.

HOAGLAND, H. *Hormones, brain function and behavior*. New York: Academic Press, 1957.

HUDSON, D. B., VERNADAKIS, A., AND TIMIRAS, P. S. Regional changes in amino acid concentration in the developing brain and the effects of neonatal administration of estradiol. *Brain Research*, 1970, **23**, 213–222.

KAWAKAMI, M., AND KUBO, K. Neuro-correlate of limbic–hypothalamo–pituitary gonadal axis in the rat: Changes in limbic–hypothalamic unit activity induced by vaginal and electrical stimulation. *Neuroendocrinology*, 1971, **7**, 65–89.

KAWAKAMI, M., AND SAWYER, C. H. Neuroendocrine correlates of changes in brain activity thresholds by sex steroids and pituitary hormones. *Endocrinology*, 1959, **65**, 652–668.

KAWAKAMI, M., AND TERASAWA, E. Studies on brain activity and conduction of synapsis in estrous cats and rabbits. *Journal of Physiological Society of Japan*, 1965, **27**, 86.

KAWAKAMI, M., AND TERASAWA, E. Effect of electrical stimulation of the brain on ovulation during estrous cycle in the rats. *Endocrinologia Japonica*, 1970, **17**, 7–13.

KAWAKAMI, M., KOSHINO, T., AND HATTORI, Y. Changes in the EEG of the hypothalamic and limbic system after administration of ACTH, SU4885 and ACh in rabbits with special reference to neurohumoral feedback regulation of pituitary adrenal system. *Japanese Journal of Physiology*, 1966a, **16**, 551–569.

KAWAKAMI, M., SETO, K., AND YOSHIDA, K. Influence of the limbic system on ovulation and on progesterone and estrogen formation in rabbit's ovary. *Japanese Journal of Physiology*, 1966b, **6**, 254–273.

KAWAKAMI, M., TERASAWA, E., TSUCHIHASHI, S., AND YAMANAKA, K. Differential control by sex hormones of brain activity in the rabbit and its physiological significance. In J. Pincus, K. Nakao, and J. L. Tait (Eds.), *Steroid dynamics*. New York: Academic Press, 1966c, pp. 237–302.

KAWAKAMI, M., SETO, K., TERASAWA, E., AND YOSHIDA, K. Mechanisms in the limbic system controlling reproductive function of the ovary with special reference to the positive feedback. *Progress in Brain Research*, 1967, **27**, 69–102.

KAWAKAMI, M., SETO, K., TERASAWA, E., YOSHIDA, K., IYAMOTO, T., SEKIGUCHI, M., AND HATTORI, Y. Influence of electrical stimulation and lesion in limbic structure upon biosynthesis of adrenocorticoid in the rabbit. *Neuroendocrinology*, 1968, **3**, 337–348.

KAWAKAMI, M., TERASAWA, E., AND IBUKI, T. Changes in multiple unit activity of the brain during the estrous cycle. *Neuroendocrinology*, 1970, **6**, 30–48.

KLUVER, H., AND BUCY, P. Preliminary analysis of functions of the temporal lobes in monkeys. *Archives of Neurology and Psychiatry*, 1939, **42**, 979–1000.

KOIKEGAMI, H., YAMADA, T., AND USEI, K. Stimulation of amygdaloid nuclei and periamygdaloid cortex with special reference to its effects on uterine movements and ovulation. *Folia Psychiatrica Neurologica Japonica*, 1954, **8**, 7–31.

KORANYI, L., AND ENDRÖCZI, E. Influence of pituitary–adreno-cortical hormones on thalamo-cortical and brainstem limbic circuits. In D. de Wied and J. A. W. M. Weijnen (Eds.), *Progress in brain research*. Vol. 32. Amsterdam: Elsevier, 1970, pp. 120–130.

KORANYI, L., BEYER, C., AND GUZMAN-FLORES, C. Effect of ACTH and hydrocortisone on multiple unit activity in the forebrain and thalamus in response to reticular stimulation. *Physiology and Behavior*, 1971, **7**, 331–335.

KRIEGER, D. T., AND RIZZO, F. Serotonin mediation of circadian periodicity of plasma 17-hydroxycorticosteroids. *American Journal of Physiology*, 1969, **217(6)**, 1703.

KRIEGER, D. T., OSSOWSKI, R., FOGEL, M., AND ALLEN, W. Lack of circadian periodicity of human serum FSH and LH levels. *Journal of Clinical Endocrinology and Metabolism*, 1972, **34(4)**, 619–623.

LAWTON, I. E., AND SCHWARTZ, N. B. Pituitary–ovarian function in rats exposed to constant light: A chronological study. *Endocrinology*, 1967, **81**, 497.

LEYENDECKER, G., AND SAXENA, B. B. Taggesschwankungen der Konzentrationen von follikelsti-mulierendem (FSH) und luteinisierendem (LH) Hormon im menschlichen Plasma. *Klinische Wochenschrift,* 1970, **48,** 236.

LONG, J. A., AND EVANS, H. M. The oestrus cycle in the rat and its associated phenomena. *Memoirs University of California,* **6,** 1922.

MANDELBROD, I., AND FELDMAN, S. Effects on sensory and hippocampal stimulation on unit activity in the median eminence of the rat hypothalamus. *Physiology and Behavior,* 1972, **9,** 565–572.

MASON, J. W. The CNS regulation of ACTH secretion. In *Reticular formation of the brain.* Boston: Little, Brown, 1958, pp. 645–670.

McCLINTOCK, J. A., AND SCHWARTZ, N. B. Changes in pituitary and plasma follicle stimulating hormone concentrations during the rat estrous cycle. *Endocrinology,* 1968, **83,** 433.

McGOWAN-SASS, B. K. Differentiation of electrical rhythms and functional specificity of the hippocampus of the rat. *Physiology and Behavior,* 1973, **11(2),** 187–194.

McGOWAN-SASS, B. K., AND EIDELBERG, E. Habituation of somatosensory evoked potentials in the lemniscal system of the cat. *Electroencephalography and Clinical Neurophysiology,* 1972, **32,** 373–381.

MENA, F., AND BEYER, C. Induction of milk secretion in the rabbit by lesions in the temporal lobe. *Endocrinology,* 1968, **83(3),** 618–620.

MOBERG, G. P., SCAPAGNINI, U., DE GROOT, J., AND GANONG, W. F. Effect of sectioning the fornix on diurnal fluctuations in plasma corticosterone levels in the rat. *Neuroendocrinology,* 1971, **7,** 11–15.

NAKADATE, G., AND DE GROOT, J. Fornix section and adrenocortical function in rats. *Anatomical Record,* 1963, **45,** 338.

NANKIN, H. R., AND TROEN, P. Overnight patterns of serum luteinizing hormone in normal men. *Journal of Clinical Endocrinology and Metabolism,* 1972, **35(5),** 705–710.

OJEMAN, G. A., AND HENKIN, R. I. Steroid dependent changes in human visual evoked potentials. *Life Sciences,* 1967, **6(2),** 327–334.

OKADA, F. The maturation of the circadian rhythm of brain serotonin in the rat. *Life Sciences,* 1971, **10,** 77–86.

PAGANO, R. R., AND LOVELEY, R. H. Diurnal cycle and ACTH facilitation of shuttlebox avoidance. *Physiology and Behavior,* 1972, **8,** 721–723.

PAWEL, M. A., SASSIN, J. F., AND WEITZMAN, E. D. The temporal relation between HGH release and sleep stage changes at nocturnal sleep onset in man. *Life Sciences,* 1972, **2(1),** 587–591..

QUAY, W. B. Differences in circadian rhythms in 5-hydroxytryptamine according to brain region. *American Journal of Physiology,* 1968, **215,** 1448–1453.

RETIENNE, K., AND SCHULZ, F. Circadian rhythmicity of hypothalamic CRF and its central nervous regulation. *Hormones and Metabolic Research,* 1070, **2(4),** 221.

RISS, W., BURSTEIN, S., AND JOHNSON, R. W. Hippocampal or pyriform lobe damage in infancy and endocrine development of rats. *American Journal of Physiology,* 1963, **204,** 861–866.

RONNEKLEIV, O. K., KRULICH, L., AND McCANN, S. M. An early morning surge of prolactin in the male rat and its abolition by pinealectomy. *Endocrinology,* 1973, **92(5),** 1339–1342.

SASSIN, J. F., FRANTZ, A. G., WEITZMAN, E. D., AND KAPEN, S. Human prolactin: 24-hour pattern with increased release during sleep, *Science,* 1972, **177,** 1205.

SAUL, G. D., AND FELD, M. The limbic system and EST seizures. In J. Wortis (Ed.), *Recent advances in biological psychiatry.* Vol. 3. New York: Grune and Stratton, 1961.

SCAPAGNINI, U., MOBERG, G. P., VANLOON, G. R., DE GROOT, J., AND GANONG, W. F. Relations of brain 5-hydroxytryptamine content to the diurnal variation in plasma corticosterone in the rat. *Neuroendocrinology,* 1971, **7(2),** 90–96.

SCHREINER, L., AND KLING, A. Behavioral changes following rhinencephalic injury in the cat. *Journal of Neurophysiology,* 1953, **16,** 643–659.

SHEALY, C. N., AND PEELE, T. L. Studies on amygdaloid nucleus of the cat. *Journal of Neurophysiology,* 1957, **20,** 125–139.

SHIRLEY, B., WOLINSKY, J., AND SCHWARTZ, N. B. Effects of a single injection of an estrogen antagonist on the estrous cycle of the rat. *Endocrinology,* 1968, **82,** 959–968.

SLUSHER, M. A. Effect of chronic hypothalamic lesions on diurnal and stress corticosteroid levels. *American Journal of Physiology,* 1964, **206(5),** 1161–1164.

SLUSHER, M. A. Effects of cortisol implants in the brainstem and ventral hippocampus on diurnal corticosteroid levels. *Experimental Brain Research,* 1966, **1,** 184–194.

SNYDER, S. H., AXELROD, J., AND ZWEIG, M. Circadian rhythm in the serotonin content of rat pineal gland:. Regulating factors. *Journal of Pharmacology and Experimental Therapeutics,* 1967, **158,** 206–213.

TAKEBE, K., KUNITA, H., SAWANO, S., HORIUCHI, Y., AND MASHIMO, K. Circadian rhythms of plasma growth hormones and cortisol after insulin. *Journal of Clinical Endocrinology,* 1969, **29,** 1630–1633.

TAYLOR, A. N., MATHESON, G. K., AND DAFNY, N. Modification of the responsiveness of components of the limbic midbrain circuit by corticosteroids and ACTH. In C. H. Sawyer and R. A. Gorski (Eds.), *Steroid hormones and brain function.* Berkeley: University of California Press, 1971.

TERASAWA, E., AND KAWAKAMI, M. Effects of limbic forebrain ablation on pituitary gonadal function in the female rat. *Endocrinologica Japonica,* 1973, **20(3),** 277–290.

TERASAWA, E., AND TIMIRAS, P. S. Electrical activity during the estrous cycle of the rat: Cyclic changes in limbic structures. *Endocrinology,* 1968a, **83(2),** 207–216.

TERASAWA, E., AND TIMIRAS, P. S. Electrophysiological study of the limbic system in the rat at the onset of puberty. *American Journal of Physiology,* 1968b, **215(6),** 1462–1467.

TERASAWA, E., AND TIMIRAS, P. S. Cyclic changes in electrical activity of the rat midbrain reticular formation during the estrous cycle. *Brain Research,* 1969, **14,** 189–198.

TERASAWA, E., KAWAKAMI, M., AND SAWYER, C. H. Induction of ovulation by electrochemical stimulation in androgenized and spontaneously constant-estrous rats. *Proceedings of the Society for Experimental Biology and Medicine,* 1969, **132,** 497–501.

TIMIRAS, P. S., WOODBURY, D. M., AND BAKER, D. H. Effect of hydrocortisone acetate, desoxycorticosterone acetate, insulin, glucagon and dextrose, alone or in combination, on experimental convulsions and carbohydrate metabolism. *Archives Internationales de Pharmacodynamie et de Therapie,* 1956, **105,** 450–467.

TIMIRAS, P. S., VERNADAKIS, A., AND SHERWOOD, N. Development and plasticity of the nervous system. In N. S. Assali (Ed.), *Biology of gestation.* Vol. 2. New York: Academic Press, 1968, pp. 261–319.

VELASCO, M. E., AND TALEISNIK, S. Release of gonadotropins induced by amygdaloid stimulation in the rat. *Endocrinology,* 1969a, **84,** 132–139.

VELASCO, M. E., AND TALEISNIK, S. Effect of hippocampal stimulation on the release of gonadotropin. *Endocrinology,* 1969b, **85(6),** 1154–1159.

VERNADAKIS, A., AND WOODBURY, D. M. Effects of cortisol on electroshock seizure thresholds in developing rats. *Journal of Pharmacology and Experimental Therapeutics,* 1963, **139,** 110–113.

VERNADAKIS, A., AND WOODBURY, D. M. Effects of cortisol on maturation of the central nervous system. In *Influence of hormones on the nervous system.* Proceedings of the International Society of Psychoneuroendocrinology, Brooklyn, 1970, Basel: Karger, 1971, pp. 85–97.

WANG, G. H. Relation between "spontaneous" activity and estrous cycle in the white rat. *Comparative Psychology Monographs,* 1923, **1,** 1–27.

WOODBURY, D. M. Effect of hormones on brain excitability and electrolytes. *Recent Progress in Hormone Research,* 1954, **10,** 65–107.

WOOLLEY, D., AND TIMIRAS, P. S. The gonad–brain relationship: Effects of female sex hormones on electroshock convulsions in the rat. *Endocrinology,* 1962a, **70,** 196–209.

WOOLLEY, D., AND TIMIRAS, P. S. Gonad–brain relationship: Effects of castration and testosterone on electroshock convulsions in male rats, *Endocrinology,* 1962b, **71,** 609–617.

14

The Hippocampus and Regulation of the Hypothalamic–Hypophyseal–Adrenal Cortical Axis

CAROL VAN HARTESVELDT

1. Introduction

It has been established since 1950 that the hormones of the adrenal cortex are regulated by the central nervous system. The adrenal cortex is stimulated by adrenocorticotropic hormone (ACTH) carried through the circulatory system from the adenohypophysis, which is activated by corticotropin releasing factor (CRF) brought via the hypophyseal portal vessels from the median eminence of the hypothalamus. Many structures of the limbic system project to the hypothalamus and are therefore potential candidates for modulating adrenocortical hormone secretion. The hippocampus has been implicated in such a role since the late 1950s. The results of much of the research in this area have led many authors to conclude that the hippocampus inhibits hypothalamic mechanisms controlling the secretion of adrenal corticosteroid hormones. However, a close examination of the existing data reveals that this generalization is probably misleading oversimplification since (1) there is more than one functional system which has access to the final common path of adrenocortical steroid hormone secretion and (2) different methods of hippocampal disruption give results which lead to different conclusions about its function. In this chapter, the data relating to the role of the hippocampus in the regulation of this hor-

CAROL VAN HARTESVELDT • University of Florida, Gainesville, Florida.

mone system will be critically analyzed using these two points as an organizational framework.

2.   Diurnal Rhythms and Responses to Stress: Different Functional Systems

Release of the hormones of the hypothalamic–pituitary–adrenal cortical axis in many mammals fluctuates with a circadian rhythm (Critchlow, 1963). Superimposed on these cycles are increases in the adrenal corticosteroids elicited by a great variety of stimuli, including both those which cause tissue damage and those which do not (Mason, 1968). Evidence that circadian changes and stress-induced changes of adrenal corticosteroid hormones are mediated by independent neural systems which converge on the median eminence of the hypothalamus comes from studies of (1) development, (2) neuroanatomy, and (3) feedback suppression of corticosteroid secretion.

Regarding development, an increase in CRF activity in the hypothalamus in response to stress can be detected by the seventh day after birth in the rat, while the circadian rhythm of hypothalamic CRF content appears only during the third week of life (Hiroshige and Sato, 1971).

Experiments in which neural inputs to the hypothalamus have been disrupted also indicate independent mediation of circadian rhythms and of stress-induced adrenal corticosteroid responses. In order to discover the sources of effective inputs to the median eminence which regulate circadian rhythms of the adrenal corticosteroid hormones, CRF, ACTH, or plasma corticosteroids have been measured in animals with partial or total deafferentation of the hypothalamus. In rats with total deafferentation of the median eminence, a basal level of plasma corticosterone can be maintained but there are no circadian changes in level (Halasz et al., 1967; Palka et al., 1969; Ondo and Kitay, 1972). These circadian changes can be disrupted if afferents to the median eminence from an anterior direction, but not from dorsal or posterior directions, are severed (Slusher, 1964; Halasz et al., 1967). Thus signals from a neural circadian pacemaker or clock to the hypothalamus are mediated via a specific anterior pathway.

The median eminence also receives neural inputs which elicit an adrenocortical response following stressful stimuli. Some stimuli act directly on the hypothalamus; the neurally isolated median eminence can still respond (although not always maximally) to ether, histamine, Escherichia coli endotoxin, insulin-induced hypoglycemia, or a large dose of formaldehyde (Makara et al., 1970). However, for an adrenocortical response to other stressors such as noise and vibration, surgical trauma, or a small dose of formaldehyde (Makara et al., 1969) or to photic or acoustic stimulation (Feldman et al., 1972), some neural pathways to the median eminence must be intact. Knife cuts severing inputs from the lateral hypothalamic region to the median eminence significantly suppress the adrenocortical response to these stimuli (Makara et al., 1969; Feldman et al., 1972). Feldman et al. (1972) have suggested that the lateral cut severs the input from the medial forebrain bundle and thus disrupts communications between the hypothalamus and both forebrain and midbrain limbic regions. The anatomical evidence, then, indicates that the signals for regulation of diurnal rhythms and for stress-induced responses of the adrenal corticosteroid hormones reach the median eminence via anterior and lateral directions, respectively.

The neural input to the median eminence is in part determined by levels of circulating adrenal corticosteroid hormones; that is, these hormones play a part in central nervous system feedback loops. Experiments in which adrenocortical secretion was suppressed by administration of corticosteroids have provided further evidence that diurnal and stress-induced changes in adrenal corticosteroid hormones are neurally mediated by independent systems. The circadian rise in plasma corticosterone can be suppressed by small doses of corticosteroids without affecting the stress response (Zimmerman and Critchlow, 1967). Using a different experimental design, Hodges (1970) found that after chronic betamethasone (a synthetic corticosteroid) suppression the diurnal rhythm of plasma corticosterone recovered sooner than stress-induced increments in plasma corticosterone concentration. Since different procedures were used, these results are not necessarily contradictory, and both sets of results again indicate that the neural mechanisms governing circadian rhythm and stress-induced changes in adrenal corticosteroid hormones are dissociable. In addition, the sites of feedback action of the corticosteroids in both cases must have been in areas removed from the final common path for adrenocortical control in the median eminence of the hypothalamus.

The study of hormone feedback mechanisms has led to further elucidation and differentiation of the neural systems governing corticosteroid responses to stress. According to the classical view, high levels of corticosterone in the plasma exert an inhibitory influence on stress-induced ACTH release (Dallman and Yates, 1968; Mess and Martini, 1968), while low levels of corticosterone stimulate release of ACTH (Gemzell et al., 1951; Barrett et al., 1957). However, the pituitary–adrenocortical system is far more complex. Dallman and Yates have shown that the degree to which corticosteroids inhibit release of CRF and ACTH depends on the nature and strength of the stressing stimulus, as well as the animal's recent past history with regard to stress. These authors discovered that there is a group of stimuli which can elicit CRF and ACTH release even in the presence of the synthetic corticosteroid, dexamethasone. Thus they concluded that there must be at least two classifications of stressful stimuli. Type I stimuli, or corticosteroid-sensitive stimuli, lead to responses of the adrenocortical system which can be inhibited by high levels of plasma corticosteroids. Type II stimuli are corticosteroid resistant; the adrenocortical system responds to these stimuli regardless of level of corticosteroids in the plasma. With prior stressful stimulation, the adrenocortical response to a type I stimulus may become less corticosteroid sensitive, or even corticosteroid resistant. These results suggest that at some location in the central nervous system there are at least two different pathways, one corticosteroid sensitive (type I) and the other corticosteroid resistant (type II), which activate secretion of CRF in the median eminence in response to different stressful stimuli.

The evidence summarized above indicates that more than one functional neural system has access to the final common path of the hypothalamic–pituitary–adrenal cortex system. Any particular neural structure might participate in none, one, or several of these functional systems.

3. The Hippocampus and Regulation of Circadian Rhythm of Corticosteroid Hormones

In most mammals, plasma levels of adrenal corticosteroid hormones fluctuate in a circadian rhythm normally synchronized by light (Critchlow, 1963), although the time of peak and trough hormone levels varies with the species. Since circadian periodicity of the hypothalamic content of CRF is maintained in adrenalectomized and hypophysectomized rats (Hiroshige and Sakakura, 1971; Takebe et al., 1972), endogenous corticosteroid feedback on the central nervous system is not a critical factor in this process; a neural input to the median eminence must be of major importance. As discussed above, this path probably enters the median eminence from an anterior direction. Considerable attention has been focused on limbic system contributions to this path.

The results of several experiments on animals with hippocampal lesions or hippocampal corticosteroid implants have led to the conclusion that these manipulations result in high "resting" or "basal" levels of adrenal corticosteroid hormones. This conclusion implies an alteration in the pattern or level of diurnal fluctuation of these hormones. In early studies on animals with hippocampal lesions, only one or two hormone measurements were taken over a 24-h period, and changes in corticoid levels were found. When measured at one time during the day, plasma corticosteroid levels were found to be higher in rats with electrolytic lesions of ventral hippocampus (Knigge, 1961) and cats with large aspirative hippocampal lesions (Fendler et al., 1961) than controls. Taking two measures of 17-OH-CS levels in the urine in the monkey, during either the day or the night, Mason (1957) found that both bilateral ablation of the hippocampus and bilateral fornix section resulted in a drop in the normally high daytime values, and a slight increase in the normally low evening values. However, the results of several more recent experiments indicate that the hippocampus plays no critical role in the modulation of diurnal adrenocortical activity. Both Coover et al. (1971) and Kearley et al. (1974) found that plasma corticosteroid levels in rats with large aspirative lesions of the hippocampus were not significantly different from those of controls when measured in the morning near trough time.

In order to properly study the role of the hippocampus in the diurnal fluctuation of the adrenal corticosteroid hormones, a sufficient number of measures must be taken over a 24-h period to locate both the times and the magnitudes of the peak and trough corticosteroid levels in the brain-damaged animals, since either the time phase or the shape of the corticosteroid curve may have been altered. Wilson and Critchlow (1973/1974) measured plasma corticosterone in rats with large aspirative hippocampal lesions at three 10-h intervals and found that these animals had normal diurnal cycles. Lanier et al. (1975) measured plasma corticosterone levels in rats with dorsal, ventral, and near-total aspirative hippocampal lesions at four 6-h intervals, and found them not significantly different from controls at any time. Jackson and Regestein (1974) found no differences in plasma cortisol measured at four 6-h intervals in rhesus monkeys with anterior, posterior, or near-total hippocampal lesions. Therefore, neither the whole hippocampus nor parts of it are necessary for the production of normal diurnal cycles of plasma corticoids.

The disparity between the results of the early and more recent studies of the role of the hippocampus in diurnal adrenocortical activity may be due to several factors. First, the results of the study by Fendler *et al.* (1961) are not applicable to the depiction of "basal" diurnal levels since blood was withdrawn from the adrenal veins of anesthetized cats over a 1-h period. The effects of the anesthesia or the prolonged venisection may have contributed to the increase in plasma corticoids in the cats with hippocampal lesions. Second, the method of making the lesion may influence the results. Knigge (1961) made electrolytic lesions of ventral hippocampus and found an increase in corticosterone, while Lanier *et al.* (1975) made aspirative lesions of approximately the same region and found no change in corticosterone level. It is possible that abnormal activity in the tissue remaining after the electrolytic lesion could alter hypothalamic mechanisms controlling adrenocortical secretion. Third, time after surgery may be important, as indicated in the studies on fornix section reported below.

3.1. Fornix Section

The results of fornix section on adrenal corticosteroid hormone secretion are relevant to the present discussion since the fornix is a major efferent pathway of the hippocampus. In Mason's (1959) further studies of fornix section in the monkey, urine samples were collected more frequently and it was discovered that the corticosteroid rhythm was not flattened or obliterated in these animals, as might have been concluded from the results of the previous two-sample study (Mason, 1957), but that the peak of hormone secretion had been shifted to an earlier time. In the rat, Nakadate and De Groot (1963) found that bilateral fornix section resulted in equivalent morning and afternoon plasma corticosterone levels. More recently, Moberg *et al.* (1971) bilaterally sectioned the fornix in the rat and measured plasma corticosterone levels at six points in the 24-h cycle. The group data for these rats show that corticosterone levels remained intermediate between the normal peak and trough levels throughout the day, with a slight elevation at the time of the normal corticosteroid peak; there was no evidence for a phase shift in the cycle. However, Wilson and Critchlow (1973/1974) found that bilateral fornix section had no effect on diurnal plasma corticosterone cycles when measured at two or three points in the cycle at 8–11, 29–32, or 75 days after surgery.

The implications of all the findings above must be interpreted in light of the results of a recent study by Lengvari and Halasz (1973). These authors measured plasma corticosterone levels in fornix-sectioned rats in the morning and evening, and found significant diurnal variations at 3 wk, but not 1 wk, after surgery. Since corticosteroid hormones were measured at 1–2 wk by Mason (1957, 1959), Nakadate and De Groot (1963), and Moberg *et al.* (1971), it is possible that the changes they found after fornix section reflected only a temporary disruption of diurnal cycling. Thus the fornix would not be considered a critical neural structure for the production of a diurnal corticosteroid rhythm; and if the hippocampus has a critical role in the regulation of diurnal corticosteroid rhythms, it is not via the fornix. However, it is possible that signals carried by the fornix might have a modulating influence on this rhythm since the data of Lengvari and Halasz reveal a slight decrease in peak values and a slight increase in trough values of corticosterone in fornix-sectioned rats even at 3 wk after surgery. The temporary nature of the effect of fornix section serves as a reminder of one weakness of the lesion technique. The effects of surgical removal of part of the brain cannot directly indicate the way that structure functions in the intact organism, but rather what the rest of the brain can do without it. Reorganization or recovery of function may occur over time.

3.2. Hippocampal Steroid Implants

Another method used to study the role of the hippocampus in adrenal corticosteroid secretion is implantation of corticosteroid hormones into the hippocampus. An implication of this method is that a corticosteroid feedback mechanism has a function in maintaining the diurnal cycle of these hormones. While adrenalectomized and hypophysectomized animals can maintain a diurnal rhythm of hypothalamic CRF content, this rhythm is shifted several hours earlier (Hiroshige and Sakakura, 1971; Takebe et al., 1972). Thus the hormone feedback may maintain the temporal aspects of the rhythm rather than the magnitude of the peak and trough levels. It is particularly unfortunate that in all of the studies to date in which steroids have been implanted in the hippocampus only one or two measures over the 24-h period have been taken. Implanting corticosterone into areas CA2 and CA3 (but not into area CA1) of rabbit hippocampus increased the incorporation of ^{14}C-labeled acetate into corticosterone in adrenal homogenates when measured 3 wk later (Kawakami et al., 1968b). Implanting cortisone into rat hippocampus (Knigge, 1966) resulted in an increase in plasma corticosterone, measured at one (unspecified) time of day. Implanting cortisol in the ventral but not dorsal hippocampus of rat resulted in an increase in adrenal corticosteroid production measured at one unspecified time of day (Bohus et al., 1968); and in an increase in plasma corticosterone levels measured at the normal trough time, and a decrease in these values when measured at the normal peak time (Slusher, 1966). Whether such changes represent temporary or permanent changes cannot be determined from these studies.

The effects of implanting steroids in the hippocampus are not simply due to diffusion of the hormone to the median eminence, since implanting steroids at that location results in a decrease in plasma corticosterone levels at both peak and trough times (Slusher, 1966). The increase in plasma corticosteroids in animals with hippocampal hormone implants could indicate a positive feedback role for the hippocampus. However, the apparent similarity of the effects of electrolytic hippocampal lesions and hippocampal steroid implants on "basal" corticosteroid levels might suggest that the steroid implant was functionally ablating or inactivating one region of the hippocampus in a nonspecific, nonphysiological manner. Further research is needed to evaluate these possibilities.

3.3.   Summary

The present data indicate that the hippocampus does not have a critical role in the regulation of the diurnal cycle of adrenal corticosteroid secretion. The temporary nature of the effects of fornix section (Lengvari and Halasz, 1973) suggests that if the hippocampus has mechanisms for diurnal adrenocortical regulation they are duplicated elsewhere in the brain. The effects of electrolytic lesions and steroid implants in the hippocampus suggest that, although the structure may not be necessary for normal diurnal adrenocortical regulation, manipulations of hippocampal function can affect it. The apparent flattening of the diurnal corticosteroid curve following hippocampal steroid implantation (Slusher, 1966) suggests a change in both excitatory and inhibitory signals from the hippocampus. Furthermore, these signals may be regionally localized since steroid implants in ventral but not dorsal hippocampus are effective (Bohus *et al.*, 1968; Slusher, 1966). Since the hippocampus is not necessary for the normal diurnal corticosteroid rhythm, the effects of electrolytic lesions and steroid implants may be mediated via a mechanism in the hippocampus normally regulating corticosteroid response to stress.

4.   *The Hippocampus and Regulation of Corticosteroid Responses to Stress*

4.1.   *Electrical Stimulation of Hippocampus*

The hippocampus has been strongly implicated in the adrenocortical response to stress, mainly by evidence from experiments in which the hippocampus was stimulated electrically. With continuing experimentation, many experimental parameters and conditions have been identified which determine the outcome of hippocampal stimulation with respect to secretion of adrenal corticosteroid hormones. The effects of hippocampal stimulation on the secretion of these hormones have both a regional specificity within the hippocampus and a temporal pattern. Mason (1957) found that while stimulation of the hippocampus of rhesus monkey had no effect or even slightly elevated plasma 17-OH-CS levels during 90 min of stimulation these levels were significantly depressed 24–48 h later. In the *encephale isole* cat, stimulation of the uncus of the hippocampus resulted in a decrease in adrenal venous corticosteroids at 15–20 min after stimulation with recovery to prestimulation levels after about 1 h; on the other hand, stimulation of the dentate gyrus led to an immediate increase in adrenal effluent corticosteroids followed by a rapid return to baseline levels (Slusher and Hyde, 1961). In human patients with psychomotor epilepsy, stimulation of the hippocampus led to a decrease in plasma 17-OH-CS at 5–30 min after stimulation, with recovery to baseline levels at about 30–60 min (Mandell *et al.*, 1963; Rubin *et al.*, 1966). In the rabbit, stimulation of areas CA2 and CA3 of the dorsal hippocampus produced an increase in the incorporation of [1-^{14}C]acetate into corticosteroid hormones 10 min after stimulation, but a decrease when measured 240 min later; on the other hand, stimulation of the fascia dentata led to no significant changes in incorporation of acetate into corticosteroids when measured 10 min after stimulation, but a decrease at 240 min after stimulation (Kawakami *et al.*, 1968*a*). Thus both regional and temporal parameters must be taken into account when describing the modulatory role of the hippocampus. Of special interest is the long duration of the inhibitory effects after stimulation, up to even 48 h later, and its underlying mechanism.

Although stimulation of the hippocampus has had regionally and temporally variable effects on adrenocortical secretion under "resting" conditions in many experiments, all available data indicate that low-frequency stimulation of the hippocampus suppresses or eliminates the adrenocortical response to stress. The earliest report of this effect was made by Porter (1954), who found that stimulation of the uncus of monkey hippocampus inhibited the eosinopenic response to surgical stress. Later, Mason (1957) showed that hippocampal stimulation in the monkey increased the threshold for ACTH release elicited by electrical stimulation of the hypothalamus. Endröczi *et al.* (1959) have found that, while resulting in no change in measures of adrenocortical activity under resting conditions, hippocampal stimulation greatly attenuated this activity in response to several stressors in several species. They found that hippocampal stimulation prevented the adrenal ascorbic acid decrease occurring after painful electrical stimulation of the leg or subcutaneous injection of formalin or epinephrine in the rat; prevented the drop in lymphocyte count in response to subcutaneous epinephrine, histamine, or formalin in dog and rabbit; and prevented the rise in corticoid level in venous blood after subcutaneous injection of histamine or formalin in the cat. Further investigation (Endröczi and Lissák, 1962) revealed that only hippocampal stimulation at low frequencies (12 or 36 Hz) prevented the increase in adrenocortical activity following stress, while stimulation at progressively higher frequencies (120 Hz) had little effect or significantly increased the adrenocortical response to stress. (In all the studies described here, the inhibitory effects have been found using low frequencies of stimulation.) In more recent work, Kawakami *et al.* (1968a) found that while stimulation of the hippocampus of the rabbit resulted in an increase of adrenocortical activity under resting conditions the same stimulation prevented the rise in incorporation of $[1-^{14}C]$acetate into corticosterone and 17-OH-CS, and prevented the rise in plasma corticosterone after immobilization stress. Similarly, stimulation of the dorsal hippocampus of the rat (Dupont *et al.*, 1972) greatly attenuated the rise in plasma corticosterone in response to nicking the tail or 20-min exposure to cold.

The conclusions to be drawn from the results of hippocampal stimulation on the adrenal corticosteroid stress response depend in part on the interpretation of the way that electrical stimulation of a structure reveals its function. It is generally assumed, at least implicitly, that low-frequency, low-current stimulation enhances the normal function of a structure, while higher levels of stimulation disrupt it or activate additional structures. In the studies described above, low-frequency stimulation of the hippocampus led to a decrease in stress-induced adrenocortical activity; therefore, most authors have concluded that the hippocampus normally inhibits the secretion of these hormones. Two possible kinds of inhibitory mechanisms have been proposed (Endröczi *et al.*, 1959). First, the hippocampus might tonically inhibit the hypothalamohypophyseal system, and this inhibition would normally be released or overcome by the effects of stressors. The lack of any major inhibitory effects of hippocampal stimulation under resting conditions does not support this hypothesis. Second, the hippocampus might exert inhibitory actions only during stressful stimulation to limit magnitude or duration of adrenocortical secretion, or after prolonged stress. Results of ablation of the hippocampus provide bases for further exploration of these ideas.

4.2. Ablation of Hippocampus

If the hippocampus exerts a tonic inhibition on hypothalamic mechanisms underlying the control of adrenal corticosteroid secretion, then ablation of the hippocampus would be expected to release these mechanisms and result in elevated adrenocortical activities, in both resting and stress conditions. The effect of hippocampal ablation on "resting" adrenocortical activity was considered with regard to the diurnal corticosteroid rhythm. Although fornix section, interrupting the signals from the hippocampus, may temporarily disrupt the diurnal cycle, there is no evidence to suggest a continued high output of corticosteroid hormones. The lack of significant changes in the adrenal gland several months after hippocampal ablation also supports this conclusion (Knigge, 1961; Fregly and Van Hartesveldt, unpublished data). In addition, in every study in which waking, unanesthetized animals with hippocampal lesions have been exposed to well-defined, carefully administered stressors, they either have not differed from control animals or have secreted less adrenal corticosteroid hormones. Kim and Kim (1961) found that unilateral adrenalectomy led to approximately the same decreases in adrenal ascorbic acid in rats with dorsal hippocampal ablation and nonoperated rats several hours after operation. Coover et al. (1971) found that rats with hippocampal ablation were not different from controls with respect to plasma corticosterone levels measured after exposure to a novel environment, water deprivation, or ether stress and venisection. Similarly, Kearley et al. (1974) found that both male and female rats with hippocampal lesions had the same increase in plasma corticosterone as controls following ether stress and cardiac puncture. Further studies in my laboratory have shown that rats with hippocampal ablation have the same increases in plasma corticosterone as control animals in response to exposure to a novel environment, foot shock, and ether stress. Knigge (1961) found that hippocampal lesions in rats did not alter the peak value of the plasma corticosterone response to 4 h of physical immobilization. Fornix section in rats has no effect on plasma corticosterone in response to several minutes' exposure to electrical shock (Nakadate and De Groot, 1963); and corticosterone increase in response to ether stress and cardiac puncture is suppressed at 8–11 but not 29–32 days after fornix section (Wilson and Critchlow, 1973/1974).

Ablation of the hippocampus, then, does not result in the increase in magnitude of stress-induced adrenal corticosteroid hormones predicted by the hypothesis that this structure exerts phasic or tonic inhibition on this system. However, there is a methodological problem in all the results reported here. In every study, the measure of central nervous system influences on the median eminence and pituitary gland has been indirect—the response of the adrenal gland has been measured. The relationship between level of adrenal corticosteroid secretion and level of ACTH released is linear only up to the point at which the response of the adrenal gland approaches a maximum. Then further increases in ACTH do not elicit greater secretion of the adrenal cortical hormones. Thus, under conditions of stress for which adrenal secretion is maximal, release of central inhibitory mechanisms might increase ACTH output, but this effect could not be measured by adrenocortical secretion: the ACTH level itself must be measured. As yet, no direct measurements of ACTH have been made in animals with hippocampal lesions.

Using measurements of adrenocortical hormones, an inhibitory influence of the hippocampus on stress-induced activation of the hypothalamohypophyseal system could be shown if the response of the adrenal cortex were submaximal under the test conditions. One way to accomplish this would be to subject the subject to a minimal stress which would elicit only a small adrenocortical response. The high "resting" levels of adrenocortical hormones reported as a result of hippocampal lesion (Knigge, 1961) or steroid implant (Knigge, 1966) might represent an increased response to a low-level stress in these animals, as has been observed in animals with septal lesions (Seggie *et al.*, 1973). In addition, exposure to a novel environment has resulted in a slight increase in adrenal cortical secretion in rats with dorsal (Bohus *et al.*, 1968) or large aspirative (Coover *et al.*, 1971) hippocampal lesions.

A submaximal adrenocortical response to stress can also be produced by central nervous system lesions. Bilateral lesions either of the amygdala or of the midbrain reticular formation in the rat greatly attenuate the increase in plasma corticosterone in response to etherization and heart puncture; hippocampal lesions have no effect. However, when hippocampal lesions are made in combination with either of these lesions, the normal stress-induced response results (Knigge and Hays, 1963).

Finally, since many central nervous system depressants suppress ACTH release, increased levels of adrenal venous corticoids in cats with hippocampal lesions under phenobarbital (Fendler *et al.*, 1961) may have been observable due to a submaximal adrenocortical response in the control animals. In contrast to the evidence that the hippocampus inhibits secretion of adrenal corticosteroid hormones are some hints that the hippocampus may have a facilitatory influence on stress-induced adrenocortical activity. Knigge's results (1961) may be interpreted with reference to change from baseline levels. Knigge (1961) found that rats with hippocampal lesions reached the same levels of plasma corticosterone after immobilization, but started from a higher baseline level. Thus, in terms of change from baseline, the control animals increased corticosterone levels more than the animals with the lesions, with respect to both absolute amount of increase and percent increase over baseline. The controls also maintained this increase over their baseline levels for a longer period of time. These results could be interpreted to suggest that the hippocampus normally facilitates both magnitude and duration of the adrenocortical response to stress. Much stronger evidence for a facilitatory role was produced by Kawakami *et al.* (1968a), who found that after fornix section in the rabbit no adrenocortical responses to immobilization stress could be observed. Finally, several studies indicate that under certain conditions hippocampal lesions result in *less* of an increase in adrenal corticosteroid hormones, leading to the conclusion that the hippocampus may facilitate adrenocortical responses to some stimuli. Kim and Kim (1961) found that chronic stress, consisting of unilateral adrenalectomy followed by repeated skin incisions, led to an increase in ascorbic acid content of the adrenal gland, presumably reflecting increased stimulation by ACTH; rats with hippocampal ablation had the smallest increases in adrenal ascorbic acid. Jackson and Regestein (1974) showed that in rhesus monkeys continuously confined in a primate restraint chair those with near-total removal of the hippocampus had lower plasma cortisol levels than controls. Coover *et al.* (1971) found that while rats with hippocampal lesions made more operant responses than controls in the first session of extinction they had lower corticosterone levels when measured directly after the session. Iuvone and Van Hartesveldt (1975) found that in rats habituated to an open field those with hippocampal lesions crossed more squares but had lower corticosterone levels than controls when measured directly after the session. The results of these studies emphasize the importance of considering the particular "stressful" stimulus employed, rather than assuming that all stresses have the same effect on the nervous system.

4.3.   Hippocampus and Adrenocortical Feedback Mechanisms

To this point, the discussion of the role of the hippocampus in the regulation of stress-induced activation of adrenocortical secretion has been focused on the signals which the hippocampus may send to the hypothalamus. However, since it is well known that this hormone system has feedback action on the central nervous system, it is legitimate to consider whether the hippocampus receives and processes information concerning plasma levels of corticosteroid hormones. It is fairly certain that at least some of the cells receptive to the adrenal corticosteroid hormones lie outside the hypothalamus. The fact that steroid suppression of circadian and stress-induced adrenocortical changes can be dissociated indicates that the site of action of these hormones lies outside the median eminence. Further evidence has been provided by Hodges (1970), who found that immediately after corticosterone treatment, when plasma corticosteroid levels were high, there was no inhibition of ACTH release in response to stress. After 16 h, plasma corticosterone levels had returned to normal, but the secretion of ACTH in response to stress was suppressed. Both the time delay before pituitary inhibition and the time delay of recovery of ACTH release suggest that there is a corticosteroid receptor region less readily accessible to the systemic circulation than the hypothalamus.

The hippocampus is a good candidate for an extrahypothalamic receptor site for feedback inhibition by the adrenal corticosteroid hormones. In the rat, corticosterone is taken up in greater concentrations and retained longer in the septum and hippocampus than in other regions of the brain (McEwen *et al.*, 1968, 1969, 1970; McEwen and Weiss, 1970; Gerlach and McEwen, 1972; Knizley, 1972). Cells whose nuclei retain corticosterone, as well as cells in which corticosterone is bound to the cytosol proteins (Stevens *et al.*, 1971), are concentrated in the hippocampus. These data suggest that the hippocampus may play an important role in feedback loops of the adrenal corticosteroid hormones, or may mediate other effects of these hormones on the central nervous system.

Electrical activity in the hippocampus changes in response to adrenocortical hormones, suggesting that the uptake of corticosteroids into cells in the hippocampus may be translated into changes in cell firing rates. Hydrocortisone administered intravenously to rabbits leads to an increase in slow waves recorded from the hippocampus (Kawakami *et al.*, 1966), while administration of corticosterone intraperitoneally to the rat leads to a decrease in the firing rate of single cells in the dorsal hippocampus (Pfaff *et al.*, 1971). However, since Steiner *et al.* (1969) were unable to locate any steroid-sensitive cells in the dorsal hippocampus of the rat by direct iontophoretic application of corticosteroids, it is possible that the changes in electrical activity which have been observed in the hippocampus as a result of corticosteroid administration were secondary to changes in activity elsewhere in the brain.

In examining the possible participation of the hippocampus in feedback regulation of the adrenal corticosteroid hormones, mediation of the effects of ACTH should also be considered. Independent of the corticosteroid feedback loop, ACTH plays a part in a short feedback loop; that is, ACTH influences the central nervous system directly and not via adrenal hormones (Motta *et al.*, 1970). In a number of conditions, including long-term hypophysectomy, adrenalectomy, hypophysectomy plus adrenalectomy, and ACTH treatment of animals with both hypophysectomy and adrenalectomy, there is an inverse relationship between plasma ACTH concentration and hypothalamic CRF activity (Seiden and Brodish, 1971). These results indicate that ACTH has an inhibitory influence on the activity of the median eminence. Furthermore, the effects of stresses which induce release of ACTH even in the presence of high levels of dexamethasone (and thus act via type II or corticosteroid-resistant pathways) can be blocked by administration of a combination of dexamethasone and ACTH. Administration of ACTH alone can block stress-induced increases in endogenous ACTH secretion, although this effect depends on the stimulus, time after pretreatment, and the strain of the animal (Dallman and Yates, 1968). Since ACTH administration does not block endogenous ACTH secretion in response to all stimuli, the receptive areas of the brain must be located in elements that functionally precede the CRF-secreting neurons. Certainly there are many areas of the brain which show changes in electrical potentials after administration of ACTH; these include areas of the hypothalamus, the midbrain reticular formation, amygdala, hippocampus, septum, and areas of the cortex (Motta *et al.*, 1970). In the hippocampus, ACTH increases fast-wave electroencephalographic activity (Kawakami *et al.*, 1966) and increases the firing rates of single units (Pfaff *et al.*, 1971), effects opposite to those found following administration of the adrenal corticosteroid hormones. Thus the hippocampus may participate in central neural mediation of the effects of both hormones of the hypothalamic–pituitary–adrenal cortical complex. To date, direct effects of ACTH on the hippocampus have not been examined.

4.4. *Hippocampal Steroid Implants*

Data from studies on hormone uptake, as well as electrophysiological recording, have led to the hypothesis that the hippocampus directly receives and processes information regarding plasma levels of adrenal corticosteroid hormones. Kawakami *et al.* (1968*b*) suggested that this function of the hippocampus may account for the difference in the results of hippocampal stimulation under stressful and nonstressful conditions: level of adrenocortical hormones may "set" or "tune" the hippocampus to respond to the same stimuli in different ways. To directly test the above hypothesis, corticosteroids have been implanted in the hippocampus and adrenocortical activity has been observed under several experimental conditions, with the assumption that this procedure reveals the effects of high plasma corticosteroid levels on the particular structure implanted.

The interpretation of at least two of these studies is complicated by increased basal levels of adrenocortical activity; with respect to determining changes in response to stress, different conclusions may be drawn depending on whether the measure examined is absolute or percent change over baseline, or final absolute level of adrenocortical activity. For example, using cortisone implants in rat hippocampus, Knigge (1966) found that basal corticosterone levels were higher in these animals (19.0 mg/100 ml) than in controls 6.4 μg/ml). Exposure to combined etherization and heart puncture increased levels in implanted animals by 6.6–25.6 μg/100 ml (134% above baseline), and in control animals by 8.6–15 μg/100 ml (234% above baseline). In a study with a similar design, Bohus et al. (1968) examined the effects of cortisol implants in dorsal and ventral hippocampus of the rat 15 min after transfer to a new environment. The cholesterol implants employed as a control procedure resulted in an increase in stress-induced corticosteroid production when placed in the dorsal hippocampus; cortisol implants had no further effect. Cholesterol implants in ventral hippocampus had no significant effect on stress-induced corticosteroid production, but cortisol implants increased these levels above controls significantly. However, calculating from the mean values of the groups as shown on the figures, controls and rats with ventral hippocampal cortisol implants had a similar increment over baseline in stress-induced corticosteroid production in vitro (9.7 and approximately 11.5 μg/100 ml/h, respectively) and reached the same percent increase over baseline value (181%).

In both studies described above, rats with hippocampal steroid implants had higher basal levels and higher stress-induced levels of adrenocortical activity than controls, and a similar absolute increment of activity above baseline. These data clearly do not fit the classical characteristics of a negative feedback system, in which a high level of corticosteroid hormones would be expected to suppress further hormone secretion. In terms of change over baseline, these implants had little effect. In terms of absolute levels of corticosteroid secretion, hippocampal implants increased corticosteroid secretion. If it is assumed that the implanted corticosteroid is acting in a physiological manner, this result could be interpreted as evidence that the hippocampus participates in a positive feedback loop; that is, that the presence of high corticosteroid level results in activation of excitatory mechanisms in the hippocampus with respect to adrenocortical activity.

In the two studies described above, the authors attempted to determine whether the hippocampus mediates the effects of *high* levels of circulating corticosteroid hormones on adrenocortical responses to stress. Using a contrasting approach, Davidson and Feldman (1967) investigated whether the hippocampus mediates excitatory effects of *low* plasma corticosteroid levels. They implanted dexamethasone in the dorsal hippocampus, performed unilateral adrenalectomy, and measured the weight of the adrenal gland 9–10 days later. In the normal animal, removal of one adrenal gland lowers plasma corticosteroid levels, resulting in an increase in ACTH secretion and compensatory hypertrophy of the remaining adrenal. Animals with dexamethasone implanted in the hippocampus had less compensatory hypertrophy than normals. Thus it could be concluded that low corticosteroid concentrations activate excitatory mechanisms in the hippocampus.

4.5. Summary

The finding that electrical stimulation of the hippocampus can block or prevent the usual adrenocortical response to stress has led to the conception of the hippocampus as a structure with a regionally organized, phasic inhibitory influence on adrenocortical secretion. Unfortunately, the conditions under which the hippocampus exerts this inhibitory influence in the normal animal have not yet been identified. Investigation of the role of the hippocampus using other techniques has been complicated by methodological problems. With respect to the ablation technique, ablation of an inhibitory structure would be expected to result in a release of further excitation and greater stimulation of the adrenal cortex; however, at high levels of excitation, adrenocortical secretion reaches a ceiling level and cannot accurately reflect the amount of ACTH released. Thus the finding that hippocampal ablation leads to no changes in peak level of adrenal corticosteroid hormones secreted following stress is not meaningful information. Clearly, direct measurements of ACTH in animals with hippocampal ablation are needed. In addition to magnitude of response, latency and duration would be worthwhile parameters to measure. The fact that hippocampal lesions decrease corticosterone secretion in response to some stimuli suggests that this structure has a facilitatory influence on adrenocortical activity in some situations. With respect to steroid implantation, the results of these studies have not yet indicated the nature of the participation of the hippocampus in feedback loops, as suggested by hormone uptake measures and electrophysiological data. The methodological problem with this technique is suggested by the great similarity between many of the results of steroid implantation and small electrolytic lesions: implanted steroids may be acting in a nonphysiological manner. More investigation of the effects of these steroids on cells in the region of implantation is needed.

Perhaps the most important goal of this chapter has been to emphasize the complexity of both the hypothalamic–pituitary–adrenal cortical axis and the role of the hippocampus in its regulation. First, adrenocortical secretion is a response which can be elicited in several ways, which are probably mediated by different central nervous system mechanisms. Diurnal cycles may be regulated by different underlying mechanisms than stress-induced responses, and even the stress mechanisms may be differentiable on the basis of severity or duration of the stressor. Thus, from the point of view of regulation, the hypothalamic–pituitary–adrenal cortical axis reflects the activity not of one neural system but of many. Second, the hippocampus itself does not have a "unitary" function with respect to control of adrenocortical secretion; it is regionally organized, and probably has both excitatory and inhibitory properties. The most meaningful contributions to knowledge in this area will come from studies designed to investigate these complexities.

5. References

BARRETT, A. M., HODGES, J. R., AND SAYERS, G. The influence of sex, adrenalectomy, and stress on blood ACTH levels in the rat. *Journal of Endocrinology*, 1957, **16**, xiii.

BOHUS, B., NYAKAS, C., AND LISSÁK, K. Involvement of suprahypothalamic structures in the hormonal feedback action of corticosteroids. *Acta Physiologica Academiae Scientiarum Hungaricae*, 1968, **34**, 1–8.

COOVER, G. D., GOLDMAN, L., AND LEVINE, S. Plasma corticosterone levels during extinction of a lever-press response in hippocampectomized rats. *Physiology and Behavior*, 1971, **7**, 727–732.

CRITCHLOW, V. The role of light in the neuroendocrine system. In A. V. Nalbandov (Ed.), *Advances in neuroendocrinology*. Urbana: University of Illinois Press, 1963, pp. 377–402.

DALLMAN, M. F., AND YATES, F. E. Anatomical and functional mapping of central neural input and feedback pathways of the adrenocortical system. In V. H. T. James and J. Landon (Eds.), *The investigation of hypothalamic–pituitary–adrenal function*. Cambridge: University Press, 1968, pp. 39–72.

DAVIDSON, J. M., AND FELDMAN, S. Effects of extrahypothalamic dexamethasone implants on the pituitary–adrenal system. *Acta Endocrinologica*, 1967, **55**, 240–246.

DUPONT, A., BASTARACHE, E., ENDROCZI, E., AND FORTIER, C. Effect of hippocampal stimulation on the plasma thyrotropin (THS) and corticosterone responses to acute cold exposure in the rat. *Canadian Journal of Physiology and Pharmacology*, 1972, **50**, 364–367.

ENDRÖCZI, E., AND LISSÁK, K. Interrelations between paleocortical activity and pituitary–adrenocortical function. *Acta Physiologica Academiae Scientiarum Hungaricae*, 1962, **21**, 257–263.

ENDRÖCZI, E., LISSÁK, K., BOHUS, B., AND KOVACS, S. The inhibitory influence of archicortical structures on pituitary–adrenal function. *Acta Physiologica Academiae Scientiarum Hungaricae*, 1959, **16**, 17–22.

FELDMAN, S., CONFORTI, N., AND CHOWERS, I. Effects of partial hypothalamic deafferentations on adrenocortical responses. *Acta Endocrinologica*, 1972, **69**, 526–530.

FENDLER, K., KARMOS, G., AND TELEGDY, G. The effect of hippocampal lesion on pituitary–adrenal function. *Acta Physiologica Academiae Scientiarum Hungaricae*, 1961, **20**, 293–297.

GEMZELL, C. A., VAN DYKE, D. C., TOBIAS, C. A., AND EVANS, H. M. Increase in formation and secretion of ACTH following adrenalectomy. *Endocrinology*, 1951, **49**, 325–336.

GERLACH, J. L., AND McEWEN, B. S. Rat brain binds adrenal steroid hormone: Radioautography of hippocampus with corticosterone. *Science*, 1972, **175**, 1133–1136.

HALASZ, B., SLUSHER, M. A., AND GORSKI, R. A. Adrenocorticotrophic hormone secretion in rats after partial or total interruption of neural afferents to the medial basal hypothalamus. *Neuroendocrinology*, 1967, **2**, 43–55.

HIROSHIGE, T., AND SAKAKURA, M. Circadian rhythm of corticotropin-releasing activity in the hypothalamus of normal and adrenalectomized rats. *Neuroendocrinology*, 1971, **7**, 25–36.

HIROSHIGE, T., AND SATO, T. Changes in hypothalamic content of corticotropin-releasing activity following stress during neonatal maturation in the rat. *Neuroendocrinology*, 1971, **7**, 257–270.

HODGES, J. R. The hypothalamus and pituitary ACTH release. *Progress in Brain Research*, 1970, **32**, 12–18.

IUVONE, P. M., AND VAN HARTESVELDT, C. Locomotor activity and plasma corticosterone in rats with hippocampal lesions. In preparation.

JACKSON, W. J., AND REGESTEIN, Q. R. Hippocampectomy in rhesus monkeys: Effects on plasma cortisol during two stressful conditions. Paper presented at the Society for Neuroscience Fourth Annual Meeting, St.Louis, October 1974.

KAWAKAMI, M., KOSHINO, T., AND HATTORI, Y. Changes in the EEG of the hypothalamus and limbic system after administration of ACTH, SU-4885 and ACH in rabbits with special reference to neurohumoral feedback regulation of pituitary–adrenal system. *Japanese Journal of Physiology*, 1966, **16**, 551–569.

KAWAKAMI, M., SETO, K., TERASAWA, E., YOSHIDA, K., MIYAMOTO, T., SEKIGUCHI, M., AND HATTORI, Y. Influence of electrical stimulation and lesion in limbic structure upon biosynthesis of adrenocorticoid in the rabbit. *Neuroendocrinology*, 1968a, **3**, 337–348.

KAWAKAMI, M., SETO, K., AND YOSHIDA, K. Influence of corticosterone implantation in limbic structure upon biosynthesis of adrenocortical steroid. *Neuroendocrinology*, 1968b, **3**, 349–354.

KEARLEY, R. C., VAN HARTESVELDT, C., AND WOODRUFF, M. L. Behavioral and hormonal effects of hippocampal lesions on male and female rats. *Physiological Psychology,* 1974, **2,** 187–196.

KIM, C., AND KIM, C. U. Effect of partial hippocampal resection on stress mechanism in rats. *American Journal of Physiology,* 1961, **201,** 337–340.

KNIGGE, K. M. Adrenocortical response to stress in rats with lesions in hippocampus and amygdala. *Proceedings of the Society for Experimental Biology and Medicine,* 1961, **108,** 18–21.

KNIGGE, K. M. Feedback mechanisms in neural control of adenohypophyseal function: Effect of steroids implanted in amygdala and hippocampus. *Abstracts of the Second International Congress on Hormonal Steroids,* Excerpta Medica International Congress Series No. 111, 1966, p. 208.

KNIGGE, K. M., AND HAYS, M. Evidence of inhibitive role of hippocampus in neural regulation of ACTH release. *Proceedings of the Society for Experimental Biology and Medicine,* 1963, **114,** 67–69.

KNIZLEY, H., JR. The hippocampus and septal area as primary target sites for corticosterone. *Journal of Neurochemistry,* 1972, **19,** 2737–2745.

LANIER, L. P., VAN HARTESVELDT, C., WEIS, B., AND ISAACSON, R. L. Effects of differential hippocampal damage upon rhythmic and stress-induced corticosterone secretion in the rat. *Neuroendocrinology,* 1975, in press.

LENGVARI, I., AND HALASZ, B. Evidence for a diurnal fluctuation in plasma corticosterone levels after fornix transection in the rat. *Neuroendocrinology,* 1973, **11,** 191–196.

MAKARA, G. B., STARK, E., PALKOVITS, M., REVESZ, T., AND MIHALY, K. Afferent pathways of stressful stimuli: Corticotrophin release after partial deafferentation of the medial basal hypothalamus. *Journal of Endocrinology,* 1969, **44,** 187–193.

MAKARA, G. B., STARK, E., AND PALKOVITS, M. Afferent pathways of stressful stimuli: Corticotropin release after hypothalamic deafferentation. *Journal of Endocrinology,* 1970, **47,** 411–416.

MANDELL, A., CHAPMAN, L. F., RAND, R. W., AND WALTER, R. D. Plasma corticosteroids: Changes in concentration after stimulation of hippocampus and amygdala. *Science,* 1963, **139,** 1212.

MASON, J. W. The central nervous system regulation of ACTH secretion. In H. H. Jasper, L. D. Proctor, R. S. Knighton, W. C. Noshay, and R. T. Costello (Eds.), *Reticular formation of the brain.* Boston: Little, Brown, 1957, pp. 645–670.

MASON, J. W. Psychological influences on the pituitary–adrenal cortical system. *Recent Progress in Hormone Research,* 1959, **15,** 345–389.

MASON, J. W. A review of psychoendocrine research on the pituitary–adrenal cortical system. *Psychosomatic Medicine,* 1968, **30,** 576–607.

MCEWEN, B. S., AND WEISS, J. M. The uptake and action of corticosterone: Regional and subcellular studies on rat brain. *Progress in Brain Research,* 1970, **32,** 200–210.

MCEWEN, B. S., WEISS, J. M., AND SCHWARTZ, L. S. Selective retention of corticosterone by limbic structure in rat brain. *Nature (London),* 1968, **220,** 911–912.

MCEWEN, B. S., WEISS, J. M., AND SCHWARTZ, L. S. Uptake of corticosterone by rat brain and its concentration by certain limbic structures. *Brain Research,* 1969, **16,** 227–241.

MCEWEN, B. S., WEISS, J. M., AND SCHWARTZ, L. S. Retention of corticosterone by cell nuclei from brain regions of adrenalectomized rats. *Brain Research,* 1970, **17,** 471–482.

MESS, B., AND MARTINI, L. The central nervous system and the secretion of anterior pituitary trophic hormones. In V. H. T. James (Ed.), *Recent advances in endocrinology.* Boston: Little, Brown, 1968, pp. 1–49.

MOBERG, G. P., SCAPAGNINI, U., DE GROOT, J., AND GANONG, W. F. Effect of sectioning the fornix on diurnal fluctuation in plasma corticosterone levels in the rat. *Neuroendocrinology,* 1971, **7,** 11–15.

MOTTA, M., PIVA, F., AND MARTINI, L. The role of "short" feedback mechanisms in the regulation of adrenocorticotropin secretion. *Progress in Brain Research,* 1970, **32,** 25–32.

NAKADATE, G. M., AND DE GROOT, J. Fornix transection and adrenocortical function in rats. *Anatomical Record,* 1963, **145,** 338.

ONDO, J. G., AND KITAY, J. I. Pituitary-adrenal function in rats with diencephalic islands. *Neuroendocrinology,* 1972, **9,** 72–82.

PALKA, Y., COYER, D., AND CRITCHLOW, V. Effects of isolation of medial basal hypothalamus on pituitary–adrenal and pituitary–ovarian functions. *Neuroendocrinology,* 1969, **5,** 333–349.

PFAFF, D. W., SILVA, M. T. A., AND WEISS, J. M. Telemetered recording of hormone effects on hippocampal neurons. *Science,* 1971, **172,** 394–395.

PORTER, R. W. The central nervous system and stress-induced eosinopenia. *Recent Progress in Hormone Research,* 1954, **10,** 1–27.

RUBIN, R. T., MANDELL, A. J., AND CRANDALL, P. H. Corticosteroid responses to limbic stimulation in man: Localization of stimulus sites. *Science,* 1966, **153,** 767–768.

SEGGIE, J. A., BROWN, G. M., UHLIN, I. V., SCHALLY, A., AND KASTIN, A. J. Effect of septal lesions on plasma levels of corticosterone, growth hormone, prolactin and melanocyte stimulating hormone before and after stimulation. Paper presented at the Society for Neuroscience Third Annual Meeting, San Diego, November 1973.

SEIDEN, G. AND BRODISH, A. Physiological evidence for "short-loop" feedback effects of ACTH on hypothalamic CRF. *Neuroendocrinology,* 1971, **8,** 154–164.

SLUSHER, M. A. Effects of chronic hypothalamic lesions on diurnal and stress corticosteroid levels. *American Journal of Physiology,* 1964, **206,** 1161–1164.

SLUSHER, M. A. Effects of cortisol implants in the brain stem and ventral hippocampus on diurnal corticosteroid levels. *Experimental Brain Research,* 1966, **1,** 184–194.

SLUSHER, M. A., AND HYDE, J. F. Effects of limbic stimulation on release of corticosteroids into the adrenal venous effluent of the cat. *Endocrinology,* 1961, **69,** 1080–1084.

STEINER, F. A., RUF, K., AND AKERT, K. Steroid-sensitive neurons in rat brain: Anatomical localization and response to neurohumours and ACTH. *Brain Research,* 1969, **12,** 74–85.

STEVENS, W., GROSSER, B. I., AND REED, D. J. Corticosterone-binding molecules in rat brain cytosols: Regional distribution. *Brain Research,* 1971, **35,** 602–607.

TAKEBE, K., SAKAKURA, M., AND MASHIMO, K. Continuance of diurnal rhythmicity of CRF activity in hypophysectomized rats. *Endocrinology,* 1972, **90,** 1515–1520.

WILSON, M., AND CRITCHLOW, V. Effect of fornix transection or hippocampectomy on rhythmic pituitary–adrenal function in the rat. *Neuroendocrinology,* 1973/1974, **13,** 29–40.

ZIMMERMAN, E., AND CRITCHLOW, V. Effects of diurnal variation in plasma corticosterone levels on adrenocortical response to stress. *Proceedings of the Society for Experimental Biology and Medicine,* 1967, **125,** 658–663.

15

Hippocampal Protein Synthesis and Spike Discharges in Relation to Memory

SHINSHU NAKAJIMA

1. Introduction

In the past 10 years or so, a great number of experiments have been conducted to examine the role played by the hippocampus in the formation and retention of memory. Earlier studies in this field were no doubt inspired by a clinical observation of deficits in recent memory in man after resection of the temporal lobe (Scoville and Milner, 1957). However, the accumulated results of animal experiments, particularly those on rats and mice, now form a body of knowledge which requires separate consideration. The scope of this chapter is therefore limited to the studies on nonprimate species.

Many experiments in this field take the form of training–experimental treatment–testing. The training (original learning) is usually the acquisition of a relatively simple task, such as position or brightness discrimination, active or passive avoidance of foot shock, or conditioned suppression of feeding or drinking. A variety of experimental treatments have been given to the hippocampus, ranging from electrical stimulation to injection of chemical substances. If the performance of a task at the time of testing (retention test) is impaired as a result of an experimental treatment, the treatment is considered to have interfered with the memory of the task. A basic requirement for this type of experimental design is that the treatment not directly interfere with the ability to perform the task.

SHINSHU NAKAJIMA • Dalhousie University, Halifax, Nova Scotia, Canada. This research is supported by National Research Council of Canada Grant A0233.

This design is most adequate in demonstrating retrograde amnesia, and widely used to study the amnesic effect of electroconvulsive shock (ECS). Memory is disrupted if ECS is given within a short period of time after original learning, but not if it is given past this period. The demonstration that the treatment has no effect if it is given at a certain point in time is important: it serves to ensure that the ability to perform the task is not impaired by the treatment. Therefore, a conclusion that memory has at least two phases (one vulnerable and the other invulnerable to the treatment) is inherent in the experimental design.

It is more difficult to demonstrate a complete disruption of memory. If performance in the retention test is impaired regardless of the time interval between training and treatment, then it becomes necessary to demonstrate that the impairment is not due to a defect in the sensory or motor system, or to the absence of motivation to perform the task. Until these possible alternative interpretations are ruled out, one cannot conclude that the impairment of performance results from a loss of memory.

In some experiments, a treatment may be initiated earlier, that is, before the original learning of a task. The design of treatment–training–testing is frequently used when the experimenter is interested in the phase of memory immediately after learning, or when it takes some time for the treatment to become effective. A good example can be found in the studies using drugs that are absorbed slowly. In this experimental design, it is important to ensure that the treatment does not influence the performance at the time of training. If it does, then it should be demonstrated that the influence is present equally at the time of training and at the time of testing.

In the following sections, experiments are reviewed in the light of these methodological criteria.

2. Macromolecular Synthesis in the Hippocampus

2.1. Macromolecular Theory of Flexner

The idea that protein synthesis in the temporal area of the brain may be related to memory arose from an experiment conducted by Flexner et al. (1963) using an antibiotic substance, puromycin. They trained mice to move into one of the arms of a Y-maze within 5 s to avoid foot shock. They then injected puromycin directly into the brain, and tested the animals for retention of the maze task 3 days after injection. Performance in the retention test was impaired if the drug had been injected into the temporal area within 3 days of original training, but not if it had been injected more than 4 days after the training. Combined injections into the temporal, frontal, and ventricular sites impaired later performance even if the injections were made more than 4 days after original training.

Since puromycin suppresses the biosynthesis of proteins, the experimenters suggested that the impairment of performance may be a loss of memory due to the suppression of protein synthesis in the brain (Flexner et al., 1963). In later experiments, they confirmed that the bilateral temporal injection of puromycin indeed suppressed protein synthesis in the hippocampus and entorhinal cortex to a level below 10% of the normal level and that the combined injections (temporal, frontal, and ventricular injections) resulted in a similar suppression in the entire brain (Flexner *et al.*, 1964, 1965*a,b*).

Another aspect of the effect of puromycin on the performance of a maze task was discovered by Barondes and Cohen (1966). They injected puromycin into the temporal area of the mouse brain and 5 h later trained the animals for position discrimination in a Y-maze. It should be noted that the task used in their experiment was different from the one used by Flexner, although the apparatus was similar. Flexner and associates (Flexner *et al.*, 1962; Flexner and Flexner, 1969*a*) trained mice to make correct choice of the arm and to avoid foot shock within 5 s, whereas Barondes and Cohen (1966) trained the animals to make correct choice of the arm regardless of avoidance or escape—a more conventional procedure.

Barondes and Cohen (1966) found that the drug-injected mice learned the task but retained it only for 15 min or so. The animals tested 45 min after the training made a large number of errors, and those tested 3 h later showed no sign of retention. On the basis of their findings and those of Flexner *et al.* (1963), they concluded that there are three successive phases of memory. The first phase is unaffected by puromycin injection, the second phase can be destroyed by puromycin injection into the temporal area, and the third phase is destroyed only by more extensive injection of puromycin into the entire brain.

In 1967, Flexner and associates reviewed the related studies and proposed a macromolecular theory of memory, which assumed that memory is retained in the brain as a self-sustained system of macromolecular biosynthesis. A learning experience triggers the synthesis of one or more species of messenger ribonucleic acid (mRNA), which alters the rate of synthesis of "essential proteins." The essential proteins have two functions. One is to change synaptic characteristics so that the transmission of impulses is facilitated in certain neurons, resulting in a change of behavior. Such a change of behavior is considered as an expression of memory. The other function is to counteract (or to produce a substance that counteracts) the genetic repressors which formed a template for the mRNA that initially influenced the synthesis of essential proteins.

In other words, the essential proteins facilitate the synthesis of mRNA, which in turn facilitates the synthesis of the essential proteins, and the mRNA-protein cycle continues in a self-sustained fashion. If protein synthesis is suppressed by a drug, such as puromycin, then the essential proteins would be depleted and the memory would not be expressed in behavior. Also, if the drug effect lasts long enough, the essential species of mRNA would be depleted, resulting in a permanent loss of memory (Flexner *et al.*, 1967).

There are other interesting features of the effects of puromycin. Its behavioral effect is antagonized by an additional injection of physiological saline into the sites where puromycin was injected (Flexner and Flexner, 1967; Rosenbaum *et al.*, 1968), but the saline antagonism does not occur if puromycin is injected shortly before or immediately after learning of the maze task (Flexner and Flexner, 1968*a*). Saline injection into the frontal neocortical area can antagonize the effect of puromycin injection into the temporal area, provided that the saline follows puromycin after more than 5 days, but not if it follows within 3 days (Flexner and Flexner, 1969b). The puromycin effect is also antagonized by the presence of monovalent or divalent cations except Na^+ (Flexner and Flexner, 1969a). Intraperitoneal or subcutaneous injection of adrenergic drugs also antagonizes the puromycin effect (Roberts et al., 1970).

Flexner and associates tried to explain these antagonistic effects by assuming two separate mechanisms. On the one hand, puromycin suppresses protein synthesis as proposed earlier, and on the other hand it forms an abnormal peptide. Flexner and Flexner (1968b) injected radioactive puromycin into the temporal area of mice and found that peptidylpuromycin remained in a wide area of the brain for more than 58 days. They proposed that peptidylpuromycin temporarily suppressed the expression of memory.

The explanation may be summarized as follows. Memory is first stored as the self-sustained system of macromolecular synthesis in the hippocampus–entorhinal cortex area. Puromycin can completely abolish memory at this stage by disrupting the self-sustained system, and subsequent injection of saline cannot restore the lost memory (Flexner and Flexner, 1968a). Within a few minutes after learning, the substrate of memory begins to spread to the entire neocortex, and completes the migration within a few days. Puromycin injection into the temporal area alone can no longer abolish memory. However, it can suppress the expression of memory because peptidylpuromycin now binds to the adrenergic synaptic sites and blocks neuronal function. The cations (other than Na^+) prevent the binding of peptidylpuromycin by occupying the anionic sites of neuronal membrane; physiological saline restores the expression of memory by removing peptidylpuromycin bound to the synaptic sites.

A statement that "the expression of memory is suppressed" is misleading. Apparently, what is "suppressed" is not the "expression of memory" but the ability to perform the task. This point is clear from the results of an experiment by Flexner and Flexner (1968a). They injected puromycin into the temporal area of mice and 5 h later trained the animals for the avoidance–discrimination task. The animals were unable to perform avoidance response. The results indicate that, as far as avoidance response is concerned, puromycin interferes with the ability to perform it. It is important to note that Flexner and associates trained their mice on two tasks simultaneously: on shock avoidance and on position discrimination. If the effect of puromycin on position discrimination had been examined separately from its effect on avoidance response, the results may have been less confusing.

2.2. Enhancement of Macromolecular Synthesis

Congruent with the macromolecular theory of Flexner and associates, Hydén and Lange (1968) found an increase in the rate of protein synthesis during learning. They injected radioactive leucine into the lateral ventricle of rats, trained them on a "transfer of handedness task," and examined the incorporation of the amino acid into protein in the pyramidal neurons of the CA3 area in the hippocampus. The task was to retrieve food pellets with the unpreferred forepaw. The rats trained on this task showed a large increase in the rate of protein synthesis as compared with control animals that continued to use the preferred paw. One interesting finding was that the rats retrained on the same task 14 days later showed a similar increase of protein synthesis, but those retrained 30 days later did not (Hydén and Lange, 1970). The experimenters suggested that the information acquired 30 days previously may have been stored outside the CA3 area. This interpretation is somewhat similar to the theory of Flexner and associates.

A similar increase of protein synthesis was found by Beach *et al.* (1969) in the CA1 area during one-way active avoidance training in the rat. Bowman and Strobel (1969) found an increase of RNA synthesis in the hippocampus of rats during a successive series of position reversals in a Y-maze.

The increase in the rate of macromolecular biosynthesis appears to indicate an increase in neuronal activity. Berry (1969) found in a monosynaptic preparation of *Aplysia* that the rate of RNA synthesis is proportional to the number of action potentials. A similar relation seems to hold in the rat hippocampus. Many of the hippocampal neurons, particularly those in the CA3 area, show a systematic change, either increase or decrease in the firing rate, in the course of acquisition of bar-pressing response for food or water (Olds *et al.*, 1969) and also during a classical conditioning of tone–food pairings (Olds and Hirano, 1969). Therefore, the overall increase of macromolecular synthesis appears to suggest that the hippocampal neurons, on the average, increase their firing rate during learning (an inhibition requires the firing of inhibitory neurons).

However, the fact that learning is accompanied by neuronal excitation and biochemical synthesis in the hippocampus does not necessarily indicate that learning is dependent on them. Learning may still be possible without excitation of hippocampal neurons. Indeed, rats can learn and retain a wide variety of tasks after complete ablation of the hippocampal formation (e.g., Douglas, 1967).

3.   Effects of Epileptogenic Substances

The macromolecular theory of memory, as proposed by Flexner and associates, is based on two assumptions. One is that the behavioral impairment signifies a loss of memory, and the other is that the impairment is caused by the suppression of macromolecular biosynthesis.

The validity of the first assumption was questioned by Deutsch (1969). His criticism was directed to the way in which memory was assessed by Flexner and associates. In their experiments (Flexner *et al.*, 1963, 1967), an animal's performance in an original learning session (before the drug injection) was compared with that in the relearning session (after injection). "Such a score could produce a serious overestimate of the amount of amnesia if the rate of learning is slowed down by the drug" (Deutsch, 1969, p. 85).

The second assumption, that the behavioral effect is caused by the biochemical effect, was challenged by the findings on the electrographic effect of puromycin injection. Cohen *et al.* (1966) found a depression of hippocampal electrical activity in an acute preparation 5 h after puromycin injection into the hippocampus. The depression they recorded may have been a postictal depression, because the mice injected with puromycin into the temporal area had a lower threshold for pentylenetetrazol convulsion (Cohen and Barondes, 1967). They suspected that the behavioral impairment observed after puromycin injection may be due to the abnormal electrical activity in the hippocampus, and concluded that puromycin is not an adequate drug to use in the study of macromolecular basis of memory.

3.1. Actinomycin D

If memory is critically dependent on macromolecular synthesis in the hippocampal area, the suppression of RNA synthesis may also be effective in disrupting memory. Thus Agranoff et al. (1967) injected actinomycin D into the brain of goldfish in order to suppress the synthesis of RNA. They trained the fish to avoid electrical shock and then injected actinomycin into the brain either immediately, 1 h, or 3 h later. The fish injected within 1 h after the original training demonstrated more failures of avoidance in a later test than those injected after 3 h.

Nakajima (1969) injected actinomycin D into the hippocampal region of the rat. Three days after injection, the Nissl substance completely disappeared from the hippocampal neurons, indicating that the synthesis of new RNA had been suppressed and RNA-dependent protein synthesis was absent. About the same time, the EEG record of the drug-injected animals began to show epileptiform spike discharges in the hippocampus.

Under these conditions, the animals were trained to make position discrimination in a T-maze to escape or avoid foot shock (the failure in avoidance was not counted as an error). The drug-injected animals made more errors than normal animals, but they eventually reached a criterion of learning. However, 1 day after the original learning, the actinomycin-injected animals made a large number of errors again in relearning the same task. The rats with electrolytic lesions in the hippocampal area showed a similar deficit in learning the task, but once they had learned it they retained it well. Nakajima (1969) suggested that the presence of the spike discharges, rather than the absence of macromolecular synthesis, is a more likely cause of the relearning deficit.

A subsequent study (Nakajima, 1972) revealed that the impairment of performance was not ascribable to a loss of memory, for the following reasons. When the animals were tested within a few days after actinomycin injection, their performance was only slightly impaired, regardless of when the task had been learned originally. When they were tested more than 4 days after injection, the performance was severely impaired, regardless of whether the task had been learned previously or not. The onset of the behavioral impairment coincided with the onset of the epileptiform discharges in the hippocampus.

It is clear that the behavioral effect of actinomycin is the impairment of ability to perform the task, and that the impairment is caused by the malfunction of the hippocampus.

3.2. Puromycin

The notion that the behavioral effect of puromycin may be mediated by the abnormal hippocampal activity, rather than the suppression of protein synthesis (Cohen
et al., 1966; Cohen and Barondes, 1967), was tested by examining the biochemical,
electrographic, and behavioral effects produced by the same injection.

Nakajima (unpublished) injected puromycin into the hippocampus of the rat
(100 μg to both the dorsal and ventral hippocampus) and examined the suppression
of protein synthesis in the brain by radioautographic method. Protein synthesis was
almost completely suppressed in the hippocampus–entorhinal cortex area for a 3-h
period starting at 5 h after the drug injection (Fig. 1A). The entorhinal cortex began
to resume protein synthesis in 24 h (B), and the hippocampus also recovered from the
suppression in 48 h except in the vicinity of the injection sites, where signs of necrosis
appeared (C).

Russell (1969) examined the electrographic aspect of puromycin injection. Spike
discharges first appeared in the hippocampus during injection of the drug solution
(Fig. 2). The effect was probably due to the mechanical distortion of the tissue, because similar discharges were also observed during the injection of physiological
saline. The hippocampal activity of the saline-injected animals returned to normal in
a few hours, whereas the puromycin animals showed a depression of EEG amplitude
as reported by Cohen *et al.* (1966). Epileptiform discharges began to appear about 20
h after puromycin injection, and continued for several days thereafter.

To examine the behavioral effect of puromycin, Russell (1969) trained the rats
on a shock-motivated position discrimination task in a T-maze. When the original
training was conducted 5 h after puromycin injection into the hippocampus and the
retention test 8 h after injection—i.e., when protein synthesis was absent in the hippocampus–entorhinal cortex area—the maze performance of the puromycin animals
was not significantly different from that of the saline control animals. When the
original training was given 24 h and the retention test 48 h after injection, the
puromycin animals showed impaired performance in the retention test. That is, the
performance of the task was normal when protein synthesis was suppressed, but impaired when spike discharges appeared in the hippocampus.

The impairment of discrimination appears to be related to the fact that the
puromycin-injected rats were relatively slow in running and received more foot
shocks. To test this possibility, Russell (1969) gave an inescapable foot shock just
prior to the retention test. The puromycin-injected animals became totally incapable
of making any discrimination, but the behavior of the saline control animals was not
affected.

Nakajima (1973) reexamined the behavioral, biochemical, and electrographic effects of puromycin injection in mice. The animal species (Swiss albino mice), the
drug dosage (90 μg to each injection site), and the injection site (bitemporal) were
identical to those in the previous studies (Flexner *et al.*, 1963; Barondes and Cohen,
1966). However, there were differences in the procedure. The failure of avoidance was
not counted as an error. The drug was injected very slowly in order to avoid the
flowback of the solution along the outside of the injection needle. In some animals,

the drug was injected through a previously implanted cannula to avoid anesthesia and skull puncture. Animals showing histological signs of bilateral asymmetry of injection were eliminated from the group.

An autoradiographic experiment showed complete suppression of protein synthesis in the hippocampus, entorhinal cortex, and overlying neocortex during a period between 5 and 8 h after the drug injection. Recovery was virtually complete within 24 h (somewhat faster than in the rat). The electrographic results are shown

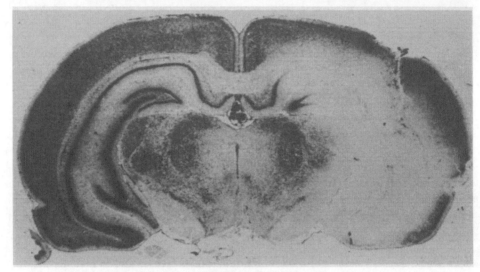

FIG. 1(A)

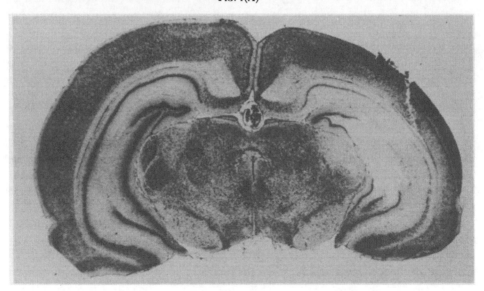

FIG. 1(B)

FIG. 1. Autoradiograms of the rat brain injected with puromycin into the right hippocampus 5 h (A), 24 h (B), and 48 h (C) before intraperitoneal injection of radioactive leucine.

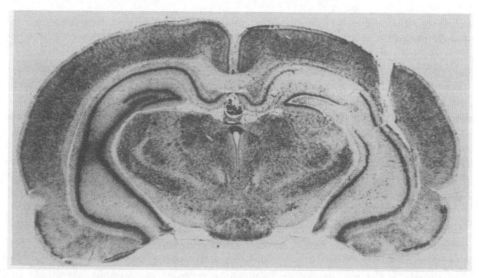

Fig. 1(C)

in Fig. 3. The initial spike discharges disappeared in a few hours, and depression prevailed in the hippocampus thereafter. Conspicuous spike discharges appeared on a generally depressed background at irregular intervals 3 days after injection (somewhat slower than in the rat).

The performance of a T-maze position discrimination task, which had been learned originally 5 h after the injection, was not impaired if the animals were tested 8 h after injection. The animals tested 24 h after injection made some errors, and those tested 3 days after injection made a large number of errors. For that matter, the performance on the third day after injection was very poor regardless of whether the task was originally learned 7 days before, 1 day before, or 5 h after injection. There was no gradient of the behavioral effect dependent on the interval between training and injection (Nakajima, 1973).

The results did not agree with the findings of Flexner et al. (1963), in which the animals trained more than 4 days before puromycin injection did not show the behavioral deficit. One of the possible causes may be the asymmetry of injection in their experiment. Too rapid injections may have resulted in a leak of the drug solution from either one of the bilateral injection sites, and may have biased the animals to turn consistently to one direction in the maze.

The experiments of Russell (1969) and Nakajima (1973) indicate that the behavioral effect of puromycin occurs when spike discharges are observed in the hippocampus, but not when protein synthesis is suppressed in the hippocampus–entorhinal cortex area. Furthermore, the behavioral effect is better explained as an interference with the performance at the time of retention test than as a loss of memory.

3.3. Penicillin

A similar relation between the hippocampal spike discharges and the performance of shock-motivated response has been reported by Olton (1970). He injected penicillin unilaterally into the hippocampus, septal area, or entorhinal cortex. Although penicillin does not suppress the biosynthesis of macromolecules as much as actinomycin or puromycin, the unilateral injection assured the normal macromolecular synthesis in the uninjected hemisphere. The hippocampal epileptiform discharge was most pronounced after penicillin injection into the hippocampus, and less after injection into the other structures. Learning of a two-way active avoidance task was impaired in the animals with hippocampal injection, but not in the others.

Schmaltz (1971) injected pencillin into the hippocampus of rats either 8 days before or 8 days after the acquisition of two-way avoidance response. The group trained before the injection did not show any deficit in the performance of the task 8 days after injection, whereas the group trained after injection demonstrated the deficit as reported by Olton (1970). It is dangerous to conclude from these results that penicillin disrupted the memory of a newly acquired response while leaving the well-consolidated memory intact. At the initial stage of two-way avoidance learning, even a normal animal receives foot shocks much more frequently than at a later stage. Had the animals of the preinjection training group received inescapable foot shocks just prior to the retention test, they would have performed as poorly as the postinjection group. It may be recalled that, in an aforementioned study (Russell, 1969), a combination of puromycin injection into the hippocampus and inescapable foot shock made the performance of a position discrimination task virtually impossible.

The disruptive effect of penicillin found in the shock avoidance situation does not seem to be generalizable to other situations. Penicillin did not produce any effect on the acquisition of position discrimination (Means *et al.*, 1972) or brightness dis-

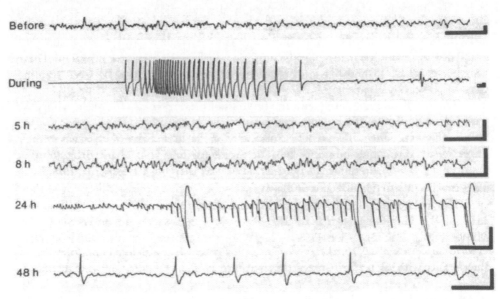

Fig. 2. Hippocampal EEG before, during, and after puromycin injection into the rat hippocampus. Calibrations: 1 s and 500 μV.

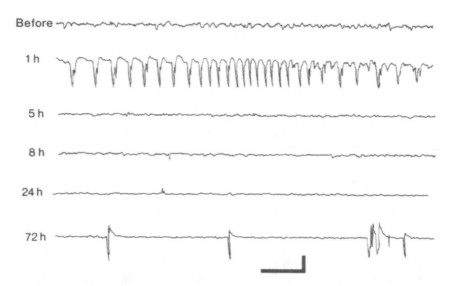

FIG. 3. Hippocampal EEG before and after puromycin injection into the mouse hippocampus. Calibrations: 1 s and 500 μV.

crimination (Woodruff and Isaacson, 1972) reinforced with food. Woodruff *et al.* (1974) found that daily injection of penicillin into the hippocampus retarded the acquisition of a brightness discrimination in a two-bar Skinner box. The results likely indicate an interference with the visual system, because the drug injected into the overlying neocortex produced more retardation.

The behavioral deficits produced by penicillin injection into the hippocampus are, like those produced by actinomycin and puromycin, better explained as the interference with either sensory, motor, or motivational mechanism. At present, there is no evidence that epileptiform discharges in the hippocampus affect memory.

4. Disruptive Stimulation of the Hippocampus

4.1. Potassium Chloride

A burst of spike discharges can be produced in the hippocampus by injecting potassium chloride (KCl). The discharges are followed by a depression of electrical activity, which slowly spreads over the entire hippocampus in one hemisphere, in a manner similar to the cortical spreading depression of Leão. If the concentration of KCl is high, a second and a third burst of discharges may occur.

Grossman and Mountford (1964) produced bilateral spike discharges in the hippocampus of the rat with crystalline KCl and examined the effect on brightness discrimination. The application of KCl increased the latency of responding, but it did not interfere with the choice behavior itself. Hirano (1965) injected a KCl solution immediately after training rats on passive avoidance and found an impairment in later retention test. Avis and Carlton (1968) trained rats to drink water and then presented a tone followed by a foot shock for four trials in the absence of water. On the next day, they anesthetized the animals, injected a KCl solution into the hippocampus, and recorded hippocampal EEG. Four days after injection, they tested the animals by presenting the tone while the animals were drinking water. Those animals which had the hippocampal spreading depression for more than 5 min failed to show a conditioned suppression of drinking.

Whether or not these behavioral deficits indicate retrograde amnesia is debatable. Huges (1969) injected a KCl solution into the rat hippocampus either 1, 3, 7, or 21 days after giving them a session of tone–shock conditioning similar to the one used by Avis and Carlton (1968). The animals failed to show the conditioned suppression regardless of the time interval between conditioning and injection. Kapp and Schneider (1971) used a similar behavioral test. They applied KCl or NaCl crystals to the rat hippocampus 10 s after a session of aversive conditioning training in one group and 24 hr after in another group. The conditioned suppression did not occur in either group when they were tested 4 days later.

The fact that the impairment was unrelated to the time interval between the conditioning and injection suggests that the effect of KCl was proactive. Probably, the neurons in a wide area of the hippocampus suffered from high concentrations of KCl and became functionally defective for a long period of time. A high concentration of KCl (e.g., 25%) is useful in producing spreading depression in the neocortex, which is protected by the dura mater. For the hippocampus, less than 1% is sufficient (Bureš et al., 1960). Hirano (1965), who injected 2 µl of 1.11% KCl (isotonic) solution, may have produced hippocampal depression without creating a long-lasting dysfunction. Unfortunately, he did not test at any time interval other than "immediately after." In the other studies, where a 25% solution or crystals were directly applied to the hippocampus, the presence of long-lasting dysfunction is certainly to be expected.

If the behavioral effect of KCl injection is caused by the hippocampal dysfunction, it may disappear when the neurons recover their normal metabolism. In support of this inference, there are a number of experimenters who observed the recovery of normal behavior long after KCl application. In a previously mentioned study, Kapp and Schneider (1971) reported the recovery of conditioned suppression 21 days after the application of crystalline KCl. Carlton and Markiewicz (1973) found a similar recovery 10 days after 25% KCl application. The period in which conditioned suppression was absent is probably the period when the hippocampus was in the state of temporary lesion. The failure to suppress drinking in the presence of a conditioned stimulus falls into the same category of deficits as those demonstrated by the animals with hippocampal lesions, namely a failure to withhold responses (Kimble, 1968).

4.2. Acute Lesions

A time-dependent gradient of impairment has been reported with a mechanical injury inflicted on the hippocampus. Bohdanecka et al. (1967) inserted a 26-gauge hypodermic needle through the skull into the hippocampus of unanesthetized mice. Bilateral puncture of this kind either 1 h before, immediately after, or 1 h after a foot shock experience resulted in a loss of passive avoidance 24 h later. The puncture 6 h before or 6 h after the shock did not have any effect. Similar results were reported by Dorfman *et al.* (1969). It is likely that the puncture produced bursts of injury discharges and spreading depression in the hippocampus, although EEG was not recorded in either study.

Uretsky and McCleary (1969) performed a more drastic surgical intervention. They first trained cats on a one-way active avoidance task and then isolated the hippocampus by transecting the fornix and the entorhinal cortex. The animals failed to perform the avoidance task a few weeks later if the surgery was performed 3 h after original training, but not if it was done 8 days after the training. The surgical operation must have produced massive injury discharges in the efferent fibers from the hippocampus and severe spreading depression in the neocortex, but EEG was not recorded.

The temporal gradient of the behavioral deficits in these experiments suggests that the effect is retrograde amnesia. Probably, what was critical for the amnesic effect was the injury discharges and spreading depression, and not the loss of hippocampal tissue, because the size of lesions need not be large (Bohdanecka *et al.*, 1967).

4.3. Cholinergic Stimulation

Disruptive effect of hippocampal stimulation with cholinergic substances has been studied by Deutsch and associates. The findings are well summarized in a review paper (Deutsch, 1971). The results of their experiments are, however, difficult to interpret because of the lack of proper control groups.

For example, Deutsch *et al.* (1966) trained rats to discriminate brightness to escape from foot shock and injected diisopropylfluorophosphate (DFP) into the hippocampus. The interval between the training and injection was either 30 min or 3, 5, or 14 days, and the retention test was conducted 1 day after the injection. The 3-day group performed better than the 5-day or 14-day group. However, there was no control group injected with the vehicle of DFP (peanut oil) and tested at these intervals, and only one control group was tested 15 days after the training without injection.

Implicit in the design of this experiment is an assumption that if a group of control animals show good retention after 15 days they must have had good memory throughout the retention period. However, the assumption has been invalidated by Huppert and Deutsch (1969). They partially trained rats on a brightness discrimination task and then retrained the animals after various intervals ranging from 30 min to 17 days. The animals retrained 7 or 10 days after the initial training took fewer trials to reach a criterion of learning than those retrained after shorter or longer intervals. Therefore, the effect of drug injection cannot be evaluated without having a complete set of control groups with and without vehicle injection.

5.   Retrograde Amnesia by Electrical Stimulation

5.1.   Stimulation of the Hippocampus

A number of experiments have been conducted using electrical stimulation of the hippocampus to produce retrograde amnesia. The hippocampus is rather susceptible to electrical stimulation, and relatively short (3–10 s) stimulation easily elicits a train of afterdischarges lasting for more than 30 s, followed by a postictal depression. Typically, electrical stimulation has been given shortly after a training trial, in a way similar to a posttrial electroconvulsive shock (ECS).

Stein and Chorover (1968) stimulated the rat hippocampus for 3 s immediately after every trial of food-reinforced maze learning and found an increase of errors in a later test. Brunner *et al.* (1970) administered two 1.5-s trains of stimulation within 10 s after a stepdown passive avoidance training and observed a failure of avoidance 24 and 48 h later. Lidsky and Slotnick (1970) gave their mice one 8-s stimulation within 92 s after a foot shock, which was contingent on feeding, and observed a failure in later performance of passive avoidance. There was a gradient of the disruptive effect: the shorter the interval between the shock and stimulation, the more pronounced the failure.

Contrary to these reports of disruptive effects, there are a few studies reporting facilitation. In the study of Stein and Chorover (1968) mentioned above, if the hippocampal stimulation was given only once, after the first trial of maze learning, later learning of the same task was facilitated. Erickson and Patel (1969) also observed facilitation of passive avoidance learning by stimulating the hippocampus for 3 s after every trial. The discrepancy in the results may come from the difference in the duration and severity of the hippocampal afterdischarge. It is possible that the disruptive effect occurs only when the stimulation is strong enough to produce a substantial duration of afterdischarge.

More recent studies, however, suggest that the hippocampal afterdischarge is not a critical factor for the disruptive effect. McDonough and Kesner (1971) first determined the threshold for afterdischarge by stimulating the hippocampus of cats for 5 s. They then trained the animals on a passive avoidance task and stimulated the hippocampus at a subthreshold intensity. All of the animals stimulated within 5 s of training failed to show the avoidance response 4 days later; the stimulation was less effective when applied 30 or 60 min later. Shinkman and Kaufman (1972) administered four trains of 10-s stimulation to the rat hippocampus within 1 min after each foot shock and observed a failure in learning of conditioned suppression. They monitored the hippocampal electrical activity during the training period and found that a weak stimulation insufficient to produce hippocampal afterdischarge was still effective in disrupting the acquisition of conditioned suppression. They suggested that there is a critical neural circuit within the hippocampus which is responsible for the behavioral effect. Either direct stimulation of the critical circuit or indirect excitation of it by generalized afterdischarges gives rise to the disruptive effect.

Generally speaking, the disruptive effect of posttrial hippocampal stimulation is similar to that of ECS. That is, the effect is maximal when the stimulation is given immediately after a training trial, and becomes smaller as the time interval between the trial and stimulation increases. These characteristics suggest that the disruptive effect of hippocampal stimulation may well be a retrograde amnesia similar to the one produced by ECS. An interesting finding is that the amnesic effect of ECS is weaker in rats with hippocampal lesions (Hostetter, 1968). The hippocampus, therefore, appears to be responsible for at least some aspects of the ECS effect, but certainly not for all aspects.

5.2. Stimulation of Other Structures

The fact that a stimulating electrode is placed in the hippocampus does not necessarily mean that the hippocampus is the structure responsible for the behavioral effect. Hippocampal stimulation may excite the structures that are outside the hippocampus but intimately connected to it (Elul, 1964). For example, stimulation of the hippocampus elicits afterdischarges in the amygdala, which may in turn produce an amnesic effect.

Goddard (1964) stimulated the amygdala immediately after every foot shock and found that the acquisition of conditioned suppression was retarded. Similar amnesic effects were observed in one-way and two-way active avoidance learning situations, but not in an appetitively motivated maze learning situation (Goddard, 1964). Kesner and Doty (1968) stimulated various structures in the limbic system of the cat and found that the amygdaloid stimulation was more effective than hippocampal stimulation in producing the amnesic effect. They concluded that the structure that plays a critical role in the mnemonic process is the amygdala, not the hippocampus. Bresnahan and Routtenberg (1972) tested the effect of very weak (5 μA) stimulation on the retention of a stepdown passive avoidance response. The rats learned the response while stimulation was applied to the medial nucleus of the amygdala, but failed to retain it 24 h later. Those with electrodes in other nuclei of the amygdala or in the hippocampus retained the response.

The effect of stimulation of the midline nuclei of the thalamus was studied by Mahut (1962). Rats stimulated immediately after every trial of maze learning reinforced with food made more errors than control animals without stimulation. Those stimulated 165 s after the training trial did not show any impairment.

The caudate nucleus is another structure that has been studied in the context of retrograde amnesia. Wyers et al. (1968) found that stimulation of this nucleus with a single pulse of 0.5 ms duration produced a characteristic spindle of electrical activity in the caudate nucleus itself and other related structures. Caudate stimulation of this kind applied within 30 s after a foot shock abolished a passive avoidance response 1 day later. A similar stimulation of the hippocampus was less effective: the passive avoidance response was disrupted only if hippocampal stimulation was given within 1 s of the shock trial. The experimenters concluded that the occurrence of the caudate spindle is more critical for the amnesic effect than the hippocampal afterdischarges. The caudate stimulation, however, may have excited the stria terminalis and produced afterdischarges in the amygdala.

5.3. Kindling

It has been generally thought that electrical stimulation excites neural tissue without leaving a permanent effect, provided that the intensity of stimulation is within the physiological range. A brain stimulated yesterday remains more or less the same today, and so it will be tomorrow. Actually, such an assumption is wrong. There is no reason to assume that the brain does not "remember" electrical stimulation, no matter how weak and how short may it be, as long as it excites the neurons that are involved in the formation of memory.

In 1967, Goddard discovered that very weak stimulation of the brain, insufficient to produce any behavioral effect, can eventually trigger convulsions if repeated for many days. A more extensive report of the phenomenon, which Goddard named "kindling" (Goddard *et al.,* 1969), showed that there are many structures in the brain, particularly in the limbic system and basal ganglia, where convulsions can be kindled. The intensity of stimulation need not be high, the duration need not be long. An important point is that the stimulation has to be repeated. On the first day of stimulation, the electrographic record may show no effect of stimulation. After a few repetitions, the same intensity of stimulation begins to produce an afterdischarge followed by a postictal depression. The afterdischarge becomes longer in duration and larger in amplitude, contraction of the facial muscles begins to appear, and finally bilateral clonus of the forelimbs develops. Once an epileptogenic focus is kindled in one structure in the limbic system, convulsions can easily be triggered by weak stimulation of other structures in the limbic system.

The phenomenon of kindling clearly demonstrates that electrical stimulation leaves an extremaly long-lasting and widespread change of neuronal excitability. It should be noted that if the hippocampus or amygdala is stimulated in order to determine a threshold for behavioral or electrographic effect, the effect of stimulation accumulates over trials and makes the entire limbic system more susceptible to further stimulation.

For example, similar stimulation of the amygdala may (Goddard, 1964) or may not (Lidsky *et al.,* 1970) produce an amnesic effect, depending on, probably, the number of stimulations applied before a formal behavioral test. It may well happen that a hippocampal stimulation does not produce any behavioral or electrographic effect at the time when these effects are examined, yet the same stimulation subsequently applied in a learning situation may produce afterdischarges in the hippocampus, amygdala, or elsewhere in the limbic system, and give rise to an amnesic effect.

6. Mechanisms Underlying the Amnesic Effect

A wide variety of experiments have been conducted to evaluate the involvement of the hippocampus in the formation and retention of memory. The effects of various experimental treatments may be summarized as follows:

1   At present, there is no evidence that the suppression of macromolecular synthesis in the hippocampus and entorhinal cortex disrupts memory. Recent studies

indicate that memory is not affected by the suppression of protein synthesis by puromycin. However, the findings do not deny the possibility that the biosynthesis of protein may be involved in the formation and retention of memory. The findings only indicate insufficiency of evidence for such a possibility.

2. Injection of epileptogenic substances such as puromycin, actinomycin, and penicillin into the hippocampus impairs the performance of discrimination and active avoidance tasks. The onset of the impairment is closely related to the onset of epileptiform spike discharges in the hippocampus. The impairment of performance appears to result from the interference with sensory, motor, or motivational mechanisms, rather than the disruption of memory.

3. Profound spreading depression in the hippocampus results in an impairment of withholding responses such as passive avoidance and conditioned suppression. The impairment is similar to that demonstrated by animals with hippocampal lesions, and probably caused by a long-lasting dysfunction of the hippocampal neurons. When the hippocampus recovers normal function, the behavioral impairment disappears.

4. Electrical, and possibly chemical, stimulation applied to the hippocampus immediately after a training trial produces retrograde amnesia in a fashion similar to the way ECS does. Prolonged excitation of the hippocampal afferent fibers by injury may have the same amnesic effect. There is a clear temporal gradient of the effect: the amnesic effect is most pronounced if the stimulation is applied within a few seconds of training trial, weaker with a longer delay, and practically nonexistent after a few hours of delay. The structure responsible for the amnesic effect may be the hippocampus itself, or it may be some other structure in the limbic system.

It is not clear at present how hippocampal stimulation gives rise to retrograde amnesia. One of the possible mechanisms, which has not been investigated sufficiently, is the involvement of the hippocampus in the control of pituitary–adrenal hormones. It has been known that the structures in the limbic system facilitate or inhibit the secretion of adrenocorticotropic hormone (ACTH) (Magni *et al.*, 1966), and ACTH in turn stimulates the adrenal cortex to produce corticosteroids. These hormones influence aversively motivated behavior in a complicated fashion (Bohus, this volume; Davidson and Levine, 1972; DiGiusto *et al.*, 1971). Recently, evidence has been accumulated that there are neurons in the hippocampus that are sensitive to the pituitary and adrenocortical hormones (Gerlach and McEwen, 1972; McEwen *et al.*, 1969, 1970; Pfaff *et al.*, 1971; Steiner *et al.*, 1969). Thus the hippocampus may be viewed as a window through which the hormones modulate the activity of the central nervous system.

In this regard, a recent study (Nakajima, 1975) is relevant. It was found that the amnesic effect of cycloheximide in a passive avoidance situation is mediated by the adrenocortical hormones. Subcutaneous injection of cyclohexmide suppressed cerebral protein synthesis in adrenalectomized mice as well as in unoperated control animals. A step-through passive avoidance task learned under these conditions was not retained by the unoperated mice, but the adrenalectomized animals retained it well. Also, a supplementary injection of hydrocortisone or corticosterone after a training trial antagonized the amnesic effect of cyclohexmide in unoperated animals. There was a temporal gradient of antagonism: hydrocortisone antagonized the amnesic effect of cycloheximide if injected within 30 min after the training trial, but not after more than 3 h. These results indicate that the amnesic effect of cycloheximide is due to the suppression of protein synthesis in the adrenal gland, which is essential for the production of corticosteroids (Ferguson, 1968; Garren et al., 1965, 1966). The findings not only speak against the macromolecular hypothesis of memory but also emphasizes the significance of hormonal effect in learning of aversively motivated tasks.

Cottrell (1975) has found that the target of the adrenal hormones in their antiamnesic effect is in the hippocampus. She trained rats on a one-trial passive avoidance task. The animals subcutaneously injected with cycloheximide before training showed no sign of avoidance at all in a test conducted 7 days later, indicating the amnesic effect of the drug. The effect of cyclohexmide was antagonized by injecting 13 μg of hydrocortisone succinate bilaterally into the hippocampus through implanted cannulas within 5 min of training. Similar injection into the amygdala was not effective. It appears that, in normal animals, an experience of foot shock leads to the secretion of adrenocortical hormones, which in turn modulate the activity of the hippocampal neurons. Later performance of the avoidance task seems to depend on this hormonal modulation: if the hormonal secretion is blocked by cycloheximide, the avoidance response is absent in a later test. Electrical or chemical stimulation of the hippocampus may produce an amnesic effect by disrupting the hormonal modulation of the hippocampus. Obviously, this is only one of many possible mechanisms.

Different aspects of behavioral impairment, all of which have been considered as memory deficit, seem to be based on different mechanisms. Some of them may in fact be related to memory, others are not; some are related to the functions of the hippocampus, others may not be. An important task at present is to investigate mechanisms underlying each of these different aspects of "memory deficit."

ACKNOWLEDGMENTS

The author wishes to express his gratitude to Dr. S. Corkin and Dr. G. V. Goddard for their critical reading of the manuscript.

7. References

AGRANOFF, B. W., DAVIS, R. E., CASOLA, L., AND LIM, R. Actinomycin D blocks formation of memory of shock-avoidance in goldfish. Science, 1967, 158, 1600–1601.

AVIS, H. H., AND CARLTON, P. L. Retrograde amnesia produced by hippocampal spreading depression. Science, 1968, 161, 73–75.

BARONDES, S. H., AND COHEN, H. D. Puromycin effect on successive phase of memory storage. Science, 1966, 151, 594–595.

BEACH, G., EMMENS, M., KIMBLE, D. P., AND LICKEY, M. Autoradiographic demonstration of biochemical changes in the limbic system during avoidance training. Proceedings of the National Academy of Sciences U.S.A., 1969, 62, 692–696.

BERRY, R. W. Ribonucleic acid metabolism of a single neuron: Correlation with electrical activity. Science, 1969, 166, 1021–1023.

BOHDANECKA, M., BOHDANECKY, Z., AND JARVIK, M. E. Amnesic effects of small bilateral brain puncture in the mouse. Science, 1967, 157, 335–336.

Bowman, R. E., and Strobel, D. A. Brain RNA metabolism in the rat during learning. *Journal of Comparative and Physiological Psychology*, 1969, **67**, 448–456.

Bresnahan, E., and Routtenberg, A. Memory disruption by unilateral low level, sub-seizure stimulation of the medial amygdaloid nucleus. *Physiology and Behavior*, 1972, **9**, 513–525.

Brunner, R. L., Rossi, R. R., Stutz, R. M., and Roth, T. G. Memory loss following post-trial electrical stimulation of the hippocampus. *Psychonomic Science*, 1970, **18**, 159–160.

Bureš, J., Burešová, O., and Weiss, T. Functional consequences of hippocampal spreading depression. *Physiologia Bohemoslovenica*, 1960, **9**, 219–227.

Carlton, P. L., and Markiewicz, B. Studies of the physiological bases of memory. *Progress in Physiological Psychology*, 1973, **5**, 125–153.

Cohen, H. D., and Barondes, S. H. Puromycin effect on memory may be due to occult seizure. *Science*, 1967, **157**, 333–334.

Cohen, H. D., Ervin, F., and Barondes, S. H. Puromycin and cycloheximide: Different effects on hippocampal electrical activity. *Science*, 1966, **154**, 1557–1558.

Cottrell, G. A. Effect of corticosteroid injection into limbic structures on cycloheximide-induced passive avoidance deficits. M. A. thesis, Dalhousie University, 1975.

Davidson, J. M., and Levine, S. Endocrine regulation of behavior. *Annual Review of Physiology*, 1972, **34**, 375–408.

Deutsch, J. A. The physiological basis of memory. *Annual Review of Psychology*, 1969, **20**, 85–104.

Deutsch, J. A. The cholinergic synapses and the site of memory. *Science*, 1971, **174**, 788–794.

Deutsch, J. A., Hamburg, M.D., and Dahl, H. Anticholinesterase-induced amnesia and its temporal aspects. *Science*, 1966, **151**, 221–223.

DiGiusto, E. L., Cairncross, K., and King, M. G. Hormonal influences on fear-motivated responses. *Psychological Bulletin*, 1971, **75**, 432–444.

Dorfman, L. J., Bohdanecka, M., Bohdanecky, Z., and Jarvik, M. E. Retrograde amnesia produced by small cortical stab wounds in the mouse. *Journal of Comparative and Physiological Psychology*, 1969, **69**, 324–328.

Douglas, R. J. The hippocampus and behavior. *Psychological Bulletin*, 1967, **67**, 416–442.

Elul, R. Regional differences in the hippocampus of the cat. II. Projections of the dorsal and ventral hippocampus. *Electroencephalography and Clinical Neurophysiology*, 1964, **16**, 489–502.

Erickson, C. K., and Patel, J. B. Facilitation of avoidance learning by post-trial hippocampal electrical stimulation. *Journal of Comparative and Physiological Psychology*, 1969, **68**, 400–406.

Ferguson, J. J. Metabolic inhibitors and adrenal function. In K. W. McKerns (Ed.), *Functions of the adrenal cortex*. Vol. 1. New York: Appleton, 1968, pp. 463–478.

Flexner, J. B., and Flexner, L. B. Restoration of expression of memory lost after treatment with puromycin. *Proceedings of the National Academy of Sciences U.S.A.*, 1967, **57**, 1651–1654.

Flexner, J. B., and Flexner, L. B. Puromycin: Effect on memory of mice when injected with various cations. *Science*, 1969a, **165**, 1143–1144.

Flexner, J. B., and Flexner, L. B. Studies on memory: Evidence for a widespread memory trace in the neocortex after the suppression of recent memory by puromycin. *Proceedings of the National Academy of Sciences U.S.A.*, 1969b, **62**, 729–732.

Flexner, J. B., Flexner, L. B., Stellar, E., de la Haba, G., and Roberts, R. B. Inhibition of protein synthesis in brain and learning and memory following puromycin. *Journal of Neurochemistry*, 1962, **9**, 595–605.

Flexner, J. B., Flexner, L. B., and Stellar, E. Memory in mice as affected by intracerebral puromycin. *Science*, 1963, **141**, 57–59.

Flexner, L. B., and Flexner, J. B. Intracerebral saline: Effect on memory of trained mice treated with puromycin. *Science*, 1968a, **159**, 330–331.

Flexner, L. B., and Flexner, J. B. Studies on memory: The long survival of peptidyl-puromycin in mouse brain. *Proceedings of the National Academy of Sciences U.S.A.*, 1968b, **60**, 923–927.

Flexner, L. B., Flexner, J. B., Roberts, R. B., and de la Haba, G. Loss of recent memory in mice as related to regional inhibition of cerebral protein synthesis. *Proceedings of the National Academy of Sciences U.S.A.*, 1964, **52**, 1165–1169.

Flexner, L. B., Flexner, J. B., de la Haba, G., and Roberts, R. B. Loss of memory as related to inhibition of cerebral protein synthesis. *Journal of Neurochemistry*, 1965a, **12**, 535–542.

Flexner, L. B., Flexner, J. B., and Stellar, E. Memory and cerebral protein synthesis in mice as affected by graded amounts of puromycin. *Experimental Neurology*, 1965b, **13**, 264–272.

Flexner, L. B., Flexner, J. B., and Roberts, R. B. Memory in mice analyzed with antibiotics. *Science*, 1967, **155**, 1377–1383.

Garren, L. D., Ney, R. L., and Davis, W. W. Studies on the role of protein synthesis in the regulation of corticosterone production by adrenocorticotropic hormone *in vivo*. *Proceedings of the National Academy of Sciences U.S.A.*, 1965, **53**, 1443–1450.

Garren, L. D., Davis, W. W., Crocco, R. M., and Ney, R. L. Puromycin analogs: Action of adrenocorticotropic hormone and the role of glycogen. *Science*, 1966, **152**, 1386–1388.

Gerlach, J. L., and McEwen, B. S. Rat brain binds adrenal steroid hormone: Radioautography of hippocampus with corticosterone. *Science*, 1972, **175**, 1133–1136.

Goddard, G. V. Amygdaloid stimulation and learning in the rat. *Journal of Comparative and Physiological Psychology*, 1964, **58**, 23–30.

Goddard, G. V. Development of epileptic seizures through brain stimulation at low intensity. *Nature (London)*, 1967, **214**, 1020–1021.

Goddard, G. V., McIntyre, D. C., and Leech, C. K. A permanent change in brain function resulting from daily electrical stimulation. *Experimental Neurology*, 1969, **25**, 295–330.

Grossman, S. P., and Mountford, H. Learning and extinction during chemically induced disturbance of hippocampal functions. *American Journal of Physiology*, 1964, **207**, 1387–1393.

Hirano, T. Effects of functional disturbances of the limbic system on the memory consolidation. *Japanese Psychological Research*, 1965, **7**, 171–182.

Hostetter, G. Hippocampal lesions in rats weaken the retrograde amnesic effect of ECS. *Journal of Comparative and Physiological Psychology*, 1968, **66**, 349–353.

Huges, R. A. Retrograde amnesia in rats produced by hippocampal injections of potassium chloride: Gradient of effect and recovery. *Journal of Comparative and Physiological Psychology*, 1969, **68**, 637–644.

Huppert, F. A., and Deutsch, J. A. Improvement in memory with time. *Quarterly Journal of Experimental Psychology*, 1969, **21**, 267–271.

Hydén, H., and Lange, P. W. Protein synthesis in the hippocampal pyramidal cells of rats during a behavioral test. *Science*, 1968, **159**, 1370–1373.

Hydén, H., and Lange, P. W. Brain cell protein synthesis specifically related to learning. *Proceedings of the National Academy of Sciences U.S.A.*, 1970, **65**, 898–904.

Kapp, B. S., and Schneider, A. M. Selective recovery from retrograde amnesia produced by hippocampal spreading depression. *Science*, 1971, **173**, 1149–1151.

Kesner, R. P., and Doty, R. W. Amnesia produced in cats by local seizure activity initiated from the amygdala. *Experimental Neurology*, 1968, **21**, 58–68.

Kimble, D. P. Hippocampus and internal inhibition. *Psychological Bulletin*, 1968, **70**, 285–295.

Lidnsky, A., and Slotnick, B. M. Electrical stimulation of the hippocampus and ECS produce similar amnesic effects in mice. *Neuropsychologia*, 1970, **8**, 363–369.

Lidsky, T. I., Levine, M. S., Kreinick, C. J., and Schwartzbaum, J. S. Retrograde effects of amygdaloid stimulation on conditioned suppression (CER) in rats. *Journal of Comparative and Physiological Psychology*, 1970, **73**, 135–149.

Magni, G., Motta, M., and Martini, L. Control of adrenocorticotropic hormone secretion. In L. Martini and W. F. Ganong (Eds.), *Neuroendocrinology*. Vol. 1. New York: Academic Press, 1966, pp. 297–370.

Mahut, H. Effects of subcortical electrical stimulation on learning in the rat. *Journal of Comparative and Physiological Psychology*, 1962, **55**, 472–477.

McDonough, J. H., and Kesner, R. P. Amnesia produced by brief electrical stimulation of amygdala or dorsal hippocampus in cats. *Journal of Comparative and Physiological Psychology*, 1971, **77**, 171–178.

McEwen, B. S., Weiss, J. M., and Schwartz, L. S. Uptake of corticosterone by rat brain and its concentration by certain limbic structures. *Brain Research*, 1969, **16**, 227–241.

McEwen, B. S., Weiss, J. M., and Schwartz, L. S. Retention of corticosterone by cell nuclei from brain regions of adrenalectomized rats. *Brain Research*, 1970, **17**, 471–482.

MEANS, L. W., WOODRUFF, M. L., AND ISAACSON, R. L. The effect of a twenty-four hour intertrial in-
  terval on the acquisition of spatial discrimination by hippocampally damaged rats. *Physiology and
  Behavior*, 1972, **8,** 457–462.
NAKAJIMA, S. Interference with relearning in the rat after hippocampal injection of actinomycin D.
  Journal of Comparative and Physiological Psychology, 1969, **67,** 457–461.
NAKAJIMA, S. Proactive effect of actinomycin D on maze performance in the rat. *Physiology and Behavior*,
  1972, **8,** 1063–1067.
NAKAJIMA, S. Biochemical disruption of memory: Re-examinations. In W. B. Essman and S. Nakajima
  (Eds.), *Current biochemical approaches to learning and memory*. New York: Halsted, 1973, pp.
  133–146.
NAKAJIMA, S. Amnesic effect of cycloheximide in the mouse mediated by adrenocortical hormones. *Journal
  of Comparative Physiological Psychology*, 1975, **88,** 378–385.
OLDS, J., AND HIRANO, T. Conditioned responses of hippocampal and other neurons. *Electroencepha-
  lography and Clinical Neurophysiology*, 1969, **26,** 159–166.
OLDS, J., MINK, W. D., AND BEST, P. J. Single unit patterns during anticipatory behavior. *Electroen-
  cephalography and Clinical Neurophysiology*, 1969, **26,** 144–158.
OLTON, D. S. Specific deficits in active avoidance behavior following penicillin injection into hippocampus.
  Physiology and Behavior, 1970, **5,** 957–963.
PFAFF, D. W., SILVA, M. T., AND WEISS, J. M. Telemetered recording of hormone effects on hippocampal
  neurons. *Science*, 1971, **172,** 394–395.
ROBERTS, R. B., FLEXNER, J. B., AND FLEXNER, L. B. Some evidence for the involvement of adrenergic
  sites in the memory trace. *Proceedings of the National Academy of Sciences U.S.A.*, 1970, **66,**
  310–313.
ROSENBAUM, M., COHEN, H. D., AND BARONDES, S. H. Effect of intracerebral saline on amnesia produced
  by inhibitors of cerebral protein synthesis. *Communications in Behavioral Biology*, 1968, Part A, **2,**
  47–50.
RUSSELL, P. J. D. The effect of intrahippocampal injection of puromycin in the rat. M. A. thesis,
  Dalhousie University, 1969.
SCHMALTZ, L. W. Deficit in active avoidance learning in rats following penicillin injection into hip-
  pocampus. *Physiology and Behavior*, 1971, **6,** 667–674.
SCOVILLE, W. B., AND MILNER, B. Loss of recent memory after bilateral hippocampal lesions. *Journal of
  Neurology, Neurosurgery, and Psychiatry*, 1957, **20(11),** 11–19.
SHINKMAN, P. G., AND KAUFMAN, K. P. Post-trial hippocampal stimulation and CER acquisition in the
  rat. *Journal of Comparative and Physiological Psychology*, 1972, **80,** 283–292.
STEIN, D. G., AND CHOROVER, S. L. Effects of post-trial electrical stimulation of hippocampus and cau-
  date nucleus on maze learning in the rat. *Physiology and Behavior*, 1968, **3,** 787–792.
STEINER, F. A., RUF, K., AND AKERT, K. Steroid-sensitive neurons in the rat brain: Anatomical localiza-
  tion and responses to neurohumours and ACTH. *Brain Research*, 1969, **12,** 74–85.
URETSKY, E., AND MCCLEARY, R. A. Effect of hippocampal isolation on retention. *Journal of Compara-
  tive and Physiological Psychology*, 1969, **68,** 1–8.
WOODRUFF, M. L., AND ISAACSON, R. L. Discrimination learning in animals with lesions of hippocampus.
  Behavioral Biology, 1972, **7,** 489–501.
WOODRUFF, M. L., KEARLEY, R. C., AND ISAACSON, R. L. Deficient brightness discrimination acquisition
  produced by daily intracranial injections of penicillin in rats. *Behavioral Biology*, 1974, **12,** 445–460.
WYERS, E. J., PEEKE, H. V., WILLISTON, J. S., AND HERZ, M. J. Retroactive impairment of passive avoid-
  ance learning by stimulation of the caudate nucleus. *Experimental Neurology*, 1968, **22,** 350–366.

Index

415